RELIABILITY TECHNOLOGY
Theory & Applications

RELIABILITY TECHNOLOGY

Theory & Applications

European Reliability Conference - REL-CON '86
Copenhagen, Denmark, June 16-20, 1986

EUREL + DIF + I-S

Edited by

J. MØLTOFT and **F. JENSEN**
The Engineering Academy of Denmark
Lyngby, Denmark

1986

NORTH-HOLLAND
AMSTERDAM ● NEW YORK ● OXFORD ● TOKYO

ISBN: 0 444 70039 0

Publishers:
ELSEVIER SCIENCE PUBLISHERS B.V.
P.O. Box 1991
1000 BZ Amsterdam
The Netherlands

Sole distributors for the U.S.A. and Canada:
ELSEVIER SCIENCE PUBLISHING COMPANY, INC.
52 Vanderbilt Avenue
New York, N.Y. 10017
U.S.A.

Library of Congress Cataloging-in-Publication Data

European Reliability Conference (1986 : Copenhagen,
    Denmark)
    Reliability technology.

    Includes index.
    1. Reliability (Engineering)--Congresses.
I. Møltoft, J. (Jørgen), 1940-   . II. Jensen,
Finn, 1937-   . III. Title.
TA169.E974 1986   620'.00452   86-9016
ISBN 0-444-70039-0 (U.S.)

PRINTED IN THE NETHERLANDS

Foreword

Our society is becoming more and more dependent on the proper functioning of increasingly complex technological products. Failure of a product will have implications ranging from consumer irritation to enormous economic losses or even the endangering of human lives. From video recorders to telecommunication systems, from automobiles to satellites, actual experience has over the past years demonstrated to everyone the importance of reliability in technology.

This first European international reliability week will bring together reliability and design engineers for the presentation and discussion of reliability technology and methods — advanced as well as basic — from the field of electronics, electro-mechanical, and mechanical engineering as well as from systems engineering and management.

The organizers of this conference — the Danish Engineers' Post-Graduate Institute (DIEU) — have utilized the experience gained from the highly successful EUROCON '82 Conference on Reliability. Together with the Engineering Academy of Denmark (DIA), they have carefully structured the programme so that experts as well as newcomers to the field can participate with maximum benefit. In doing so, the organizers have enjoyed the whole-hearted assistance of leading European experts in reliability engineering.

The help given from all these persons is gratefully acknowledged, as well as the sponsorship given to REL-CON Europe '86 by **EUREL** — the Convention of National Societies of Electrical Engineers of Western Europe, **DIF** — the Danish Society of Chemical, Civil, Electrical and Mechanical Engineers, and **I-S** — the Society of Engineers of Denmark.

Jørgen Møltoft   Finn Jensen

THE TECHNICAL PROGRAMME COMMITTEE

Erling Andersen
ElektronikCentralen, Denmark

Heinz H. Frey
BBC Brown, Boveri & Company, Ltd., Switzerland

Wolfgang Gerling
Siemens AG, Federal Republic of Germany

Roland Goarin
CNET, France

Örjan Hallberg
RIFA AB, Sweden

Finn Jensen
Finn Jensen Consultancy, Denmark

Valter Loll
TERMA Elektronik AS, Denmark

Jørgen Møltoft (Chairman)
The Engineering Academy of Denmark, Denmark

Patrick D.T. O'Connor
British Aerospace, United Kingdom

Emiliano Pollino
CSELT SpA, Italy

Leo M. Van Tetterode
Philips International B.V., The Netherlands

SESSION CHAIRMEN

Harold Ascher
Naval Research Laboratory, USA

D.S. Campbell
Loughborough University of Technology, United Kingdom

Albert Ferris-Prabhu
IBM General Technology Division, USA

Ove Stig Jørgensen
Post & Telegraph, Long Line Office, Denmark

Barry Popplewell
Pye Unicam Ltd., United Kingdom

George Rzevski
Kingston Polytechnic, United Kingdom

Arne Salomon
Radiometer A/S, Denmark

Kurt Stochholm
TERMA Elektronik AS, Denmark

Wayne Tustin
Tustin Institute of Technology, USA

Table of Contents

Part 1
INDUSTRIAL
CASE HISTORIES

KEYNOTE ADDRESS

Reliability Technology — Theory & Applications
J. Møltoft and F. Jensen (Editors)
© Elsevier Science Publishers B.V. (North-Holland), 1986

MODERN QUALITY & RELIABILITY ASSURANCE SYSTEMS

E. FUCHS
Director, Quality Assurance Center
AT&T Bell Laboratories
Holmdel, New Jersey, USA

Ever since the 1920's, when Walter Shewhart of AT&T Bell Laboratories laid the foundations of statistical quality control, AT&T has been a leader in quality technology. In this paper, we discuss the new AT&T quality system and the role that this technology is playing in it. In particular, we emphasize the statistical quality and reliability assurance techniques employed as essential elements of the technology.

1. INTRODUCTION

On January 1, 1984, AT&T underwent a process known as divestiture in which the company implemented a transition from public utility to competitive enterprise. The divestiture process divided AT&T, then a corporation with $150 billion in assets and $70 billion in operating revenues, into eight smaller companies. Seven of the companies are regional holding companies that own the Bell operating telephone companies. The eighth company is the new AT&T, a company which in 1984 had $40 billion in assets and $33 billion in operating revenues. Before divestiture, AT&T was restricted to the telecommunications business. It was vertically integrated; the clients for the telecommunications products and services it developed were its own operating entities. After divestiture, AT&T can take its technology and products to whatever markets it deems attractive.

Before divestiture, the AT&T quality system was inward-looking. The supplier of products and the customers for most of them had the same bottom line. Therefore, the central focus of the AT&T quality system was to minimize life-cycle costs for its products.

The new AT&T deals with its customers at arms length. Therefore, it faces the task of identifying potential customers and their needs, marrying the needs with technology and selling products in competitive markets. This requires two important additional goals: satisfying diverse populations of customers and contributing to wider cost-price margins. A good quality system is an effective way to accomplish these complementary objectives.

2. QUALITY STRATEGY

The dimensions of our quality and reliability assurance systems are: leadership, the use of an effective product realization process, application of modern quality and reliability technology and the organizational structure that puts leadership, accountability and resources where they are needed.

2.1 Leadership

You may have seen the recent issue of Quality Progress magazine with the picture of the president of AT&T on the cover. He is convinced that quality and reliability are central issues, spanning all of our corporation's critical strategies. Moreover, he recognizes that our company's priorities need to be national priorities. Therefore, he accepted the position of Chairman of U.S. National Quality Month. He is convinced that his leadership is critical to the success of our quality initiative. How did he become a believer? He saw AT&T's quality and reliability costs and the opportunities to reduce those costs. The amounts that can be saved are in the hundreds-of-millions of dollars. Those are large numbers, even for a corporation the size of AT&T. The results of our customer contacts and surveys also clearly indicated the role of quality as a market differentiator. Once he saw the facts his leadership was assured.

2.2 What is quality

I think it useful to be sure that we have a common understanding of what we mean by "quality." AT&T president Jim Olson talks about the "new definition of quality." He too understands that the traditional meaning, conformance to requirements, is outmoded as a total definition of quality. It begs the issue of what requirements?, whose requirements? The key point is that quality is in the eyes of the beholder and the beholder that counts is the customer. The end-use customer for our products and services is critically important, to be sure; it is he who holds the ultimate power of the purse. But, each of us has a customer and a supplier. It is very important that everyone understands this quality chain concept. Professor David Garvin of the Harvard Business School has published a definition of quality that illustrates its multi-faceted nature (Figure 1). The definition is useful because it at once comprehends the several customer views of quality: performance, reliability, "fits and finishes," esthetics, etc., as well as the traditional manufacturer's definition, relating to conformance. This comprehensive view is fast becoming the new suppliers' definition of quality, and is the one that is operative in AT&T.

2.3 Product realization process

AT&T has developed an intergrated quality system that views the quality chain from the point of view of the corporation. This requires a quality plan for each element of the business. That quality plan is but a facet of a more comprehensive plan called the product realization process.

The product realization process is a business plan that integrates the sequential and overlapping steps in the life cycle of products and services. In order for a business to operate successfully, the information and processes involved in every step must be coordinated and orchestrated. For instance, an interesting view of

the central portion of the product realization process is the well-known Hewlett-Packard diamond (Figure 2). This is a view of manufacturing that takes in account its interactions with its adjacent elements. In many industries, the interactions are more important than the labor, and manufacturing becomes an information management activity.

In order to produce high quality product at low cost, the quality system must address development functions, including design, engineering, simulation and test. It must include the purchasing, scheduling, production management and materials management functions. And it must include material handling, robotics and process and assembly controls. Information generated at each step affects decisions and performance of every other step. One cannot have a functioning quality system unless the right information is generated when needed, is provided where needed, and is used.

3. QUALITY AND RELIABILITY TECHNOLOGY

We will now focus on quality and reliability technology to support the needs of our business and the means for delivering or transferring the technology to the user communities. We begin by identifying the technical functions performed at each step of the quality system (Figure 3). Space does not permit discussion of all these elements. We have selected three for elaboration in this paper: reliability modeling, reliability data analysis and robust process design. These three topics are interesting topics that are related to the subject of this meeting.

3.1 Reliability modeling

Reliability is the ability of an item to perform its defined function over time. The set of engineering tasks that are performed during product development to insure the reliability of product is called reliability engineering.

Traditionally reliability engineering has involved the setting of product reliability objectives, such as the familiar one for electronic switching systems of no more than 2 hours of downtime in 40 years, and then designing the product to achieve these objectives. Designs are evaluated for their ability to meet these requirements by using probabilistic modeling tools and empirical data about part reliability. These data are usually codified in a database such as those based on U.S. Military Standard 217D. In AT&T this traditional view of reliability engineering is complemented by a more modern viewpoint compatible with the zero defect quality philosophy.

At AT&T it is not enough to demonstrate that a proposed product design is capable of meeting its reliability objectives; it must also be demonstrated that all reasonable effort has been made to prevent failure, even if in the process the reliability objectives are surpassed. This emphasis on the prevention of failure comes to AT&T from its long time involvement in the design and construction of ultrareliable systems such as submarine cable systems and electronic switching systems.

The AT&T approach to reliability engineering calls for the extensive use of qualitative reliability engineering techniques such as Failure Mode, Effect and Critical Analysis (FMECA) and Fault Time Analysis (FTA).* These tools are used to make sure that all causes of failure are accounted for and as far as possible prevented.

When it comes to the use of quantitative reliability tools based on probability models, the AT&T approach to reliability engineering demands that these tools be used not only for reliability prediction and measurement but also for the identification of weak links in the product design. These weak links can then be eliminated by, for instance, the use of redundancy, higher levels of component integration, better grade parts, parts derating, and parts screening.

To facilitate the use of probability modeling for the identification of weak links, AT&T has built a software tool called System Used for Prediction and Evaluation of Reliability (SUPER). This tool is fully integrated with AT&T's computer aided engineering and design tools and with AT&T's in-house parts reliability database. Thus, SUPER is easy to use routinely as part of the overall product realization process.

Since SUPER is commonly used to identify weak links, SUPER's user interface has been designed to facilitate comparisons among alternative product design options. In particular, tradeoff and sensitivity studies are very easy to do. With SUPER it is possible to gauge the effect on product reliability of changing, for example, the product architecture or the assumed shape of the parts failure rate function. SUPER derives its ability to handle a wide variety of assumptions from its numerical subroutines and the number crunching capabilities of today's minicomputers. SUPER's numerical subroutines are based on sophisticated techniques from computational probability and numerical analysis.

Among SUPER capabilities is the ability to gauge the effect of infant mortality on product reliability, including the effect of replacing failed parts with new spares. SUPER's ability in this area is heavily dependent on information about the infant mortality behavior of most of the parts used by AT&T contained in AT&T's internal parts reliability database. For these parts, this information is far more extensive than that in US MIL-STD-217D.

With SUPER it is possible to realistically model the effect on product reliability of different temperature and electrical stresses on parts. When the user provides SUPER with a list of components or subsystems being stressed, together with appropriate times and stress conditions, SUPER will analyze the system reliability under the new conditions.

---

* The concept of the FTA was originated in 1961 by H. A. Watson of AT&T Bell Laboratories as a means to evaluate the safety of the Minuteman Lauch Control System.

SUPER can model maintained systems. For example, when the repair policy is entered into SUPER, it will calculate system availability as a function of time in addition to the usual steady state value. The expected downtime and the expected number of failures in an arbitrary interval of time are two other maintained system reliability figures of merit that SUPER can calculate when appropriate repair policy information is provided.

SUPER also has some capabilities to model the effect of different spare policies on the availability, expected downtime, and expected number of failures of products involving redundancy.

Other capabilities of SUPER include guidance and help on-line, a forms-driven or conversational screen display and high-quality graphical output capabilities.

An interesting future enhancement to SUPER now in the planning stage is the use of artificial intelligence techniques to suggest changes that enhance product reliability.

3.2 Reliability data analysis

Empirically determined parts failure rates are the starting point of any product reliability model. Hence, these models can be only as accurate as the data and the statistical methods form which the parts failure rates are estimated. Even when the parts failure rates are accurate, product reliability predicted from the reliability models has to be verified due to the great number of assumptions involved in most reliability models.

In general, reliability data arises throughout the product realization process, for example: during parts qualification and certification; during accelerated testing of critical parts; during product design and development; during prototype testing; during manufacturing process design and vendor qualification efforts; during factory testing; and during field tracking and warranty analysis.

Since very high reliabilities are frequently found in the electronic industry, it is usually very expensive to obtain reliability data. For this reason, at AT&T we have developed a software system called Software Tool for the Analysis and Presentation of Reliability Data (STAR) that uses the most efficient statistical techniques to extract all the available information in reliability data.

STAR can handle all commonly occurring reliability data. For example, it can handle data that is either censored or truncated. Censoring means that exact failure times are known for very few, if any, units and for all other units it is only known that the failure times fall somewhere in an interval of time. Truncation means that it is not known how many units started at time zero, since only those that survived the first t_0 hours could be observed. Notice that these special characteristics of reliability data make the conventional statistical techniques inappropriate. For example, what is the average of data where some failure times are known, but for some other units it is only known that they survived 1000 hours?

STAR provides confidence intervals for all the estimates that it provides. This is a very useful feature because it prevents the common mistake of making product decisions on the basis of inadequate data. Confidence intervals also help to determine the appropriate sample size for a reliability experiment; thus eliminating the extra cost of unnecessarily large sample sizes.

For a tool like STAR to be effective it has to be made available to the engineers who collect most of the reliability data. For this reason, STAR provides guidance to the nonstatistician on the proper analysis of reliability data. For example, STAR guides the user through the following three-step strategy which has proven useful in analyzing reliability data:

(1) Non-parametric or model-free analysis to look at the data without restrictive model assumptions.

(2) Graphical methods to choose the best parametric model.

(3) Fitting the best parametric model and then using this fit to answer important reliability questions.

STAR also guides the user through the analysis of accelerated data.

STAR is a powerful tool even for the experienced statistical analyst since it contains many advanced features. For example, STAR can be used to analyze reliability data in which there are several explanatory variables. This is done through the use of the proportional hazards model.

Yet it is not the most advanced features of STAR that have contributed most to the increased reliability of AT&T products. It has been the widespread use of STAR's simplest features. Features such as STAR's capability to quickly and simply analyze reliability data with no explanatory variable and STAR's capability to generate presentation and report quality reliability plots. The widespread use of STAR throughout AT&T is the direct result of STAR's ease of use and of a widely attended three-day workshop in which AT&T engineers learn about STAR and the statistical principle on which it is based.

3.3 Robust design

A new quality engineering method called "parameter design," was pioneered by Professor Genichi Taguchi, who is a former director of the Japanese Academy of Quality, and a Deming Award winner. Parameter design is presently the target of interest, debate, research, analysis, extension and use. Within AT&T, parameter design is used to determine the most cost-effective way to manufacture products while maintaining high quality standards. In brief, the parameter design methods help engineers to find the best settings for the process variables, those settings that produce products that meet the product design specifications in the face of component, environmental and manufacturing variations.

Often, process engineers must consider many parameters that interact in nonlinear ways to affect each processing step. To illustrate the complexity of such process design problems, consider a one-step process that has nine parameters, where each paramenter is a variable that can take on two or three values or settings. There are thousands of combinations of settings. In order to seek the optimum settings in the traditional way, via sequential, change-one-thing-at-a-time experiments, more than 6,000 experiments must be conducted. No one can afford that many tests. Alternatively, a detailed model of the interactions that lends itself to analytic or simulation optimization is required. Such models almost never exist. Therefore, in the past, optimization of this kind was hardly ever done.

To circumvent these problems AT&T is using design of experiment techniques. The process engineer identifies the pertinent variables and the practicable range of their settings. Then the engineer selects a small number of judiciously chosen process conditions using experimental design methods. This allows many factors to be studied simultaneously. Thus, a tractable number of experiments is almost always possible. The allowable number of experiments can be defined in terms of experimentation time or cost. There is usually a designed experiment that does not exceed reasonable time or cost constraints. Of course, the smaller the experiment, the more the possibilities of error. For this reason, once the results of the experiments have been analyzed and a "best" set of conditions have been predicted, one or more confirming experiments are conducted to validate the results. As an illustration of the efficiency of the experimental design approach consider the example described above, where there are more than 6,000 combinations of settings where usually less than 50 combinations of settings will suffice. When these efficiently designed experiments are used there is usually no way of knowing whether the settings determined by analysis of the experimental data is the global optimum. What is always known is whether the results are significantly better than before and this is usually enough.

While engineers and other process designers have used some of Taguchi's experimental design methods for parameter design successfully, mathematical statisticians and some other theorists have been disturbed by the lack of rigorous theoretical underpinnings for his methods. The lack of firm theoretical understanding stands in the way of our ability to embed the approach in software tools that can be made widely available for use by members of our design community who are not trained in statistics, without the consulative support of our limited supply of expert statisticians.

For these reasons, we have a considerable amount of research underway to enhance our theoretical understanding of the use of experimental design methods in parameter design. Some of this work has been reported in the literature (References).

3.4 Technology transfer

The kinds of quality technology described above are valuable to AT&T only if they can be widely and efficiently transferred to the user communities. Five strategies are important here. First, we develop software-based tools or support systems to disseminate as many of the methods as is possible. Second, we develop education and training packages to convey the methods and tools. Third, we are implementing an advanced Technology Transfer Laboratory to perfect the courses and software tools. Fourth, we have cooperative relationships with our corporate and R&D education and training organizations to deliver the courses to the user communities. Fifth, we work with systems engineers, developers and managers of our industrial automation and information systems to integrate our tools into our CAD, CAM, and information movement and management systems.

4. ORGANIZATION

To effectively channel resources and technology to achieve desired ends, an appropriate organizational structure is required. This structure must facilitate the flow of leadership and resources downward from the top, the flow of information upward and the transfer of technology from its sources to the user communities. Figure 4 characterizes our organizational structure to meet these objectives. At the top, the corporate management is supported by a small quality policy staff. The responsibilities of this staff include formulation of policy, allocation of corporate resources and dealing with inter-line-of-business needs and issues.

Each line of business (LOB) has a quality manager, whose responsibility it is to plan the quality systems for the LOB, provide the expertise and support to facilitate implementation, and to support the education and training needs of the LOB.

Each LOB has one or more quality assurance groups whose responsibility it is to help determine appropriate quality metrics, collect the data and report the results.

We have a Quality Assurance Center whose responsibility it is to do the research, systems engineering and development of generic quality systems and quality technology, and to transfer the technology to the user communities. The user communities, of course, have the responsibilities to tailor generic quality systems to their particular needs and to implement them and make them work.

Finally, we have education and training organizations whose job it is to help the business entities to be sure that each member of AT&T knows what his job is and how to do it.

REFERENCES

(1) AT&T Technical Journal, Special Issue on Quality Theory and Practice, (scheduled for March, 1986 publication).

(2) Genichi Taguchi and Yu-In Wu, Introduction to Off-Line Quality Control, (available in English through American Supplier Institute, Inc., Detroit, MI, USA).

(3) Raghu N. Kackar, "Off-Line Quality Control, Parameter Design, and the Taguchi Method," given at the Conference on Frontiers of Industrial Experimentation, Mohonk Mountain House, New Paltz, N.Y., (published in <u>Journal of Quality Technology</u>, October, 1985).

BIOGRAPHY

Edward Fuchs is Director of the Quality Assurance Center at AT&T Bell Laboratories, Holmdel, N.J., USA. He is responsible for quality and reliability technology for AT&T Bell Laboratories and for quality systems support to the AT&T Lines of Business. Mr. Fuchs joined Bell Laboratories in 1960 as a Member of the Technical Staff and assumed his present position in 1976. Mr. Fuchs received a BSEE from Polytechnic Institute of Brooklyn and a MSEE from New York University. He has studied statistics at Rutgers University and economics at New York University. He is a member of ACM and ASQC and a senior member of IEEE. He was a recipient of the National Science Foundation Fellowship.

FIGURE 1

DIMENSIONS OF PRODUCT QUALITY

- **PERFORMANCE (PRIMARY CHARACTERISTICS)**
- **FEATURES ("BELLS AND WHISTLES")**
- **RELIABILITY (FREQUENCY OF FAILURE)**
- **CONFORMANCE (TO SPECIFICATION)**
- **DURABILITY (SERVICE LIFE)**
- **SERVICEABILITY (SPEED OF REPAIR)**
- **AESTHETICS ("FITS AND FINISHES")**
- **PERCEIVED QUALITY (INTANGIBLES)**

TAKEN FROM <u>PRODUCT QUALITY:</u>
<u>AN IMPORTANT STRATEGIC</u>
<u>WEAPON</u>, DANIEL A. GARVIN, 1984

FIGURE 2

THE MANUFACTURING BUSINESS

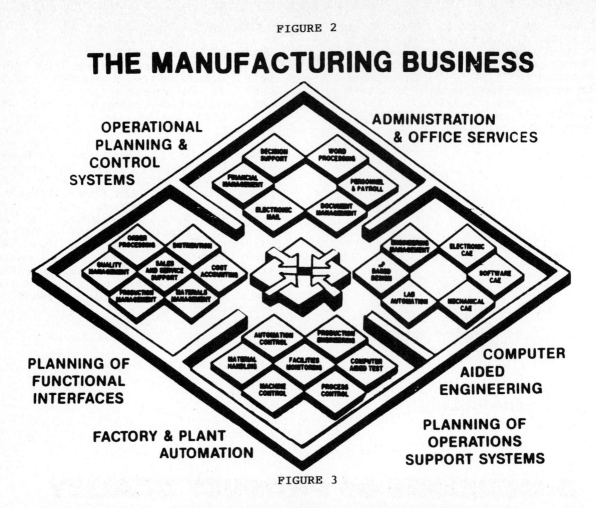

OPERATIONAL PLANNING & CONTROL SYSTEMS

ADMINISTRATION & OFFICE SERVICES

PLANNING OF FUNCTIONAL INTERFACES

COMPUTER AIDED ENGINEERING

FACTORY & PLANT AUTOMATION

PLANNING OF OPERATIONS SUPPORT SYSTEMS

FIGURE 3

QUALITY FUNCTIONS FOR THE PRODUCT REALIZATION PROCESS

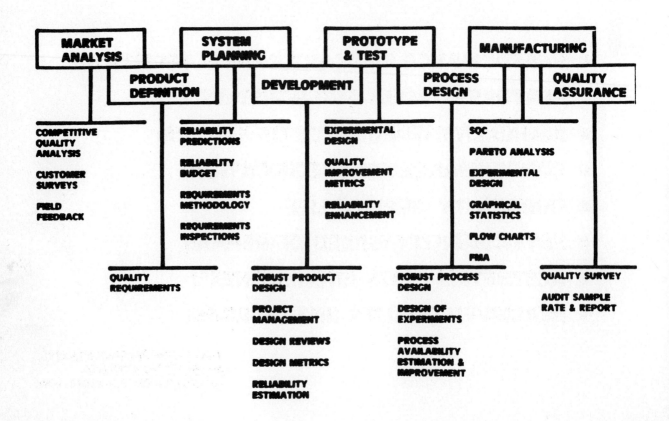

FIGURE 4

THE AT&T QUALITY SYSTEM

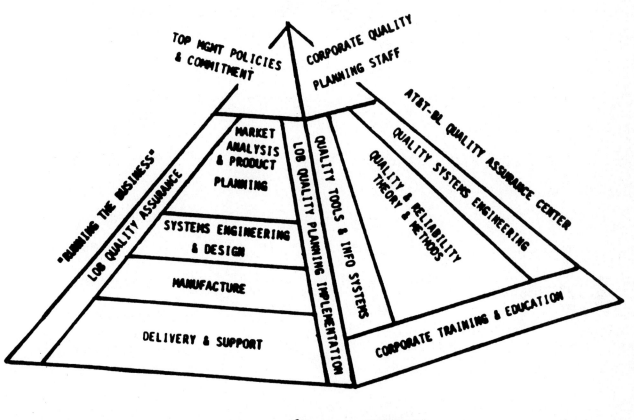

FIELD FAILURE STUDIES

Reliability Technology — Theory & Applications
J. Møltoft and F. Jensen (Editors)
© Elsevier Science Publishers B.V. (North-Holland), 1986

FIELD FAILURE DATA FROM REPAIRABLE SYSTEMS
A DANISH CASE STUDY

Lars Rimestad, Claus Kjærgaard and Jørgen Møltoft
The Engineering Academy of Denmark
Department of Electrical Engineering

For repairable systems the traditional probability tools (such as hazard rate, cumulative distribution functions, Weibull plotting etc) are not applicable. This paper describes a simple and easy-to-use mathematical model and demonstrates its usefulness on actual field data for electronic systems.

1. INTRODUCTION

Large electronic systems are not discarded when they fail but repaired and put to use again. What is the probability of failure after the second, third or tenth failure of the same piece of equipment? This depends very much on the previous history of the apparatus; was it a "rogue" component that failed? How many times has the system failed before? Did repair weaken it? There are a thousand questions like this that have to be answered in order to give a well-founded estimate of the probability of the next failure. This is impossible and the mathematics would be extremely heavy and tiresome. Obviously a short-cut is needed.

This paper offers a very simple model for evaluating the failure pattern for repairable systems from actual field failure data. The use of the model is demonstrated on data from 8 different electronic systems.

This is part of the results of a current research project at The Danish Engineering Academy. The research project, named RELIABILITY UNDER USER CONDITIONS, is supported by The Danish Council of Technology and 3 major companies cooperate by supplying data on their products.

2. THE MATHEMATICAL MODEL

In reliability it is useful to have a variable that tells you when to expect the failures. For non-repairable parts or systems this variable is the probability of failure at time t, on the condition that the system has survived until then. This is called the hazard rate h(t). For repairable systems there is no such condition because the system may have failed one or even more times before time t. In this case the variable can be named the FAILURE INTENSITY I(t) and it is easily estimated from actual field data by:

$$I(t) = \frac{k}{N(t) \ \Delta t}$$

where k is the number of failures in the time interval Δt and N(t) is the number of systems in use at that time (Δt is assumed to be so short that N(t) does not change). This will nearly always be a very fluctuating estimate, so therefore an integration is recommended.

The integral of I(t) is called the Mean Cumulative Number of Failures per system, M(t).

$$M(t) = \int_0^t I(t)dt$$

M(t) can be estimated by numerical integration, but it is much easier to use the recursive formula

$$M(t_i) = M(t_{i-1}) + \frac{1}{N(t_i)}$$

where t_i is the time until i'th failure.

The Weibull process (not to be confused with the Weibull distribution of times-to-first-failure) occurs when I(t) takes on the shape of

$$I(t) = \frac{\beta}{\eta} \left(\frac{t}{\eta}\right)^{\beta-1}$$

In that case the mean cumulative number of failures per system can be expressed as

$$M(t) = \left(\frac{t}{\eta}\right)^{\beta}$$

Plotting M(t) versus t on a double logarithmic paper will give a straight line with slope β, as can be seen by the logarithmic transformation

$$\log M(t) = \beta \log t - \beta \log \eta$$

It is also seen that η is the value of t that gives M(t) = 1.

3. CALCULATION OF TIME

It is important to note that the time t is the operational time. Downtime due to repair, transport and storage is not included. This may not be entirely correct since some failure mechanisms develop with calendar time. Furthermore, repair can be more or less harmful to the system that ought to be aged correspondingly. However, these minor objections are ignored because it is believed that the majority of the failure mechanisms (and the most serious) are developing only when the system is in use.

Figure 1 shows an example of the event history of 5 systems in calendar time. Figure 2 is the corresponding representaion in operational time where all downtimes have been left out. Figure 3 is the M-curve calculated for the same example. Note that it is possible for M to be larger than 1.0 unlike the cumulative distribution of the well known Weibull function for times-to-first-failure.

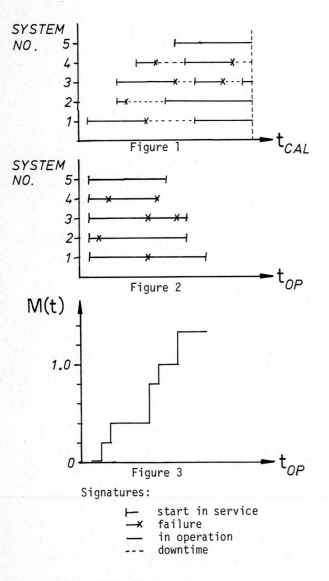

Signatures:

$\vdash$	start in service
$\rightarrow\!\!\times$	failure
—	in operation
- - -	downtime

4. HOW TO READ THE M-GRAPH

Very often the failures of an electronic system are not evenly distributed in time - they may gather in a few groups. This is seen as humps in the failure intensity and these humps are detected by steep inclines on the M-graph. See figure 4.

When the M-graph is a straight line it is valid to assume a Weibull process and the slope will be the shape parameter β of the failure intensity.

If β is larger than unity, the failure intensity is increasing while a β smaller than one means a decreasing I(t). In the special case where β is equal to one, the failure intensity becomes constant, $I = 1/\eta$.

Upon seeing an S-curved M-graph (such as figure 4) the reader may wonder if this indicates some

sort of a bimodal Weibull-process - perhaps, but we do not yet know how to describe this mathematically.Therefore this concept cannot be used - yet.

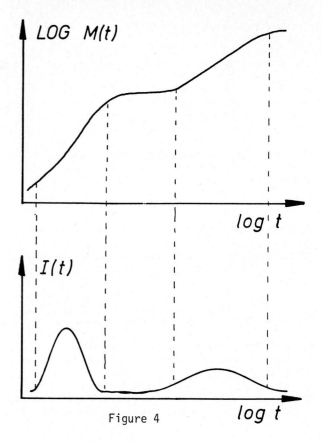

Figure 4

5. ABOUT THE DATA

The investigated electronic equipment is a large installation, containing between 16 and 28 modules of 16 different types.

The 8 module types presented here consist of printed circuit boards of high complexity. They are mounted in slots, and when failure occurs the failed module is easily replaced with a working module. Repair is done at the company plant after an incircuit test. The repaired module is later used for replacement after a failure in another installation. This means that the installation soon will contain modules of different age, so therefore the analysis of the failure intensity has to be done on module level, not installation level.

It takes a lot of recording of data to keep track of the events of every module, and incompleteness of data is inevitable.

However, the investigation included a large number of installations and thence a much larger number of modules, so the M-graphs will be representative of the behaviour of systems under user conditions.

6. COMMENTS ON THE GRAPHS

The failure data for the 8 investegated module types (systems) have been collected over a few years. The graphs of M versus operational time (in days) have been calculated by computer and they are appended to this article.

From the graphs the systems can be devided into two groups:

a) Those with an M(t)-curve that is more or less a straight line, suggesting (but not proving) a single Weibull-process.

b) Those with an M(t)-curve with one or several steep inclines, indicating humps in the failure intensity.

The systems number 3 and 5 are examples of type a. Both are very complex circuits and components of several types are known to fail. Yet there are no very significant humps in the failure intensity. β is less than unity so the failure intensity is decreasing.

For the other systems there seem to be humps. It is generally not yet known which components are the causes of the humps.

7. RELAYS

The systems number 4, 5 and 6 contain large amounts of relays of a certain kind. A Weibull plot of a few percent of these relays was obtained at an earlier stage of the research project (see figure 5). This shows a hump in the hazard rate around 10 days. This agrees well with systems 4 and 6 that also have a hump at approximately 10 days. Further investigation indeed did show that the 10-day hump of system 4 was caused by the relays. For system number 6 too little information was obtainable about the type and position of the components that failed.

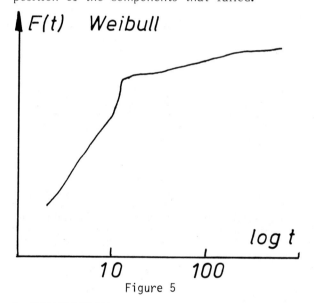

Figure 5

8. CONCLUSION

A mathematical model for treating field failure data on repairable systems has been evolved and demonstrated on a set of real-life field data.

This model will

1) reveal humps in the failure intensity,

2) determine whether the failure intensity is decreasing, constant or increasing, and this tells you whether "Things are getting better or worse",

3) make the basis for investigations of the dominating failure mechanisms. However, this can only be done when you have information about the component-type and -position for every failure (or most of them).

As indicated by the example of the relays, it can be very useful to combine the use of this model for repairable systems with the conventional probability tools (Weibull plotting etc) used on component level.

Finally, one very important result has come out of this work, and that is the fact that humps do exist - not only in the hazard rate for components, but also in the failure intensity for repairable electronic systems.

REFERENCES

{1} Gisela Härtler: "Graphical Weibull analysis of repairable systems". Quality and Reliability Engineering International. Vol. 1 (1985).

{2} F. J. Brunner: "Praktische Zuverlässigkeitsanalyse des reparierbaren Systems 'Farhzeug' ". QZ-Qualitätstechnik, Heft 6 (1984).

ACKNOWLEDGEMENTS

The authors wish to acknowledge the financial support given to this project by the Britisk Ministry of Defence and the Danish Council of Technology. The scientific and technical expertise supplied by our research partners at the Loughborough University of Technology is also gratefully acknowledged.

Thanks are also due to Svend Aagesen who participated as a research fellow in the first stages of this programme and did much of the initial work to help the project under way.

AUTHORS

Lars Rimestad and Claus Kjærgaard received their engineering degrees in 1984. They are now employed as research fellows at the Danish Engineering Academy on contracts sponsored by the British Ministry of Defence and the Danish Council of Technology.

Jørgen Møltoft is chairman of the technical programme committee of the RELCON '86 CONFERENCE. He is an associate professor at the Danish Engineering Academy and his main topic of research and teaching has for the last decade been reliability technology. From 1979 to 1980 he was visiting professor at Loughborough University of Technology, U.K., researching reliability.

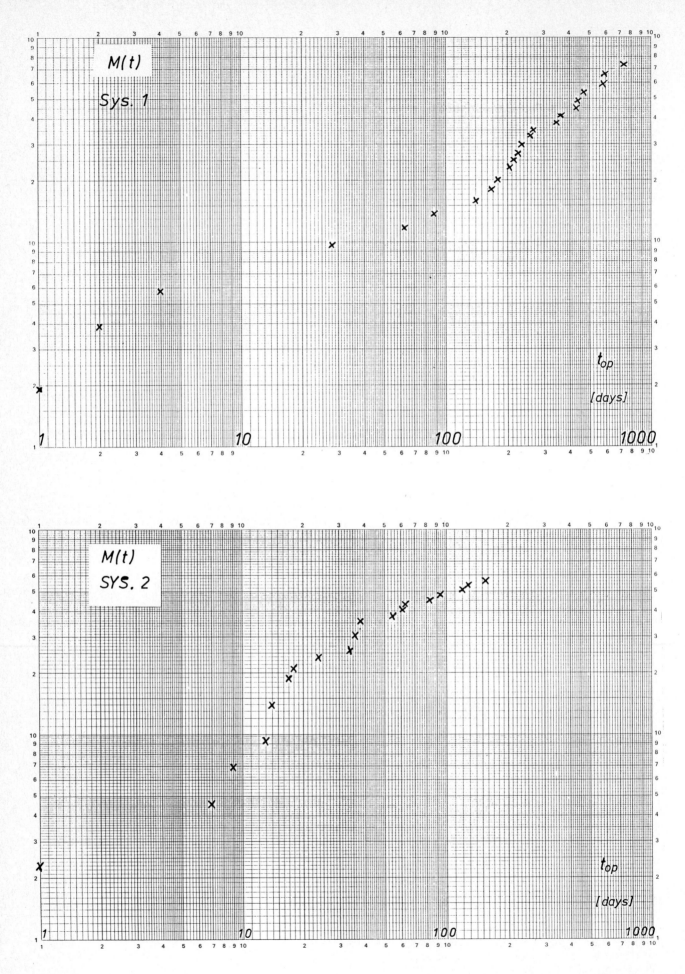

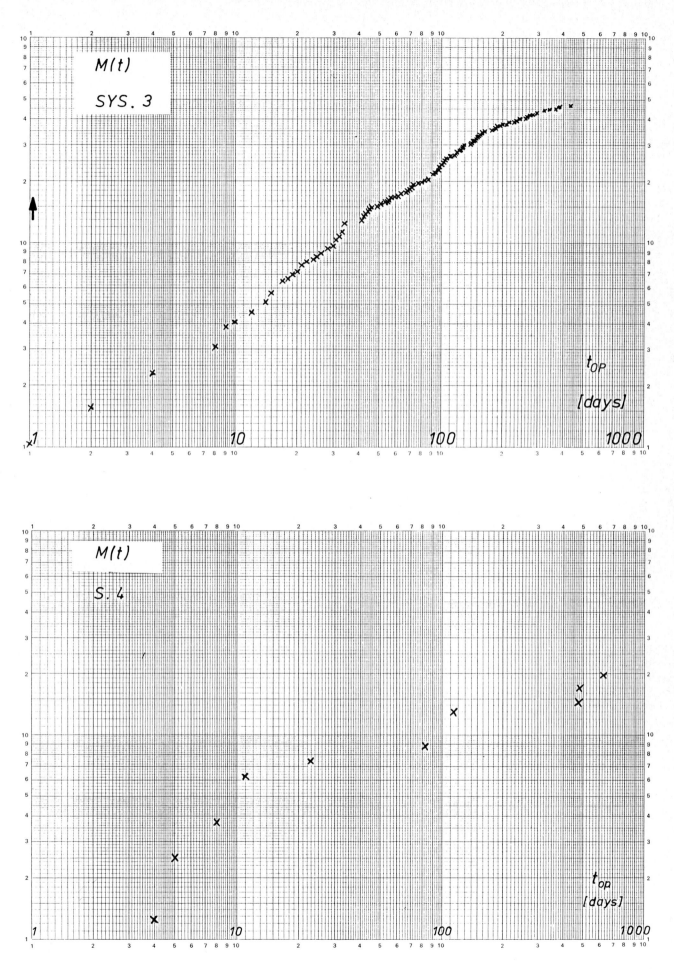

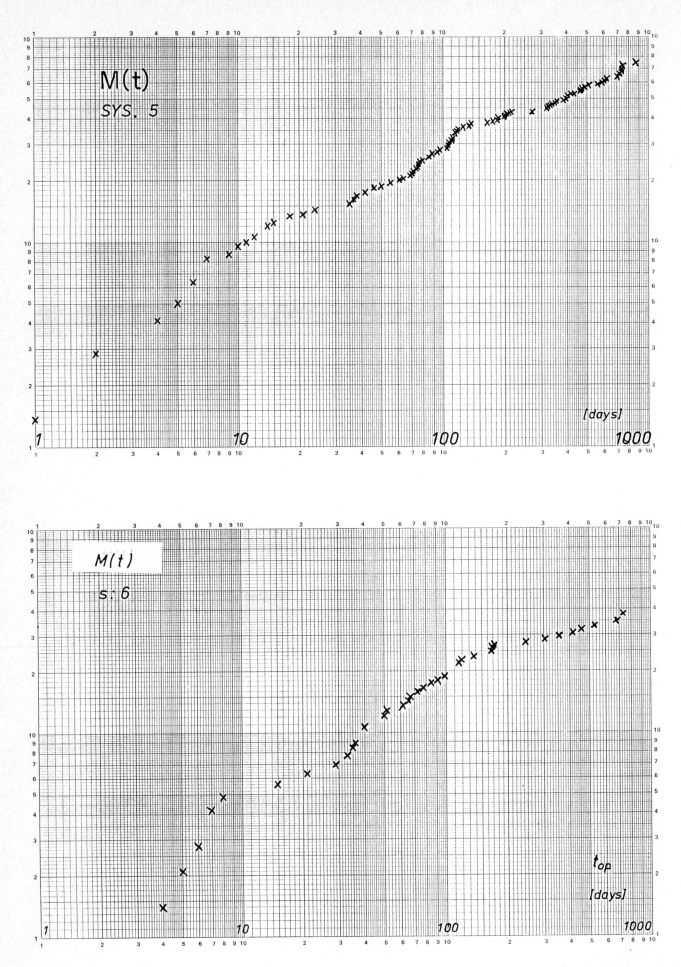

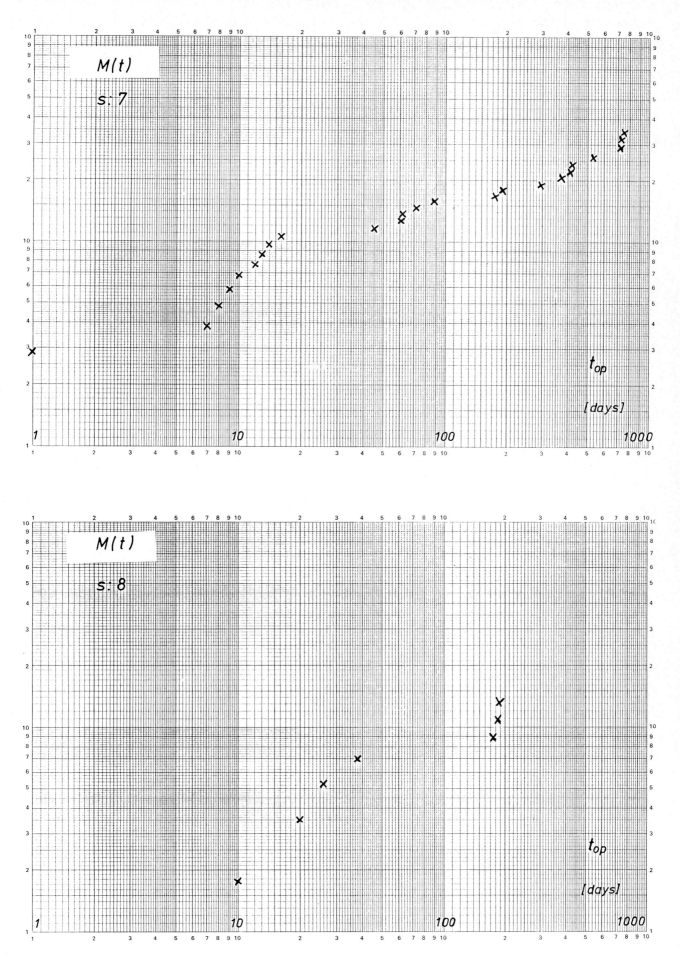

Reliability Technology — Theory & Applications
J. Møltoft and F. Jensen (Editors)
© Elsevier Science Publishers B.V. (North-Holland), 1986

SURVEILLANCE OF AN EFT/POS TERMINAL SYSTEM

Leif Kjoeller   and   Preben Rahtgen

PKK, The Danish Banks' and Savings Banks'
Debit and Credit Card Company Ltd.

Lautrupbjerg 10, DK-2750 Ballerup, Denmark.

A nationwide electronic funds transfer system is described with emphasis on total system reliability and security. To prevent even minor defects from deferring overall system performance, an extended and highly automatized surveillance system is required. The end-to-end system solution concerning reliability and security based on collection of automatically reported statistics and failures is described with the EFT/POS (Electronic Funds Transfer/Point Of Sale) terminal as the central element.

1.0 INTRODUCTION.

In recent years payment by means of debit- and credit cards has been a rapidly growing trade. In order to stay in the business of handling money-transfers the Danish banks and savings banks a few years ago decided jointly to develop a nationwide card payment system, the Dankort, and for that purpose set up the company PKK.

When designing such a system it is absolutely essential to consider the two aspects Security and Reliable Performance.

Security is considered the most important factor as the system is transferring real money in its transaction handling. Reliable performance is also of the greatest importance in order to maintain a 24 hours accessability and thereby assure acceptance and confidence from both retailers and cardholders.

2.0 SYSTEM ARCHITECTURE.

The system was designed as a total project between PKK, the Dankort company, and the chosen central computer supplier, the Danish PTT, the telephone companies, the chosen terminal manufacturer, the retailers, the consumer bodies, the government and a number of consultants.

The central computer is owned and maintained by PKK, and the communication network and the EFT/POS terminals are delivered and maintained by the telephone companies.

The communication network is based upon the existing two separate networks, the Public Telephone switching Network and the Nordic Data Network, Datex.

In order to achieve the cheapest possible terminals and the lowest cost on communication, the terminals are utili-

zing the local telephone lines to the nearest telephone exchange only. At the telephone exchange a network concentrator serves as concentrator for up to 200 terminals as well as it functions as protocol converter between the asynchronous terminal modem connection and the synchronous X.21 Datex network. This network again constitutes numerous input lines to the front-end of the central computer at PKK.

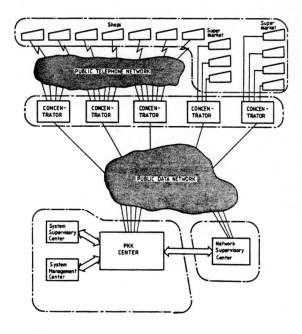

Fig. 1 The Dankort topology.

The network is planned to serve at least 60,000 terminals scattered all over the country and in conjunction with the central computer it is able to handle up to 700 transactions per second at peaks.

The system architecture is based upon a high degree of built-in reliability:
- The terminals have built-in self-diagnostic routines at power-up, checking the terminal and its externally connected equipment.
- The danish telephone network is generally a highly reliable communication means with a low MTTR.
- The Nordic Data Network has alternative routing between switching points and maintains a 24 hours surveillance. The Datex centres constantly checks all ports (connection points) for being alive and provides numerous error codes when a data connection is in progress.
- The input lines to the central front-end are dualized and are physically separated.
- The network concentrators have a built-in diagnostic program and provide performance statistics and alarms to local or remote surveillance terminals.
- The front-end and central computers are designed as a fault tolerant (n+1) redundant system where all data are stored in a dualized mode. The total system accessability is designed to be better than 99.5% between six o'clock in the morning and midnight and better than 85% for the remaining part of the day. Furthermore the central computer is equipped with several features regarding transaction processing performance, error detection and alarm generation.

The time in which the central computer - including the front-end - system is not accessable is calculated from the formula:

$$To = \frac{1}{M} \cdot \sum_{i=1}^{M} \frac{1}{Ni} \sum_{j=1}^{Ji} Tij + Tp$$

where

To = total non accessable time

M = number of active communication control units in the front-end system

Ni = number of active communication ports in communication control unit i

Ji = number of non active (faulty) communication ports in communication control unit i

Tij = drop-out time of communication port j in communication unit i

Tp = drop-out time of main processor

The number of active communication ports is equal to the total number of ports in the system's configuration table.

3.0 THE TERMINAL CONCEPT.

A trade situation can be divided into a registration phase and a payment phase. In the latter the card payment terminal plays the role as the central element providing a clean interface to the registration equipment, the cash register, and acts as the birthplace of the EFT transaction as well as the final communicator and registry of the transaction result.

In order to make certain that the registered result is valid and data have not been tampered with during transmission, a comprehensive number of steps have been taken to assure end-to-end security and control.

First of all the standard DES encryption algorithm is utilized between the terminal and the computer centre in order to protect data if being monitored on the open public telephone lines. Further the encryption gives a very good control of communication quality, as even a change or substitution of a single bit will result in a completely distorted and not understandable information after decryption at the receiving point.

Secondly other forms of end-to-end control measures have been taken in the transaction, such as including a terminal identification number unique to each terminal and generation of a random number for each transaction transmitted in the encrypted part of the transaction. This number is only returned to the terminal if the transaction was authenticated and verified in the centre. Otherwise the random number is substituted with an error code describing why the transaction was not approved.

Further a daily balance counter is incorporated in the terminal's securitymodule. The balance of the day will only be updated if the complete transaction is approved in the terminal upon controlling the answer from the centre for the mentioned parameters. The returned new balance (old terminal balance plus purchase amount) is verified for being the same as calculated and expected in the terminal itself.

Also the date in each received answer will be compared to that stored in the terminal when 'logged-on' by the retailer. The dates must match to ensure validation of the transaction on the correct banking day.

In case the transaction is not authenticated in the terminal, e.g. due to a faulty answer, the balance will not be updated and a flag will be set in the terminal status field, which will be transmitted to the centre in the following transaction.

Transactions approved and logged in the centre will only be cleared with the retailers' and cardholders' banks if the transaction is succeeded by a transaction from the same terminal containing correct balance and terminal status.

TERMINAL   CENTRE

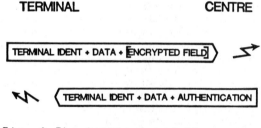

Fig. 2 The terminal and its communition records.

4.0 THE TERMINAL ELEMENTS

The Dankort terminal is, contrary to most other EFT/POS terminals, designed to be operated by the cardholder alone. This means that only two other elements are necessary to form a complete system at the retailers - an electronic programmable cash register terminal and a connection unit. The cash register provides the purchase amount for the transaction, contains the operator keys and display and performs the printing of receipt and journal information. The connection unit constitutes the power supply and modem for the terminal. In case a suitable cash register is not available its functions may be handled by a retailer unit, which is delivered as a module of the Dankort terminal. Together with this and the connection unit the three units form a free standing terminal system.

Housed in an ergonomically and neatly designed cabinet the Dankort terminal holds a microprocessor that provides the function control of the built-in elements, as well as it controls the communication to both cash register/retailer unit and modem/telephone connection unit. The built-in elements are:

- a manually operated cardreader (ISO track 2),
- a display containing 10 fixed guidance messages and a 10-digit numeric field,
- four large function keys,
- a sound alert,
- the security module with encryption functions and integrated PIN-code (Personal Identification Number) keypad.

The function control keeps a firm control of the functions in the connected retailer unit and provides all the records for the receipt- and journal-printers formatted for direct printing.

5.0 THE SECURITY MODULE

For reliability and security reasons a tamper resistant and battery backed-up security module has been developed and incorporated in the terminal. Primarily this module consists of a processor able to perform encryption functions according to the NBS Data Encryption Standard (DES).

The most secret information part of an EFT/POS transaction is considered to be the PIN-code, which has to be keyed in by the cardholder on the 10-digit numeric keypad. In order not to have this PIN floating on 'open wires' in the terminal the keypad is integrated with the encryption processor in the module enabling the code to be encrypted immediately upon entering. When all data for a transaction are available, a second encryption takes place forming the entire encrypted field of the transaction.

The encrypted datafield of the transaction consists of several encryption blocks chained to each other in a way that any change in any part of the field will affect the entire decryption function and give a meaningless result.

The encrypted information are all sensitive data, i.e. amount, balance, random number, PIN-code and certain information from the magnetic card.

The secret keys used for the encryption are stored temporarily in the security module together with a number of other information such as counters, balance, date, random number etc. All these data will vanish if the module is tampered with, as well as if the built-in battery expires. The secret keys may be changed upon request in a special key-exchange scheme and the remaining data only relates to either a log-on/ log-off period or to a single transaction.

The random number is generated within the module, transmitted in the encrypted field and returned as a transaction acknowledgement. The security module will validate a number of returned data and finally compare the received random number with the stored original, and thereafter constitute the final tranaction result to the function control processing.

The reliability of the terminal is ensured by means of the battery backed-up data in the security module, enabling the terminal to hold its data when e.g. the mains power is interrupted. The calculated back-up time of the battery is more than 5 years.

The integrated keypad is designed to withstand more than 300,000 operations of each key. As no terminal is expected to generate more than 200 transactions per day this durability will be sufficient for the terminal's life time.

6.0 THE TERMINAL ERROR COUNTERS.

The terminal contains a number of error counters which are incremented as different errors are detected by the terminal:

- The PIN failure counter is incremented by receiving a certain errorcode from the centre indicating that a false PIN was entered.
- The communication error counter is updated when either the terminal does not manage to get a call tone from the telephone network or a data carrier from the network concentrator within the time-out periods, or if it detects an error in the communication to the concentrator.
- The internal communication failure counter is incremented when a communications error is detected on the interface to externally connected equipment, i.e. a cash-register.
- The soft error counter is incremented when the terminal detects an error from which it is able to recover.
- The card reading failure counter is incremented when the third attempt to read a card fails.

Each errrortype has an individual number of retries, e.g. 3 for communication, before it is treated as an error. The error will be registered in the journal and displayed to the operator, who has to make a soft reset of the terminal in order to continue operation. A detected and registered error will update the appropriate error counter.

The limit for each counter is set to assure detection of serious errors, i.e. if too many errors of a type occur during a day. If the limit has been reached, an 'alarm' will be set in the terminal status field in the transaction following the necessary soft reset.

7.0 AUTOMATIC SURVEILLANCE.

The error counter values are reported to the centre in the log-off transaction at the end of a day or a period. If a terminal has had a number of errors, causing the limits to be reached, the centre will know this from the information in the terminal status field. According to the seriousness the centre may react differently upon this information and then e.g. cause an alarm in the centre or answer the terminal with an errorcode that forces it to log-off and thereby transmit all counter values for further investigation.

As every transaction is logged, it is also possible over a longer period to create statistics on the development of deteriorating transmission lines or internal behavior of the terminal.

As the terminal in each transaction informs about the previous transaction result and response time, it is possible to detect an increasing response time for the individual terminal or if any error flag is set.

The transaction processing in the centre will detect which error counter is overrun and generate the appropriate alarm to the surveillance centre. In severe cases, i.e. too many PIN-failures, the transaction processing will block the terminal record against futher use of the terminal. The following investigation then will show whether the detected PIN-failures were caused by an attempt to defraud or by a malfunction in the keypad.

Every alarm generated in the centre is logged in a special alarm file for later processing. In this way the life cycle of an individual terminal can be followed and thereby give an indication of the reliability of the terminals.

As the terminal informs about the response time seen from the terminal, statistics can be calculated to determine whether the network load is too heavy in a special geographical area, or at special times of the day, or if the terminal simply does not receive an answer within the time-out period.

If the centre detects that a terminal is having communication difficulties, the centre itself generates a test-transaction, calls the relevant network concentrator and tests, by receiving the transaction itself, if the concentrator is alive and if the communication to the centre is OK. If this test shows an error the centre will generate another test-transaction which will receive its answer direct from the concentrator and thereby determine which part of the communication network is malfunctioning.

Each transaction is validated after decryption for not having any values out of range. The decryption of the transaction will also detect errors not found by block check and parity control and spot severe errors in the security module.

8.0 COLLECTED DATA.

From data collected during the initial test period commencing November, 1984 the following examples of statistic presentation have been taken.

One of the more interesting measurement was calculation of response time seen from the terminal as function of time of day.

Shown in fig. 3 the distribution of reported response time is parted in 4 columns - response time less than 3 seconds (33), between 3 and 5 seconds (55), between 5 and 15 seconds (00) and no response received (XX). It is seen that in the hour commencing at eight o'clock the response time is increasing. This is due to the increased load in the system as every retailer has to log-on the terminal each morning. The no response column does not show in the figure as it in no case accounts for more than 0.5 % of the transactions.

9.0 CONCLUSION.

Although the total number of terminals installed by January, '86 was no more than 1.000, the gained experience clearly shows the necessity of automatic surveillance and collection of data.

Early terminals showed defects in the card reader. These card reading failures were reported to the center by the terminal thereby making it possible to correct the problem before it became too troublesome to the retailers and cardholders.

The testperiod did, not surprisingly, also show that the patience of the human being is not as great as that of a machine. This was accounted for by establishing a manned customer service centre dealing with incoming calls, like complaints and investigations from both retailers, cardholders and banks.

As the knowledge of the total system has improved all major problems have been detected by the built-in automatic surveillance system.

```
100 -

 90 -
     33      33              33              33
 80 -33      33              33      33      33
     33      33              33      33      33
 70 -33      33      33      33      33      33
     33      33      33      33      33      33
 60 -33      33      33      33      33      33
     33      33      33      33      33      33
 50 -33      33      33      33      33      33
     33      33      33      33      33      33
 40 -33      33      33      33      33      33
     33      33      33      33      33      33
 30 -33      33      33      33      33      33
     33      33      33      33      33      33
 20 -33      33      33  00  33      33      33
     33      33  00  33  00  33      33      33
 10 -33  00  33  00  335500  33  00  33  00  33  00
     335500  33  00  335500  335500  335500  335500
     +   +   +   +   +   +   +   +   +   +   +   +   +   +

     6:00    7:00    8:00    9:00    10:00   11:00   12:00
                                              TIME OF DAY

DISTRIBUTION OF REPORTED RESPONSE TIME
```

Fig. 3 Example of statistics presentation.

REFERENCES

(1) National Bureau of Standards, Data Encryption Standard (FIPS pub. no. 46, 1977).

(2) Davies, Donald W., The Security of Data in Networks (IEEE cat. no.EH0183-4, 1981)

(3) Rahtgen, P. and Kjoeller L., The Dankort Terminal and its Interfaces (PKK, Copenhagen, 1985)

BIOGRAPHY.

Preben Rahtgen (born 1944) is projectcoordinator of PKK and has an extensive background in computers and electronic cash registers. Prior to joining PKK in 1982 he was employed by NCR, Denmark, as rework and development supervisor.

Leif Kjoeller (born 1951) is systems designer of PKK. He received the B.Sc. degree in electrical engineering from the Danish Engineering Academy in 1974. Prior to joining PKK in 1984 he was employed by Audilux, Denmark, as technical manager.

Reliability Technology — Theory & Applications
J. Møltoft and F. Jensen (Editors)
Elsevier Science Publishers B.V. (North-Holland), 1986

INFLUENCE OF CLIMATIC CONDITIONS ON RELIABILITY OF ELECTRONIC COMPONENTS USED IN TELECOMMUNICATIONS EQUIPMENT

Alain LELIEVRE and Roland GOARIN
Centre National d'Etudes des Telecommunications
Centre de Fiabilité
BP 40, 22301 Lannion, FRANCE

A study of a group of French public telephone exchanges was conducted in 1984 and 1985 (operation CLIMEX). It revealed the significant influence of climatic conditions on the reliability of equipment.

The great number of observations made enabled numerous data to be collected. Sophisticated statistical processings were made using data processing means, in particular data analysis methods. Prominence was thus given to the various parameters affecting the reliability of components (temperature, hygrometry, location of the board, air conditioning process...).

A physical analysis of the failure of components complemented the statistical study. It resulted in taking practical measures to improve the field reliability of a number of telephone exchanges.

KEY WORDS
Reliability in the field, electronic components, telecommunications, telephone exchanges, data analysis.

1. INTRODUCTION

This study was devoted to two types of electronic telephone exchanges, which were called A and B. Eleven exchanges A and four exchanges B were concerned. The exchanges A connected about 200 000 subscribers. Data were collected from July 1984 to October 1985. The exchanges B connect 70 000 subscribers, and data were collected from November 1983 to September 1984.

2. INTEGRATED CIRCUITS

2.1. CMOS Integraded Circuits

2.1.1. Telephone exchanges A

The reliability of 9 types of CMOS Integrated Circuits was studied in telephone exchanges A. Table 2 gives the results according to the rooms - main or remote buildings - and the position of boards in frames - top or bottom.

Exchanges A	Exchanges B
Data collected from July 1984 to October 1985 Various climatic conditions according to rooms and supports	Two steps : - step 1 : from November 1983 to February 1984 - step 2 (normal climatic conditions) : from March to September 1984. A third of the supports were ventilated during step 1, not any during step 2.

Table 1
Climatic conditions during the observation of the exchanges

Various parameters affect the reliability of components (temperature, hygrometry, location of the board, air conditioning process...) in the two types of exchanges. Two steps must be distinguished for the exchanges B : various climatic conditions were produced during step 1 (temperature between 5°C and 25°C, hygrometry up to 95 per cent), but these conditions were regulated during step 2 (table 1).

The locations of all the components on the boards and frames of each type of telephone exchanges were perfectly recorded. The name of manufacterer and the date of manufacture were also well known for each component in exchanges B. All this information was filed for data processing.

The boards involved are situated in four racks numbered 2, 3 7 and 8 downwards. As the frames are air-conditioned, hygrometry decreases and temperature increases from the bottom up (figure 1). Temperature and hygrometry are controlled in main buildings by means of air conditioning.

Figure 2 shows a comparison of the predicted and observed failure rates of ICs. Predicted rates are calculated in accordance with the CNET Reliability Data Handbook for electronic components reliability [6]. These rates particularly depend on the complexity of ICs. The observed reliability of ICs is on average lightly lower than that predicted. The difference is particularly important in IC 22100 and IC 4028. These two components were more thoroughly investigated by means of failure analysis.

A. Lelièvre and R. Goarin

	Population	Predicted failure rates[(*)]	Number of failures	Observed failure rates						
				Total	Main buildings			Remote buildings		
					TF	BF	Tot.	TF	BF	Tot.
IC 22100	70520	146.0	136	244.	300.	309.	305.	125.	305.	226.
IC 4094 B	52890	103.6	35	84.	44.	103.	77.	51.	113.	86.
IC 4585 B	17630	58.5	4	29.	0.	0.	0.	44.	34.	38.
IC 4040 B	8810	114.5	1	14.	0.	0.	0.	44.	0.	19.
IC 4512 B	17630	56.6	16	115.	0.	257.	145.	44.	153.	105.
IC 40257 B	8810	70.8	4	57.	0.	0.	0.	44.	102.	76.
IC 4028 B	8810	64.9	19	273.	266.	412.	348.	174.	305.	248.
IC 4001 B	8810	22.3	6	86.	266.	0.	116.	0.	136.	76.
IC 4011 B	8810	22.3	5	72.	266.	0.	116.	0.	102.	57.
TOTAL	202750	101.0	226	147.	158.	182.	171.	79.	190.	141.

Table 2 - Telephone exchanges A
Reliability of Integrated Circuits

TF : top of frames (racks 2 and 3)
BF : bottom of frames (racks 7 and 8)
Failure rates are given in 10^{-9}/h. (Fit, i.e failure in time)

* Using the CNET Reliability Data Handbook

		IC 22100	Other ICs	Mean
Main buildings	Top of frames	300.	60.	158.
	Bottom of frames	309.	98.	182.
	Total	305.	81.	171.
Remote buildings	Top of frames	125.	44.	79.
	Bottom of frames	305.	81.	190.
	Total	226.	65.	141.
Main buildings	Standard ventilation	301.	75.	139.
	Non-standard ventilation	411.	106.	193.
Remote buildings	Natural convection	245.	61.	114.
	Top of frames ventilation	366.	147.	210.
Main buildings	Direct cooling	206.	32.	82.
	Cold water cooling	557.	175.	284.
Remote buildings	No cooling	270.	71.	128.
	Heat pump cooling	195.	89.	119.
Remote buildings	Convector heating	246.	70.	120.
	Radiator heating	175.	53.	88.
	No heating	264.	76.	130.
Remote buildings	Dehumidifiers	253.	121.	159.
	No dehumidifier	273.	73.	130.
Main buildings	Humidifiers	392.	92.	178.
Remote buildings	No humidifiers	260.	77.	129.
Remote buildings	Ventilation	257.	86.	135.
	No ventilation	259.	64.	120.
Mean		244.	71.	147.

Table 3 - Telephone exchanges A
Reliability of ICs according to climatic conditions
Failure rates are expressed in fits (10^{-9}/h.).

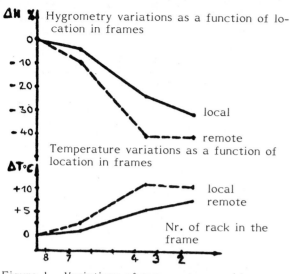

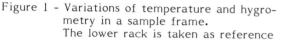

Figure 1 - Variations of temperature and hygrometry in a sample frame.
The lower rack is taken as reference

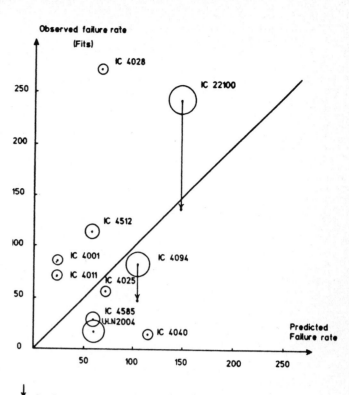

Figure 2 : Telephone exchanges A
Reliability of Integrated Circuits

The area of each circle is proportional to the IC population

The variation of the failure rate of ICs with time shows that the reliability is worst in summer – the warmest season – and in October, because of humidity (figure 3).

So it is interesting to examine the influence of the location of boards in rooms and frames - according to different climatic conditions on the reliability of ICs. Table 3 shows that this reliability is better at the bottom than at the top of frames. The difference is greater in remote than in main buildings. Fig. 1 shows that this variation of hygrometry and temperature in bottom racks (7 and 8) and in top racks (2 and 3) in a typical exchange. The decrease in hygrometry in frames from the bottom up explain the better reliability of ICs in racks 2 and 3, specially in remote buildings.

The failure rate of ICs is higher in main buildings than in remote buildings because of air-conditioning (table 2 and 3). Non-standard ventilation in main buildings corresponds to a bad reliability of ICs, and natural convection in remote buildings correspond to low failure rates. In the same way, the reliability of ICs is better in remote buildings without humidifiers than in main buildings with humidifiers. Nevertheless dehumidifiers do not improve the reliability of ICs. So the harmful effect of humidity was revealed, but using air-conditioning to control hygrometry does not reduce the failure rate of ICs.

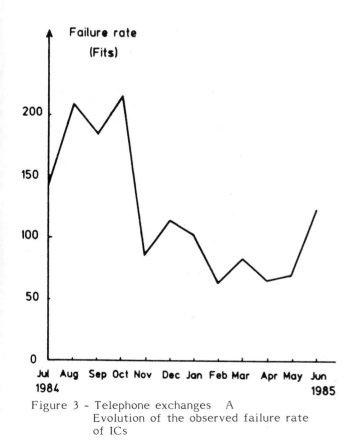

Figure 3 - Telephone exchanges   A
Evolution of the observed failure rate of ICs

		Top of frames	Bottom of frames	Total
Main buildings	Observed failure rate	300.	309.	305.
	Confirmed failure rate	150.	180.	167.
Remote buildings	Observed failure rate	125.	305.	226.
	Confirmed failure rate	54.	171.	118.

Table 4 - Reliability of IC 22100 according to the position of boards in frames (in 10^{-9}/h.).

	Hermetic	Non hermetic	Corrosion	No corrosion
Confirmed by electric measurements	24	24	30	18
Not confirmed	41	2	3	40

Table 5 - Failure analysis of 91 IC 22100 replaced

2.1.1.1. Reliability of IC 22100

The IC 22100 is a CMOS LSI Integrated Circuit encapsulated in a ceramic can. This 16 pins IC is an element of a cross matrix in subscribers boards. The boards studied in telephone exchanges A had a total of 70520 IC 22100. 136 IC 22100 were replaced from July 1984 till June 1985 : the observed failure rate was 244 fits. The predicted failure rate calculated in accordance with the CNET Reliability Data Handbook is 146 fits (table 1).

The influence of climatic conditions on the reliability of ICs was demonstrated above, particularly according to the difference in hygrometry between the top and the bottom of frames. This influence is confirmed by table 4. The confirmed failure rate was calculated after having measured each replaced component.

Table 4 revealed that the difference between the top and the bottom of frames is greater for the failure rates confirmed by electric measurements. The mean confirmed failure rate (130 fits) is better than the predicted rate (146 fits).

136 IC 22100 were replaced during the period concerned. Hermeticity test and physical failure analysis were carried out on 91 of them. The results are given in Table 5.

The failures of non hermetic replaced components are confirmed by electrical measures. 63 % of hermetic replaced components do not correspond to real failures. Helium testing reveals hermeticity defects in fifty per cent of really failed IC 22100.

The main physical failure cause is corrosion [1, 2]. 3 among 43 good components were corroded, though the functioning of these circuits had not yet been affected.

Components showing corrosion are more numerous than non hermetic ICs. Bad hermiticity is the main cause of corrosion, but internal corrosion also occurs, independently of climatic conditions.

Discrimination between hermetic and non hermetic ICs gives the following results :

Non hermetic 26 $\longrightarrow$ confirmed failures 24 $\longrightarrow$ corrosion 17 (71 %) No corrosion 7

Hermetic 65 $\longrightarrow$ confirmed failures 24 $\longrightarrow$ Corrosion 13 (54 %) No corrosion 11
Non confirmed failures 41

Among the 30 components which are really failing, the number of corroded components is almost the same, whether these components are hermetic (13) or not (17). Corrosion may depend on climatic conditions but in some cases it may be an intrinsic defect of the integrated circuits.

2.1.1.2. Reliability of IC 4094

35 IC 4094 were replaced during the period concerned. The observed failure rate (84 fits) is better than the predicted rate (103.6 fits) for this CMOS Integrated Circuit. Electrical tests and physical failure analysis were carried out on 32 IC 4094. The results are given in table 6.

13 failures of ICs out of 32 could not be confirmed by electrical tests.

12 among 17 bad components are not hermetic and corrosion appears in 13 of them.

Electrical test		Hermeticity test		Failure analysis			
		hermetic	non-hermetic	non revealed	corrosion	contamination	other
Not confirmed	13	12	1	13	0	0	0
Parametric failures	4	2	2	0	2	1	1
Catastrophic failures	13	3	10	0	11	1	1
Non measurable	2	2	0	2	0	0	0
Total	32	19	13	15	13	2	2

Table 6 - Telephone exchanges A
Analysis of 32 IC 4094 replaced

		4001 B	4051 B	4052 B	4066 B	4512 B	Total
Populations		4584	27154	4584	4583	8864	49769
STEP 1 ventilated frames	observed failure rates	1670.	1520.	1000.	1670.	350.	1290.
	confirmed failure rates	1340.	1410.	1000.	1670.	350.	1200.
STEP 1 non-ventilated frames	observed failure rates	760.	660.	1050.	1620.	500.	760.
	confirmed failure rates	760.	580.	1050.	1240.	250.	640.
STEP 2	observed failure rates	760.	580.	130.	540.	140.	470.
	confirmed failure rates	490.	530.	130.	360.	90.	400.
Predicted failure rates		23.	55.	110.	50.	57.	57.

Table 7 - Telephone exchanges B
Reliability of Integrated Circuits
Observed failure rates and confirmed rates
after electrical tests

2.1.2 Telephone exchanges B

The reliability of 5 types of CMOS Integrated Circuits was studied in telephone exchanges B. These ICs came from four manufactures: RTC, SGS, Motorola and RCA. Figure 4 shows that the reliability of these components strongly depends on climatic conditions. The mean failure rate is higher during step 1 (fluctuating temperature and hygrometry) than in step 2 (normal climatic conditions). During step 1 the reliability is better in non-ventilated than in ventilated frames. Motorola ICs have a good reliability during the step 2, and they are not much influenced by variations of climatic conditions. On the other hand, the integrated circuits from the other manufacturers, have higher failure rates and are very sensitive to climatic environment.

Electrical tests were made on each component replaced. Results are given in table 7. The improvement of reliability corresponding to normal climatic conditions is observed in the five types of integrated circuits. This improvement is corroborated by the confirmed failure rates calculated after electrical tests.

The influence of climatic conditions is confirmed by the figure 5, showing an increase in the failure rate of ICs in summer.

Two integrated circuits are used in telephone exchanges A and B : IC 4001 and IC 4512. The observed failure rate of these two components is better in exchanges A during the periods concerned.

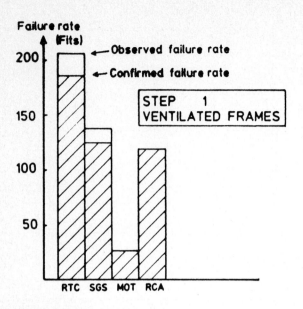

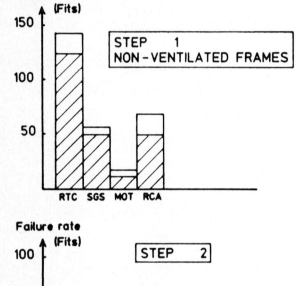

Figure 4 - Telephone exchanges B
Reliability of CMOS Integrated Circuits
according to manufacturers

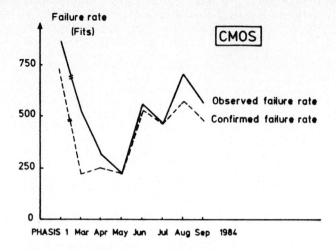

Figure 5 - Telephone exchanges B
Evolution of the reliability of CMOS ICs

Hermiticity test and physical failure analysis were made on 174 CMOS integrated circuits, and failure was confirmed. The results are given in table 8.

Only 3 defects on non-hermetic ICs among 112 are not revealed by physical failure analysis, and 106 components show corrosion. 58 % of hermetic ICs are also corroded.

The integrated circuits from RCA are the less reliable ones (figure 4). Most of those which were replaced were not hermetic (36 among 52) and showed corrosion (50 among 52).

Corrosion is the main cause of failure for the five types of IC studied. The table 9 compares the causes of failure during step 2 and step 1, in ventilated and non-ventilated frames.

Corrosion is the main cause of defects during the two steps. Figure 5 shows that normal climatic conditions (step 2) correspond to a good reliability of ICs in April and May. But the failure rate increases in summer because of corrosion.

Hence a high proportion of confirmed ICs failures during step 2 are due to corrosion.

2.2. TTL Integrated circuits

One type of TTL IC was followed up in exchanges A : the ULN 2004, which is an analogic interface. Only six components were replaced during the period concerned. The failure rate (17 fits) was much lower than predicted (59.5 fits).

Five types of logic TTL-LS ICs were studied in exchanges B. 9 components confirmed in failure were analysed. The causes of failure were fusion and breakdown. The main confirmed failure rate was 64 fits (predicted rate : 25 fits).

3. OTHER COMPONENTS

3.1. Transistors

Common silicon transistors were studied in exchanges A (2N2222, 2N2369...). 247000 transistors were installed on the boards concerned. 55 were replaced, and the mean observed replacement rate was 28 fits (predicted rate : 8 fits).

3.2. Photocouplers

Photocouplers SL 5501 were used in exchanges A and B. The observed failure rates are 14 fits in exchanges A and 10 fits in B (predicted rate : 50 fits).

	non revealed defects	light corrosion	corrosion	strong corrosion	fusion	impurities	Total
Hermetic	22	11	16	9	3	1	62
Non hermetic	3	18	24	64	2	1	112
Total	25	29	40	73	5	2	174

Table 8 - Telephone exchanges B
Failure analysis of 174 integrated circuits

	STEP 1		STEP 2
	ventilated frames	non-ventilated frames	
Hermetic	9	27	26
Non-hermetic	17	21	74
Total	26	48	100
Non-revealed defects	5	9	11
Corrosion	21	36	85
Fusion	0	2	3
Impurities	0	1	1

Table 9 - Telephone exchanges B
Failure analysis of ICs according to climatic conditions

3.3. Relays

3.3.1. Exchanges A

The relays studied in exchanges A are "European 4 Form C" relays ("break before make"). 105 had to be raplaced during the period concerned. The mean observed replacement rate was 47 fits (predicted failure rate : 50 fits).

The table 10 shows that the best reliability of relays is found in remote buildings and particularly at the top of frames, where hygrometry is the lowest. The influence of climatic conditions is confirmed by figure 6, which shows that the failure rate of relays is higher in summer.

Only 18 % of relays replaced are confirmed in failure by electrical tests.

3.3.2. Exchanges B

Miniature "2 Form C" relays were used in exchanges B. Most of them had been manufactured by ITT. The table 11 shows the influence of climatic conditions on the reliability of relays. The failure rate confirmed by electrical tests is only 20 fits during step 2 (non-ventilated frames, normal temperature and hygrometry).

Most of the relays confirmed in failure had too high contact resistances. Some of them presented pollution of contacts by silica or impurities.

The same phenomena were observed on relays from exchanges A.

4. STATISTICAL PROCESSING BY FACTORIAL ANALYSIS

4.1. Methodology

Factorial Analysis is a family of statistical methods based on geometrical criterion. These techniques permit to make a graphical representation on plans (factorial plans) of the information contained in a rectangular table [3, 4, 5].

Among these methods the Factorial Analysis of Correspondences is used to describe correspondence tables. Let us consider a population with k items described by a set of p characteristics. The items may be for instance failed components, a characteristic I being the date of failure and J the number of the rack where the boards concerned are located. Each characteristic may be split up in several attributes, for example m months for the date and n racks for the position in the frame. So it is possible to establish a binary table with k lines (k components) and m in columns (m for the months, n for the racks). A component on a rack j which fails during the month i is then described by a line with 1 in the columns i and m + j, and with O in the other columns.

Factorial Analysis is based on the following principle : each line of the table may be represented

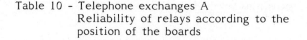

	top of frames	bottom of frames	Total
Main buildings	37.	84.	63.
Remote buildings	29.	52.	42.

Table 10 - Telephone exchanges A
Reliability of relays according to the
position of the boards

	STEP 1		STEP 2
	ventilated frames	non-ventilated frames	
Number of repla-cements	10	20	28
Observed failure rate	110.	70.	40.
Confirmed failure rate	80.	30.	20.

Table 11 - Telephone exchanges B
Reliability of relays according to climatic
conditions

Figure 6 - Telephone exchanges A
Evolution of the observed failure rate
of relays

by a point in a m + n dimensions space according to the m + n columns of the table. This set of points is projected on one or several plans on which the deformation is minimal. These factorial plans are defined by main axes of inertia. Each factorial axis must be orthogonal to the preceding axes and extract the maximum of inertia from the set of points.

The representation of the table is particularly satisfactory when the proportion of inertia extracted by the first axes is important. In this case the two or three first axes are enough to correctly condense the information contained

in the table. These axes determine two or three factorial plans.

In the same way the columns of the table may be represented on plans. It is proved that the axes calculated from the lines or the columns of the table are identical. So components and attributes may be simultaneously represented on the same factorial plans.

4.2 Application to reliability

Factorial Analysis may be extended to a table in which lines and columns represent several caracteristics with numerous attributes. Suppose for instance the table K has k lines corresponding to k climatic conditions in exchanges A. The table has n columns which correspond to the n types of components on the board which is the most used in exchanges A. K (i, j) is the failure rate of the component j in the climatic conditions i. K is formed of 33 lines (33 types of climatic conditions) and 6 colums which correspond to the whole board and to 5 sets of components:

- IC 22100
- other ICs
- relays
- transistors
-photocouplers.

To obtain a binary table of correspondence each column of table K has been split up in three as follows (if m is the mean and s the standard deviation of the column) :

- column 1 : failure rates $\lambda < m-s/2$ (for ex. fig 7 : board-1)
- column 2 : $m-s/2 \leqslant \lambda < m+s/2$ (for ex. Fig. 7 : board-1)
- column 3 : $\lambda \geqslant m + s/2$ (for ex.: board -3).

So each column of the table K is split up in three columns which contain 1 if λ is in the correspon-

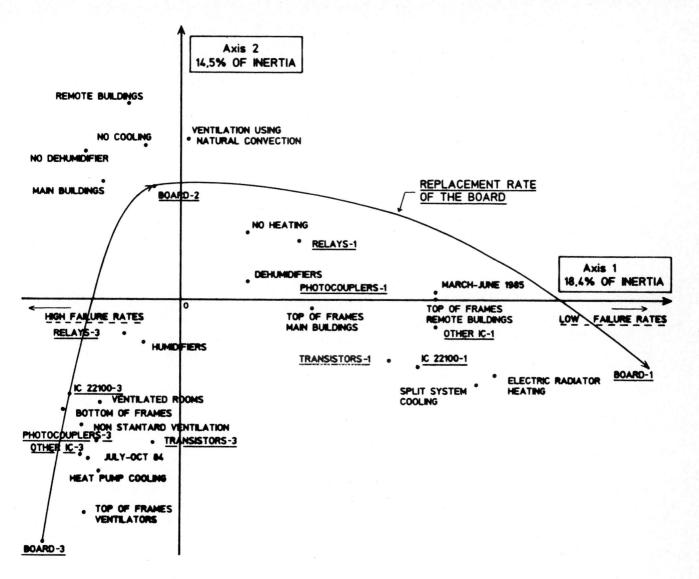

Figure 7 - Telephone exchanges A
Statistical processing by factorial analysis : the first factorial plan

For ex. : IC 22100-1 indicated the failure rates λ of IC 22100 when
λ < m-s/2 (m is the mean, s the standard deviation)
IC 22100-3, the failure rates λ when λ ⩾ m + s/2

ding interval and O otherwise. The binary table obtained has 33 lines and 18 columns. It has been processed by means of Factorial Analysis.

The main factorial plan represents the third of the inertia extracted from the table (figure 7). The less interesting points are in the central region of the plan. They have not been represented for clarity reasons. The factorial plan is explained by first observing the periphery of the projected set of points, where the most typical attributes which best characterize some groups of climatic conditions can be found.

Factorial Analysis permits to represent lines and columns of the table on the same plans. So climatic conditions and reliability of components may be correlated. The curve on figure 7 joints the points representing bad, medium and good reliability of the whole board. Low failure rates are represented by points placed towards the right extremity of the first axis. So the best reliability of all the groups of compo-

ponents was obtained during the period from March to June 1985, in the top racks, specially in remote buildings (low hygrometry). The best reliability was obtained in buildings with electric radiator heating, split-system cooling and sometimes dehumidifiers.

High failure rates correspond to the southwest region of the plan. The components the more influenced by bad climatic conditions seem to be integrated circuits and photocouplers. High failure rates were observed from July to October 1984, in the bottom racks, in buildings with top of frame ventilators, heat-pump cooling or non-standard ventilation. This proves that bad climatic conditions have a bad effect on reliability of components, particularly integrated circuits.

Factorial Analysis was also applied to telephone exchanges B. That permitted to graphically display most of the important statistical information and to describe the differences between steps 1 and 2.

5. CONCLUSIONS

The influence of climatic conditions on the reliability of components in telephone exchanges,
was widely proved in this study. Correlations
between environmental conditions and reliability
were precisely described by means of Factorial
Analysis of Correspondences. So it is proved
that air-conditioning has a bad effect on reliability of integrated circuits due to corrosion
when the hygrometry rate is too high. The reliability of components is nevertheless acceptable
with controlled climatic conditions. For most
of the components, the observed failure rates
generally correspond to the predicted rates,
in accordance with the CNET Reliability Data
Handbook.

This was confirmed by calculations made for
the whole of France from July 1984 to June
1985. Results show an additional increase in
the reliability of components in the French public
telephone exchanges during this period.

BIOGRAPHY

Alain LELIEVRE received his Master's degree
and Doctorate in Mathematics from the University of Rennes. He is working for the Reliability
Centre at C.N.E.T. Lannion. His main field of
interest is the application of mathematical statistics to reliability problems.

Roland GOARIN is the Head of the Reliability
Centre at C.N.E.T. Lannion. He is more particularly in charge of the Reliability Data Handbook
for electronic components. He received his Master's degree in Physics and Doctorate in electronics from the University of Rennes. He is the
chairman of the Reliability Committee of E.O.Q.C.
and a member of the I.E.C. Committee 56 (Reliability and Maintainability).

REFERENCES

[1] Lelièvre, A., Monfort, M.L. and Berger, C.,
    Fiabilité de composants utilisés dans les
    centraux téléphoniques E10 B, 4è. Colloque
    Intern. sur la Fiabilité (Perros-Guirec, France,
    mai 1984).

[2] Archambault, C., Goarin, R., Lelièvre, A.,
    A., Monfort, M.L. and Petibon, M.,
    Influence de l'environnement hygrométrique
    sur la fiabilité des équipements et des com-
    posants dans les unités de raccordement
    d'abonnés téléphoniques, 2è. Colloque Nat.
    sur la Thermique et l'Environnement (Perros-
    Guirec, juin 1985).

[3] Benzecri, J.P.,
    L'analyse des Données, Vol. 2 : L'analyse
    des correspondances (Dunod, Paris, 1976).

[4] Goarin, R.,
    Application de l'analyse des correspondançes
    à l'étude de la fiabilité des composants
    électroniques, Congrès Nat. de Fiabilité
    (Perros-Guirec, France, 1972).

[5] Lelièvre, A.,
    Etude de la fiabilité d'un transistor à effet
    de champ As Ga pour hyperfréquences par
    des méthodes d'analyse factorielle et de
    classification automatique, Microelectronics
    and Reliability 4 (Pergamon Press, Oxford,
    1983).

[6] Recueil de Données de Fiabilité du C.N.E.T.,
    Vol. 1 : Composants électroniques (Lannion,
    1983).

Reliability Technology — Theory & Applications
J. Møltoft and F. Jensen (Editors)
© Elsevier Science Publishers B.V. (North-Holland), 1986

COMPONENT FAILURE RATES: FIELD EXPERIENCE VERSUS LABORATORY TESTS AND EXISTING MODELS

Ferdinando FEDERICI, Giorgio TURCONI
ITALTEL - Società Italiana Telecomunicazioni
Cascina Castelletto, I-20019 Settimo Milanese, Italy

The execution of reliability prediction for electric components does not involve only the application of existing models: it must be noted in fact that, to be able to apply the models, the components user must fulfil all the quality procedures which assure on one hand the reaching of the required quality levels, on the other hand the possibility of maintaining them constant in time.
This of course in all the manufacturing phases: component qualification, quality and reliability periodic verifications, electrical and environmental incoming inspection tests.
The purpose of this paper is to show the way followed by Italtel for the solution of this problem consisting in using models adjusted by the data coming from field and is based on a complete Quality Assurance system for components.

1. THE PROBLEM OF RELIABILITY PREDICTION

The execution of reliability prediction in electronics is meant for obtaining, during the design phase, the required information to carry out the choices referring to circuits systems structure and logistics, of course besides controlling the respect of the existing specifications objectives.

Therefore it comes out the necessity that prediction calculations must be based on data and reference models which are reliable in themselves: the calculated values infact have to be considered as realistically reacheable objectives, thus leaving to the testing activities (qualification) the duty to point out possible anomalous conditions to be removed in time to assure the reaching of such objectives.

Moreover, it must be emphasized the fact that employed models have not only to guarantee the data internal coherency, as it could be acceptable to make comparative evaluations of different possible solutions, but have to provide absolute reliable values: design alternatives based on different technological choices may be evaluated only if the different models describing them supply real values.

2. SOURCE, VALIDITY AND LIMITS OF THE MODELS

A first consideration which must be done is that the reliability models for the solution of complex systems, whether they are based on Fault Tree, Markov or simulation techniques, refer and get validity from the failure rate values allocated to the system blocks.

In electronics the most complete reference as regards the failure rate models for components is the MIL-HDBK-217 which supplies, within the assumption of constant failure rates (exponential distribution of times to failure), the failure rate value as function of electrical, environmental and quality parameters.

The validity of the handbook lays on the large amount of experimental data taken into account in the definition of models yet this is also its limit owing to the difficulty of following rapidly the technological evolution of components.

Generally, in fact, the new technologies are evaluated in a conservative way.

As regards the structure of models, they include a certain number of factors, each one being function of a certain characteristic (electrical, environmental, quality); the different factors are bound together by sum or product and through a scale parameter they give the component failure rate.

To apply the models of HDBK-217 in another field different from the military one and, as far as we are concerned, in Telecommunications (TLC), a lot of problems comes out for the identification of the reference parameters (environmental and quality conditions).

This difficulty in identification occurs in particular as regards the quality level: for the users of components in TLC field, which surely apply quality procedures (qualification/screening)', usually none of the levels listed in HDBK-217 exactly corresponds to its own test procedure and, on the other hand, to each quality level is correlated a moltiplicative factor of the base failure rate, therefore its value depends directly on the coice of the quality factor which can cover a two decades interval', as shown in Table 1.

3. INTEGRATION OF MODELS WITH EXISTING USER DATA

The solution of the problem can be obtained through the correlation of models to the real data by introducting some suitable corrective factors in order to fit the models to the components user.

This can be made on condition that the user himself satisfies a certain number of requirements:

a/ He must fulfil quality procedures to cover all production phases as regards the components;

b/ These procedures have to guarantee the intrinsic quality and reliability of the components (qualification/screening) and of the processed components (qualification and control procedures of the manufacturing processes);

c/ The components may be used only after getting through the qualification tests;

d/ The real reference data for the correction of models have to refer to components without failure systematicity of anomalous behaviour, this thanks to the fact that the prediction values which can be obtained form models must be considered as objectives, and therefore really obtainable, but only after removing any failure cause external to those being intrinsic in the use of components;

e/ The hypothesis of constant failure rate must be guaranteed, on one hand by removing the infant mortality (through the screening at component level - if necessary - and at equipment/system level)', on the other hand by adopting the necessary measures in case of components which may show problems regarding the aspect of useful life (for example the Integrated Circuits in plastic case).

Consequently the integration between the models and the experimental data is carried out following two ways:

- one for assuring to the components an uniform quality level and therefore the same behaviour as regards quality/reliability;

- the other for the determination of coefficients and factors to be applied to the general models in order to turn them into models applicable to the components user.

4. MODELS MODIFICATION

There are different ways of intervention on the models according to the type of modification one wants to introduce:

a/ Intervention on the functions of the model;

b/ Intervention on the parameters (additives or multiplicative) of the model.

Source: MIL-HDBK-217

INTEGRATED CIRCUITS

Quality Level	S	B	B-0	B-1	B-2	C	C-1	D	D-1
Quality Factor	0.5	1	2	3	6.5	8	13	17.5	35

TRANSISTORS

Quality Level	JANTXV	JANTX	JAN	Lower	Plastic
Quality Factor	0.12	0.24	1.2	6	12

Table 1

Generally the a/ intervention is adopted when the modifications are substantial', for example for extending the validity range of a model (fit for a more limited ambit) because the technological evolution of components should be followed.

The b/ intervention is carried out for fixing a landmark for the models when the correlation with the existing experimental data is performed.

For consolidated components normally it is enough to intervene in this second way, while for the most recently introduced ones it could be better to operate in a deeper way: it must be considered infact that the models available in literature (and among them the HDBK-217 ones) are made public some years later than the time of manufacturing of the examined components and therefore often they are not up-to-date if compared with the rapid technologic evolution.

5. FIELD DATA

As mentioned, the examination of the data observed in field is essential in the acquisition of the elements necessary for the adaptation of models.

Nevertheless it must be specified that, even if speaking of data coming from field, we refer to the data originated by the repair centres where the maintenance at 3rd level is carried out (repair by replacement of components).

Consequently it is right to call of Removal Rate (measured in Rit=removals/billion hours) instead of Failure Rate (measured in Fit=failures/billion hours).

Even excluding, for statistical reasons, contemporary failures, often during repair more than one component is replaced therefore, in general:

$$Rit \geq Fit$$

this is due to the fact that the fault locations procedures are planned with a view to minimize the total repair cost, so it may occur that it is preferable to replace a group of components including the failed one, instead of spending time for isolating the single damaged component (particularly where the fault locations procedures are not performed by Automatic Test Equipment).

Moreover it may happen that the failure of a component causes the failure of another component connected with the first one (dependent failure).

Through the exact determination of the

real failure rate can be carried out only fulfilling the failure analysis of all the components replaced during repairs, nevertheless in the presence of a wide number of data and under the hypothesis of random events it is possible to make an evaluation of the failure rate by introducing the Mean Removal Rate, that is by assigning 1/N failures to each of the N components replaced during a repair.

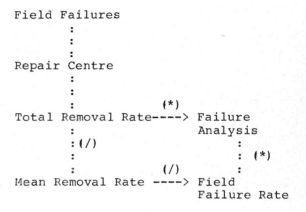

(/) This is the way now in use
(*) Way of future development

6. MEASURED REMOVAL RATES

The examination of the Mean Removal Rate (RIT) values measured in field will be now performed.

The sample that we show here refers to a PCM transmission system which has reached a field steady behaviour.

The data relate to an observation period covering the two years 1984-1985.

The reference environment conditions are the following:

- Installations on permanent racks;

- Racks are placed in masonry buildings without direct opening towards outside;

- The effect of the power dissipation of the equipment is such as to keep a constant temperature of the components at an average value of approx. 40°C;

- The value of the electric stress applied to the components ranges between 10% to 50% (reference value = 20%) compared to the max. applicable value.

Only data referred to components which has cumulated more than one billion hours x component have been taken into account.

Component	Sample Hours x Comp. (10^9)	Mean Removal Rate (RIT) $(10^{-9} h^{-1})$
Film Resistors (2% tol)	68.3	0.2
Plastic capacitor	18.4	2.0
Tantalum capacitor	3.4	1.9
Ceramic capac. (multilayer)	10.2	0.3
Signal diode	9.7	0.9
Zener diode	19.2	0.9
Rectifier diode	16.2	0.9
PNP transistor	7.3	2.5
NPN-PNP Darlington trans.	6.6	4.2
I.C. LSTTL SSI/MSC Cerdip.	8.9	5.1
I.C. CMOS SSI/MSI Cerdip.	1.7	14.2
Inductor (ferrite)	11.9	0.9
Audio transformer	6.1	2.8

Table 2

In the end, a special interesting point is the comparison between the values calculated according to the HDBK-217, the data coming from field and the corrective factors used for adapting the models (Table 3).

The meaning of the terms used in Table 3 is the following:

Lambda base:
Failure rate calculated in accordance with HDBK-217 for conditions listed in Section 6 excluding Quality Factor.

Lambda MIL:
Failure rate calculated as above under the following conditions with regard to Quality Factor.

Resistor and Capacitors: Level P
Discrete Semiconductors: Level JAN
Integrated Circuits: Level B
Inductive devices: Level MIL

RIT:
Mean Removal Rate.

K1:
Corrective factor applicable to Lambda base into models for obtaining field values

The stated results, even if they cannot be generalized to other users being referred to the Quality Assurance (QA) system adopted in Italtel, are a real example of the method illustrated in this paper.

Component	Lambda base $(10^{-9} h^{-1})$	Lambda MIL $(10^{-9} h^{-1})$	RIT $(10^{-9} h^{-1})$	K1
Film resistor (2% tol)	0.87	0.26	0.2	0.22
Plastic capacitor	1.27	0.38	2.0	1.56
Tantalum capacitor	5.79	1.73	1.9	0.33
Ceramic capac.(multilayer)	0.85	0.25	0.3	0.35
Signal diode	0.20	0.30	0.9	4.40
Zener diode	0.66	1.98	0.9	1.35
Rectifier diode	0.30	0.45	0.9	2.93
PNP transistor	0.58	0.70	2.5	4.26
NPN-PNP Darl. trans.	1.25	1.50	4.2	3.34
I.C. LSTTL SSI/MSI	3.60	3.60	5.1	1.4
I.C. CMOS SSI/MSI	4.16	4.16	14.2	3.4
Inductor (ferrite)	2.08	6.24	0.9	0.43
Audio transformer	2.08	6.24	2.8	1.34

Table 3

7. EXAMPLE OF ITALTEL COMPONENT QUALITY ASSURANCE PROCEDURE: INTEGRATED CIRCUITS IN PLASTIC PACKAGE

QA System adopted in Italtel for Integrated Circuits in plastic package is structured on:

a/ Technological analysis;
b/ Device Qualification tests;
c/ Incoming inspection Quality and Reliability tests;
d/ Requirements for manufacturing processes applied to I.C. in plastic package.

The purpose of a/ and b/ tests is to select components in order to assure quality and reliability level; c/ tests assure the maintaning of the required level; d/ tests assure that manufacturing processes do not affect the intrinsic quality and reliability of components.

Test activity flow is shown in details in the following tables.

TECHNOLOGICAL ANALYSIS
:
: --Package related tests--
/o/Class transition temperature
: of the plastic material
:
/o/Water extractable ionic
: substances
:
/o/Water absorbtion test
:
: --Dice related test--
/o/Glassivation integrity test
: (MIL-STD-883, Meth. 2021)

REQUIREMENTS FOR MANUFACTURING PROCESSES APPLIED TO I.C. IN PLASTIC PACKAGE
:
/o/Preheating (Duration and
: max temperature)
:
/o/Soldering (Duration and
: max Temperature)
:
/o/Cooling (Temperature rate)
:
/o/Cleaning (Cleaning agent,
: Cl- residual contamination,
: Ultrasonic frequency and
: water absorbtion)

DEVICE QUALIFICATION TESTS
:
:
/+/Visual and mechanical
: inspection (MIL-STD-883,
: meth. 2009.5 - 2015.3
/o/Dimensions (MIL-STD-883,
: meth. 2016)
:
/-/Static and dynamic
: electrical test at 25°C
: (MIL-STD-883,
: meth. 3000 & 4000)
:
/o/Static and dynamic
: electrical test at Tmax
: and Tmin (MIL-STD-883,
: meth. 3000 & 4000)
:
/o/Solderability
: (IEC 68-2-20)
:
/+/
:
/o/Resistance to soldering
: heat (IEC 68-2-20)
:
/+/
/-/
:
/o/Thermal shocks-liquid
: to liquid (MIL-STD-883,
: meth. 1011.5)
:
/+/
:
/o/Thermal cycles
: (MIL-STD-883, meth. 1010.5)
:
/+/
/-/
:
/o/Damp heat 85°C/85% RH with bias
: (CECC 90000 - 4.6.2)
:
/+/
/-/
:
/o/Operating life at 125°C
: (MIL-STD-883, meth. 1005.4)
:
/+/
/-/
:
/o/ESD sensitivity
: (MIL-STD-883, meth. 3015)
:
/-/

INCOMING INSPECTION QUALITY AND RELIABILITY TESTS

```
:
:
/o/Week code verification (*)($)
:
/o/Visual and mechanical
:  inspection (MIL-STD-883,
:  meth. 2009.5-2015.5)
:
/o/Dimensions (MIL-STD-883,
:  meth. 2016)
:
/o/Functional, static and dynamic
:  electrical test
:  (Individual device spec.)
:
/o/Solderability
:  (IEC 68-2-20)
:
/o/Resistance to soldering
:  heat, only for SMD (+)
:  (IEC 68-2-20)
:
/o/Damp heat with bias (§)(+)
```

(+) Sample size independent from lot size

(*) Week codes not older than 12 months

($) up to 5000 dev.: 2 week codes max, up to 15000 dev.: 3 week codes max, more than 15000 dev.: 4 week codes max

(§) Accelerated: 85°C
                 85% RH, 1000 hrs
or HAST   :130°C
                 85% RH,   50 hrs

8. CONCLUSIONS

The paper shows the approach used by Italtel for the solution of the problem concerning the evaluation of prediction failure rate for the components in TLC field.

The most important points are the following:

- Use of models derived from MIL-HDBK-217 modified on the ground of the data coming from field;

- Evaluation of field data through the Mean Removal Rate;

- Use of a complete Quality Assurance system for components in order to assure a quality level constant in time as a condition of applying prediction models.

9. REFERENCES

(1) MIL-HDBK-217D, Reliability Prediction of Electronic Equipment, DoD USA, 1982

(2) F.H. Reynolds, Measuring and modelling integrated circuits failure rates, EUROCON 82, 32-44

(3) J. Moltoft, Behind the "bathtub" curve, a new model and its consequences, Microelectronics and Rel, Vol 23, N. 3, 1983, 489-500

(4) C.M. Reyerson, The Reliability Bathtub Curve is Vigorously Alive, IEEE Annual Proc. on Reliability and Maintainability, 1982, 313-316.

BIOGRAPHIES

Mr. Ferdinando Federici (born 1947) received his degree in physics at Università degli Studi in Milan in 1971. He has been with Italtel since 1972 as responsible in qualification and reliability evaluation of electronic components of new technology. In 1983 he became the responsible for incoming inspection of components and materials. Since 1985 he has been responsible for all the activities related to the components Quality and Reliability Assurance.

Mr. Giorgio Turconi (born 1946) received his degree in electronics engineering (telecommunications) in 1971 by Politecnico in Milan. In 1973 he was employed by Italtel and became reliability responsible for military equipment. In 1978 he undertook the responsible for reliability methodologies with the Central Reliability Department dealing with reliability and maintainability predictions, LCC, logistic support and failure data base on complex systems. Since 1982 he has also been responsible for Quality Control methodologies and data collection procedures.

MECHANICAL AND ELECTROMECHANICAL RELIABILITY

Reliability Technology — Theory & Applications
J. Møltoft and F. Jensen (Editors)
© Elsevier Science Publishers B.V. (North-Holland), 1986

RELIABILITY EVALUATION OF INDUCTION GENERATORS AND
WIND ELECTRIC ENERGY CONVERSION SYSTEMS

S. S. Venkata
R. Natarajan
M. A. El-Sharkawi
Energy Group
Department of Electrical Engineering
University of Washington
Seattle, WA 98195

N. G. Butler
Bonneville Power Administration
Portland, Oregon

Abstract: Present practice in the U. S. is to use commercially available
induction motors as generators in wind energy applications. As generators,
these machines are exposed to hostile atmospheric and operating conditions
and hence their life expectancy is tremendously affected. Due to the lack
of prolonged operating experience on these generators in such applications,
the reliability and availability of these machines are yet to be established. However, these generators are expected to be safe, reliable, and
require minimal maintenance. Therefore, there is a need to estimate the
reliability indices of these generators as well as the Wind Energy Conversion System (WECS) in order to ensure a near optimum system performance.
We devote the first part of the paper to the reliability assessment of the
induction generator. We present the realibility studies of the WECS, with
induction generator treated as a component, in the second part of the
paper. We also demonstrate the application of the fault tree methodology
for this for this investigation.

1. INTRODUCTION

Several wind farms comprising of
clusters of intermediate-size induction
generators are emerging in some parts
of U. S. In each farm, there are hun-
dreds to thousands of machines most of
which are equipped with induction generators in the range of 50 to 1,000-kW.
Usually these machines are interconnected to the respective utility grid.
We, at the University of Washington,
are currently investigating the feasi-
bility of commercially available
induction motors as generators for wind
energy conversion systems. The primary
objectives of this project, sponsored
by the Bonneville Power Administration,
are: a) to find schemes for improving
the generator performance, and b) to
investigate effective ways to reduce
the reactive power drain on the utility
system.

To assure efficient and effective
use of electric energy, the wind turbine system must be designed to ensure
reliable, efficient, and cost-effective
operation. While these machines are
intended to survive about 30 years in
the normal industrial environment, can
they have similar life span if they are
exposed to continously varying
speed/load operation and hostile environmental conditions? There is hardly
any established historical data available to confidently determine the reliability and life expectancy of these
wind generators. Therefore, there is a
need to estimate the reliability
indices of these generators, as well as
the WECS, in order to aim at their
optimum system performance.

We describe the reliability analysis of the induction generator systematically in the following Section. In
Section 3, we present a similar analysis for the WECS with the induction
generator treated as a system element.
Section 4 concludes the paper.

2. INDUCTION GENERATOR RELIABILITY ANALYSIS

In any WECS, the induction generator is driven by a wind turbine and
the input mechanical power is applied
to the machine shaft. The output electrical power is deliverd from the generator stator terminals to a
distribution network through a circuit-breaker and a step-up transformer.
Suitable fuses are installed in all the
three phases of the generator as a measure of overload protection. Func-
tionally, the generator is expected to
deliver maximum possible real power in
the expected range of wind speed, while
keeping the system voltage within the
allowable regulation. From the reliability point of view, such a generator
can be modeled as 27 electro-mechanical
components listed in Table 1.

Table 1: List of Three-Phase
Induction Generator Components

1. Supply fuse in phase A
2. Supply fuse in phase B
3. Supply fuse in phase C
4. Terminal of phase A
5. Terminal of phase B
6. Terminal of phase C
7. Terminal board
8. Stator frame
9. Laminated stator iron assembly
10. Phase A of stator winding

11. Phase B of stator winding
12. Phase C of stator winding
13. Slot insulation
14. Shaft
15. Key
16. Rotor lamination assembly
17. Rotor bars
18. Rotor end-rings
19. Fan
20. Fan cover
21. End shield (non-drive end)
22. Bearing (non-drive end)
23. Bearing cover (non-drive end)
24. End shield (drive end)
25. Bearing (drive end)
26. Bearing cover (drive end)
27. Assembly studs

The status of components 1 to 3 dictate the supply conditions; 4 to 13 form the stator assembly; 14 to 20 consist of the rotor assembly and 20 to 27 represent the final assembly. we can now identify the following undesirable modes of operation in the generator{2}. In fact, these modes are in accordance with the concepts presented in IEEE STD-500{3} and references{4,5}.

2.1 Undesirable Modes of Operation

Mode A - Incipient Failure: Any imperfection in the generator can cause incipient failure and this can lead to degraded or catastrophic failure if no corrective action is taken. Normal aging of the machine, excessive mechanical noise, vibration, overheating due to reduced cooling, or minor deviations in the output voltage due to increased voltage drop at the contacts are some of the causes leading to this mode of operation.

Mode B - Degraded operation: In this mode, the generator fails to perform within the expected specifications, such as reduced output from its terminals. This mode affects the energy produced by the generator and can manifest itself in the following three ways:

Sub-mode B1: One line open
Sub-mode B2: Line-to-line short circuit
Sub-bode B3: One-line-to ground fault

We claim that these three modes usually result due to abnormal operating conditions.

Mode C - Catastrophic failure: This mode implies that the generator totally fails to perform its intended function. Three-phase fault in the generator, major mechanical component failure such as break in the shaft, or the bearings are some of the causes leading to this mode.

2.2 Procedure for Reliability Assessment

We carried out the reliability analysis using the fault tree technique. This analysis assumes that each component follows exponetial distribution model. This means that each of the mechanical and electrical components has constant hazard rate over their useful life period{4-9}. This assumption has been established for most of the engineering systems. A summary of the key reliability equations for this model are given in Appendix A-1. We adopted the following procedure for the analysis.

Step 1: Identify all possible causes leading to a particular mode of operation.

Step 2: Draw the corresponding fault tree in which the top event is clearly identified.

Step 3: Identify all the minimal cutsets corresponding to each cause mentioned in step 1.

Step 4: Evaluate the probability of the top event occuring, using the hazard rate of each component occuring in the set of minimal cutsets.

Step 5: Repeat steps 1 to 4 for each mode or sub-mode of undesirable operation.

Step 6: Determine the reliability indices from the results of Step 5 above as described later in this Section.

Appendix A-2 explains the fault tree methodology for a small hypothetical system once the fault tree is constructed. For further explanation, see references{3-5,8,9}.

The analysis procedure for the induction generator can be explained as follows:

Step 1:

Mode A- This mode could occur due to one of the following reasons.

a. High contact resistance in fuses (1,2,3) due to accumulation of dirt, moisture and other environmental factors.

b. High contact resistance between phase terminals (4,5,6) and terminal board(7).

c. Stator frame or laminated assembly overheating (8,9) due to reduced cooling.

d.   Contamination in end winding (10,11,12) or slot insulation(13).

e.   Dirt accumulation in fan, and fan-cover (19,20). This reduces the effective generator cooling.

Step 2:   The resulting fault tree of the induction generator for Mode A is shown in Fig.1.

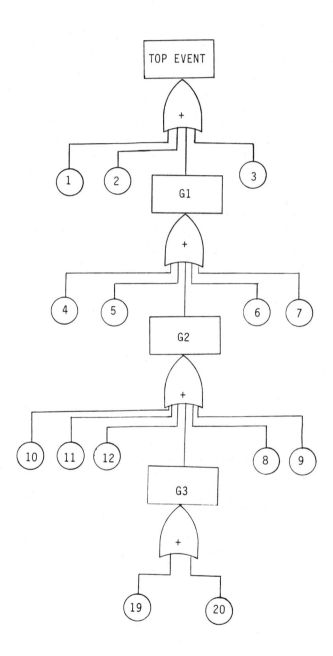

Fig. 1   Fault Tree for the Incipient Failure (Mode A) of the Induction Generator

Step 3:   The minimal cutsets for Mode A failure are: 1,2,3,4,5,6,7,8,9,10,11,12,19,20.

Step 4:   we can now evaluate the reliability of the top event due to Mode A failure, which is "the probabilities of incipient failure of the induction generator occuring":

$$P(\text{top event}) = Q_A = \sum_{i\epsilon\ \text{minimal cutsets}} q_i \qquad \text{--(1)}$$

Where q_i = i^{th} component failure probability index given by equation (A2) as:

$$q_i = (1 - \epsilon^{-\lambda_i t}) \qquad \text{--(A2)}$$

Where λ = i^{th} component failure rate usually listed in number of failures per million hours in standard reference manuals{3,4}. The summation in equation (1) results due to the fact that all basic events are mutually exclusive (use of logical OR gates). We consider "t" as one year or 8760 hours. If equation (A2) is approximated as:

$$q_i \cong \lambda_i \qquad \text{--(2)}$$

then, equation (1) is reduced to

$$Q_A = \sum_{i\epsilon\ \text{minimal cutsets}} \lambda_i \qquad \text{--(3)}$$

Equation (3) can also be expressed as

$$Q_A = 1 - \epsilon^{-\lambda_A t} \qquad \text{--(4)}$$

Where λ_A represents the number of failures per year due to mode A failure of the generator. Knowing Q_A, λ_A can be obtained.

$$\text{Where,}\ Q_A \cong \lambda_A \qquad \text{--(5)}$$

By knowing λ_i for each component, Q_A or λ_A for mode A failure can be found. Table 2 lists maximum, minimum, and median failure rates of all 27 components listed in Table 1. We assume these values to be equally valid for all the three modes of undesirable operation. The Mean-time-To-Failure (MTTF) for mode A failure is:

$$(MTTF)_A = \frac{1}{\lambda_A} \qquad \text{--(6)}$$

Table 2: Failure Rates of Induction Generator
          Components

Component	Failure Rate/10^6 Hours	
	Minimum - Maximum	Median
1,2,3	0.009 - 0.300	0.030
4,5,6	0.050 - 0.271	0.070
7	0.270 - 1.000	0.615
8,9	0.150 - 0.700	0.175
10,11,12	0.270 - 1.000	0.615
13	0.011 - 0.720	0.500
14,15	0.080 - 0.350	0.200
16	0.015 - 0.700	0.175
17,18	0.350 - 0.900	0.450
19	0.210 - 0.800	0.225
20	0.010 - 0.019	0.018
21	0.015 - 0.700	0.175
22	0.035 - 1.719	0.815
23	0.139 - 0.350	0.200
24	0.015 - 0.700	0.175
25	0.035 - 1.719	0.815
26,27	0.139 - 0.350	0.200

Step 5:

Mode B- The causes contributing to
this mode of failure are:

a. Open fuse (1,2,3) in supply lines.

b. Unacceptable levels of contamination in the terminal board(7), leading to the short-circuit of one or more
stator phases.

c. Mechanical deformation of the stator frame (8) and lamination assembly
(9) due to excessive mechanical stresses.

d. Phase winding (10,11,12) or insulation failure (13) due to the accumulation of dirt, moisture and/or
corrosion.

e. Slight deformation of shaft (14)
and key (15) due to fatigue.

By following similar procedural
steps as for mode A, we can now identify the minimal cutsets for each
sub-mode of B as:

Sub-mode B1:1,4,7,(8 AND 9),(10 AND
13),(14 AND 15),19
Sub-mode B2:(4 AND 5),7,(8 AND 9),
(10 AND 11 AND 13),14,19
Sub-mode B3:7,9,(12 AND 13),(14 AND
15),19

Here "AND" implies logical AND operation.

Mode C- In a similar logical manner, we can identify the following
causes for the catastrophic failure of
the induction generator:

a) Catastrophic failure of all fuses; 1
   AND 2 AND 3.
b) Failure of all terminals; 4 AND 5 AND 6.
c) Failure of all phase windings; 10 AND
   11 AND 12.
d) Failure of components 14,15,16,17,
   18,19,22,25.

In this condition, the componenet(s)
concerned could be replaced and the
generator can be put back to work. However, in this analysis the repair condition is not considered. Also, the
secondary failure of some of the components are not considered. The minimal
cutsets for mode C can be listed as:
(1 AND 2 AND 3),(4 AND 5 AND 6), (10
AND 11 AND 12), 14, 15, 16, 17, 18, 19,
22, 25.

Step 6: By following the same
procedure outlined for mode A, we can
determine all the reliability indices
for other modes B and C; expect for
mode B, the overall λ_B will be

$$\lambda_B = \lambda_{B1} + \lambda_{B2} + \lambda_{B3} \qquad --(7)$$

The combined failure rate
for the induction generator will
then be

$$\lambda_G = \lambda_A + \lambda_B + \lambda_C \qquad --(8)$$

and the MTTF of the generator is:

$$(MTTF)_G = 1/\lambda_G \qquad --(9)$$

2.3 Results and Discussions

The fault tree methodology developed is applied to a 50-kW, 480V,
three-phase, induction generator.
Table 3 shows the failure rates and the
MTTF for all the modes under consideration. The combined failure rate and
the MTTF for the induction generator
are given in row 7 of the table. The
table also shows the corresponding values for standard induction motors and
synchronous generators.

Table 3: Ranges of Failure Rates and the MTTF of Induction Generator

Mode	Failure Rate(Failure/year)		Mean-Time-To-Failure(years)	
	Range	Median Value	Range	Median Value
A	0.0695 - 0.0144	0.0293	14.39 - 69.26	34.04
B1	0.0322 - 0.0067	0.0115	31.02 - 148.96	86.60
B2	0.0286 - 0.0052	0.0118	34.88 - 191.69	84.43
B3	0.0293 - 0.0056	0.0119	34.15 - 180.14	83.75
B	0.0901 - 0.0175	0.0342	11.09 - 57.14	29.23
C	0.0743 - 0.0047	0.0312	13.45 - 209.59	32.04
Combined	0.2239 - 0.0366	0.0946	4.27 - 27.32	10.57
Induction motor(3)	0.0914 - 0.0061	0.0166	10.94 - 161.14	60.15
Synch. Generator	0.3628 - 0.0021	0.0394	2.75 - 473.48	25.36

Further, the calculated reliability indices, $R_G=1-Q_G$ for the induction generator are given in Table 4.

Table 4: Reliability of Indicies of the Induction Generator

	Reliability Index, R_G	
Mode	Range	Median Value
A	0.9328 - 0.9857	0.9712
B	0.9138 - 0.9826	0.9663
C	0.9345 - 0.9971	0.9712
Combined	0.7977 - 0.9657	0.9114
Std. Motor	0.9126 - 0.9939	0.9835
Syn. Gen	0.6957 - 0.9979	0.9613

Based on the analysis, the following conclusions can be made:

a) The calculated failure rate for the induction generator is higher than the corresponding values for a standard induction motor or synchronous generator as shown in Table 3. The corresponding MTTF is also lower. This is to be expected since the wind generator operates in a more hostile environment, which was taken into consideration in selecting some of the component failure rates. However, we caution that further validation of the analysis is required before placing confidence in these results, which serve as a guide in improving the generator design for wind turbine applications.

b) Based on the results presented, we identify the following critical components of the generators. They are: bearings, phase windings, terminals, terminal board, slot insulation, and fan. we therefore, recommend that these components be given additional care in the design and/or maintenance stages.

3. RELIABILITY ANALYSIS OF WECS

The WECS under consideration consists of the generator, transmission gears, and the turbine rotor as illustrated in Fig.2. The generator is connected to the utility grid through a three phase circuit-breaker and step-up transformer. In order to correct the power factor close to unity, an Adaptive Power Factor Controller (APFC) is used{1} at the terminals. Appropriate braking mechanisms are present at the turbine shaft.

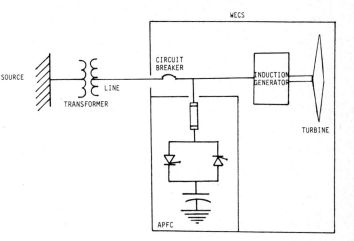

Fig. 2 Line Diagram of the WECS with Induction Generator

3.1 Modeling of Failure Modes

The WECS can be modeled as 6 components or sub-sytems listed in Table 5.

Table 5: List of WECS Components

1. Turbine
2. Brake
3. Belt
4. Circuit-breaker
5. Induction generator
6. APFC

The failure modes are identified below, in accordance with IEEE-STD 500{3}.

a) Mode A: Incipient Failure
b) Mode B: Degraded operation
c) Mode C: Catastrophic failure

Mode A-Incipient failure: Any minor problems in the WECS can cause incipient failure and this can lead to degraded or catastrophic failure if no corrective action taken. Any one of the following causes may lead to this mode of undesirable operation.

a) Lack of lubrication in the turbine bearings(1).
b) Brake failure(2).
c) Belt slippage(3).
d) Circuit-breaker malfunction(4).
e) Overheat in generator(5).

The minimal cutsets for this mode of operation are: 1,2,3,4,5.

Mode B- Degraded operation: A failure which is both gradual and partial. Such a failure does not stop all functions but compromises a function. The compromise may involve by any combination of reduced, increased or spurious output of voltage, current, frequency, or power output. Such an operation may be deterimental to the machine, to the system to which it is interconnected, or to the operating personnel. In time, such a failure may develop into a complete failure. Any one of the following causes are responsible for this mode of failure.

a) Brake and circuit-breaker failure(2 AND 4).
b) Belt and circuit-breaker failure(3 AND 4).
c) Generator and circuit-breaker failure(5 AND 4).
d) APFC and circuit-breaker failure(6 AND 4).

The minimal cutsets are: (2 AND 4), (3 AND 4), (5 AND 4), (6 AND 4).

Mode C- Catastrophic failure: A failure which is both sudden and complete. With this failure the WECS stops functioning and requires immediatate corrective action. This is an important issue since this affects the availability of the generator to supply power to the utility. Any one of the following causes are responsible for this mode of failure.

a) Failure of the turbine(1).
b) Brake failure (2).
c) Failure of belt(3).
d) Circuit-breaker failure(4).
e) Generator failure(5).

The minimal cutsets for this mode of failure are: 1,2,3,4,5.

As before, in order to evaluate the reliability indices of the WECS, the failure rates of the components listed in Table 5 are identified in Table 6.

Table 6: Failure Rates of WECS Components or sub-systems

Component or Sub-System	Failure Rate/10^6 Hours		
	Minimum – Maximum		Median Value
1(a,b,c)	0.115	– 2.700	1.175
2(a,b)	0.011	– 0.720	0.500
2(c)	0.010	– 0.700	0.400
3(a,b,c)	0.200	– 0.800	0.600
4(a,b)	0.320	– 0.900	0.700
4(c)	0.010	– 0.850	0.600
5(a)	0.014	– 0.069	0.029
5(b)	0.017	– 0.090	0.034
5(c)	0.004	– 0.074	0.030
6(a,b,c)	0.620	– 0.900	0.770

(a) - failure rate for incipient failure
(b) - failure rate for degraded failure
(c) - failure rate for catastrophic failure

Since each component is more of a subsystem, it makes sense to identify its failure rate based on the individual mode of failure, which we have taken into consideration in selecting the data in Table 6.

3.2 Results and Discussions

We have applied the methodology to a specific WECS which employs the 50-kW induction generator mentioned in the Section 2.3. In Table 7 the calculated failure rates and the MTTF of the WECS are given.

Table 7: Calculated Failure Rates and the MTTF of WECS

Mode	Failure Rate(Failure/year)		Mean-Time-To-Failure(years)	
	Range	Median Value	Range	Median Value
A	0.0456 – 0.0058	0.0286	21.99 – 171.21	33.84
B	0.1082 – 0.0191	0.0424	9.21 – 52.41	23.55
C	0.2572 – 0.0032	0.0332	3.88 – 312.75	3.04
WECS	0.4110 – 0.0281	0.1042	2.43 – 35.57	9.59

Also the calculated reliability indices of the WECS are listed in Table 8.

Table 8:   Reliability indices of
WECS

Mode	Reliability Index	
	Range	Median Value
A	0.9555 - 0.9994	0.9712
B	0.8974 - 0.9810	0.9584
C	0.7734 - 0.9967	0.9673
WECS	0.6629 - 0.9722	0.9010

4. CONCLUSIONS

The fault tree methodology is applied to study the reliability of an induction generator first and the WECS in which the generator is deployed. It must be emphasized that the failure rates used in computing the probability of the top event are conservative and hence the analysis can be taken as a guide line only. The environmental factors are included in the failure rates of most of the components in this paper. However, in wind power applications, the following additional factors take greater importance:

a) Corrosion (in the coastal areas)

b) Adverse weather (blowing sand in the desert and extreme weather conditions of the mountains)

The probability of shut down due to the hazardous operation can be minimized by proper preventive maintenance practices, in which case, it is necessary to include the repair time of the components in the analysis.

5. ACKNOWLEDGEMENTS

The first three authors gratefully acknowledge the support of Bonneville Power Administration for this study, funded under contract number BP22849. We also express our appreciation to M. C. Wehrey and R. J. Yinger of Southern California Edison Company for providing necessary support for the work reported in this paper. We also thank C. C. Liu, S. V. Vadari, and M. Chen, University of Washington, for their valable input and suggestions to the work.

6. REFERENCES

1. M. A. El-sharkawi, S. S. Venkata, T. J. Williams, and N. G. Butler, "An Adaptive Power Factor Controller for Three Phase Induction Generators", IEEE Trans. on PA&S, Vol.PAS-104(7), Dec.1984, pp.1827-1831.

2. L. J. Rejda, and K. Neville, "Industrial User's Handbook of Insluation for Rewinders", Elsevier, New York, 1977.

3. IEEE Std-500, Reliability Data, 1984.

4. E. J. Henley, H. Kumamoto, "Reliability Engineering and Risk Assessment", Prentice Hall, 1981.

5. A. M. Polovko, "Fundamentals of Reliability Theory", Academic Press, New York.

6. S.S. Venkata, M. Chinnarao, E. W. Collins, E. U. Ibok, "Reliability of Mine Electrical Power System Transients Proctection, Reliability Investigation, and Safety Testing of Mine Electrical Power System", Vol.II, Grant No: GO 144137, West Virginia University, WV 26506, Aug,1979.

7. M. Chinnarao, E.U. Ibok, E.W. Collins, and S.S. Venkata, "Safety Analysis of Underground Coal Mine Power Equipment", Conference Record, IEEE-IAS Annual Meeting, 1978, pp.296-301.

8. N. H. Norman, "Fault Tree Handbook; NUEREG-0492", 1978.

9. E. K. Stanek, S. S. Venkata, "Mine Power System Reliability", Presented in IEEE Industry and Applications Society Annual Meeting, Toronto, Oct 1985.

APPENDIX A

A-1 Reliability Indices

With expontential distribution model, and a constant failure rate of a mechanical component, the conditional probability of survival in an interval of time $(t + \Delta t)$, given that it has survived up to time t, is:

$$r(\Delta t) = e^{-\lambda \Delta t} \qquad --(A1)$$

and the probability of failure is

$$q(\Delta t) = 1 - e^{-\lambda \Delta t} \qquad --(A2)$$

which is independent of t. This property implies that a used component is essentially as good as new, and therefore, from eq.(A2) the probability of failure for any given duration of operation, say a period of 6 hours, is simply

$$q(t = 6) = q = 1 - e^{-6\lambda} \qquad --(A3)$$

where λ is expressed as number of failures per hour. Then r(t), the probability of the device performing successfully for the intended period of time is given by:

$$r(t) = (1-q(t)) = e^{-\lambda t} \qquad --(A4)$$

And the Mean Time To Failure (MTTF), the expected time to fail for a given component is:

$$MTTF = \int_{0}^{\infty} r(t) \, dt = \frac{1}{\lambda} \qquad --(A5)$$

A-2 Reliability Assessment by Fault Tree Technique

The procedure for the detemination of the top event, knowing the minimal cutsets is given below.

a) Let Y_i = 1 if the basic event i occurs
   and, Y_i = 0 otherwise

b) Let $\bar{Y}$ = (Y1, Y2, Y3 , . . Yn) be the vector of basic event outcomes. Define

$\Psi(\bar{Y})$ = 1 if the top event occurs
        = 0 otherwise

c) Enumerate the minimal cutsets of the fault tree. Let there be k minimal cutsets and each cutset contain same Y_i.

d) Then $\Psi(\bar{Y})$ is given by

$$\Psi(\bar{Y}) = \coprod_{s=1}^{k} \prod_{i \in K_s} Y_i \qquad --(A6)$$

and P(top event) = E{$\Psi(\bar{Y})$} $\qquad --(A7)$

Also, $\coprod$ means union of events

$\prod$ means intersection of events

and E{$\Psi(\bar{Y})$} is the expected value of $\Psi(\bar{Y})$.

Steps a) - d) is applied to a hypothetical system, the fault tree for which is shown in Fig. A1.

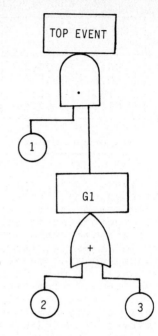

Fig. A1 Fault Tree for the Example

The minimal cutsets are (1 AND 2) OR (1 AND 3). Therefore, from eq.(A6),

$$\Psi(\bar{Y}) = \coprod_{s=1}^{2} (Y1Y2, Y1Y3)$$
$$= 1 - (1 - Y1Y2)(1 - Y1Y3)$$
$$= Y1(Y2 + Y3 - Y2Y3) \qquad --(A8)$$

making use of the idempotency condition condition that Y_i^2 = Y_i in the product of Y1Y2Y1Y4.

From (A7),
P(top event) = E{ (Y1(Y2 + Y3 - Y2Y3)}
$$= q1(q2 + q3 - q2q3) \quad (A9)$$

Knowing the numercial value of the failure rates (λ's) the probability of the top event can be evaluated.

Reliability Technology — Theory & Applications
J. Møltoft and F. Jensen (Editors)
© Elsevier Science Publishers B.V. (North-Holland), 1986

SOME MAXIMUM LIKELIHOOD RELIABILITY ESTIMATES FROM WARRANTY DATA
OF CARS IN USER OPERATION

Pasquale ERTO and Maurizio GUIDA
University of Naples - CNR Istituto Motori
Naples, Italy

Usually the warranty data of cars report only the component and failure codes and the mileage interval in which the failure occurred. No information is directly provided about the number of all cars which cover the various mileage intervals without failures, and hence no direct knowledge is available about the population to which the observed failures must be referred. In this situation, only approximate estimation procedures are up till now used, that is procedures which are generally based on some "a priori" (and subjective) evaluation of car distribution versus mileage intervals. Instead, this work proposes the use of some rigorous maximum likelihood estimators which enable one to estimate both the failure and car distributions versus mileage. The method has been successfully used in real-life cases which are partially reported in an illustrative example included in this work.

1. INTRODUCTION

Today, a new car model must ensure a specified reliability level since its initial launching on the market, on pain of obscuring the company image which will be restored with difficulty   by   subsequent   improvements. Generally, the reliability targets can be achieved by means of a good design, many preliminary life tests on components or subsystems, and a quality control policy. Nevertheless, in the case of mass production, it is essential to verify constantly that these targets are really fulfilled in service. In fact, the in-service reliability level may result different from the expected one, mainly owing to faults of production process and/or to unforeseen stresses induced by real operating enviroment.

Then the manifacturer has a pressing need to collect and analyze field data in order to detect the causes of a possible discrepancy between the in-service and the expected reliability, and so to adopt immediately the necessary corrective actions. Obviously, since the cars are products whith wide range of operating enviroments and users, one should have a great many of manufactured units under monitoring during their entire life to be confident in the measure of their reliability. But it is quite evident the impractability of a such a policy.

Thus the approach generally undertaken consists in monitoring only the units belonging to homogeneous samples of limited size (e.g. a taxi fleet), and/or controlling the repair operations of manufactured units during the warranty period. Other sources of information, like the number of spare parts sold, are sometimes used, but they are less informative and will not be considered here.

The monitoring of a vehicle fleet allows to collect information about the entire life of the product, taking into account both early and wear-out failures. Moreover, in many cases, these fleets are subjected to a more intensive use than the normal one (see taxi), and this makes possible to obtain measures of reliability in a relatively short time. Neverthless these measures are generally valid only for the operating enviroment and use of the particular sample chosen, and, with difficulty, can be extended to other situations. Besides they cannot take into account the impact of improvements that were subsequently made.

The use of warranty data makes available timely information at low cost for reliability evaluations. Obviously, this data are of truncated type, i.e. they take into account mainly the impact of early failures, but their use has the advantage of allowing quickly both to choose corrective actions and to check the effectiveness of those just adopted.

2. RELIABILITY ANALYSES FROM WARRANTY DATA

Usually, warranty data contain only the following information:

- vehicle type code
- building date
- component and defect code
- mileage to failure.

In a formal statistical language the warranty data are failure observations from a truncated sample with suspended items. Thus, to carry out a reliability analysis both the number of failures and the number of suspensions for each mileage interval are

required. Obviously the warranty data give no
mileage information about those vehicles which
are sold and reach the end of each mileage
interval without any claim being made. Thus no
direct information is available about the
population to which the reported number of the
failed items must be referred. Therefore, all
the known procedures used by automotive
industries [1],[2],[3] require the a priori
estimation (often arbitrary) of the vehicle
distribution versus mileage, in order to
partition the total number of non-failed
vehicles under warranty into mileage intervals.
Note that this process may regard also the
vehicles under special maintenance contract
including the warranty period too, or only a
part of the vehicles under warranty, as all
vehicles from a particular production unit, or
all vehicles produced in a given period, etc..

This work proposes a method which does not
need any a priori estimation, enabling one to
estimate simultaneously both the mileage and
the failure distribution functions. In the
next section we discuss a special case occurred
in a real-life situation, in which the proposed
method of analysis was successfully used.

3. A REAL-LIFE RELIABILITY ANALYSIS FROM SOME INCOMPLETE WARRANTY DATA

3.1 The Available Data Set

The failure data normally refer to about
forty different components (or parts) of some
car model; the kilometers to failure are
typically grouped into equal width lifebands,
each of 10,000 kilometers; all vehicles under
consideration are sold during the same year in
which repairs are made. In a case from real
life, the cars sold in this year were 498 and
the total number of warranty claims required to
the manifacturer were 70. Furthermore,
irrespective of the parts involved, this number
were distributed over the lifebands as follows:

Lifebands (km/1000)

	0-10	10-20	20-30	30-40	> 40
number of claims	55	11	3	1	0

The characteristic that makes peculiar a
case as this one is that no age (from selling
date) distribution and no distribution of
covered kilometers are given for the fleet
under consideration. Hence it is not possible
to allocate the non-failed units in each
lifeband.

To overcome this difficulty we propose the
following approach, introducing an estimation
procedure which involves at the same time both
the failure and the kilometer distributions.

3.2 The proposed method of analysis

Let be Tf the random variable (r.v.)
representing the "kilometers to failure", and
G(t)=Prob(Tf<t) its unknown distribution
function. Moreover, let be F(t) the
probability that a car, sold during the year of
observation, doesn't exceed the kilometers t,
till the end of this year.

The experimental context under analysis is
equivalent to a progressively suspended
sampling, and more exactly a sampling in which
some of the items under life-testing have their
test randomly suspended before failure. Thus,
a r.v. representing the "kilometers to
suspension", say Ts, is defined, with the
unknown distribution function F(t)=Prob(Ts<t).
Hence, it follows that:

Probability that an item fails before t
kilometers = Prob(Tf<t and Ts>Tf) =

$$= \int_0^t [1 - F(x)] \, dG(x),$$

and:

Probability that an item is suspended before t
kilometers = Prob(Ts<t and Tf>Ts) =

$$= \int_0^t [1 - G(x)] \, dF(x),$$

being Tf and Ts independent r.v..

Assuming that G(t) and F(t) are of
exponential type, that is G(t) = 1-exp(-a*t)
and F(t) = 1-exp(-b*t), the above probabilities
become:

Prob. that an item fails before t =
= a{ 1 - exp[-(a+b)t] }/(a+b)

Prob. that an item is suspended before t =
= b{ 1 - exp[-(a+b)t] }/(a+b).

Some comments on the assumption of the
exponential model for G(t) are required:
warranty data are essentially data on early
failures, hence, from a theoretical viewpoint,
the kilometers to failure should have a
decreasing failure rate. Then, as an example,
a Weibull model with shape parameters less than
1 should be more sound. However, experimental
results have shown that, in situations similar
to the present one (see [1], [3]), the shape
parameter of the Weibull distribution is very
close to 1. Hence it appears to be no

practical advantage in using a model more complex than the exponential one for $G(t)$.

To explain the choice we made for $F(t)$, we use, in a qualitative manner, some information given in [2] about the kilometer distribution versus age (from selling date) for a fleet of european cars. In the above reference, information on the two 5 percent tails of the distributions at 3, 6, 9 and 12 month of car age is available. We have extrapolated this information assuming for each of these distributions a Weibull model which had the same two 5 percent tails. In fig. 1 the kilometer distributions (respectively at 3, 6, 9 and 12 months of age) are reported using a Weibull probability paper. Then, we derived the "compound" kilometer distribution which corresponds to car ages (from selling dates) uniformly distributed over the range of 12 months. As can be seen from the same figure, this distribution results to be very close to an exponential one. Even if this result cannot be considered as a decisive proof, nevertheless the exponential model appears to be at least as the preferential candidate for $F(t)$ in the present situation.

3.3 Estimation procedure

The definition of the probabilities of both failure and suspension allows to construct the likelihood function, say L, for a sample arising from the experimental situation under study. In fact, let be:

- N = total number of cars under observation

- n_i = number of failures in the i-th lifeband (T_i, T_{i+1})

- m = number of lifebands

- $n = \sum_{i=1}^{m} n_i$ = total number of failures,

the likelihood function, L, is proportional to:

$$L \propto \left\{ \int_0^\infty [1 - G(x)] \, dF(x) \right\}^{N-n} *$$

$$* \prod_{i=1}^{m} \left\{ \int_{T_i}^{T_{i+1}} [1 - F(x)] \, dG(x) \right\}^{n_i} =$$

$$= [b/(a+b)]^{N-n} \prod_{i=1}^{m} \{a \,[\exp(-(a+b)*T_i) +$$

$$- \exp(-(a+b)*T_{i+1})] / (a+b)\}^{n_i}.$$

The values of a and b, which maximize this function (given a sample with known N and n_i) are the well-known maximum likelihood estimates of the unknown parameters a and b.

Reparametrizing for convenience the likelihood function in terms of $a=a$ and $c=a+b$, and equating to zero the partial derivatives of $\ln(L)$, the following equations result:

$$a/c = n/N$$

$$\sum_{i=1}^{m} n_i \frac{[T_{i+1} \exp(-c*T_{i+1}) - T_i \exp(-c*T_i)]}{[\exp(-c*T_i) - \exp(-c*T_{i+1})]} = 0$$

To solve this last non-linear equation in c, an iterative method is required (e.g. Newton-Raphson method). An initial temptative value, c^*, can be easily found taking into account that the probability that an item fails in the i-th interval has the non-parametric estimate n_i/N. Thus, as an example, for the first lifeband $(T_1=0, T_2)$ it results:

$$\text{Prob } (0< T_f <T_2 \text{ and } T_s>T_f) =$$
$$= [1 - \exp(-c*T_2)] \, a/c = n_1/N,$$

and, being from the first likelihood equation $a/c = n/N$, an initial temptative value follows:

$$c^* = [- \ln(1 - n_1/n)]/T_2$$

then the estimates a and $b=(c - a)$ can be obtained without any further difficulty.

3.4 Practical example and comments

The maximum likelihood method was applied to the sample of warranty claims from real life, given in a previous section. The following estimates of the unknown parameters a and b were found:

$$a = 2.11 * 10^{-5} \text{ km}^{-1}$$

$$b = 12.93 * 10^{-5} \text{ km}^{-1}$$

The model chosen appears to fit at an extremely high degree of accuracy the experimental data. In Table 1 the observed and the estimated numbers of failures for each lifeband are reported.

TABLE 1
Lifebands (km/1000)

no. of claims	0-10	10-20	20-30	30-40	> 40
observed	55	11	3	1	0
estimated	54	12	3	1	0

The estimated average number of kilometers for
the fleet under test is:

$$1/b = 7734 \text{ km}$$

which is a very plausible value for a fleet of
cars which ages are presumably distributed
almost uniformely over 12 months.

REFERENCES

[1] VIKMAN, S. and JOHANSSON, B. "Some
    experiences with a programmed Weibull
    routine for the evaluation of field test
    results". EOQC II European Seminar on
    Life Testing and Reliability, Torino 1971,
    p 70.

[2] TOTH-FAY, R. "Evaluation of field
    reliability on the basis of the
    information supplied by the warranty".
    EOQC II European Seminar on Life Testing
    and Reliability, Torino 1971, 71-78.

[3] TURPIN, M. P. "Application of computer
    methods to reliability prediction and
    assessment in a commercial company".
    Reliability Engineering, 3 (1982),
    295-314.

Address: CNR Istituto Motori - Piazza Barsanti
           e Matteucci - 80125 Napoli - Italy

BIOGRAPHIES

P. Erto, Associate Professor in "Reliability
Theory" at the University of Naples, member of
the IEEE Reliability Society, was born in 1946.
He received a Dr. degree in Mechanical
Engineering and post-doctoral degree in
Computer Science from University of Naples.
Since 1974 he has given formal courses of
"Reliability Theory" in the University of
Calabria and of "Reliability and Quality
Control" in the University of Naples. He has
published more than 30 papers, many in
mechanical reliability and related statistical
topics, in journals that include 'Statistica',
'IEEE Transactions on Reliability',
'Reliability Engineering' and 'Quality and
Reliability Engineering International'.

M. Guida, Head of Dept. of Statistics and
Reliability of the "Istituto Motori" of Naples,
was born in 1948. He received Dr. degree in
Mechanical Engineering from University of
Naples. He is engaged in the study of the
practical and theoretical problems of the high
mechanical reliability evaluation from few
and/or accelerated tests. His papers appeared
in journal as 'IEEE Transactions on
Reliability', 'Reliability Engineering' and
'Quality and Reliability Engineering
International'.

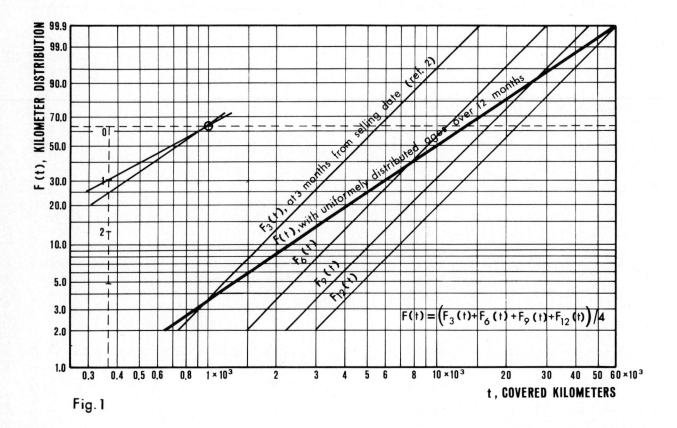

Fig. 1

Reliability Technology — Theory & Applications
J. Møltoft and F. Jensen (Editors)
© Elsevier Science Publishers B.V. (North-Holland), 1986

THE INFLUENCE OF THE OPERATING
ENVIRONMENT ON BUS RELIABILITY

Tomasz NOWAKOWSKI
Institute of Machine Design and Operation
The Technical University of Wrocław, Poland

The problem of the dependence of technical object reliability on environment interactions was discussed on the example of licence-manufactured bus Jelcz PR 110 U. The "continous test" method was used to estimate several reliability indicators for bus, its functionsl subsystems and its elements. The tests were realized in five towns in Poland. The mathematical model of vehicle operating conditions was discussed and some parameters of this model were shown. The bus reliability dependence on operating conditions was estimated by the modified multiple regression method. Some results of this influence were described.

1. INTRODUCTION

First buses Jelcz PR 110 U were manufactured under Berliet licence in 1976. From that moment their reliability is tested by Institute of Machine Design and Operation of the Technical University of Wrocław. Field reliability test of the pilot lot of the buses was realized in three big towns. Preliminary test results make surprize because of their differences between towns. In the proffesional literature the operation conditions in towns are regarded as identical and it is not suggested to distinquish them. Thus, several investigations were carried out (technical and social) to identify the problem of influence of the operating conditions on bus reliability [1, 4].

2. BASIC PRINCIPLES

The correlation model was used to solve the problem. The research object was regarded as a "black box" but the analysis was complited for bus as the whole or for vehicale functional subsystems or for bus elements.
It was assumed that the vehicle is forced by operating conditions factors $x_1(t),\ldots, x_s(t)$ and nonmeasurable disturbances $z_1(t),\ldots, z_u(t)$ (fig.1).
Outputs $y_1(t),\ldots, y_p(t)$ are the system responses. One of the outputs means reliability of vehicle $R(t)$. Thus the model of the influence of the operating conditions on vehicle reliability can be expressed by function:

$$R(t) = f\left[x_1(t),\ldots, x_s(t), z_1(t),\ldots, z_u(t)\right] \qquad (1)$$

To solve the problem it is necessary:

(i) to estimate the reliability of the bus, its subsystems and its elements

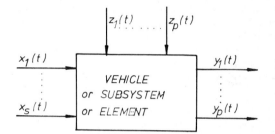

Fig.1. Problem principles.

for different operating conditions,

(ii) to evaluate the characteristics of the operating conditions for several environments,

(iii) to estimate the function (1).

3. RESEARCH OF BUS RELIABILITY

3.1. Test method

The "continous test" method was realized in five operating environments (it means in five towns) to estimate the reliability of Jelcz PR 110 U bus.

The "continous test" method is described by four elements [3]:

$$M_{ct} : \langle P_{ct}, Q, \lambda_o, T_t \rangle \qquad (2)$$

where: P_{ct} - continous test parameters
   Q - load, which forces wear out of the bus,
   λ_o - operating intensity,
   T_t - total test time.

"Continous test" method parametres consists of:

$$P_{ct} : \langle I, Z\, v_j, Z_t \rangle \qquad (3)$$

where: I - sample size,

$Z(v_j) = Z_j(0) - Z_j(t_0)$ - margin of the operating potential for vehicle j which came to the sample set at the moment t_0,

$Z_t = Z_j(t_0) - Z_j(t_t)$ - operating potential of vehicle j which is planed to use from the moment t_0 till test end at the moment t_t.

The "continous test" parameters P_{ct} (avarage values of $Z(v_j)$) are shown in table 1.

Table 1
Continous test parameters

Measure	I	$Z(v_j)$	Z_t
Measure	unit	km	km
Total sample	110	8202	100 000
Environment sample	10-50	4375--15398	100 000

The load of the buses Q and the operating intensity λ_0 were characterised for specific environment (λ_0 =122÷470 km/day). The total time of the test amounted T_t = 210 ÷ 821 days.

Informations about bus failures were collected on the special designed forms [2]. They were verified and computer processed.

The bus was devided into twenty functional subsystems. The series reliability configuration for bus and subsystems was assumed and independence of the failures.

3.2. Test results

As a result of data processing, it was obtained (among others):

(i) distribution functions of mileage to the first failure and mileage between failures for bus and subsystems,

(ii) expected value and standard deviation of the distributions.

The number of element failure data allows only to compare failures amount between environments. For this purpose, the failure factor k_e was used.

$k_e = n_e/n_s$ (4)

where: n_e - number of element e failures in specific environment,
n_s - number of subsystem failures (element e belongs to this subsystem).

Fig.2 shows the reliability function confident intervals of the bus (in expotential distribution paper). In the third environment, the Weibull distribution (with shape parameter β = 1.25) was found as the best fitted one.

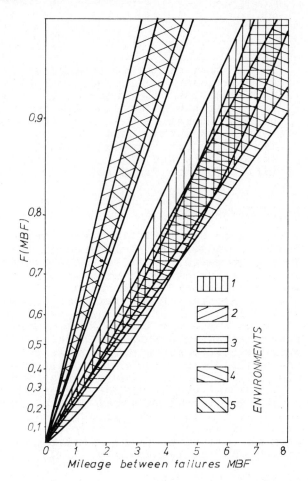

Fig.2. Confident intervals of bus reliability function

In table 2, the type of the distribution functions of some subsystems is compared. The confident intervals of expected values of this functions are shown on fig.3. Fig.4 shows the failure factors for chosen elements.

Table 2
Reliability function types

System	Environment				
	1	2	3	4	5
Cluch	W_l	W_l	W_l	W_m	E
Brakes	W_l	W_m	E	W_m	E
Suspension	E	E	E	W_l	E
Vehicle	E	E	W_m	E	E

Markings:

E - expotential distribution

W_l - Weibull distribution ($\beta < 1$)

W_m - Weibull distribution ($\beta > 1$)

According to these examples, the variability of the bus reliability in different operating conditions seems to be significant.

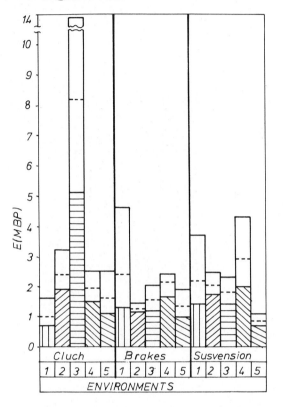

Fig.3. Confident intervals of mean MBF of subsystems.

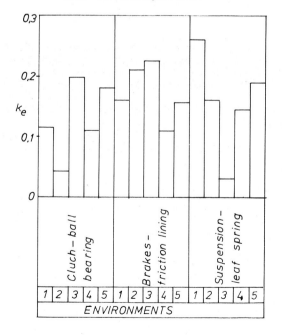

Fig.4. Failure factors of bus elements.

4. RESEARCH OF THE OPERATING CONDITIONS

Operating conditions (operating environment, surrounding) mean variable phisical conditions in which the technical object performs. The classification of the operating condition factors could be made on the basis of various criterions [4]. In this paper two criterions were considered (fig.5).

(i) Subiective factors resulted from a consious or an unconsious activity of a man.

(ii) Factors conditioned by object performing (influence on the object only during its operating).

<u>OPERATING CONDITIONS</u>

<u>SUBIECTIVE FACTORS</u>   <u>OBIECTIVE FACTORS</u>

- Psycho-physical features of drivers
- Psycho-physical features of servis personnel

Factors non-conditioned by object performing

- Servis organization
- Operational use organization
- Atmospheric factors
- Subgrade

Conditioned by object performing factors

- Traffic conditions
- Vehicle load
- Route geometry
- Route microgeometry

Fig.5. Vehicle operating conditions.

The research of the subjective factors was complited [1] and the factors non-conditoned by object performing were recognized [4]. It was found that first of all the difference of bus operating conditions occures in bus motion conditions and in bus load.

4.1. Model of bus motion conditions

The describtion of vehicle motion in spacial coordinate system X, Y, Z was a basis for analysis of bus motion conditions (fig.6). It was assumed that

it is possible to regard the bus motion
conditions as the stochastic processes
$X(t)$, $Y(t)$, $Z(t)$, $\phi(t)$, $Y(t)$, $E(t)$.
The solution of the problem was bounded
to stationary and ergodic processes.
The motion processes were described by
the first and the second moments of
probability distributions. The town
traffic properties, the horizontal
cross-section and longitudinal section
of streets were considered. The harmful
motions were neglected. Thus, the bus
motion conditions were described by the
matrix W:

$$W = \begin{vmatrix} E(X) & D^2(X) & R_{xx}(\tau) & R_{xz}(\tau) & R_{x\varepsilon}(\tau) \\ E(Z) & D^2(Z) & R_{zz}(\tau) & R_{z\varepsilon}(\tau) \\ E(E) & D^2(E) & R_{\varepsilon\varepsilon}(\tau) \end{vmatrix} \quad (5)$$

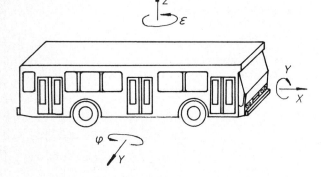

Fig.6. Vehicle motion coordinate
       system.

4.2. The test of bus operating conditions

The special equiped test bus was used
in five towns to estimate the traffic
conditions and the bus load. In order
to get the basis for operating conditions comparance, four criterions were
considered [6]:

(i) the criterion of the test route
choise,
(ii) the criterion of the test period
choise,
(iii) the criterion of the test time
choise,
(iiii) the criterion of the test bus
driver choise.

The test bus (fig.7) was equiped in
the measurement system which can record 14 values, among others:
- distance covered,
- travelling speed,
- steering angle,
- bus load.
Also the kind and the state of the road
surface were encoded.

4.3. Test results

Test out-comes were obtained as functions:

$$x_i = x_i(t) \quad (6)$$

where: i - number of operation condi-
            tions factor.

The analysis of the test out-comes were
complited in order to obtain probability
distribution of the factor x_i

$$F(x_i) = P(X_i > x_i) \quad (7)$$

Fig.7. Test bus cabin.

and to obtain probability distribution
of time t or distance l under the
fixed value of random event X_i:

$$F(t_{X_i(t)=x_i}) = P(T > t/X_i(t) = x_i) \quad (8)$$

Fig.8 shows the distribution functions
of travelling speed $F(v)$ which were
approximated by the composite Weibull
distribution.

In the analysis of the test results was
considered:

(i) the verification of the hypotesis
H_0^1: $E(x_i) = E(x_j)$,

(ii) the verification of the hypotesis
H_0^2: $F_i(x) = F_j(x)$,

(iii) the verification of the hypotesis
that the processes $X(t)$, $Z(t)$, $E(t)$,
$Q(t)$ were stationary,

(iiii) the verification of the hypotesis that the processes $X(t)$, $Z(t)$, $E(t)$
were ergodic.

The quantitative results of H_0^1 and H_0^2
verification were illustrated by factor

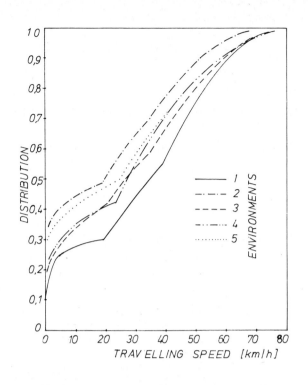

Fig.8. Distributions of the bus travelling speed.

k_H (fig.9):

$$k_H = (A[H_O^1] + A[H_O^2] / (R[H_O^1] + R[H_O^2]) \qquad (9)$$

where: $A[H_O^1]$, $A[H_O^2]$ – number of accepted hypotesis,
$R[H_O^1]$, $R[H_O^2]$ – number of rejected hypotesis.

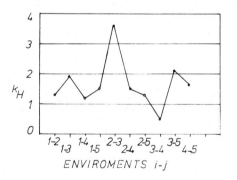

Fig.9. K_H – factor values.

The hypotesis that the bus load process Q(t) is stationary was rejected, but there were no reasons to reject this hypotesis for the bus motion processes. The analysis of the autocorrelation function of the stationary processes as the ergodic ones.

5. EVALUATION OF THE INFLUENCE

The influence of the operating conditions on the object reliability could be evaluated by three classes of methods [5]:

(i) deterministic methods (operating conditions and object reliability are described by one parameter),

(ii) quasiprobabilistic methods (probability distributions are represented by the first and the second moments)

(iii) probabilistic methods (allow to predict the reliability function on the basis of distribution function of the operating conditions).

5.1. Evaluation method

The modified multiple regression method (method of add and discard) was used. The function (1) got the form:

$$y = f(x_1, \ldots, x_s, b_0, b_1, \ldots, b_k) \qquad (10)$$

It allows to predict the average habites of the object. Thus, the function (10) was computed as:

$$\bar{M} = b_0 + b_1 x_1 + \ldots + b_n x_n + b_{11} x_1^2 + \ldots b_{ij} x_i x_j +$$
$$+ \ldots + b_{kk} x_k^2 \qquad (11)$$

where: $\bar{M}$ – mean mileage between failures.

5.2. Evaluation results

The regression functions were estimated for the bus and for its subsystems. Fig. 10 shows the correlation factors of mean mileage between failures of the bus with the operating conditions factors. Fig.11 presents the correlation factors for the break system.

This is accepted to be a significant regression function for the bus:

$$\bar{M} = b_0 (-1 + 0.094\ v) \qquad (12)$$

where: v – mean value of travelling speed.

Thus, the bus reliability increases while operating in free motion (high mean travelling speed, long travelling time between stops, long segments of travelling straight on and on the bend).

This function was estimated for the break system:

$$\bar{M} = b_0 (1 - 0.013\ \bar{a}\ \alpha) \qquad (13)$$

where: $\bar{a}$ – mean value of travelling deceleration,
α – mean value of steering angle.

The reliability of the break system is relatively weak depended on traffic conditions. The failure rate of the breaks increases proportionally to the travelling deceleration and the steering angle.

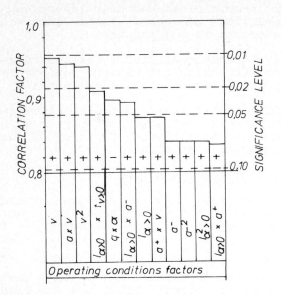

Markings:

v - mean travelling speed,

a^- - mean travelling deceleration,

$l_{\alpha > 0}$ - mean cornering distance,

$t_{v > 0}$ - mean running time,

q - mean bus load,

a^+ - mean travelling aceleration.

Fig.10. Correlation factors of bus mean MBF.

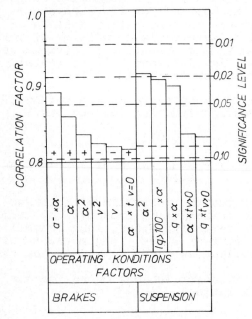

Markings:

a^- - mean travelling deceleration,
α - mean steering angle,
v - mean travelling speed,
$t_{v=0}$ - mean parking time,
$l_{q > 100}$ - mean distance of overload bus,
q - mean bus load,
$t_{v > 0}$ - mean running time.

Fig.11. Correlation factors of sub-systems mean MBF.

6. CONCLUSIONS

The presented analysis allows for quantitative estimation of the influence of the operating conditions on the bus reliability. Some discussed relationships were confirmed in the tests of next productive lots of the bus. Neverthless, the stability of the evaluated dependences remains the significant problem. The test results show some possibilities of bus reliability improvements. It was confirmed that the identification of the operating conditions on the object design level is very important.

REFERENCES

[1] Gałdzicki Z., Gidzińska D.: Analiza struktury społeczno-demograficznej pracowników obsługi autobusów Jelcz PR 110 U (Analysis of social and demographic structure of bus service workers), IOZ report no. 10/79, Technical University of Wrocław 1979.

[2] Gołąbek A., Młyńczak M., Nowakowski T., Stolarek E.: Projekt systemu zbierania informacji i przetwarzania na EMC danych o eksploatacyjnej niezawodności autobusów, samochodów ciężarowych i pochodnych (Data collection and computer processing system of cars operating reliability), IKEM report no. 549/78, Technical University of Wrocław 1978.

[3] Hebda M., Janicki D.: Trwałość i niezawodność samochodów w eksploatacji (Cars life-time and reliability in operation), WKŁ, Warszawa, 1977.

[4] Nowakowski T.: Wpływ warunków użytkowania na niezawodność pojazdu na przykładzie autobusu miejskiego PR 110 U (Operating conditions influence on bus reliability), IKEM report no. 047/80, Technical University of Wrocław, 1980.

[5] Nowakowski T.: Metody szacowania wpływu otoczenia na niezawodność urządzeń technicznych (Evaluation methods of environment influences on technical object reliability) in: Problemy niezawodności transportu - Cedzyna 83 (Transport reliability problems), NOT Kielce, 1983.

[6] Nowakowski T.: Metoda badania warunków użytkowania pojazdu (Research methods of vehicle operating conditions), in: Zeszyty Naukowe Politechniki Poznańskiej (Scientific Publications of Poznań Technical University), Mechanika no.30, 1984.

BIOGRAPHY

Tomasz Nowakowski was born in Wrocław (Poland) in 1953. 1971 he entered the Mechanical Faculty of the Technical University of Wrocław. 1976 he finished studies and obtained MS degree than he enrolled for Doctor Studies at the TUW. 1980 he obtained the degree of Doctor

of Technical Sciences. He is the research worker at the Institute of Machine Designe and Operation of the TUW.
He works on problems concerning reliability of minings of transport (buses,
trucks, rail-way cars, inland ships).
He is specialy interested in reliability field tests, reliability evaluation on design level and interactions
man - machine.

PLENUM SESSION

Reliability Technology — Theory & Applications
J. Møltoft and F. Jensen (Editors)
© Elsevier Science Publishers B.V. (North-Holland), 1986

EEC PRODUCT LIABILITY LEGISLATION AND ITS
PRACTICAL CONSEQUENCES FOR THE ELECTRONICS
INDUSTRY

Frede Ask
Association of Electronics Manufacturers in Denmark

The EEC Council Directive regarding product liability aims at protecting the consumers against damage caused by defective products. The impact of this new legislation will probably be a slightly increased number of lawsuits but not to the extent that it has been seen in the USA. Manufacturers of consumer electronics are advised to improve their quality control and make their manuals more user friendly. It should be noted that the existing national rules on product liability will remain in force together with the new EEC rules.

1. INTRODUCTION

Product liability is a fairly new legal institution. However, it is older than often believed. The first cases date back to the beginning of the industrialization late in the nineteenth century.

Rroduct liability is usually defined as the producer's liability for damages caused by defects in his products.

Product liability is not to be confused with the following legal institutions:

1.1 Liability in contracts, (for instance the buyers right to claim damages in accordance with the rules of the Sale of Goods Act).

1.2 Ordinary liability for damage caused by a tortious act.

1.3 Liability for services.

1.4 Product control, i.e. mandatory standards sanctioned by punishment (typically a fine).

None of the above mentioned legal institutions are suited for solving the problems of product liability:

1.1 and 1.3 would require a contract between the producer and the injured person. The latter might as well be a third party. Consequently, no contract exists.

1.2 would place the buyer in a less advantageous position (if he were the injured person) than that granted by the Sale of Goods Act.

1.4 only deals with the criminal aspect. The fine is independent of whether a damage actually has occurred or not. Consequently, these rules cannot solve problems concerning liability for damage.

2. EXISTING PRODUCT LIABILITY RULES

Product liability rules differ somewhat from country to country. However, some of the main features can be explained by mentioning the conditions for being held liable for damages in accordance with the legal usage in this field:

2.1 There must be a product (as opposed to service or real estate).

2.2 There must be a defect (when the product is sold).

2.2 There must be a loss (or a damage) to persons or property.

2.3 There must be a casual relationship between the defect and the damage.

2.4 There must be a basis of liability. This can be:

2.4.1 Negligence, where the burden of proof lies with the injured person.

2.4.2 Negligence, where the burden of proof lies with the producer.

2.4.3 Strict liability, i.e. the producer is liable without fault on his part.

3. WHY NEW RULES ?

In the preamble of the Council Directive of July 25th, 1985 (85/374/EEC) the following arguments are presented:

3.1 It is necessary to create an approximation of the laws of the Member States concerning product liability as the existing divergencies may:

3.1.1 Distort competition.

3.1.2 Effect the free movement of goods within the common market.

3.1.3 Entail a differing degree of protection of the consumers.

3.2 Strict liability is the sole means of creating a fair apportionment of the risks inherent in modern production.

Argument 3.1.1 seems a little weak. As a product liability case normally will be settled by a court in the country in which the damage has taken place, the business conditions are the same for all producers irrespective of their nationality. Arguments 3.1.2 and 3.1.3 are admittedly valid. Regarding argument 3.2 the opinions vary substantially. However, the trend towards a better protection of consumers is a development that the industry can hardly struggle against.

4. THE RULES OF THE DIRECTIVE

The Directive requires that Member States at the latest by July 30th, 1988 shall bring into force the new rules of the Directive. Some of the main elements are:

4.1 The basis of liability is strict liability, i. e. the producer is liable even if there is nothing at all to blame him for. Admittedly, this a very severe rule.

The present main rule is negligence where the burden of proof lies with the producer. However, during the last couple of decades the courts have administered this rule in a fairly strict manner, i.e. the requirements to the proof have become so rigid that the rule is approaching a rule of strict liability. One may then argue that it doesn't do any harm to the industry that the rule is called by its true name.

One of the positive aspects of strict liability is that the fact that a producer is held liable in a product liability case does not entail any repression as opposed to a judgement where the producer is found guilty of negligence because he cannot prove that this was not the case. Thus, it cannot be precluded that this new rule with little practical impact from a legal point of view may contribute to an improved image of the industry.

4.2 It should be noted that , according to the Directive, the burden of proof regarding the damage, the defect and the casual relationship between defect and damage still lies with the injured person.

4.3 It should also be noted that the new rules do not cover damage on property in commercial use. Here the existing national rules will apply.

4.4 The Directive permits the Member Countries to introduce a limit of 70 million ECU for the producer's total liability for damages caused by identical items with the same defect.

4.5 Finally, it should be mentioned that the producer cannot be held liable when 10 years have elapsed since the products were marketed.

5. CONCLUSION

One might ask whether there is a risk that Europe will experience a product liability crisis like that of USA. There is a good chance that the answer is no. Firstly, Eu-

ropean countries will hardly accept a law-
yer practice like the so-called "no cure,
no pay" system. Secondly, the system of
using juries of laymen to the extend that
it is done in the US is also unacceptable
to Europeans. Thirdly, the social security
system in Europe gives an injured person
an acceptable alternative to the courts.

It is incontestable that the new product
liability legislation constitutes a certain tightening of business conditions.
Especially producers of consumer electronics should be aware of the increased risk
of being held liable for damages. It is
advisable to improve quality control and
make manuals and other instructions as
user friendly as possible. However, it
would definitely be an exaggeration to use
the term "catastrophe" in this context.

BIOGRAPHY

Frede Ask, born 1931, was graduated as a
B. Sc. in Electrical Engineering in 1953
(Aarhus Technical College) and in 1971 as
a Master of Law (University of Copenhagen).

He was with Bruel & Kjaer, 1955-57 and
with Storno, 1957-73. His field of activity covered reliability investigations,
marketing and law.

Since 1973 Frede Ask has been employed by
The Association of Electronics Manufacturers in Denmark as its Director.

Frede Ask is a Member of the Danish Academy of Technical Sciences and of the
Product Liability Committee under The Federation of Danish Industries. He is a
co-author of the book: Industrial Product
Liability and Product Liability Insurance
1981 (Federation of Danish Industries).

SYSTEM SAFETY STUDIES

Reliability Technology — Theory & Applications
J. Møltoft and F. Jensen (Editors)
Elsevier Science Publishers B.V. (North-Holland), 1986

RELIABILITY AND SAFETY OF TRAFFIC SIGNALS

Klaus SCHLABBACH

Technical University Darmstadt, Department of Traffic and Transport Engineering
Petersenstrasse 15, 6100 Darmstadt, West Germany

The aim of increasing reliability and safety of traffic signals by avoiding signal indications which endanger traffic shall be achieved with the German guideline DIN 57832. Influences will be pointed out by several components of a traffic control system (controller, signals, detectors f.e.) and by extended criterions (planning process, operating process). Requirements to the producers of controllers and recommandations to the operators of signals are given.

1. INTRODUCTION

Formerly traffic control was the policemens job. But as the signals came, they had to clear the stage. Since that time the automatic followers with most elaborated technology move irresistible forward. In between it is well known that there will never be absolute safety in the technical sphere. By means of mutual understanding improbable defects are merely defined into impossible defects. For example the chance that all engines of an aircraft break down at the same time, is never taken into account, provided that all necessary services have been done. Even the breakdown of all engines is not impossible but only improbable.

The safety of traffic signals means that technical, planning or human defects do not lead to dangerous signal indications. Reliability means that all the respective components carry out their function extensive without error. Finally, the availability is the ratio of the actual working hours to the desired working hours of the complete system; in a hardly to achieve ideal situation this ratio becomes 1; wishes and reality just cover each other.

2. VDE 57832 - A SEALED BOOK?

Traffic signals are electrical equipments, at any rate they are a little bit more complicated as a hair-dryer or a vacuum cleaner. The number and the mode of monitoring systems at traffic signals are laid down in the terms of reference VDE 57832. The already mentioned dangerous signal indications ought to be prevented by:

- application of suitable circuits and working funds,
- careful production of the equipments,
- reliable erection of the complete system,
- duly operating of traffic signals and
- an experienced traffic engineering and planning.

This listing clearly shows the interconnection of the different branches like engineers, producers, building contractors and operators of a traffic signal system. Therefore it is not worth-while to increase the specific demands only for one group. It is obvious to compare this with the securing of a house against burglars. The cellar-outdoor fulfills all the security regulations and the neighbouring window is only of a thin glass.

The technical failure of a system, which can occur at any time, demands further arrangements for protecting against traffic accidents. The terms of reference VDE distinguish three modes, depending on the degree of traffic endangerment and the probability of it taking place:

A) indispensable securing measures at

- failure of red light on vehicle signals never mind if there are priority signs in this approach or not,
- failure of red light on vehicle signals infront of crossings with pedestrian signals,
- failure of red light on pedestrian signals at crossings behind signals with a green arrow for those who turn (including the "diagonal-green-arrow"),
- failure of red light on cyclist signals at placed back crossings for cyclists or when the priority sign for cyclists is presented,
- failure of red light on bus- and tram signals,
- undesired appearing of a green light on vehicle signals with a green light on incompatible vehicles, pedestrian or cyclist signals at the same time,
- premature appearing of a green light on an incompatible vehicle or a pedestrian signal with the yellow light of a vehicle signal,
- simultaneous appearing of red and green light at the same pedestrian signal with a green light on an incompatible vehicle signal.

B) conditional dispensable securing measures at

- extended yellow light on vehicle signals with green light on incompatible vehicle or pedestrian signals,
- green and yellow light on vehicle signals with green light on incompatible vehicle or pedestrian signals,
- premature appearing of a red and yellow light with the yellow light on an incompatible vehicle signal,

- extended yellow light on vehicle signals with red and green light on incompatible vehicle signals,
- fixing the signal stage longer than the intended time (stopping of signal cycle),
- reduction of intergreen time.

C) <u>dispensable securing measures at</u>

- failure of yellow light,
- failure of flashing yellow light behind/at vehicle signals,
- simultaneous green and red light on the same signal,
- reduction of the vehicle green time to less than the minimum green time.

This detailed description not at all should lead to confusion; it becomes obvious how many possible conflicts must be considered. The reader now is able to reconstruct the situation by comparing it to an intersection known to him. In a second step he had to ask if the classification into the three categories is unconditionally right or not. For example the reduction of intergreen time: avoiding this belongs perhaps to the indispensable securing measures? And why is prevention against the undesired reduction of the green time below the minimum value for the pedestrian signals not also necessary?

The basic principle of electricity "switch on first, then check" demands a fast reaction against mistakes, before the road user takes notice of it; 0.3 s cut-off-time is taken to be enough by the terms of reference VDE; this value should be further reduced with today's technology. The recognition of switching errors happens mostly by comparing independent electrical data and by this way finding out the undesired appearance, abscence or disappearance of lights. The countermeasure is then carried out automatically and in most cases it consists of cutting off the whole system by keeping in view all the conditions mentioned in the guideline for traffic signals (RiLSA). The blocking circuits, which were formerly used, prevent the switching on of undesired lights, without interrupting the further signal program. For example in the past the auxiliary signal for left-turn ("diagonal-green-arrow") would be suppressed if the incompatible pedestrian green light behind was switched on simultaneously. The green-arrow remained dark and the whole system did the job furthermore.

If there are several signals for one approach and one direction it is not clear which of them to use for monitoring a signal group. As you know, the main signal on the right of the road is repeated in some cases on the left and/or on a cantilever over the road. In May 1982 a questionaire carried out by the Road Traffic Authorities of big German cities resulted, that in 5 of 12 cases only the right signals on the main - and secondary road were monitored. In Munich f.e. only the right signals of the priority road are monitored as a rule; the failure rate therefore is only 0.4 % which means 1.5 failures per year and signal system. It is very interesting to compare this with Stuttgart, where - according to the questionaire - all the red lights are monitored; the failure rate was about five per signal system and year.

The users often have different opinions about the binding force of the terms of reference VDE. This varies from basic dissaproval ("general technical regulations of this type do not create any legal obligation") to differentiated intervals of application (renewal programs in many towns and districts). The most common indispensable securing measures are "undesired appearing of incompatible green lights (UAG)" and "undesired failure of monitored red lights (UFR)". In addition to that there is since 1980 the monitoring of the intergreen-time reduction; the countermeasure in this case is not the cutting off of the system, but in retarding the new signal to the nominal value. For sometime past preventive measures against fixing the signal stage longer than the intended time ("cycle monitoring") is usual. For example, if the advice "green light, main direction" is not repeated within 180 s, the signal system cuts off. Old systems with electro-mechanical timers yet showed these conditions ("constant red"), because the lingering gear defects were not detected below this limiting value. For lawfull procedure under such circumstances refer part 5.

3. RELIABILITY OF COMPONENTS

A signal system is built up of a different number of components depending on the type of traffic engineering problem. In all cases there are the controller, the signals with the bearing construction, the necessary communication required for the connection and the external power supply. If you need traffic-actuated control, detectors must be installed additionaly. Large controlling areas need control centers with corresponding peripheral equipment and communicating connexions with the intersections. All these parts can, either alone or under combined action, cause failures and defects and by this way reduce the reliability of the whole system. These possibilities are depicted below.

3.1 Controllers

Usually they are devided in two sections: one for the signal control and the other for switching the commands. Dependant on the used technology there are three generations of equipment:

- electromechanical parts,
- all electronic elements,
- highly integrated circuits.

The beginning of technology differs from one manufcturer to another; also mixed versions of the generations do exist. The LSI ("large-scale-integrated") circuits which are in use in Germany from 1980 have become popular under the keyword "microprocessor". The conception of hardware/software came into being along with them. Whereas hardware consists of the suitable physical parts of the equipment, software is non other than the "condensed form of all that brainware" [1], which has been invested in the analysis, diagnosis and elaboration of the traffic process. Only with these instruments it is possible to cross the narrow borders in the prevailing conception of signal programs.

The elements (condensators, transistors, flat devices, etc.) must pass a number of tests, before they are built into the equipment. The sub-contractors already must conform to certain quality standards. The point of main effort during checking the function is the climatic stress (-25° C to +40°C surrounding temperature, humidity of 100 % for a short time). That is clear, because you will have an interior temperature up to 60°C, if the equipment is in an unsuitable position under direct exposure of the sun. Secondly the equipment itself contributes a rise in temperature; most of all when the switching section runs with the low voltage of 42 V, there will be at the transformer (220 V - 42 V) a noticeable heat. Due to the highly concentrated way of packing the different devices, such heat could not be transmitted. The

result was using ventilation, which only distributed the heat uniformly all over the elements and prevented the so-called heat-nests. A cooling by sucking air from outside is almost impossible, as the pollution by dust particles and corrosive components cannot be checked.

The use of microprocessors raises the question of the reliability of the software. Even leading engineers and scientists agree, that we are hardly in a position, to check and control all defects and possible exceptional cases of computer runnings [2]. Perhaps this general statement is diminished in the field of traffic engineering by means of the long experience and detailed testing of the software by the manufacturers. Those who have intensively occupied themselves with the boundary conditions, compulsions and handicaps of algorithms and user software have however doubts about this. Quotation of a practical minded man: "Some of the software is like a banana - it ripens after

Table 1
Controller disturbances per month/number of monitoring circuits

Controller No.	UAG	UFR	Number of phases	Disturbances per month without red light		
				1980/81	1982/83	1984/85
1	17	14	11	3.2	0.8	0.1
20	26	15	14	4.5	0.7	0.3
31	8	14	10	3.3	0.2	0.1
41	25	16	13	2.0	0.2	0.5
68	21	24	13	5.0	0.2	0.6
69	7	7	7	0.7	1.3	0.2
88	27	18	13	3.4	0.8	0.4
104	7	6	7	0.2	0.4	0.6

Table 2
Controller disturbances per year (without red light)

Type	disturbances per year					
	1980	1981	1982	1983	1984	1985
all controllers	6.6		6.5	4.5	4.1	3.3
pedestrian signals	0.9		0.8	1.0	1.2	1.2
intersections	8.7		8.9	6.1	5.3	4.2
under this BX61 ⎫ electro-	9.0		6.9	6.0	3.6	4.4
BX62 ⎭ mechanical	10.0		7.8	6.2	7.6	3.9
BX E ⎫ electronic	6.7		9.3	6.4	3.9	4.8
BE ⎭	6.0		3.5	9.0	4.8	4.8
SBM LSJ-circuits	-		11.3	3.1	4.8	3.8

having come into the hands of the customer".
Only seldom the question is of <u>traffic endangering defects,</u> which should lead to the cut off
of the whole system; quiet often there are
<u>traffic hindering defects,</u> for example a signal
program or a special part of it, which is in
not the intended way optimised to the local
traffic situation. On the other hand it should
also be mentioned that the important inter-
green-timemonitoring was only possible by the
development of suitable software devices. The
latest equipments also waste up other securing
measures with the help of software; they consist of two microprocessors which carry out the
control and monitor functions by job sharing
and counter checking [3].

The hard- and software are considered as a
system and tested together, for example by
"burning-in" (12 hour operation, 80°C) with a
following comparison of software data.

From the German speaking part of the world
there are very few publications available regarding the reliability of old and new equipments during permanent operating. A research
paper [4] - already from 1974 - states overall
12 to 25 disturbances per year and system.
However it is presumed that controller defects
and red light disturbances, which will be explained later, are included. Informations from
abroad - as in [5] with only three or in[6]
with 2 to 8 disturbances per system and year -
cannot be transfered, as system configuration,
monitoring - and maintenance regulations cannot
be retraced. Reservations against overall statistics are so long justified till differentiated causes of defects are listed. This is
confirmed by the experience with the 140 systems installed in Darmstadt. Table 1 shows the
disturbances per month, the number of green-
green monitoring (UAG) and the monitored red
lights (UFR) for some intersections with controllers in LSJ-technology. The number of researched intersections is too small for making
valid the theoretically desired assertion, that
the number of disturbances increases with the
number of green-green monitorings. The high
values in the years 1980 - 81 are explained by
the fact that the service and maintenance staff
had less experience with the new technology,
which resulted in delayed repair actions. But
all these controllers are situated at important
intersections, which means that they were not
to be left without signals for a longer period
due to traffic safety. That is the reason why
the systems were switched on again and again,
so that we had failures one after another. For
example in march 1981 the controller no. 41 cut
off 9 times due to the same defect in the
performance part; figure 2 shows the controller
disturbances in the long run of all signal
systems in Darmstadt classified to the type of
intersection and the type of controller technology. In general the disturbances happened like
points of main effort (1/3 of the disturbances
concentrated itself only at 10 % of the controllers). If these points are first of all
recognized, it is most of all possible to reduce the disturbance to a great extent and
thereby increase the availability.

3.2 Signals

As important as the controller itself, where
the commands are generated and executed, is the
end of the information chain: the electrical
impulses are transformed into a visible form
for the road users by the bulb, reflector and
lens in the signal. Already during the construction of a system care should be taken of
visibility and recognition of the signals
(size, repetition, contrast-glave, angle of
inclination, sunrays etc.). During the years
the environment can essentially change: sign-
posts will be put up, hedges and trees grow,
glowing advertising boards will be installed or
changed. Today the signals are made up of syn-
thetic products (polycarbonates). They must be
protected against "unpermitted usage" and
should be moisture - and dust proof. According
to [7], dust on the lens reduces the lightin-
tensity by about 45 %. This should be an important tip, while considering the traffic safety
and maintenance.

Traffic endangering situations can also be
created by phantom lights; this means the
lighting of a signal - most frequently by sun
rays - without the signal bulb being switched
on. For relief are suggested special dispension
panes or reflectorforms as well as additional
installations (segments, phantomcrosses) .

Most of all the duration of bulbs life is
important, as the failure of monitored red
lights leads to the cut off of the system
(compare part 2). At the same time an extended
exchange of the bulbs can mean a considerable
saving in the servicing costs. In the Federal
Republic of Germany it is usual to use a normal
alternating current of 220 V for light bulbs.
This historic founded development is not optimal in regard to luminous intensity and shed-
ding the load. In the early sixties the cities
of Munich and Düsseldorf developed an alternative by using the low voltage 42 V. These bulbs
achieve the necessary luminous intensity with a
lower electrical performance than the 220-V-
bulbs, and in addition they also contain a
second filament, which in case of a defect
would switched on automatically. This idea has
proved itself at the railways for almost sixty
years. A similar thing can be imitated in the
220-V-technique by repeating the red pedestrian
signals respectively by the application of a
double reflector. The latter has in one signal
chamber two bulbs by using a divided mirror; in
the case of one bulb fusing the other one would
still give light. The 12-V-technique is widely
used abroad. By installing a circular central
transformer and adjusting the signal securing
units this technique could also be combined
with the locally available 220-V-technique. In
table 3 the different potentials needed for
bulbperformances in order to give the same
luminous intensity is shown according to [8].

Table 3: Bulbs: Potential, performance and duration of life by the same luminous intensity; *) mean duration, **)minimum duration (failure rate 2 %)

Potential [V]	Performance [W]	Duration of bulbs life [h]
10...12	24	7400*)
40...42	40	3000**)
220...240	75	3000**)

Experiences from abroad show us that the duration of bulbs life is strictly combined with the maintenance strategy (preventive or re-establishing according to [9]). The optimal failure rate for bulbs under consideration of the service costs is given in 10 for Australian conditions, as 2.3 % for 3360 and 15.9 % for 4200 lighting hours. The advantage of the low-voltage bulbs is explained with the higher exploitation of light and the smaller luminary, which has a better efficiency with respect to the luminous flux.

In Austria a special low voltage optic has been developed (10 V/50 W-halogen) and has been researched about its longterm influence on traffic safety [11]. Along a 3.4 km long throughroad eight intersections were equiped with the new signals. Although the accidents in the stretches between the intersections rose slightly, the accidents in the signal area could be reduced from before 29 to afterwards 16 per year; the traffic volume remained unchanged (30000 CPD). A remarkable part of this decrease had such accident types, which could be positively influenced by a better recognition of the new signals: Rear-end collisions went down from 12 to 7 per year and accidents due to red-light-neglects from 4 to 2 per year.

3.3 Means of communication

The cables connecting the controller with the signals are normally passed through PVC-tubes. In contrary to a second possibility (cables directly in earth), in the first case a completion of wires is possible quickly and without greater expenses. According to the terms of regulation VDE 57832 each signal column should be cabled seperately ("star cabling"). Disturbances of the communication means are mostly done during underground works due to carelessness or ignorance. A break of the wire leads to an UFR-situation in a very short time; a touching of wires ("short circuit") in general shows itself as an UAG-disturbance. At the moment we only can presume about the long term duration of cable isolations life; present estimations are about 50 years.

In exceptional cases (f.e. securing of construction sites) the connection between controller and signals is provided wireless. The securing of failing signals needs a two way wireless system, which is in fact available, but has not found enough resonance by the user. Apart from that it is in no case possible to get a cut-off-time of 0.3 s, because one wireless frequency can only transfer a certain volume of data. The cut-off-times usual today are about 0.7 s. Inspite of all these limitations, "wireless signals" are an important help to reduce the accident risks at conflict points, which would be otherwise for a certain time without signals.

3.4 Power supply

The confidence which we have developed with regard to the power supply of electrical systems is at most called in question by timid natures during thunderstorms at the summertime. Though a total failure of the power supply is only seldom, the secondary disturbances of it can lead to the cut off of the controllers. These secondary disturbances can be:

- short time breakdown of current of more than 0.3 s,
- voltage fluctuations of more than +10 % and - 15 % or
- frequency fluctuations of more than 0.5 %.

Such deviations occur already when switching over inside the network or when large scale consumers in the neighbourhood are switched on. In case of lightening in the surroundings of a signal system an increasing in the voltage to about more than 10.000 V can happen. The consequences can be limited with the aid of overvoltage conductors. The additional question of securing the software during failure of power supply arises in the case of controllers with microprocessors. As long as the software is stored in PROM, there are no problems. But if you use RAM you have to support it by balancing batteries; otherwise RAM-data disappear never to be run again within milliseconds after the failure of power supply.

3.5 Detectors

In the case of fluctuating and unforeseeable traffic volumes, optimal signal programs are often traffic-actuated. For this the detection of the different traffic flows with induction loops, pedestrian sensors and/or overhead systems contacts is necessary. All these installations can be disturbed and by this way give wrong or no message. Some of the manufacturers offer for new equipments detector programs, which monitor on-line the plausibilities (for example 2 detectors for one traffic stream) or comparative values. In the case of mistakes it should automatically be switched over to an alternative program and at the same time the area control center should be informed. Thereby we have to distinguish traffic endangering and traffic hindering impacts of the mistakes. Indirect traffic endangers can occur if special required phases for pedestrians, cars or turning trams are not set in operation. Through proper preventive measures such situations could be avoided. Traffic hindering situations always occur, if the malfunction of detectors does not lead to an optimal signal program (for example no green period modification in the main direction; repeated setting of special phases). Let us state that the traffic-actuated

components do not have any effect on the reliability of a signal system.

3.6 Controlling centers

Amongst the controlling of the traffic flow,
which is however getting more decentralized,
the traffic control computer takes over the
functions of monitoring and managing the system
network. A questionaire among 21 German towns
of different sizes has been taken [12]. The
following disturbances inside the network had
been registered by the computer:
- break of cables   18 cities
- failure of controller   18 cities
- failure of power supply   15 cities
- undesired failure of monitored red
  light (UFR)   12 cities
- undesired appearance of green
  light (UAG)   11 cities
- monitoring of cycle time   7 cities
- monitoring of intergreen-time   4 cities
- monitoring of minimum green time   3 cities

In the meantime traffic control computers are
considered to be very reliable; as an example
the table 4 shows the MTBF-data of special
parts of traffic control computers according to
[13]:

Table 4: Traffic control computer - mean time
         between failures

system element	MTBF [years]
central processing unit	5
32-K- storage	5
I/0- processor	23
line printer	10
cable multiplexer	8
tape controlling unit	5

Additionally we have the cut-off-times during
preventive maintenance, about 2 hours in 3
months. It is interesting to notice the reserve
facilities which are provided for, in case of
failure/cut off the traffic control computer.
Named under [12] are:
- double computer hot-line standby   2x
- double computer normal standby   6x
- area-/master controller   7x
- emergency program in the junction
  controller   15x

An appropriate job sharing between the traffic
control computer and the subordinate installations under normal operating conditions and
even more in case of disturbances causes, that
the road user can hardly recognize, if the
central control is operating or not. Beyond
that the traffic control computer aids for
- effective and cheap service organisation,
- reduction of failure times,
- systematic quality control and
- reducing the sources of error.

The dialogue between the traffic control computer and the connected signal systems can be
made with parallel data transmission (each
information is refered to one wire) or serial
data transmission. Serial means the data is

transformed by modem and will be transmitted
through only 2 or 4 wires; at the end of the
transmitter line a reversible action takes
place, so that the data can be used in its
original form. The reliability of this type of
transmission is secured by control signals
("paritybits") or by repeating the data tele-
grams.

4. INFLUENCES IN RELIABILITY AND SAFETY BY EXTENDED CRITERIONS

It is very important to do an expert traffic
engineering before the installation of a controller as well as afterwards a regular operating. Refering to the previously discussed
components of traffic control systems, these
both fields form the extended surrounding area
and are discussed with the most important safety aspects.

4.1 Planning process

Already during the first designing of intersections many aspects like number of lanes, length
of lanes, traffic islands, chanelling and
crossing of pedestrians and cyclists have to be
taken into consideration; often times even
contradicting points of view have to be considered. Traffic engineers are supported by the
guidelines of the German Road and Traffic Research Association. Additionally everybody has
more or less experiences from his everyday
work. A comprehensive work involving great
responsibility is the calculation of suitable
signalprograms. As a part of it the calculation
of the intergreen-times (between incompatible
traffic streams), which is important for the
traffic safety, has also to be determined. The
guidelines for signals (RiLSA) lay down in
details the assumptions for acceleration and
speed and for the calculation processes which
are built upon these assumptions. Recently
simplified ways have been put into discussion,
so for example as in [14], which based upon
experiments with the theory of probability.
Those methods achieve a comparable and satisfactory standard of safety. Such efforts have
to be encouraged to spread out. By the use of
software like CROSS or KNOTEN a greater reliability beyond the calculation of the intergreen-times is possible. Those "software tools"
are available from consultants and manufacturers. If not, the distribution of the work (program development and program testing) to different persons of the staff helps in reducing
mistakes during the planning process.

Even when signal programs are developed and
calculated corresponding to the state-of-the-
art they can still include latent (this means
veiled) dangers [15], "if the assumptions under
certain structural realities lead to constella-
tions of signals, which do not consider the
actual behaviour of a pedestrian satisfactorily." The interested reader will find more details in the literature given at the end; at
this point it is adequate just to mention it as
an example to show how far a broad discussion
of the safety aspects at traffic signals reaches out.

4.2 Operating

The planning, construction and operating of a signal system should form an unit. This claim, from teaching and research often times accentuated, is very seldom respected in the practice. Even if there are interpretation conflicts concerning certain details, but in general the requirements for operating are satisfactorily described in the terms of regulation RiLSA[16] and VDE 57832 [17] (- perhaps they are too perfect? -); they only had to be noticed. Today even here computer-added-design is offered to a great extent (for example IPSYS, SBS, AMPEL).

To operate duly it takes more than fussy keeping of control books and monitoring household lists, for example:
- component related decision about re-establishing or preventive maintenance strategy,
- service organisation and planning of staff assignment,
- sectionalised fault analysis according to the technique of the system,
- development of devices with greater reliability and reduced maintenance,
- clear and definite points of intersections between the concerned authorities.

The scetched operating management is necessary, if we want to change the first mentioned linear sequence (planning, constructing, operating) into a constant ameliorative circular process, and thereby increase the availability of signal systems.

5. LEGAL ASPECTS

The behaviour of the road users in the face of traffic signs is disunited. On the one side there are signs (for example those which regulate the parking facilities) which are constantly and purposely disobeyed. On the other side the green light is looked upon as an absolute right of way, which is not associated with by the jurisdiction.

Basically it remains a fact, that the cut off or failure of signal systems depicts an admissible situation for the traffic flow at an intersection. In this case a substitute system in the form of the road traffic laws guarantees the traffic flow, which can be lifted to a higher level of service by additional measures (f.e. police).

In one case a priority approach did not show the red light due to a failure, though waiting duty of the yield approach was quashed by a green light. The Federal Supreme Court - already in 1970 - turned down the appeal to make the traffic authorities responsible for the liability of endangering the traffic. The Regional Court of Düsseldorf with a judgement from May 1976 added, that a town responsible for the traffic securing duty should prove that it is not guilty when in case of a failure in the signal system an accident occurs.

As already mentioned in part 2 the traffic situation is also unaccountable, if the signal indication "constant red light" is given to some traffic streams or to all road users. In such cases the principle of extreme suspicion is valid. As the Regional Court of Cologne said in April 1980, the red light is allowed to be passed with great caution. Previously an appropriate waiting time of 3 - 5 minutes should be kept up.

6. OUTLOOK

Nowadays the failures of microelectronic installations cannot any longer be diagnosed or repaired without other microprocessors. One example: In Germany all cars have to be checked within two years by a technical inspection called TÜV. It is not possible to check the brake mechanism of modern cars, which have an "anti-locking-device", with the usual cylinder test-bench. The usual speed of rotation of the test-benches is not more than 5 km/h which is too small to activate most of the anti-locking devices. That is why one relies completely on the selftest built in the system itself which can be read by sensors and a testing computer. Similarly the manufacturers of the controllers are expected to feel themselves even more than system suppliers and offer suitable user located and standardised error detecting routines. The diagnosis is desired to be so far that it is able to determine the errors to such an extent one can easily exchange the components.

The operators of signal systems should be interested in connecting higher reliability and lower service and maintenance costs. The first step could be an intensified specialised exchange of nation-wide knowledge with the aim o having comparable data of analysis. In the second step these data should be discussed in an international range. The relationship between accident risks and failures of signal systems including the securing measures of part 2 has to be revised intensively. The published literature [4] for that does not serve the purpose fully.

The authors of the terms of regulation VDE 57832 are of the opinion that "informations about quantitative reliability for example about the availability are not possible, because the number of detrimental factors is very large and those factors cannot be evaluated." Considering the 35000 signal systems installed in the Federal Republic of Germany it becomes obvious how urgent - perhaps already overdue - it is to do more fundamental research in this area.

Most of all let us never forget that the safety of traffic signals does not solely depend upon the technical devices but also on the behaviour and mutual respect of the road users.

REFERENCES

[1] Nake, F., Points of intersections between mankind and machines, in Wagenbach, P. (ed.), Kursbuch No. 75: Computer culture (Rotbuch Verlag, Berlin, 1984).

[2] Mahr, B., The government of the directions for use, in Wagenbach, P. (ed.), Kursbuch No. 75: Computer culture (Rotbuch Verlag, Berlin, 1984).

[3] Spannaus, R., Reliability of traffic signals by using micro-computers, Die Strasse 3 (1981) 84-87.

[4] Kittel, J., Research of criterias for a homogeneous securing of signals according to traffic engineering aspects, in German Road and Traffic Reserch Association (ed.), Informations about research No. 17 (Kirschbaum Verlag, Bonn, 1977).

[5] Silovski, W., Traffic management and traffic signals, in International Meeting (ed.), Traffic accidents and car damages (Wien, 1979).

[6] Blase, J.H., Computer aids to large-scale traffic signal maintenance, Traffic Engineering and Control 7 (1979) 341-347.

[7] Mitchell, M.E., The key to preventive maintenance on a traffic signal system, Traffic Engineering 11 (1975) 30-31.

[8] Schreiber, G. Technical properties of light at traffic signals, Strassenverkehrstechnik 2 (1981) 7-10.

[9] Fischer, K. and Hultsch, K.H., Ways to secure the availability of traffic signals, Die Strasse 1 (1978) 7-10.

[10] Hulscher, F.R., Reliability aspects of road traffic control signals, Traffic Engineering and Control 3 (1977) 98-101.

[11] Tesarek, H. and Zibuschka, F., The absorption optic with halogen, Strassenverkehrstechnik 2 (1981) 45-48.

[12] Schlabbach, K. Questionaire about traffic control centers, Town Planning Authority, Darmstadt (May 1982).

[13] Philipps Inc., Traffic control system ATC (Hilversum, 1982).

[14] Jakob, G., A simplified method for calculating inter-green times at traffic signals, Strassenverkehrstechnik 4 (1982) 109-113.

[15] Retzko, H.-G. and Häckelmann, P., Latent dangers to pedestrians at traffic-light facilities, Zeitschrift für Verkehrssicherheit 4 (1977) 151-156.

[16] German Road and Traffic Research Association, Guidelines for traffic signals (Kirschbaum Verlag, Bonn, 1981).

[17] German Normalizing Committee, Standardizing sheet VDE 57832 - Traffic signals (Benth Verlag, Berlin, 1985).

BIOGRAPHY

Dipl.-Ing. Klaus Schlabbach;
born 21. July 1947 in Altenkirchen;

1954-1966   elementary and grammar school,
1966-1974   study of civil egineering at the Technical University Darmstadt;
since 1974 traffic engineer with the Town Planning Authority of Darmstadt;
since 1979 always there as the head of the "Traffic actuated light control" - project group;
since 1984 aditionally part-time as a researcher with the Technical University Darmstadt;

member of the German Road and Traffic Research Association with the committees "Traffic control in urban networks" and Designing of traffic systems for pedestrians"; numerous publications about traffic control and traffic safety;

married since 1974, two daughters.

Reliability Technology — Theory & Applications
J. Møltoft and F. Jensen (Editors)
© Elsevier Science Publishers B.V. (North-Holland), 1986

A COMPARATIVE RELIABILITY STUDY OF
PROTECTION SYSTEM FOR PIPELINES IN INDONESIA

A. LEROY, Risk Analysis Group, TOTAL - CFP
G. MATOUSKOFF, Safety Systems Department, Contrôle et Prévention

INTRODUCTION

The subject of the paper is a pipeline in Indonesia constituted of three different geographic installations as described in figure 1.

TOTAL's production/installation 1
Pipeline and producer side deliver/installation 2
Customer side delivery/installation 3

The 20" pipeline of the TOTAL company installation (1 and 2) ties in to the 36" and 42" pipelines of the third party installation. The maximum allowable working pressure of the third party installation pipeline is only 947 psig* while it was decided that the 20" pipeline would be designed to operate at inlet pressures as high as 1200 psig.

Moreover there are no full flow relief valves installed between installation 1 and 2 and the third party installation so, it is apparent that operation of the 20" pipeline at pressures exceeding 947 psig could result in overpressure of the pipelines of the third party installation.

In order to protect the third party installation pipelines against any overpressure which could be caused by installation 1 compressors, TOTAL has proposed two solutions and the American Company a third one.

In order to assess the safety level of each solution, a comparative reliability study has been performed.

The different stages of method used are:

- description of the system to be studied,

- filling up of the Failure Mode and Effects Analysis tables for the system,

- fault tree drafting,

- reliability data gathering,

_____

* Pound force spare inch.

- quantification of the fault trees,

- synthesis and conclusions.

The best data sources were used to perform the calculations: viz.

- OREDA (Offshore Reliability Data Handbook): data collected on North Sea platforms,

- then the SYSTEMS RELIABILITY SERVICE data bank of the UNITED KINGDOM ATOMIC ENERGY AUTHORITY,

- then the NONELECTRONIC PARTS RELIABILITY DATA - 2 handbook (NPRD-2)

1. GENERAL DESCRIPTION OF THE SYSTEM

The gas pressurized by two compressors is delivered at TOTAL's customer through a 20" pipeline. The main equipment on both sides is described here below.

Each High Pressure module is made up of:

- One interstage scrubber (V8160 and 8260).

- One gas turbine powered compressor (T 8150/C8170 and T 8250/C8270) with a nominal output of 100 MMSCFD.

- One gas aftercooler (A8180 and A8280).

- One discharge scrubber (V8190 and V8290).

- Two pneumatically operated valves allowing the isolation of the compressor from the well heads:

  . POV 8162 (8262) opened in normal functioning.

  . POV 8163 (8263) closed in normal functioning. It is only opened for the starting of the compressor.

- One anti-surge valve FV 8191 (8291) allowing by-pass of the compressor on receipt of order from the logic cubicle.

Then gas is delivered to the 20" pipeline through the V 8310 tank and the metering unit.

20" pipeline between TOTAL and its customer is equipped with several isolation gas-operated valves (GOV 9020 to 9050).

Before entering third party's piping system gas goes through:

- two parallel tanks V9070/V 9080, a by-pass being provided,

- and a safety gas-operated valve GOV 9051.

2. PROPOSED SOLUTIONS

2.1 Installation 1

Proposals A, B and C.

- to decrease the set point of the overpressure switches PSHH 8175 and 8275, down to 1030 psig (71 bars (g)),

- to decrease the set pressure of the pressure safety valves 8192, 8193, 8194 and 8292, 8293, 8294, down to 1060 psig (73 bars (g)).

2.2 Installation 2

Proposal A:

- to equip GOV 9050 with one PSH set at 928 psig (64 bars (g)),

- to decrease the set point of GOV 9051 from 971 psig (67 bars (g)) down to 928 psig (64 bars (g)).

Proposal B:

- to equip the GOV 9040 with a PSH set at 913 psig (63 bars (g)),

- to decrease the set point of GOV 9051 from 971 psig to 913 psig (63 bars (g)),

- to decrease the set pressure of the PSV 9076, 9077, 9077, 9086, 9087 from 1305 psig (90 bars (g)) to 928 (64 bars (g)),

- in this case the by-pass of the drums V 9070 and V 9080 must be lock-closed.

Proposal C:

- to equip GOV 9050 with one PSH set at 915 psig (63,1 bars (g)),

- to decrease the set point of GOV 9051 from 971 psig (67 bars (g)) down to 925 psig (63,8 bars (g)),

- to decrease the set pressure of the PSV 9076, 9077, 9086 and 9087 from 1305 psig (90 bars (g)) to 947 psig (65,3 bars (g)), all four being on duty,

- in this case, drums V 9070 and V 9080 are both on-duty.

The tree solutions A, B, C are shown in figures 2, 3, 4.

3. QUALITATIVE ANALYSIS

3.1 Failure mode and effects analysis

The F.M.E.A. objective is to identify and analyse the effects of all the components failure modes of the system.

According to the aim of the study this F.M.E.A. does not consider the failure modes concerning the availability of the installation but only those concerning the availability of the overpressure protection system.

This F.M.E.A. applies in particular to all the mechanical, hydraulic or electrical components of the overpressure protection systems.

A detailed F.M.E.A. is established for the gas operated valves (piloting and power systems).

A classification of the severity of the failure effects has been realized as described in the table overleaf.

	EFFECTS	
CLASS	DESCRIPTION	
1	More than two levels of protection against overpressure may still be available	
2	Only two levels of protection against overpressure may still be available	
3	Only one level of protection against overpressure may still be available	
4	No protection against overpressure	

For the determination of effect classes, it is considered that the first protection against overpressure is at installation 1 and the last one at third party's installation.

The different components considered in the F.M.E.A. are:

- Compressor by-pass valve.

- Pneumatic Operated suction valve.

- Pneumatic Operated valve.

- Gas turbine drive H.P. gas compressor.

- High discharge gas pressure.

- Pressure transducer.

- Pressure safety high.

- Logic cubicle.

- Pressure safety valves.

- Gas operated valve: 9051.

  . isolation valve,

  . three ways valve,

  . H.P. filter,

  . normally open distributor 2/2,

  . normally closed distributor 2/2,

  . distributor 5 use outlet,

  . distributor 3/2,

  . normally open distributor 3/2,

  . pressure inlet and distributor 2 use outlet,

  . distributor 4/3,

  . exchangers,

  . actuator of the valve,

  . ball valve.

- Pressure safety valves 9076, 9077, 9086, 9087.

It is not possible in an article of this nature to reproduce the F.M.E.A. established for each of those components, however after studying the system, it appears that:

- Possibility of human error in testing Pressure Safety Valves at TOTAL's installation could be seriously reduced if the PSV were interlocked.

- Maintenance, tests and periodical inspection are an important factor in failure prevention.

3.2 Fault tree analysis

The objective of the fault trees is to represent on a graph, the event combinations which lead to an unwanted event. This breakdown is deductive and stops at basic events which are wholly independent.

OVERPRESSURE PROTECTION SYSTEM FAILURE FAULT TREES

Three fault trees have been drawn representing (for each proposed solution A, B and C), the overpressure protection system failure.

- SCHEME A overpressure protection system failure.

- SCHEME B overpressure protection system failure.

- SCHEME C overpressure protection system failure.

GOV FAILURE TO CLOSE FAULT TREE

A fault tree has been drawn for the unwanted event: GOV failure to close.

The tree is drawn for the GOV 9051 and it is similar for the GOVs 9040/9050.

Fault tree representing the overpressure protection system failure for solution A is shown in figure 5.

4. QUANTITATIVE ANALYSIS

4.1 Reliability characteristic

The reliability characteristic used is the unavailability known as the fractional dead time (FDT), which is the average proportion of the time that the system is in the failed state. Indeed, for a proportion system, this value is more useful than its probability of failure to close, which is calculated anyway.

Simplified formulae

If the System is tested at intervals T, we have:

$$FDT = \frac{1}{T} \int_{o}^{T} P(t)dt$$

with P(t) is probability of failure system t:

$$P(t) = 1 - e^{-\lambda t}$$
$$P(t) = \lambda t \text{ if } t \ll 1$$

For a component:

$$FDT = \frac{1}{T} \int_{o}^{T} (1 - e^{-\lambda t})dt = 1 - \frac{1}{\lambda T}(1 - e^{-\lambda T})$$

$$\text{if } T \ll 1: FDT = \frac{1}{2}\lambda T$$

For a m out of n systems:

$$\text{if } T \ll 1$$

$$\text{for 1 out of 2 } FDT = \frac{1 (\lambda T)^2}{3}$$

$$\text{for 1 out of 3 } FDT = \frac{1 (\lambda T)^2}{4}$$

Effect of tests

Two test intervals are considered:

- general overhaul each 8000 hours concerning all the components,

- complete checking test each 4000 hours concerning GOV and POV piloting systems.

4.2 Data

Reliability data come from non-redundant data sources:

- OREDA (Offshore Reliability Data Handbook)

  . Phase I (81-82): D.V. report N° 82 -0529.

  . Phase II (83-84): Confidential report given to participants to the 83-84 data collection (TOTAL OIL MARINE, TOTAL British subsidiary, is a participating company).

. SYSTEMS RELIABILITY SERVICE (S.R.S.) of
  the UNITED KINGDOM ATOMIC ENERGY
  AUTHORITY.
  S.R.S. data do not come from a BIKINI
  output (generic reliability values) but
  from a MAGPIE output, upon our request.

. NONELECTRONIC PARTS RELIABILITY DATA-2
  (NPRD2) OF THE RELIABILITY ANALYSIS CENTER
  (U.S. Air Force).
  The best suited terms used in the N.P.R.D.
  2 handbook on platform environment are the
  following:

  - Dormant (stand-by);

  - Ground fixed;

  - Ground mobile;

  - Ship sheltered.

. And from the data bank of the French Power
  Board (S.R.D.F. of the E.D.F.) for some
  particular cases.

Data on all the components which could be
compared to those of the worked out system
have been recorded for each data bank or
collection.

4.3 Common cause failure rates

Systems which employ redundant components of
protection are subject to common cause
failures. Wherever possible, common causes of
failure have been indentified and assessed.

There are, however, several factors which
cause coincident failure of redundant
components and which cannot be identified
separately. These failures are often divided
into five categories:

- Environmental effects (extraneous to the
  system):

  . dust, wetness, salty atmosphere, electric
    circuit vibrations and tightness,

  . impact on the environment of an accident
    (fire, industrial missile) on hydraulic
    pipes or wires, electro-valves or electro-
    nics.

- Design errors.
- Fabrication errors.
- Installation errors.
- Human errors.

The common cause failure rate is related to
the independent failure rate of an individual
component in a redundant system by the
expression:

$$\lambda_{cm} = \frac{\beta}{1 - \beta} \cdot \lambda_I \quad \text{with } \beta = \frac{\lambda_{cm}}{\lambda_{cm} + \lambda_I}$$

A factor of 0.1 is taken when components are
of the same design and manufacture. A difference in design and manufacture reduces the
risk signicantly, but a maximum defence
against common causes of failure could only be
achieved by complete diversity (i.e. different
actuating media etc.) and special rules
applied to operability maintainability.

A value of 0.01 is usually assumed to be the
ultimate achievable by diversity.

In view of the uncertainty in the value
ascribed to the β factor, it was taken for
calculation two values: β = 0.1
                         β = 0.05

5. CONCLUSION

Calculations are made for the two following
situations:

- One compressor running, one on stand-by -One
  vessel on duty (2 Pressure Safety Valves).

- Two compressors running - Two vessels on
  duty (4 Pressure Safety Valves).

For each running situation and for each
proposed scheme calculations are made, with
two β factors:

  β = 0.1

  β = 0.05

First, the FDT of the GOV are calculated (for
a single one i.e. $3,1 \times 10^{-2}$).

Then the final result is obtained by multi-
plicating the FDT of each installation.

Whatever the hypothesis on the common cause
failure factor β made, and the number of PSV
set on-duty at installation 2, it appears
that:

- two solutions (B and C) are apparently
  highly reliable in the same range (FDT <
  10^{-6}),

- solution A was less repliable than solutions
  B and C (FDT ≈ 10^{-6}) but as well less
  costly.

Nevertheless a safety level of 10^{-6} is not
realistic so it was suggested to consider
solution A as reliable enough to provide
protection of 36"/42" pipelines against
overpressure which could be caused by TOTAL's
installation compressors.

PRACTICAL RELIABILITY SCREENING

Reliability Technology — Theory & Applications
J. Møltoft and F. Jensen (Editors)
© Elsevier Science Publishers B.V. (North-Holland), 1986

The Practical Application of Burn - In

Hans Peter Boisen

DANFOSS A/S, Denmark

The test philosophy for making burn-in tests on mass produced complex electronic units is presented.
The markovian method for defining the burn-in time is discussed. The improvement of the reliability of the components, the design and the production process shows a decreasing number of failures during burn-in over time.

1. Introduction.

The normal definition of burn-in is:
"Burn-in is a test on units before their final application with the purpose of stabilizing the parameters of the unit and identifying early defects".

"Units" can be components, sub-systems or whole units. In the following only results from burn-in on whole units is discussed. The units are electronic motor controls used for standard AC-motors in the range from 0,75kW to 110kW. The components and sub-systems in the final product have not been burn-in tested before assembly because experiments have shown that it is not possible to reduce the number of failures during burn-in of the whole unit by making burn-in of the components or the sub-systems.

2. The burn - in test in the production.

The production flow is presented on fig. 1. All components goes through incoming inspection to stock and from here to the automatic component insert maschine. When the PCB's have been mounted, they undergo an incircuit test and after being assembled to sub-systems they undergo a functional test.
The sub-systems can now be assembled to a complete unit which undergoes a functional test When the unit has passed the functional test successfully it goes to the burn-in test and after this test it undergoes a final test before shipment to the customer.

The burn-in test panel is built up with a computer to control the test cycle and to monitor the unit continuously. The test is characterized by:

- a high load consisting of an AC motor and generator.

- high temperature during the whole burn-in test.

- a thorough test cycle.

- a continuously measuring for faults and registration of time-to-failure.

- a strong requirement to parameter drift.

The test pattern of the burn-in test cycle is change when new requirement to the units comes up.

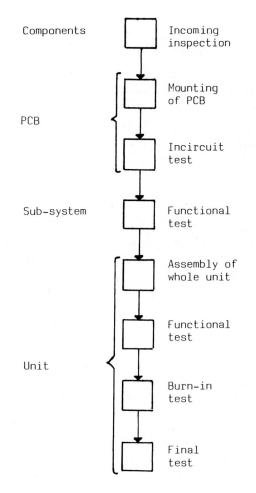

Fig. 1: The testflow in production of the units. The figure does not show the flow for repaired units.

3. Calculation of the burn - in time

Calculation of the burn-in time for a new product is based on an estimate of the burn-in time. This estimate is only made from the experience obtained through the production of similar units. Experience has shown that it is not possible to use a proper prediction model, bacause valid data for the bimodel failure distribution for the components are not available.

The optimal burn-in time is calculated on the bases of a test of the pre-production. In the first pre-production a burn-in time of approximately 6 times the estimate is used. If the number of failures during burn-in of the pre-production is too low to make a sufficient statistical analysis, the long burn-in time is used during production of the first series in current production.
To eliminate failures in the product design and the production process, all failures are analysed and corrective action is taken - if possible.

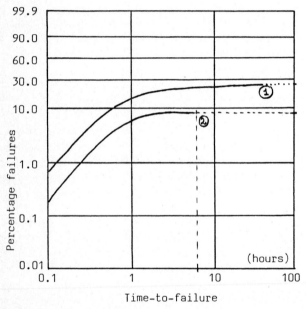

Fig. 2 Weibull curve of times-to-first failure for pre-production (1) and Weibull curve for running production (2).

The next step is a Weibul analysis during which all data are plotted.
Only times-to-first failure are analysed.

To calculate the actual burn-in time the method of Markovian burn-in is used {1}.
The method is based on the assumption that weak systems have times-to -first failure that follow an exponential distribution. To evaluate this assumption the weak population of systems is analysed. The separation from the main population is made by using Bayes´ method for separating distributions.{2}.

The weak population normally follows the exponential distribution, i.e Weibull shape parameter equal to unity. Table 1 shows a number of data for typical weak populations.

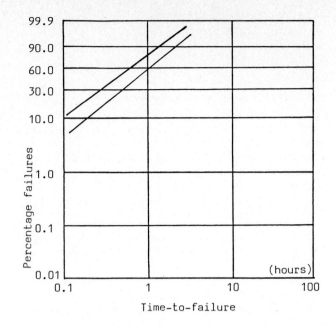

Fig. 3 Weibull plot of two typical weak populations.

	η	β	Weak population
Product A	1,1 h	1,0	11%
Product B	1,3 h	1,2	45%
Product C	1,0 h	1,0	32%
Product D	1,4 h	1,0	50%
Product E	1,4 h	1,2	4%
Product F	0,9 h	1,0	12%
Product G	0,9 h	0,9	30%

Table 1: List of the Weibull parameters for 7 different products.

To calculate the burn-in time, defined as the failure - free burn-in periode, it is necessary to calculate the number(N) of critical components defined as components having bimodal distribution of time-to-failure. In practice, N can vary between 20 to 400 depending on the complexity of the unit.

Calculating the hazard rate(λ_1) of the weak components gives a reduction of the failure rate from 0 to 35% using the formula {1}:

$$\lambda_1 = \lambda_s \cdot \frac{P_s}{P_c \cdot N}$$

P_s: Proportin of weak systems.

P_c: Proportion of weaks in the ensemble of critical components.

The final failure-free burn-in periode is
calculated using the Markovian model. The
percentage of units having one or more weak
components remaining after the burn-in is not
accepted higher than 1%. The result is a 4-8
times higher burn-in time than the characteristic lifetime of the weak population.

4. Updating the burn - in test.

Regularly the production data are analysed to
investigate changes, if any, in the failure-
pattern. If new failures arise they are
analysed - the objective is to reduce the
number of failures during burn-in and to reduce
the burn-in time. Experience has shown that it
is possible to see an imnprovement, a lower
number of units failing and a possible
reduction in the burn-in time.

A decrease in the number of failures does not
normally justify af reduction in the burn-in
time - in some cases it is necessary to have a
longer burn-in period ! Fig. 4 shows the
development during a 7 years period. In table
2. the actual Weibull parameters of the weak
population are listed.

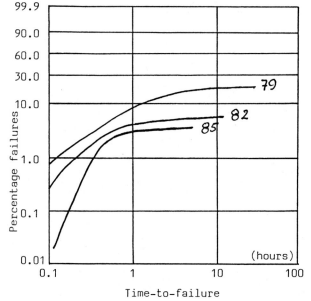

Fig. 4 Weibull plot of the same product from
three different years of production.

Production year	η		β	Weak population
79	3	h	0,7	16%
82	1	h	1	5%
85	0,2h		1,5	3%

Table 2: Development in burn-in
parameters.

5. Conclusion.

The choise of the Marvian method for analysing
burn-in data has been made because it is easy
to use and gives good results. Other methods
such as "Total Time on Test " {4} , "As good as
new", and "As bad as old" give other results
and need to be explained by a reliability
specialist. The use of the Markovian method and
the Weibull plot gives a conclusion on the
efficiency of the burn-in test easy to understand also for the the non-reliability specialist. The conclusion is that the Markovian
method using Weibull plot is a simple and
reliable way of designing burn-in tests in
practise - bringing into mind that the purpose
of our efforts is to produce products with high
quality at the lowest possible costs.

References

{1}   Finn Jensen and Niels Erik Petersen,
BURN-IN, An Engineering Approach to the
Design and Analysis of Burn-in Proce-
dures, John Wiley, 1982.

{2}   Charles Lipson and Narendra J. Sheth,
Engineering Experiments, McGraw-Hill,
1973.

{3}   Harold Ascher and Harry Feingold.
Repairable Systems Reliability: Modeling,
Inference, Miscorceptions and their
Caunses, Marcel Dekker Inc., New York 1984

{4}   Barlow, R.E., Campo, R.A., Total time on
Test Processes and Applications to
Failure Data Analysis, in Reliability and
Fault Tree Analyses, ed. Barlow, Fussel
and Sigpurvalla, Pencylvania, SIAN 1975.

Biography of the author

1980 Electrical engineer (B.Sc) from the
Engineering Academy of Denmark, Lyngby.

1980-1981 Military service at the Danish Defense
Research Laboratory.

1981-1983 Employed at Danfoss in the central
quality department.

1983-   Employed in the quality department
for Electrical drives and Controls.

1985   Degree in management (B.Com.) from
the Copenhagen, School of Economics
and Buisness Administration.

Reliability Technology — Theory & Applications
J. Møltoft and F. Jensen (Editors)
© Elsevier Science Publishers B.V. (North-Holland), 1986

ACCELERATION OF BURN-IN BY THERMAL CYCLING

Antoni DRAPELLA
Technical University of Gdańsk, 80-952 Gdańsk, Poland

This short paper describes industrial case history dealing with the problem of burn-in and screening. Because of a very long failure-free time typical burn-in can not be optimized with respect to costs. To accelerate burn-in procedure thermal cycling was applied.

0. INTRODUCTION

The loss of goodwill and markets forced producer of small portable stereo receivers to improve reliability. Field collected failure data /rather rough/ showed that a commonplace early failures infested customers. To obtain detailed information to design burn-in procedure a laboratory test was carried out. A sample of N=24 was tested under typical conditions. The result is shown in Fig. 1. A considerably long failure-free time was revealed. In a face of this fact two different ways to solve the problem were proposed.

1/ Typical burn-in optimized with respect to costs

2/ Accelerated burn-in by thermal cycling.

The first way is use-proved. However a failure-free time mentioned above stumbled engineers. In other hand thermal cycling may initiate unexpected failure mechanisms.

This paper consists of five sections. Section 2 describes a way in which a suitable life-time model was developed. In section 3 this model is involved in burn-in/cost analysis. Section 4 describes a performance of accelerated burn-in and the results obtained. Section 5 is a short conclusion.

1. NOTATIONS AND BASIC FORMULAE

t   time-to-failure
$F(t)$ cumulative distribution function
$R(t)$ survival function
$h(t)$ hazard rate function
$R(t) = 1 - F(t)$
$h(t) = -\dot{R}(t)/R(t)$
$R(t) = \exp(-((t-tf)/a)^b)$ Weibull distrib.
a,b   scale and shape parameters, resp.
tf   failure-free time
$x = \ln(t)$, $y = \ln(-\ln R)$ axes of the Weibull probability paper
E   burn-in efficiency measure
Kg   guarantee costs
Kb   burn-in costs/per hour
Ko   total costs

2. CHOOSING THE LIFE TIME MODEL

Fig. 1A shows an empirical c.d.f. plotted on the Weibull probability paper. The apparent curvature observed can be reasonably explained as due to failure-free life phenomenon. To estimate tf a method presented in [3] page 66 was used. An appropriate formula is

$$tf = t2 - (t3-t2) \cdot (t2-t1)/(t3+t1-2 \cdot t2)$$

$$/2.1/$$

Substituting t1=120 hours, t2=265 hours t3=840 hours /see Fig. 1A/ we get tf=71 hours. Fig. 1B shows adjusted c.d.f. Unfortunately this empirical c.d.f. can not be fitted by the straight line. The function being considered is the left-hand side of the S-shaped general failure pattern presented in [2]. The sample was taken from heterogeneous population. So the subsample of N1=9 must be treated separately. Fig. 1C shows an appropriate empirical c.d.f. We can see that in this case early failures follow the Weibull distribution having parameter values given below

scale parameter a = 900 hours
shape parameter b = 1.2

The survival function Ro(t) describing the whole population is a compound Weibull distribution. The value of w is 9/24.

$$Ro = w \cdot R(t, tf, a,b) + (1-w) \cdot Rs(t,?)$$

The survival function Rs(t) is ascribed to "strong" subpopulation. Parameter values remain unknown. However we assume that Rs = 1 for t < 1500 hours. This time interval overlaps burn-in time as well as guarantee time.

3. BURN-IN AND ECONOMY

To assess an efficiency of burn-in procedure the following measure proposed in [1] was applied

$$E(u) = 1 - \ln R(T/u)/\ln R(T) = 1-S2/S1$$

Fig.2 illustrates a substance of this measure. Fig. 3A shows the dependency between u and E(u) in the case being cinsidered. The following burn-in/cost model was used.

$$Ko = Kb \cdot u + Kg(1 - \exp((1 - E(u))\ln R(T)))$$

The most important problem in the optimization of burn-in is to assess the ratio Kg/Kb. A value of Kg is a sum of two costs

1/ A difference between repair costs in a factory and in a field

2/ The loss of customers

It was assumed that one failure during a guarantee period causes the loss of three potential customers. Finally the value of Kg/Kb was assessed as being about 300. Fig. 3B shows time/costs dependencies. The total costs are a strictly increasing function of burn-in time. There is no economic reason to accomplish a typical

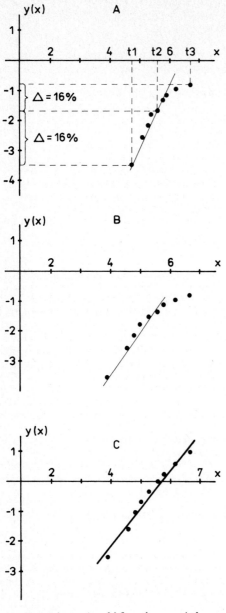

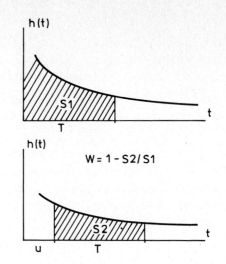

Fig. 2 Illustration of a burn-in measure
        applied.

Fig. 1. Choosing the life-time model.
        A: Inlet data. B: Failure-free time
        subtracted. C: A subsample of "weak"
        devices.

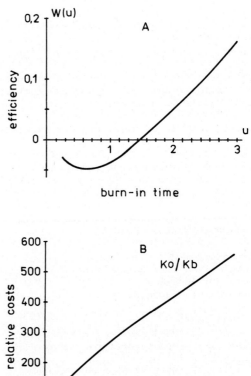

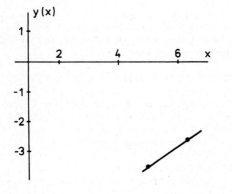

Fig. 4 . Result of post-burn-in test.

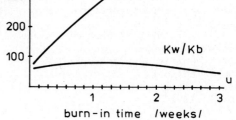

Fig. 3. Efficiency /A/ and the total costs
         /B/ as functions of burn-in time.

burn-in procedure.

4. THERMAL CYCLING

A structure of failures observed during the
test was as follows

solder joints	5 failures
semiconductors /wire bonds/	2 failures
passive components	2 failures

Mounted printed circuit boards underwent thermal cycling to destroy weak interconnections.
The test performance was as follows

sample size	24
number of cycles	5
temperature range	253 - 333 K

Four failures have been revealed after the
test. Next repaired and non failed receivers
passed the reliability test. An appropriate
c.d.f. is shown in Fig. 4. We can see that
only two failures occurred.

5. CONCLUSION

The case described above confirmed a utility of
thermal cycling as the method to screen-out
early failures accordingly to several works
earlier presented in reliability literature.

6. REFERENCES

[1] Drapella, A., A simple burn-in efficiency
    measure. Relectronic 85. 6th Symposium on
    Reliability in Electronic. Budapest,
    Hungary, August 1985.
[2] Jensen, F., Petersen, N.F., Burn-in, An en-
    gineering approach to the design and ala-
    lysis of burn-in procedures. /Wiley,
    Chichester, 1982/
[3] O'Connor, P.D.T., Practical reliability en-
    gineering. /Heyden 1981/

COMPONENT LIFE TESTING CASES

Reliability Technology — Theory & Applications
J. Møltoft and F. Jensen (Editors)
© Elsevier Science Publishers B.V. (North-Holland), 1986

ACCELERATED STRESS TESTS
OF PLASTIC PACKAGED DEVICES

Pietro BRAMBILLA and Fausto FANTINI
Telettra S.p.A.
Reliability and Quality Dept.
20059 Vimercate (Milano) - ITALY

The reliability of plastic packaged semiconductor devices was evaluated by means of accelerated temperature/humidity tests. Results of tests and failure analyses are reported and the effectiveness of the different tests with respect to failure rate and wearout evaluation is discussed. A value for the temperature acceleration of the corrosion phenomenon is derived and compared with field results.

1. INTRODUCTION

The reliability of plastic packaged semiconductor devices is of increasing importance for their application in professional equipment, due to their lower cost and easier availability with respect to their hermetic equivalents. However several doubts still exhist about these components, because of their higher sensitivity to many failure mechanisms, such as surface contamination, thermo-mechanical stresses and corrosion [1-2]. Moreover the need to use gold wires is imposed by the construction of the packages and creates the risk of "purple plague" at the wire bonds [3].

In particular, it is suspected that these failure mechanisms are of wearout type, and therefore have an increasing failure rate, and cannot be screened our by proper stresses.

A great improvement in the reliability of plastic packaged devices has been demonstrated in the least years [2], however a large dispersion still exhists in the data [4], along with the risk of the presence of bad lots.

Although in the seventies we did not qualify plastic packaged devices, we were forced to use a few TTL integrated circuits during a market crisis and we collected the results in a controlled application.

The results of the failure analysis are reported in fig. 1 and the presence of a few corrosion and "purple plague" failures can be detected, though not in a percentage so high as to drastically change the removal rate (if not the failure rate) with respect to the hermetic equivalents.

If we can consider the situation of digital bipolar IC as fairly established, much less confidence exists in the CMOS devices. This is because their very low dissipation eases the water vapour condensation on the chip surface, leading to corrosion problems [5].

Therefore we took as the main test vehicles different kinds of CMOS ICs, although other technologies were also tested [2].

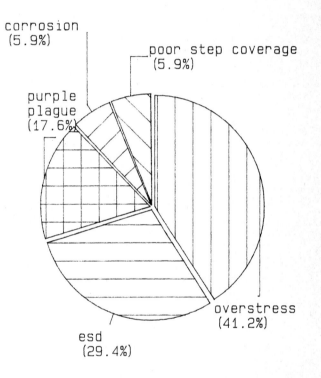

Fig. 1. Distribution of the failure mechanisms found on TTL plastic packaged ICs, that failed in the field.

In this paper we focus on the results on CMOS ICs: chapter 2 describes the tests and the samples stressed, chapter 3 reports the results of the tests, in chapter 4 these results are discussed and the conclusions are drawn.

2. SAMPLES AND TESTS

The CMOS integrated circuits were chosen as test vehicles because of their well known sensitivity to the package related reliability problems [6].

Different chip technologies were evaluated:
- CMOS 4000 devices, 6 μm Al-metal-gate with

SiO$_2$ or SiN passivation;
- HCMOS 74 HC devices, 3 μm Si-gate with SiN passivation;
- LSI memories, 3 μm Si-gate with SiN passivation.

Two types of packages were tested: the standard Dual-In-Line (DIL) and the Small-Outline (SO) packages for surface mounting or hybrid application. The tests were performed at different temperatures and constant humidity (85%), always with applied bias, which was 5 V for HCMOS and LSI and 15 V for CMOS.

The performance of higher temperature tests required a careful development of suitable test fictures, which are able to withstand at least 1000 h of test and enable easy measurements.

All the tests were run till to achieve a percentage of failures significant for deriving the parameters of the failures distribution. We also performed long-term (at least 2 years) operating life tests (OLT) in reference condition (i.e. 20°C, 50% R.H.) [6].

The complete set of tests and samples is shown in table 1.

under the surface passivation coming from the pad windows and the chip borders (fig. 3). Whilst in cathodic corroded lines no trace of contaminants was found, at least within the limits of sensitivity of the electron microprobe analysis, chlorine was detected in the positively biased corroded stripes.

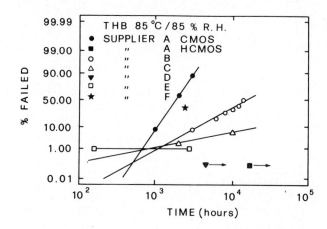

Fig. 2. Lognormal plot of the time-to-failure in THB tests.

Table 1
Duration of the performed tests in hours x 10^3

Package	DIL					SO	
Technology	CMOS			HCMOS	LSI	HCMOS	
Supplier	A	B	C	A	D	E	F
OLT	24	24	–	–	–	–	–
85°C/85% R.H.	4	10	14	14	6	4	4
125°C/85% R.H.	0.5	–	–	–	3	–	–
150°C/85% R.H.	0.15	–	–	–	–	1.8	0.15

3. RESULTS

3.1. Temperature Humidity Bias tests at 85°C 85% R.H.

In this test, generally called THB, seven different kinds of devices were stressed for, at least, 4000 hours.

In table 2 the results are analyzed in terms of lognormal distribution, which usually fits the data of accelerated tests in the best way [2,6,7]. The lognormal plots of the cumulative failure distributions are reported in fig. 2. Nearly all the devices of 4000 series from manufacturer A failed between 1000 and 3000 hours of test; the failure mode being initially the increase of leakage currents and then open circuits. The failure analysis revealed both cathodic and anodic corrosion [2].

The open metallizations were found near the pads or towards the edges of the chip, confirming the hypothesis that humidity penetrates

Quite surprisingly no humidity induced failures were obtained on devices of the HCMOS series from the same manufacturer after 4000 hours; the only failure being a fracture of the gold wire near the ball-bond.

The same result, i.e. no failures, was achieved with the LSI memories but after only 6000 hours.

A different failure mechanism appeared on HCMOS devices in SO package from manufacturer F. At 3300 hours 24% of the components abruptly failed at the functional tests, due to input voltage level (V_{IL}) degradation. The problem can be ascribed to contamination by ionic impurities, such as Na$^+$ [8]. The second supplier of these components showed only one very early failure with a large corrosion effect, but no further failures at 4000 hours.

3.2. Highly Accelerated Stress Tests (HAST)

Two test conditions were used, at 125°C and 150°C, for these highly accelerated stress

Table 2
Results of 85/85 THB test, evaluated by means of the lognormal distribution

Package	DIL					SO	
Technology	CMOS			HCMOS	LSI	HCMOS	
Supplier	A	B	C	A	D	E	F
Median life (hours)	1.8×10^3	1.5×10^4	1.4×10^6	–	–	–	–
Standard deviation	0.43	1.15	3.2	–	–	–	–
Percentage of failure at 1000 hours	8	0.8	0.6	0 (1)	0 (1)	1 (1)	0 (1)

(1) From the test data it is not possible to evaluate any distribution parameter.

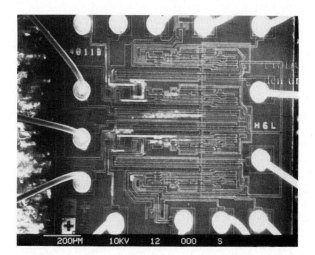

Fig. 3a. SEM photograph of a CMOS failed in the THB test.

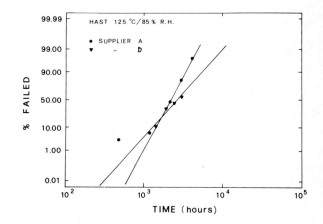

Fig. 4a. Lognormal plot of the time-to-failure in the HASTs at 125°C/85% R.H.

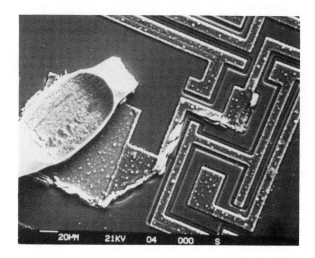

Fig. 3b. Exploited view of a corroded pad in another failed device.

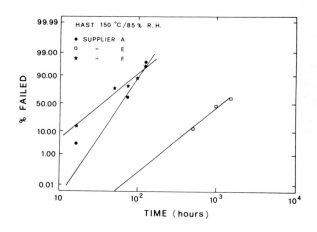

Fig. 4b. The same at 150°C/85% R.H.

tests [9].

In supplier A the main failure mechanism found was cathodic corrosion, the same as in THB test, as is confirmed also by the appearence of the lognormal plots (fig. 4) showing a similar slope, in the two HAST tests and in the THB test.

On the contrary, LSI memories did not show corrosion effects, except in one case, but failed mainly because contamination induced parameter shift. Cl and K were found also in the corroded area of fig. 5.

Fig. 5. SEM photograph of the corroded memory.

The same happened in SO devices tested at 150°C: contamination occurs first and corrosion generally follows after some time.
However the extraordinarily good results obtained with the manufacturer E must be underlined.
All these results are summarized in table 3.

(case SOT 23), did the more accelerated test, at 150°C, cause the package to swell, destroying the device by mechanical effect. No similar effects were found in SO packaged integrated circuits; on the contrary supplier E performed extraordinarily well, and SOT 23 transistors come from the same supplier.

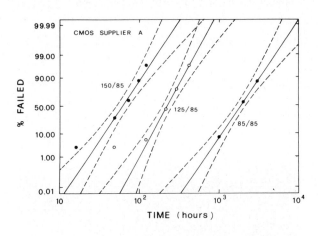

Fig. 6. Comparison of the results obtained on supplier A in the different test conditions.

Table 3
Results of the HASTs, evaluated by means of the lognormal distribution

Package	DIL		SO	
Technology	CMOS	LSI	HCMOS	
Supplier	A	D	E	F
125/85				
Median life (hours)	221	2665	–	–
Standard deviation	0.34	0.57	–	–
150/85				
Median life (hours)	61	–	1184	33
Standard deviation	0.42	–	0.78	0.78

4. DISCUSSION AND CONCLUSION

The results reported in this paper confirm both the improvement in the reliability of plastic packaged devices and the disuniformity in the resistance to the humidity stress. Thus the need for continuous monitoring of the quality of these components is also confirmed.
However for this purpose the standard 85/85 THB test is useless, due to the very long time required to obtain meaningful results. In this situation the use of the HAST is much more attractive, in fact the same failure mechanisms are found in the HAST and THB tests.
Only in the case of a small outline transistor

A good example of the correlation among the different tests was obtained with supplier A. In this case the failure distributions show a very similar behaviour at the different temperatures, with the standard deviation ranging between 0.34 and 0.43, as shown in fig. 6.
The estimated values, taking into account the three tests, are reported in table 4: the $\sigma = 0.41$ was derived by parameter analysis. This enables us easily to extrapolate to the room temperature condition, keeping the high humidity condition, by means of a pure Arrhenius relationships, as shown in fig. 7.
The activation energy is 0.68 eV, that compares well with many published data [5, 10-11].

plication, we still have no reliable way apart from non-accelerated tests. Results after 3 years of dynamic test in laboratory environment may suggest quite optimistic conclusions, however it is worth stressing that we must always face the problem of lot-by-lot variability. The use of HASTs seems to be applicable to this end.

ACKNOWLEDGEMENT

The Authors wish to thank the staff of the Component Quality and Reliability Evaluation Dept. for their help in performing tests, failure analyses and data analyses, and A. Scorzoni (LAMEL-CNR) for the SEM analyses.

REFERENCES

[1] Edfors, H.C., Reliability considerations in plastic encapsulated microcircuits (RADC Techn. Mono 71-1, 1971).

[2] Brambilla, P., Canali, C., Fantini, F., Magistrali, F. and Mattana, G., Reliability evaluation of plastic packaged devices for long life applications by THB test, Microelectron. Reliab., to be published.

[3] Gale, R.J., Epoxy degradation induced Au-Al intermetallic void formation in plastic encapsulated MOS memories, in 22nd Annual Proceeding Reliability Physics, 37-47 (IEEE, New York, 1984).

[4] Sauve, S.P. and Robinson, D.D., Field reliability data on plastic encapsulated solid-state devices, in Proceedings of ISTFA 1980, 111-124 (ATFA, Torrence, 1980).

[5] Stroehle, D., Influence of the chip temperature on the moisture induced failure rate of plastic-encapsulated devices, IEEE Trans. on CHMT, 6 (1983) 537-543.

[6] Brambilla, P., Fantini, F., Malberti, P. and Mattana, G., CMOS reliability: a useful case history to revise extrapolation effectiveness, length and slope of the learning curve, Microelectron. Reliab., 21 (1981) 191-201.

[7] Reynolds, F.H., Semiconductor component accelerated testing and data analysis, in Esaki, L. and Soncini, G. (eds.), Large Scale Integrated Circuits Technology: state of the art and prospects, 517-596 (Martinus Nijhoff, The Hague, 1982).

[8] Snow, E.H., Grove, A.S., Deal, B.E. and Sah, C.T., Ion transport phenomena in insulating films, J. of Appl. Phys., 36 (1965) 1664-1673.

[9] Gunn, J.E. and Malik, S.K., Highly accelerated temperature and humidity stress test technique, in 19th Annual Proceedings Reliability Physics, 48-51 (IEEE, New York, 1981).

[10] Merrett, R.P., Bryant, J.P. and Studd, R., An appraisal of high temperature high humidity stress tests for assessing plastic encapsulated semiconductor components, in 21st Annual Proceedings Reliability Physics, 73-82 (IEEE, New York, 1983).

[11] Gallace, L. and Rosenfield, M., Reliability of plastic-encapsulated integrated circuits in moisture environments, RCA Review, 45 (1984) 249-277.

BIOGRAPHIES

Pietro Brambilla received the degree in Electronic Physics from the University of Milano in 1970. In the same year the joined the Component Laboratories of Telettra S.p.A. in Vimercate, working on thin film deposition and Tantalum capacitor design and manufacturing. In 1975 he moved to the Reliability and Quality Department working on passive component reliability and, since 1980, he is responsible for the Semiconductor Component Reliability Group.

Fausto Fantini received the degree in Electronic Engineering from the University of Bologna in 1971. After serving in the Army, in 1973 he joined the Reliability and Qaulity Department of Telettra S.p.A. in Vimercate where he worked on the reliability of semiconductor devices. In 1979 he transferred to the Telettra Bologna plant, where he is currently responsible for the Quality Centre.
He is a member of the Institute of Electrical and Electronic Engineers (IEEE), the Italian Electrotechnical Association (AEI), the Italian Physical Society (SIF), the Italian Association for Quality Control (AICQ) and the Italian Society of Electron Microscopy (SIME). He is also Chairman of the National Technical Committee 47 "Semiconductor Devices" of the Italian Electrotechnical Commission (CEI).

Table 4

Group lognormal parametric analysis to Arrhenius Model at fixed 85% R.H.

T(°C)	Parameter	Estimate	90% confidence limits	
			lower	upper
	Standard deviation	0.41	0.30	0.62
85	Median life (hours)	1896	1548	2321
125	Median life (hours)	202	164	247
150	Median life (hours)	62	50	75
	Activation energy (eV)	0.68	0.55	0.81
	Extrapolated ML at 25°C/85% R.H. (hours)	167.000	–	–

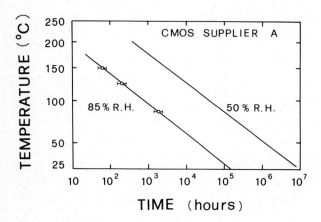

Fig. 7. Extrapolation of the results of accelerated tests by means of Arrhenius law.
Humidity acceleration was derived by reference [10].

We obtain a median life of nearly 20 years, that, associated with the very tight standard deviation, gives very low failure rates during the first years of operation, but also a very sharp increase of the failure rate, which could occur during the life of the equipment. If we also take into account the acceleration due to the humidity, the results look much better and can give some confidence in the application of plastic packaged devices (see fig. 7). This is also confirmed by the long term dynamic operating life test (OLT), performed on two thousand devices, one each from suppliers A and B, at 25°C, 50% R.H.. After nearly three years of test no failures appeared, so that a failure rate lower than 100 FIT, with 90% Confidence Level, has already been demonstrated.
Good results can also be forecasted from accelerated tests, taking into account the effect of high humidity. However this extrapolation is strongly dependent on the acceleration formula, in this case we have used the one developed at British Telecom Laboratories[10], whereas a general rule is probably unapplicable. Moreover the large differences in the distribution parameters found do not allow the observations made on supplier A to be extend to the other suppliers. Therefore each technology and supplier must be evaluated separately. From the results obtained up to now on HASTs, the following conclusions can be drawn:

a) the failure mechanisms of CMOS devices, in different packages, occurring during HAST up to 150°C, are the same as experienced in THB tests. The only exception was found with Small Outline Transistors, that did not survive the test at the highest temperature.

b) Correlations between HAST and THB tests have been demonstrated, where a clear wearout mechanism was present. However the extrapolation to operating conditions is still uncertain, so that it may be very dangerous to derive reliability targets from HASTs.

c) The effectiveness of HAST lies in the ability to make rapid comparisons between different technologies; in this respect they can replace the old THB tests.

d) New technologies, such as those employing the Silicon Nitride passivation, perform better [2], however good results were also obtained with simple Silicon Dioxide passivation. The fact that HCMOS devices generally perform better than CMOS is probably influenced by the lower applied bias.

e) To evaluate the failure rate in field ap-

Reliability Technology — Theory & Applications
J. Møltoft and F. Jensen (Editors)
© Elsevier Science Publishers B.V. (North-Holland), 1986

INTEGRATED TEST DEVICES FOR RELIABILITY EVALUATION OF NON- HERMETIC ENCAPSULATION

Sauli M. Palo and Timo J. Salo
Micronas, Inc.
P.O.B. 51
SF-02771 Espoo
FINLAND

Paul E. Collander and Raimo A. Mäkelä
Nokia Electronics
P.O.B. 780
SF-00101 Helsinki
FINLAND

The concept of integrated CMOS test devices has been adopted in a large project within the Nokia Electronics Group to assess the reliability of a number of non-hermetic encapsulations, associated materials and chip passivations as well as some assembly houses. Advantages related to the test devices are reviewed and some experiences and test results are discussed.

1 INTRODUCTION

Non-hermetic encapsulation has been studied in depth worldwide during recent years. The aim has naturally been to possess the ability to replace the more expensive hermetic packages with more cost-effective solutions even in high-reliability applications, such as telecommunications and mobile equipment.

Most published evaluation work has been made comparing standard IC's from various vendors in humid environments. Test results have varied considerably from test to test. However, useful information has been obtained for choosing an acceptable vendor for a particular technology, device family and a package type. Direct comparison of test results from different sources has usually been difficult due to somewhat vague device failure criteria and incomplete failure analysis. With the ever increasing complexity of LSI-devices the situation becomes even worse.

In a project set up within the Nokia Electronics Group to find a reliable non-hermetic encapsulation for high pin-count gate arrays and custom IC's, new solutions for suitable test devices were sought. The idea of custom designed integrated test devices came up. In fact, integrated test devices are not uncommon in developing a new IC-process, but have been less seldom used in evaluating the package itself.

2 ADVANTAGES OF TEST DEVICES

Integrated test devices offer a number of potential advantages (compared with standard IC's) for reliability evaluation, such as:

- simple structures for each parameter/failure mechanism
- straight-forward electrical measurements
- a better picture of physical phenomena
- higher stress levels possible and hence shorter test times
- failure analysis simplified/eliminated

The main drawback of test devices is that they usually have to be specifically designed and processed for their intended purpose. Furthermore, much test data is generated and therefore it is advisable to make a computer program to manipulate the data.

Once a test device has been designed, it is of great significance to be able to use the same structures periodically to monitor the mass production. It is even possible to transfer the structures to eg. five test sites on a production wafer, which can then be utilized in the incoming inspection of the wafers.

3
TEST DEVICE STRUCTURES

Due to a very tight time-schedule of the project, fully custom designed structures were not practicable. Therefore CMOS-gate array based test devices were reviewed and found to give a sufficient coverage of the most important failure mechanisms prevailing in non-hermetic devices.

The following structures could be easily wired into the test chip:

- NMOS- and PMOS- transistors with a common gate
- n+/p+ and aluminium contact resistance structures
- two aluminium metallization corrosion structures, one on periphery of the chip and the other in center
- 23-stage inverter ring oscillator
- structures for determining inverter, 2-input NAND- and NOR- gate delays
- wire bonding structure (8 internal bonds)

The above structures require a maximum of 32 bonding pads, but it is naturally possible to utilize a more limited number of structures if desired. An example of a test structure is shown in Fig. 1.

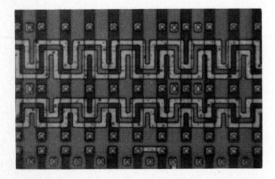

Fig. 1.
A microscope photograph of a corrosion test structure comprising of two adjacent but separate Al-conductors with minimum dimensions. During the humidity test a potential difference exists across the conductors.

4
ELECTRICAL MEASUREMENTS

A total of 31 electrical parameters were chosen to be measured at each test step. A ready-made test program and facilities at the Technical Research Centre of Finland were somewhat modified and extended to make it useful for the test devices.

The test program can be used to measure test structures both at wafer and packaged device levels. Hence test device wafers and possibly later normal production wafers incorporating a few test device chips can be comfortably evaluated for incoming inspection purposes.

5
PACKAGE TYPES AND TESTS

A number of the most interesting non-hermetic package types and associated materials on the market were chosen for the evaluation together with standard and some non-standard chip passivations. For example, both in-house and commercially assembled EPIC-packages (with epoxy blob coating) and commercial PLCC's from a number of assembly houses were tested incorporating the designed CMOS test chips with normal PSG- , LTO PSG- and silicon nitride passivations.

Environmental tests used were a normal 85C/85%RH humidity test for 1000h/2000h and a non-saturated pressure-cooker test at 125C/85%RH for 100h, both under biased conditions. The tests were preceded by temperature cycling without bias. Some monitor lots have also been burned-in at 125C for 1000h. In the future new lots encapsulated in hermetic packages will be tested in a humid environment together with the non-hermetic ones for full comparison.

6
SOME TEST RESULTS

The package groups (20 pcs each) have been tested in the humidity test 85C/85%RH but the intended computer printouts have not yet been fully completed. Figures 2 and 3 give some idea of the form of the printouts as well as some actual results. Included is the mean behaviour of the threshold voltage of an array transistor and the gate delay of a 2- input NAND- gate in the course of the humidity test.

Test step	Vt (V)			
	1.0 !	1.1 !	1.2 !	1.3 !
Character-ization	o	+	¤ ★	"
Temp. cycling		o+	★	"
168 h 85/85	o	+	★	"
500 h 85/85	o	+	★	"
1000 h 85/85	o	+	★	"
2000 h 85/85	o	+	★"	

Fig. 2.
The behavior of threshold voltage Vt of an NMOS array transistor in a number of test groups in the course of a 85C/85%RH humidity test up to 2000h.
List of symbols for test groups:

★ mean value of all test groups
" PSG chip passivation/EPIC-package
+ Si3N4   "   /   "
o LTO PSG   "   /   "
¤ PSG   "   /   "
  (commercially assembled)

Test step	2-input NAND delay (ns)			
	5.0 !	6.0 !	7.0 !	8.0 !
Character-ization			o¤ ★	"
Temp. cycling	¤		o +★	"
168 h 85/85	¤		o ★+	"
500 h 85/85			o +★	"
1000 h 85/85			o ★+	"
2000 h 85/85		o	★ +	"

Fig. 3.
The behavior of a 2-input NAND-gate in a number of test groups in the course of a 85C/85%RH humidity test up to 2000 h.
The groups and symbols used are identical to those in Fig. 2.

It can be concluded from the figures that all shown EPIC- type of packages have done very well in the test, except the one which was commercially assembled. This group was extremely weak, so that a mere temperature cycling was able to degrade it mostly beyond the specification limits. No apparent difference seems to exist in the various chip passivations, either.

Furthermore, it was observed that the Al- conductors monitoring possible corrosion were very stable in the above groups, except in the commercially assembled group. This gives further assurance that no corrosion has occurred during the humidity test.

7
PRESSURE COOKER TESTS

Unsaturated pressure cooker testing has come into use during recent years. Its main advantage is shortened test time, a typically reported acceleration factor is 5...7 compared with the standard 85/85- humidity test. However, there remains the question whether the two tests stress the components identically, ie. revealing the same failure mechanisms.

In this project a number of package groups have been tested in the conditions 125C/85%RH for 100h. The results obtained so far seem to indicate that EPIC-type of packages fail more frequently in the pressure cooker test (125C/85%RH 100h) than in the conventional 85/85 test (up to 2000h). The differences between epoxies used to protect the chip in an EPIC-package come out clearly in the pressure cooker. On the contrary, the transfer-molded PLCC-packages behave well and in an identical manner in both humidity tests. After all the tests have been completed a fuller view of the correlation of the two tests will be obtained.

It is concluded that in any case the unsaturated pressure cooker testing is valuable in a quick comparative evaluation of new packages and materials.

8
CONCLUSIONS

The idea of integrated CMOS- test device has proved to be well suited to the development and evaluation of non- hermetic encapsulations for high- reliability applications. A computerized test station as well as data handling is necessary to fully utilize the benefits of the test devices. While the conventional 85C/85%RH humidity test is probably the most reliable assessment method for package evaluation, the unsaturated pressure- cooker test is a viable alternative especially for speedy, comparative evaluation work.

BIOGRAPHIES

Sauli M. Palo received his Diploma in Electrical Engineering from the Helsinki University of Technology in 1975. After graduating he was working at the Finnish PTT in the field of semiconductor components reliability. In 1979 he joined the Nokia Electronics Group and is presently responsible for reliability assurance and evaluation of CMOS integrated circuits at the R&D Department of Micronas, Inc., a subsidiary of Nokia.

Timo J. Salo received his Diploma and Dr.Tech. in Electrical Engineering from the Helsinki University of Technology in 1965 and 1974, respectively. In 1973 he joined the Semiconductor Laboratory of the Technical Research Centre of Finland, where his group started the first MOS processing line in Finland. In 1979 he joined Vaisala, Inc. to start up the first industrial CMOS line in Finland. In 1984 he joined Micronas, Inc., where he is the Vice-President of R&D.

Paul E. Collander received a Master of Science degree in Physics from the University of Helsinki in 1972. After graduating he was a researcher at the Semiconductor Laboratory of the Technical Research Centre of Finland. In 1974 he joined Nokia Electronics, Components Department, responsible for setting up the hybrid technology laboratory. Presently he is Development Project Leader for the Microelectronics group of Nokia Electronics R&D Department. Mr. Collander has been a member of ISHM/Nordic since 1974.

Raimo A. Mäkelä is receiving his Diploma in Electrical Engineering from the University of Oulu in 1986. While studying he worked one year as a research Engineer in the R&D Department of Nokia Electronics. Presently he is working as a Product Manager in the Hybrid Production group of Nokia Electronics. Mr. Mäkelä has been a member of ISHM/Nordic since 1984.

Reliability Technology — Theory & Applications
J. Møltoft and F. Jensen (Editors)
© Elsevier Science Publishers B.V. (North-Holland), 1986

RELIABILITY INVESTIGATION ON R.F. POWER TRANSISTORS FOR MOBILE RADIO SYSTEMS

P. Ferrero, A. Gallesio, M. Liberatore, E. Pollino, D. Riva

Centro Studi E Laboratori Telecomunicazioni S.p.A.
Via G. Reiss Romoli 274, 10148 TORINO - ITALY
Phone +39 (11) 21691

A complete evaluation program on R.F. power transistors was carried out, and evidence was demonstrated of the effectiveness of Construction Analysis as a powerful tool for obtaining "reliable" information on the reliability of electronic components prior to their use.

Step-stress and life test results are provided and discussed along with failures and degradation mechanisms which were detected during failure analysis.

INTRODUCTION

Power transistor operation is somewhat critical, mostly due to the high currents and voltages involved which in turn cause junction temperature to rise to dangerous levels. Moreover R.F. devices may face particular mismatch conditions (high VSWR) and formation of hot spots is very likely to occur if safety operating limits are exceeded.

Based on the assumption of good quality electrical project our attention was focused on electromigration, and the test plan was prepared with the specific purpose to investigate the resistance of the devices to this catastrophic phenomenon.

Three device types very similar to one another as far as performance were selected to be employed in the amplifying stage of a new mobile radio system. The aims of the evaluation program were to establish their suitability for this use, to find the best choice comparing the results and to help preparing the qualification procedure.

ELECTROMIGRATION

Mass transport phenomena are a potential failure mechanism for power transistors. R.F. power transistors are generally used in the common emitter configuration, thus large amounts of current are carried by the emitter metallization paths. High current densities along with high temperature gradients and eventually structural defects may initiate a change in the diffusion coefficients of the metal ions which in turn causes a change in the rate of mass transport at some points along the conducting path. In this case the transfer of momentum between conducting electrons and ions in the direction of the current flow can lead to the formation of voids and hillocks, that is metal enhancement occurs at the anode site whilst metal depletion is observed at the cathode.

This phenomenon was widely studied for Aluminium. Black proposed an Activation Energy dependent Mean Time to Failure defined as:

$$MTF = \frac{A \exp(Ea/kT)}{J^n} \qquad [1]$$

where Ea is the Activation Energy for migration, k is the Boltzmann constant, T is the conductor absolute temperature, J is the current density and A is a pre-exponential constant which depends on the properties of the conductor.

The value of A may vary according to the grain size of the Al conductor and the eventually added metal ions, such as silicon or copper.

The generally accepted value for n is 2, but it has to be pointed out that other values have been proposed in case of strong temperature gradients along the conducting film (n=3) and by theoretical studies about contacts on bulk material (n=1).

All the transistors we are dealing with have a gold based metallization system. Gold is a very good conductor for current and heat, but it reacts very easily with silicon even at room temperature. Moreover gold does not adhere satisfactorily to silicon dioxide, and it appears quite clearly that a reliable gold based metallization system must have a barrier layer to prevent indiffusion at the contacts and an adhesive layer to prevent the detachment of the paths deposited onto the dielectric layer.

Gold exhibits a better resistance to electromigration, i.e. higher current densities can be reached without affecting the integrity of the system, and the behaviour of the metallization system can be improved adding impurities or covering the gold film with a metal or dielectric layer.

Several techniques are used to improve the quality of both Al and Au based metallization systems, but which one is the most reliable is still undecided.

CONSTRUCTION ANALYSIS

The first step of the program was a detailed construction analysis of the three device types, mostly focused on the composition of the metallization systems and passivation layers, as well as on the physical dimensions of the conducting paths with particular emphasis on the emitter bonding pads where the highest current densities are most likely to be found.

The measurements made of the sizes of the emitter bonding pads are summarized in Table 1.

The analysis of the metallization systems, made with careful sample preparation and precision cross-sectioning and polishing was mainly carried out with EDS (Energy Dispersive Spectroscopy) and AES (Auger Electron Spectroscopy). The results are the following:

Type A devices: the passivation layer is silicon nitride, the bonding film is Ti/W, and a good amount of oxygen was found through all the metallization thickness.
Type B devices: no passivation layer; the die is coated with silicone based resin, and the gold layer is sandwiched between two thin Ti/W layers and less oxygen was found at the interfaces.
Type C devices: the passivation layer is silicon dioxide, the barrier layer is Ti/W and the amount of oxygen found was about the same as in type B devices.

All three device types are manufactured according to the split-cell approach, but type C devices are composed of a much higher number of cells (32) compared with the others (8) and ballast resistors are sized not only to match temperature gradients of each individual cell, but

also of the entire die. Moreover these devices are supplied with an internal matching network.

Thus we are comparing three different design philosophies.

Type A devices are designed to prevent grain boundary diffusion of gold into the Ti/W film, which is achieved by adding a relevant amount of oxygen to saturate grain boundaries.

The mechanical constraint exerted to the gold film by the outer Ti/W layers and the lower impurity atom content of these films which enhances interactions with the gold layer can help preventing gold from migrating in type B devices.

A careful design which allows better temperature and current distributions on the die together with a thicker Ti/W bonding film is the method used to prevent indiffusion and electromigration phenomena in type C devices.

STEP-STRESS TESTS

Ten devices of each type for a total of thirty transistors have been step stressed at increasing collector current values ranging from .1 to 3.1 times the maximum rated collector current; each step lasted 48 hours with a corresponding current increase of .2 times. The junction temperature was continuously monitored and a complete electrical characterization was carried out after each step.

Five devices of each type were tested with opened packages in a suitable environment, and were carefully examined after each step to observe the degradation progress.

The failure distribution is shown in Fig. 1.

Failure analysis carried out during and at the end of the tests revealed that:
- Type A devices failed due to electromigration mostly evident at emitter bonding pads as shown in Fig. 2.
- Type B devices failed due to metallization path detachment as shown in Fig. 3.
- Type C devices failed due to ballast resistor fusion which resembles closely the appearance of PROM fusible links after programming, as shown in Fig. 4.

Step-stress test results provide a quick reference being very easy to compare and useful in the determination of the values to be applied to operating life test parameters and in the identification of the expected degradation mechanisms.

The obtained data can be also extrapolated to actual operating condition for a first rough estimation of the reliability of the devices in terms of useful life. In doing so we must of course make some basic assumptions, which are the following:
- The one and only expected degradation mechanism is electromigration, so that we can calculate the Mean Time to Failure using Eq. [1];
- We assume Ea = .8 eV and n = 1; these values are considered conservative as far as Au based metallization systems are concerned;
- The data used for extrapolation are reported in Table 2.

Substituting the data of Table 2 in Eq. [1] we obtain for the expected operating life values under nominal operating conditions:

- device A: 270,000 hours, i.e. about 30 years
- device B: 1,300,000 hours, i.e. about 150 years
- device C: no significant failures after 73,000 hours.

LIFE-TEST RESULTS

A sample of 27 devices of each transistor type have been life tested under DC bias. Each group was divided into three sub-groups and feeded with three different constant current values, that is 5, 7 and 10 Amperes, the first one being the nominal collector current value.

VCE and Ic were continuously monitored throughout the whole test, and eventually adjusted to compensate drops due to power supply instabilities.

At the end of selected cycles an electrical characterization was carried out, and it must be pointed out that no appreciable drifts were detected up to device failure.

The life-test conditions are listed in Table 3 and life-test results obtained so far are summarized in Table 4.

From Table 4 one can easily see that only the devices stressed at the highest current value and A type devices stressed at 7 Amps have completed the test, while the other devices are still operating without showing appreciable parameter drifts.

The observed degradation mechanisms are the same as those detected after the step-stress tests, i.e. electromigration for type A devices, metallization detachment for type B devices, and electromigration, but with associated pheonomena which will be described later for type C devices.

As far as type A devices are concerned by substituting in Eq. [1] the corresponding values for 7 and 10 Amps, we obtain n = 3.45, which is relatively higher than the conservative value that we assumed to extrapolate step-stress test results.

Hence the expected operating life at the maximum allowable collector current turns out to be about 30 years, which is not so different from the one derived from step-stress tests.

As for type B devices have only one failure rate corresponding to the maximum stress current (i.e. 10 Amps.). However in this case the sole degradation mechanism involved seems to be the detachment of the metallization stripes from the substrate, which is caused by Au indiffusion through the barrier layer. This phenomenon, which is temperature dependent, is well understood, and a value Ea = 1.8 eV for the activation energy is generally accepted. From the corresponding Arrhenius plot an average operating life of 600 hours at 260° C is equivalent to more than 100 million hours at 110° C, which is supposed to be the normal operating temperature. Thus type B devices exhibit a good resistance to electromigration, and suffer from a different degradation mechanism which is extremely slow to activate.

The only available data for type C devices is the mean life at the maximum applied stress, that is 5000 hours at a current density of 2.9×10^{5} at a junction temperature of 190° C, which is not enough to obtain a precise value for MTF. However at least two different degradation mechanisms have been observed, that is to say gold migration and ballast resistor opening. The latter could be easily regarded at as Ni/Cr electromigration, but requires a more detailed analysis, so it will be discussed later on. Both mechanisms are obviously catastrophic and lead to an open circuit condition. Furthermore, under the previously described test conditions, they seem to proceed in parallel.

Indiffusion at emitter contacts has been also observed which caused short circuits between emitter and base diffusions. A better understanding of all these phenomena will be possible only when the data of the tests which are not yet completed (i.e. those with a lower stress applied) become available.

However applying the same parameter values used for type A devices (Ea = .8, n = 3.45, A = exp[-30.55]) we obtain for type C devices an MTF which is much longer than the experimental value of 5000 hours. Now looking at Table 3 we see that this value was obtained at lower current density and junction temperature compared with type A and B devices even at maximum stress conditions. Hence we can say that type C devices feature an improved thermal and electrical design, which means a longer useful operating life.

FAILURE ANALYSIS

As far as type A devices are concerned the analysis carried out on the failed parts has clearly shown evidence of gold migration starting at emitter bonding pads. One of the pads at the final degradation stage can be observed in Fig. 5.

The periodic electrical measurements have demonstrated that the degradation is very fast, and a very short time elapses from the onset of the phenomenon to the failure so that is practically impossible to detect appreciable parameter drifts in the meantime.

The only failure mechanism detected for type B devices was the detachment of some metallization paths from the die (see Fig. 6).

The behavior of type C devices can be attributed to more than one degradation mechanism. The more relevant seem to be:
a) gold migration at emitter bonding pads (see Figs. 7 and 8) and ballast resistors contacts (see Figs. 9 and 10), and
b) Ni/Cr migration in emitter ballast resistors (see Figs. 11 and 12).

Both mechanisms seem to proceed in parallel.

Based on the fact that the current feeded into all the devices is kept constant through all the test, the larger the number of open emitter ballast resistor, the higher the current flowing through those which are not yet broken. When the current reaches a critical value the resistors act as PROM fusible links and start fusing (see Fig. 11). A fused resistor is shown in Fig. 13 which provides a nice example of the described phenomenon, that in this particular case is due to a mask defect which caused an electrical short between an emitter and a base finger.

Fig. 7, which was taken at the optical microscope, shows the gold depletion around the pad and the appearing of the underlying Ti/W layer. The Secondary Electron Image (Fig. 8) shows that the phenomenon starts around the wire bonding. Figs. 9 and 10 instead show gold extrusions at the ballast resistors, which means that somewhere gold must have diffused through the barrier layer. This phenomenon is illustrated in Fig. 13 which is the optical image of a cross sectioned device showing that gold has actually penetrated through the Ti/W film and goes towards the resistors.

Evidence of Nichrome migration is shown in Figs. 11 and 12; in the former, which was taken at the optical microscope, five ballast resistors belonging to the same transistor cell are shown, and a detailed examination can help understand in which order they have failed. The first to open was the longer one, that is placed right in the middle of the cell: Ni/Cr migration occurred in the direction of the electron flow and resistor depletion is evident at the site where the lower electrical potential is supposed to be found. The sudden current rise along the other emitter fingers led to partial fusion of the other resistors which were already partially depleted, and this is the reason why there the phenomenon did not proceed so slowly and regularly as in the first case, but they look broken as for a sudden catastrophic overstress.

In Fig. 12 the area where Nichrome depletion has occurred appears lighter. This can be explained as surface charge-up phenomena, since the resistor was lifted up (perhaps due to thermal expansion), as can be seen in the cross-section of Fig. 14, where the sectioned ballast resistor was not yet broken, or for the increased Nichrome thickness which enhances its Secondary Electron emissivity. The crack seems to extend beyond resistor borders, but it is actually the glassivation layer which started breaking.

CONCLUSIONS

The first conclusion that can be drawn is that any of the tested transistor types can be satisfactorily employed, provided that absolute maximum ratings are not exceeded. As far as resistance to electromigration type B devices seem to represent the best choice, since the sandwich-like (Ti/W-Au-Ti/W) metallization system has actually proven effective in inhibiting the phenomenon.

The test results show that type A devices are the most critical, but the extrapolated MTF value of 30 years is fairly good, considering that they have been DC stressed at the maximum allowed ratings, while they are supposed to be used for RF applications.

Type C devices seem to last far longer than the others, but if we look at Table 3 we see that they operate at a much safer current density value and lower temperature. As far as resistance to electromigration they are more critical than type A devices, in fact if we apply Black's law substituting the parameter values obtained for type A devices under the stress conditions at which the first data for type C devices were obtained (10 Amps), we see that the MTF value is better than the experimental one.

Now it is important to emphasize how the same kinds of results had been obtained after the first part of the evaluation program, that is from Construction Analyses and step-stress tests, which provide mostly qualitative but extremely realistic and thus immediately applicable data in a very short time. This is of the greatest importance if we consider that the technological scenario keeps on changing unceasingly and quickly.

REFERENCES

[1] J. S. Lamming: "Microwave Transistors" - from "Microwave Devices" edited by Howes & Morgan. John Wiley & Sons, 1976.

[2] A. Gallesio, C. Guasco, M. Liberatore, E. Pollino, G. Ottaviani: "Reliability investigation on gold metallized RF transistors".

[3] S. D. Mukherjee: "Interactions of metal films on semiconductors" - from "Reliability and Degradation" edited by Howes & Morgan. John Wiley & Sons, 1981.

Emiliano Pollino was born in Torino, Italy, in 1947. He received a degree in Applied Physics in 1970 from the University of Torino. In the same year he joined CSELT, where he was involved in reliability and evaluation programs on electromechanical and electronic components. Now he is in charge of the Component Reliability Department and manages all the activities regarding technological and failure analysis of electronic devices to be used in Telecommunication equipment.

Danilo Riva was born in Torino, Italy, in 1954. He received a degree in Applied Physics in 1978 from the University of Torino. In the same year he joined the Reliability Department of CSELT where he was mainly involved in the evaluation of advanced electronic devices. Now his main activity is carrying out studies on GaAs microwave and digital devices. Dr. Riva has been qualified as an Inspector of Space Components by the European Space Agency in 1981.

Alessandro Gallesio was born in Gorzegno (CN), Italy, in 1953. He received a Telecommunication Engineer degree from Pininfarina Technical High School of Torino in 1974. In 1975 he joined the Reliability Department of CSELT where he is now mainly involved in Failure Analysis of electronic devices and in the development of new analytical techniques for component evaluation.

Piergiorgio Ferrero was born in Torino, Italy, in 1938. He graduated as an Electrotechnical Engineer in 1957 at Avogadro Technical High School of Torino. He joined the Cable and Transmission Department of CSELT in 1959 where now he is mainly involved in the study and design of protection networks and shielding against electro magnetic interferences for telecom equipment.

Michele Liberatore was born in Palazzo S. Gervasio (PZ), Italy, in 1958. He took a degree in Electronics in 1981. He joined the Reliability Department of CSELT in 1976, and now is mainly involved in the characterization and evaluation of optoelectronic devices for optical fiber communications.

	Device A	Device B	Device C
Length (um)	95	75	90
Width (um)	38	35	48
Thickness (um)	2	2	2.5
X-section (cm2)	5.3x10exp(-6)	4.4x10exp(-6)	6.9x10exp(-6)
Nominal current density (A/cm2)	2.4x10exp(5)	2.8x10exp(5)	1.3x10exp(5)

Table 1: Physical dimensions of emitter bonding pads and current densities.

Man.	Test conditions Jmax [A/cm2]	Tmax ['K]	MTF [h]	Nominal values Jop [A/cm2]	Top ['K]
A	5.04x10exp(5)	483	480	2.4x10exp(5)	373
B	7x10exp(5)	543	576	2.8x10exp(5)	383
C	4.03x10exp(5)	473	624	1.3x10exp(5)	358

Table 2: Data used for extrapolation.

I [A]	Device A J [A/cm2]	Tj ['C/'K]	Device B J [A/cm2]	Tj ['C/'K]	Device C J [A/cm2]	Tj ['C/'K]
10	4.8	250/523	5.6	260/533	2.9	190/463
7	3.36	160/433	3.92	170/443	2.02	150/423
5	2.4	100/373	2.8	110/383	1.45	100/368
nom.	2.4	100/373	2.8	110/383	1.3	95/368

Table 3: Life-test conditions compared to nominal values.

I [A]	Mean Time to Failure [h] Dev. A	Dev. B	Dev. C
10	24	600	5000
7	3000	> 6500	> 6500
5	> 6500	> 6500	> 6500

Table 4: Life-test results in terms of MTF.

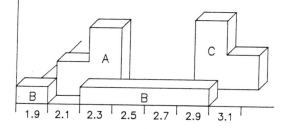

Fig. 1. Failure distribution after step-stress tests.

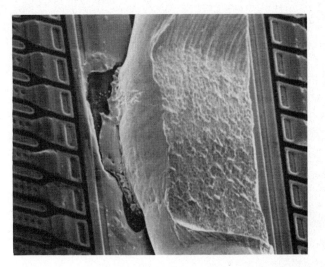

Fig. 2. Electromigration at one of the emitter bonding pads of type A devices.

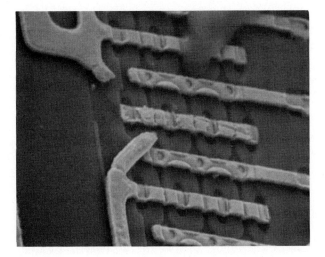

Fig. 3. Metallization path detachment on type B device.

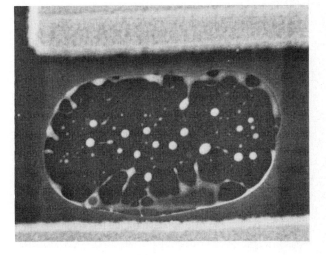

Fig. 4. Fused ballast resistor of type C device.

Fig. 5. Electromigration at one of the emitter bonding pads of type A device after life-test.

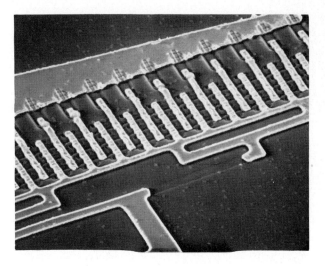

Fig. 6. Metallization path detachment on type B device after life-test.

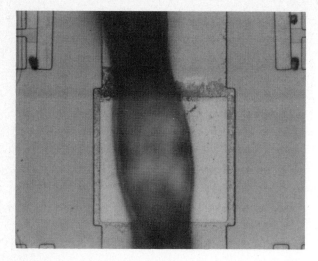

Fig. 7. Electromigration at one
of the emitter bonding
pads of type C device
after life-test.

Fig. 8. Detailed view of gold
depletion around the
bond on type C device.

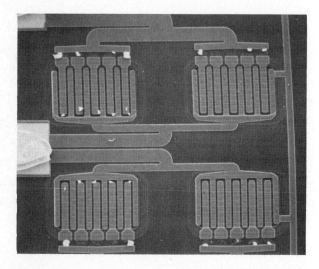

Fig. 9. Gold extrusions at the
ballast resistors of type
C device.

Fig.10. Detailed view of one of the
cells of type C device with
gold extrusions.

Fig.11. Nichrome electromigration at
the ballast resistors of
type C device.

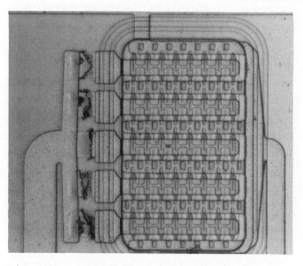

Fig.12. Detailed view of open
Nichrome ballast resistor
on one cell of type C dev.

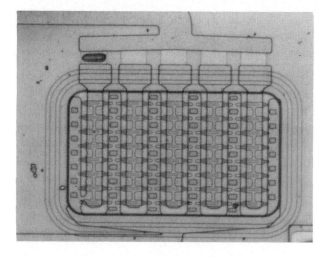

Fig.13. A mask defect caused fusion of one of the emitter ballast resistors of type C device. after life-test.

Fig.14. Cross-sectional view of one of the emitter contacts of type C device showing gold penetration through the barrier layer.

PRACTICAL RELIABILITY MANAGEMENT

Reliability Technology — Theory & Applications
J. Møltoft and F. Jensen (Editors)
© Elsevier Science Publishers B.V. (North-Holland), 1986

RELIABILITY MANAGEMENT FOR
PURCHASED FUNCTIONAL COMPONENTS

Dr. Angelo STELLA

Quality Services — I.R.E. (Industrie Riunite Eurodomestici)
Comerio (Italy)

ABSTRACT

Continuous progress in consumer maturity, strong competition on the market and the steady increase of costs paid for the consequences of down-time and failure have caused interest in reliability to be held in ever greater consideration among household appliances producers. Assuming they have achieved a good level of confidence in the state of control of their processes, the most logical means of reaching reliability assurance, is that of monitoring in a reasonable way the mass-production of main functional components. This can be done by asking suppliers to perform themselves routine life-testing of their parts, having agreed with them about test methods and conditions, together with sampling plans. These last are set considering the desired reliability level of the components and the model representing their behaviour in the field.

1. INTRODUCTION

I.R.E. (Industrie Riunite Eurodomestici) is the second European producer and among the leading Companies in the World in the field of major domestic appliances.

Every year 2.500.000 appliances and 1.800.000 compressors come out of our Italian factories, located in Cassinetta, Siena, Napoli and Trento. The Company was born in 1969 from a joint venture between the Italian Manufacturer IGNIS and its Commercial Partner PHILIPS.
Later on, in 1972, PHILIPS acquired the remaining 50 % of the company and IRE became the substantial part of the Product Division Major Domestic Appliances, within the Philips Organization.
The headquarters of the Product Division were established and still are in Comerio, 60 kilometers north of Milan.
The main commitment of the Product Division is to manufacture in its factories products which are sold and serviced by Philips National Organizations (50 all over the world), these being our direct customers.
For this transaction, assurance on quality and reliability levels of the products must be given.

In our turn, in order to be able to offer such a warranty, we have to set up continuous monitoring of the reliability of functional components, which are those mainly responsible for product failures.

This paper describes the steps in the reliability assurance program: starting from the collection of field information and setting of the targets by the producer, to the routine life-testing activity on functional components by the supplier.

2. MOTIVATION FOR RELIABILITY IMPROVEMENT

Reliability had always been one of the keys to success in consumer durables and a leading criterion of consumer choice when replacing.
It is also the basis on which the good image of many Companies has been built.
In recent years the growing of consumer maturity, due to the experience coming from use and promoted by the circulation of information on comparative tests, has led to new requirements and greater expectations.

Moreover the steep increase of costs paid for down-time and repairs during the guarantee period has emphasized the economical aspects and possibility of savings, connected with product reliability improvement.

In the field of household appliances the standstill of the market, the strengthening of the competition and the drive to lengthen guarantee periods, have all caused increased interest in reliability among producers in these last years.

3. FIELD RELIABILITY REPORTING SYSTEM

The first step in establishing a correct reliability management system, is to build a quick and reliable information system, making it possible to constantly control the situation, to plan and to verify the improvement in a dynamic way.

With this aim in mind we have in use in several Western European Countries a Quality Reporting System (in short QRS), designed to provide reliability data and to gather information on the cost of repairs activities.
The supplied information is production addressed, in the sense that reliability of the products is related to their period of production, so as to have an insight into the effects of modifications and an actual survey of the rate of the improvements.
The system is arranged in such a way that a closed park is established with the appliances whose guarantee cards are sent back by end-customers.
The information content of the guarantee card tells us the production and installation dates of the appliance.
Should a repair be performed on one of these appliances, from the jobsheet filled in by the serviceman, we can deduce the age of the machine when failure occurs and the nature of repairs, together with data for the calculation of relevant costs.

The data processed by computer gives rise to various possibilities of reporting.
The most interesting programs for the purpose of reliability improvement planning are those which give:

-       the figures concerning the total failure rates of complete products during the guarantee period
-       the contribution of single items of fault, together with their cost consequences in terms of

material, labour and travelling times (fault spectra)

4. RELIABILITY ASSURANCE PLANNING

As mentioned in the introduction, since manufacturing and servicing the products are separate tasks within our organization, the correct reliability level has to be planned together by the Product Division (manufacturer) and the National Organization (seller), in order to have a Concern-integrated approach to the matter and to attain the optimum for the total quality cost.

For this purpose, every year a quality agreement is set up, by means of which the Management of the factory manufacturing the products gives assurance to the National Organization (selling and servicing them) on the foreseen level of reliability of complete products and on expected repair rates of main sources of their failures.
On the basis of the possible existing gap between the situation arising from field reporting (QRS) and the desired level, the needed quality improvements are planned.
Should this difference exist at the end of the stated period of verification, the extra costs of the servicing activity are charged to the factory.
Therefore the system promotes a quick reaction from factory side to solve reliability problems and gives the National Organizations the right tool for budgetting, in the most accurate way, the service costs within the selling expenses.

Coming back to the reliability assurance planning, the starting point of the program is then the information on the behaviour in the field of appliances of consolidated mass-production.
Obviously, when targets have to be set, if the situation arising is unsatisfactory, the figures are corrected, fixing a different desired level.
Taking into consideration the QRS program, which splits up the total amount of failures into their main sources, usually one finds that, according to the well-known Pareto principle, a few main functional components are responsible for the most important part of the overall number of product failures.
This fact leads to the focal point of the question that is: the reliability for the appliances can be assured only if the reliability of their

critical functional components is constantly under control.

The conclusion is that we have to build up an assurance system for monitoring the reliability of such components in their mass-production.

The same procedure for calculating reliability targets is followed in the case of new products.

The reliability target for the product originates reliability targets for its critical components.

The figures are set considering the acquired experience on similar products in the field and foreseen deviations, due to the product innovation and evolution.

5. FUNCTIONAL COMPONENTS RELIABILITY MANAGEMENT

The functional components reliability assurance activity is therefore always a necessary condition for getting the assurance on the end product.

That become also a sufficient condition (i.e. exhaustive), if one assumes that the manufacturing process is completely in state of control (i.e. no failure due to assembling mistakes) and the specifications for components are adequate to the use of the product (i.e. components not overstressed).

It would also be fairly impossible and certainly uneconomical for the end-product manufacturer to monitor by himself the reliability of all critical functional components purchased externally.

The large number of items would impose such heavy structures for life testing and data processing, that it would be out of the question to bear such a burden of costs.

Therefore the most effective and successful way of facing the problem is to have it carried out by relevant suppliers.

In fact, they have a much more limited range of items to test and they have developed the expertise and specific know-how on the modes and patterns of failure for their products: so they can do it better.

A further reason, encouraging the activity, is the consideration that they could have similar requests from different customers and the possibility of satisfying at the same time the needs of all their vendees.

Moreover it can be deemed a profitable occasion to enlarge the business, getting new orders and new customers, on the wave of the favourable effect of their image.

The common interest in reliability certification among producers and in standardization of relevant tests from the supplier's side has led to the establishment of a Domestic Appliances Manufacturers Committee in the Italian Association for Quality (AICQ).

The Committee aims to define a common approach for reliability targets setting and its Technical Working Groups agree upon test methods, allowing suppliers to standardize tests facilities and reporting.

Then each Company retains the possibility of negotiating and agreeing, in a bilateral way with its vendor, on the suitable reliability level and consequent sampling plan to attain the assurance.

6. VENDOR COMMITTMENT AND CERTIFICATION

The first step in the components reliability certification process is the definition of life-testing conditions and relevant equipment.

In several cases we are dealing with components, whose running time is an important part of calendar time.

So it is usual to run into difficulties as we have to choose an accelerating factor other than time.

In setting factors like temperature, pressure, stress and so on, one has to pay attention not to go beyond the limits of distorsion of phenomena.

Moreover, in some cases the importance of interactions between the factors is so determining that they cannot be varied independently one at a time.

Then the agreement on tests methods can become the longest and most difficult phase, requiring more than one attempt and verification.

Also from the economic point of view this has to be deemed the most tactful phase, as the investments for the suitable equipment depend on test conditions.

When life-testing methods and facilities are established, the reliability target and relevant confidence level of assurance has to be fixed.

In general two different objectives are provided: the reliability at the expiry of the guarantee period and the reliability expected for useful life.

The former is in relation with service costs control, the latter with customers' satisfaction and the Company's Quality Policy.

Reliability target and its confidence level, together with a mathematical model of component behaviour in the field, are the bases for the calculation of the sampling plan.

Confidence level, being a degree of protection, is a subject for discussion and contracting with the supplier.

Greater practical troubles, on the contrary, are connected with model building, as we make broad use of electromechanical components, for which the handy exponential distribution of the reliability function is too approximate when it is not completely unreal.

In the majority of cases we assume that the distribution of times to failure is unknown and we estimate the reliability function only for a point of the curve: the target point. At the fixed time there are only two possibilities for the state of the system : either it functions or it doesn't, and the one excludes the other.

So the probability of the event is a binomial one.

In the cases where previous history allows us to have confidence in the knowledge of times to failure distribution, we make use of the Weibull model, which with its scale and shape parameters fits in well with and matches many practical situations.

Moreover, if the estimation of Beta (shape parameter) is nearly 1, things can be further simplified, assuming the costant failure rate of the exponential distribution.

This allows the maximum of flexibility as one can increase testing times and reduce the sample size with the same ratio and vice versa.

Anyhow, the sample size and the acceptance number for the failures emerge from the equations, which impose the desired reliability level being the lower confidence limit of the estimation, with the fixed degree of confidence.

As for the planning of activities, to know in advance the moment of decison gives undoubtedly some advantages, the truncation of the tests is usually preferred.

Furthermore, to obtain at the same time protection against epidemic failures, that is even more important than target verification, frequently, for setting up the operating curve of the sampling, two different reliability levels are chosen.

The acceptable reliability (whose associated confidence level is the supplier risk) corresponds with our wishes, whereas the unacceptable reliability (whose associated confidence level is the consumer risk) corresponds to what we deem epidemicity from which to defend ourselves.

This way of doing things gives the considerable advantage of a better understanding in supplier - customer relations, establishing a perfect analogy with the quality assessment at delivery, which is a more commonly practised activity.

Finally, if the sample size is too heavy in comparison with present available equipment of the supplier, who has to carry out the life-testing activity, we suggest a suitable sequential sampling plan.

In this case the answer on conformity with our requests might be given over longer periods, being the right dimension of the sample, as to have the assurance, split up into smaller sub-samples.

In any case, when the sampling plan is ready, we agree with relevant vendors the frequency of certification, that corresponds with the period foreseen for completing routine life-testing.

With stated frequency we receive life-testing results, which can be processed and accumulated to achieve a better picture of behaviours and of mathematical models representing them.

The calculation of sampling plans and data processing of results is rather complex.

As specific reliabiity software available on the market is very poor, one of the hardest points of the program has been the preparation of a suitable package for executing the abovementioned calculations by means of our Personal Computer Professional 350 Digital.

In the following pages, some expressive examples are reported.

7.   CONCLUSIONS

Today the system is running for a selected number of key functional components and its implementation is going to be extended to all of them. We are fully confident that in a few years we will get the following direct benefits:

-     absolute protection against important problems in the field
-     a powerful tool for reliability improvement planning
-     diffusion of reliability culture towards our suppliers

- increasing know-how allowing technological evolution
- capacity to cope with more and more ambitious targets

But the indirect benefits of the operations could be even more attractive:
- strengthening of brand image
- shortening of suppliers qualification terms, allowing to reduce the new products development times
- product innovation sheltered from unpleasant surprises
- higher profitability and market share.

REFERENCES

(1) F. Galetto, Affidabilita', vol. 2, C.L.E.U.P., Padova
(2) W. Nelson, Weibull Analysis of Reliabity Data with Few or No Failures, Journal of Quality Technology, Vol. 17, no. 3, July 1985, 140-146
(3) Handbook H 108, Sampling Procedures and Tables for Life and Reliability Testing, U.S. Dept. of Defense, Washington, 1960.

BIOGRAPHY

Angelo C. STELLA
CQD - Quality Services
I.R.E. S.p.A.
Viale G. Borghi 27
21025 Comerio (VA) - ITALY

A.C. Stella graduated in Aeronautical Engineering at the Politecnico of Milan in 1970.
After having taught Mechanics in the Italian Air Force Academy and Industrial Institutes, he joined IRE (a part of the Philips Organization) in 1973.
From 1973 to 1976 he was Responsible for Design Quality of Refrigeration products.
From 1976 to 1980 he worked as Quality Information and Process Auditing Responsible in the Refrigeration Factory.
Now, since 1980, he has been in charge of Quality Services (Systems and Methods for Quality and Reliability Assurance) for the whole Major Domestic Appliances Product Division.

EXAMPLES OF SAMPLING PLANS FOR THE VERIFICATION OF RELIABILITY TARGET

- EXPONENTIAL DISTRIBUTION - (TWO PARAMETERS)

TARGET   R= .98 after 1000 cycles with 90% confidence level

Type of sampling	Accept. Reliab.	Supplier Risk	Unaccept. Reliab.	Consumer Risk	Test Dur. (cycles)	Sample size	Failure acc. n.	Failure rej. n.
	.98	10%	.96	10%	1500	343	14	15
	.98	10%	.96	10%	2500	206	14	15
Single	.98	10%	.94	10%	1500	105	5	6
	.98	10%	.94	10%	2500	63	5	6
	.98	10%	.90	10%	1500	37	2	3
	.98	10%	.90	10%	2500	22	2	3

Type of sampling	Accept. Reliab.	Supplier Risk	Unaccept. Reliab.	Consumer Risk	Test Dur. (cycles)	Sample size	Failure acc. n.	Failure rej. n.
	.98	10%	.96	10%	1000	20		4*
						40		5
						60		5
						80		6
	* minimum sample size for rejection = 20 (4 failures)					100		6
						120*	0	7
						140	0	8
	minimum sample size for acceptance = 120 (no failure)					160	1	8
						180	2	9
	.98	10%	.94	10%	1000	20		3*
						40		4
						60*	0	5
	* minimum sample size for rejection = 20 (3 failures)					80	0	5
						100	1	6
Sequential						120	2	7
	minimum sample size for acceptance = 60 (no failure)					140	3	8
						160	3	8
						180	4	9
	.98	10%	.90	10%	1000	20		3*
						40*	0	4
						60	1	5
	* minimum sample size for rejection = 20 (3 failures)					80	2	6
						100	3	7
						120	4	8
	minimum sample size for acceptance = 40 (no failures)					140	5	9
						160	6	10
						180	7	11

EXAMPLES OF SAMPLING PLANS FOR THE VERIFICATION

OF RELIABILITY TARGET (ONE PARAMETER)

TARGET   R= .98 after 1000 cycles with 90% confidence level

Reliability distribution	Test duration (cycles)	Sample size	No. of failures accepted
Unknown	1000	114	0
	1000	194	1
	1000	265	2
	1000	333	3
Exponential	1500	77	0
	1500	130	1
	1500	177	2
	1500	223	3
	2500	46	0
	2500	78	1
	2500	106	2
	2500	134	3
Weibull $\beta = 0.5$	1500	93	0
	1500	157	1
	1500	215	2
	1500	270	3
	2500	72	0
	2500	122	1
	2500	167	2
	2500	209	3
Weibull $\beta = 1.5$	1500	62	0
	1500	105	1
	1500	143	2
	1500	180	3
	2500	29	0
	2500	49	1
	2500	67	2
	2500	84	3
Weibull $\beta = 2.0$	1500	51	0
	1500	86	1
	1500	117	2
	1500	147	3
	2500	18	0
	2500	31	1
	2500	42	2
	2500	53	3

```
**************************************************
*                                                *
*      RELIABILITY ESTIMATION FOR WASHING PUMPS  *
*                                                *
**************************************************
```

```
        No. OF ITEMS TESTED =     50

        No. OF FAILURES      =      3


        FAIL.No.   CYCLES    I.O.     O.N.     RANK

        -------------------------------------------------

           1       280.     1.00     1.00     1.39

           2       650.     1.00     2.00     3.37

           3       980.     1.00     3.00     5.36
```

CORREL. COEFF. FOR WEIBULL LINE = 0.99980676

SHAPE PARAMETER (Beta)   =   1.0897

CHARACTERISTIC LIFE   =   14154.1357 Cycles

          ** RELIABILITY FUNCTION **
     $R(t) = EXP((-t/\ 14154.1357)^\wedge\ 1.0897)$

LIFE-TEST DATA

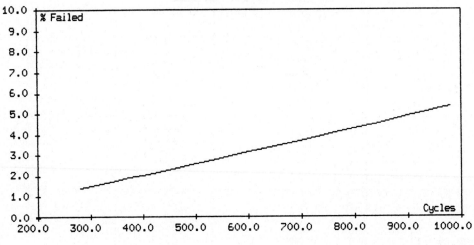

RELIABILITY R (250 CYCLES)   = 0.9878

FAILURE RATE H (250 CYCLES)   = 0.00005361

FAILURE DENSITY D (250 CYCLES)   = 0.00005295

No. OF CYCLES CORRESPONDING TO R(t)= 0.99:   207.70

:liability Technology — Theory & Applications
Møltoft and F. Jensen (Editors)
Elsevier Science Publishers B.V. (North-Holland), 1986

COMMON ERRORS IN ARM PLANS

P.P. Kirrane

Rex, Thompson & Partners, Newnhams, West Street,
Farnham, Surrey, England

ARM activities in a complex project must be correctly planned, managed and resourced if the
ARM targets and objectives are to be achieved. Many projects fail to achieve their ARM
targets and objectives because the initial ARM Plan contains fatal errors or omissions.
This paper outlines some of the more common errors and suggests how they can be avoided.

INTRODUCTION

:e achievement of Availability, Reliability
:d Maintainability (ARM) targets in a complex
oject requires a coherent and well-organised
ogramme of ARM activities. To be successful,
is programme must be carefully planned,
sourced and managed from the beginning of the
oject. However, many projects fail to
hieve their ARM targets because of fatal
rors or omissions in the initial ARM Plan.

e ARM Plan is a formal document (sometimes
sued as separate reliability and
intainability plans) defining the ARM
tivities which will take place during design
d development. In particular it should
ecisely define the following:

ARM Responsibilities
ARM Targets and Requirements
The methods by which the targets are to be
achieved.
The methods by which this achievement is to
be demonstrated
The phasing of ARM activities with other
project activities.

e ARM Plan should be produced by the
ntractor at the start of the project and may
rm part of the contractual agreement between
e contractor and the customer. The ARM plan
normally the basis for monitoring ARM
ogress throughout the project.

effective ARM plan will be tailored to suit
e ARM requirements, the timescales and
sources available and the structure and
ocedures of the organisation concerned.
nsideration of ARM needs and resources should
rm part of the initial project planning and
ade offs and therefore the ARM plan should be
epared as early as possible. This is
rticularly important with respect to ARM
sting, which will normally impinge
gnificantly on overall project costs and
mescales.

spite the recent advances in ARM prediction
d analysis techniques, testing is still the
st effective means of measuring ARM achieve-
nt and of detecting design deffects. There-
re preparation of the ARM plan will often
gin with a consideration of the needs for
sting - what types of testing are required,
en should they take place, how long should

they be, how should corrective action be
implemented.
However, testing alone is not enough. ARM
should be considered as an integral part of the
design process at all stages. If ARM is not
considered until the test phase, it is likley
that such serious ARM deficiencies will exist
in the basic design that it will be too late to
correct them. Therefore the next stage in the
preparation of the ARM Plan is to construct a
programme of ARM prediction and analysis
activities to be carried out during design and
development. Formal design review procedures
should be established to ensure that the
information obtained from these analyses is
fed back into design and that the necessary
corrective actions are taken.

The third stage in the preparation of a typical
ARM plan is to allocate responsibilities for
carrying out ARM activities. Effort will be
required from project management, quality and
design departments, etc. In complex projects
information and support required from sub-
contractors and suppliers should be
identified. A procedure for monitoring and
maintaining ARM progress will be required, and
a single focal point for ARM information should
be identified. Finally, procedures for
updating and reviewing the ARM plan itself in
the light of results obtained should be defined.

Unfortunately, many ARM plans are produced with
little thought or effort, merely to satisfy
contractual requirements, and most of these
contain errors in one or more of the above
areas. The purpose of this paper is to
describe some of the more common errors in ARM
plans and to give examples of their effects.
The paper also presents a checklist of ARM Plan
elements which may be used to assist in the
preparation of plans.

2. MANAGEMENT

The ARM Plan is a management document and
should define the procedures by which the ARM
programme is to be managed. Some of the more
common errors encountered in this aspect of the
plan are as follows:

a. Failure to specify ARM responsibilities
    The ARM Plan will require activities by
    design and quality departments as well as
    by ARM specialists. Therefore it is
    important that the responsibilities for

for carrying out these various activities should be clearly defined. In particular, an ARM co-ordinator should be identified, with overall responsibility for management and co-ordination of all ARM activities. In some cases the project manager or design manager may be nominated as ARM co-ordinator. In a typical example of the type of problems that can arise, project designers were not aware that they were responsible for providing electronic component derating and stress level information to the ARM section to support prediction work. As a result detailed stress level calculations were not recorded and had to be repeated by the ARM section - at considerable expense in terms of time and money.

b. Failure to set up procedure for monitoring ARM Progress. Ideally ARM aspects should be recognised by all concerned as having similar importance to cost and performance. In practice, however, problems in other areas (eg achieving performance targets) can cause resources to be diverted from ARM activities. This causes ARM progress to slip back, reducing the influence of ARM activities on design. It is important that the ARM plan should specify regular reviews of ARM progress, and that the review body should have the authority to force all departments to respond to ARM needs with at least the urgency that they respond to cost and performance needs. There are many examples of projects where ARM has been paid lip service at the planning stage, and has then gradually been allowed to lapse in importance until the system ARM suddenly is found to be completely inadequate at a late stage.

c. Failing to specify feedback into design. Predictions, FMEA's and FTA's are of little use unless they have some influence on the design. The ARM Plan should specify timely design reviews at which the outputs of ARM activities will be reviewed with the designers to identify aspects of the design which could be improved. In one recent case it was decided that the FMEA would not be carried out until all the detailed drawings were available, ie after the design was frozen. No Design review was specified. The FMEA identified some serious shortcomings in failure detection by system BITE, but by then it was too late to modify the design as initial systems had already been delivered. Instead the very expensive FMEA became merely a paper exercise, satisfying the plan's requirement but having no effect on ARM achievement. This example also illustrates the importance of correctly phasing ARM activities with other project activities.

d. Failure to pass on ARM requirements to sub-contractors. The design and development of major sub-systems may be sub-contracted in complex projects. Clearly if these sub-systems fail to achieve their individual ARM targets it is likely that the system will also fail to achieve its overall ARM targets. Therefore it is important that the ARM plans should clearly identify ARM targets and responsibilities for sub-contractors and that the appropriate ARM activities should be carried out during the design and development of all sub-systems. In some cases major sub-contractors should produce their own ARM plans compatible with the main project ARM plan. Failure to pass on ARM requirements commonly results in inadequate inputs to system level predictions and analyses, failure to achieve sub-system ARM targets and failure to detect this under-achievement until the sub-system has been delivered and integrated for system level reliability testing.

3. SOFTWARE

Software is an important element of many systems. Failing to consider software is probably the most common error in ARM plans. The difficulties in predicting or defining the reliability of software are well known. However, this is no excuse for ignoring software completely or for dismissing it as an area in which nothing can be done. Software can "fail" (ie cause the system failure) and does require "maintaining" (ie action to restore the system to normal operation) and these aspects should be addressed in the ARM plan. Common errors concerning software are:

a. Failing to consider software when setting ARM targets. Even though the likely level of software ARM is difficult to quantify at the design stage, some allowances should be made when setting ARM targets for the system. If this is not done, the hardware designers will inevitably design to the full targets, so that any level of software unreliability becomes unacceptable.

b. Failing to call up software quality and analysis procedures. Software "reliability" can be improved by adopting structured programming techniques, full documentation and thorough modular testing against specifications at all design stages. Software maintainability can be improved by using fully commented code, modular program construction, error trapping, etc. Most organisations have in-house software quality and design manuals which define these procedures, and these should be specified in the ARM plan. In cases where a software failure could have severe consequences (eg automatic flight control software) the program could also be analysed and validated using systems which are now becoming available such as MALPAS(1). This very powerful software tool uses directed graphs and regular algebras to detect unreachable codes, "black holes", unset variables, unused values, combinations of inputs that cause illegal outputs, etc. The techniques to achieve reliable and maintainable software are therefore available, but they require planning and resourcing from the start of a project if they are to be applied.

4. MAINTAINABILITY

As with software, the most common errors in ARM Plans concerning maintainability are errors of omission. Common examples are:

a. Failure to consider logistics structure as a whole. Many ARM Plans consider diagnosis and repair of failures at first line (ie in the field) but fail to consider the overall logistics structure for the project. In a complex system there may be several levels of repair - assemblies diagnosed as faulty at first line may be returned to a base or depot for sub-assembly level repair and the faulty sub-assemblies may then be returned to the manufacturer for component level repair. Different types of test equipment and spares will be required at each level of repair, and the analysis to identify this structure must be called up in the ARM Plan. The results of this type of error on life cycle costs may be severe - in one case a contractor set up a third level repair organisation to repair pcb's at component level even though reliability predictions indicated that only a few failures of each pcb type were likely to occur each year - it would have been more economic to throw failed pcb's away. In another case the plan did not adequately define the distinction between first and second line repair - as a result the first line equipment was designed to diagnose faults to sub-assembly level, but first line spares were provided as individual pcb's. Therefore extra time-consuming diagnostic procedures were required to indentify individual failed pcb's within sub-assemblies.

b. Failure to fully define test equipment characteristics. No realistic test equipment can be expected to accurately diagnose 100% of failures - there will always be some failures which are not detected, or some failures which cannot be isolated to an individual replaceable item etc. Therefore the ARM plan should identify design targets for test effectiveness (ie proportion of failures which are detected by the test equipment), isolatability (ie ability of test equipment to accurately diagnose failures to the correct replaceable item), false alarm rate, etc. If the designer has no such targets to work to, he is unlikely to produce test equipment which is satisfactory in all respects. For example, a test equipment used to test stored missiles annually had a false alarm rate of 2% per test. The missiles in storage were very reliable, with only about 2% failing in each year. As a result approximately 50% of missiles diagnosed as faulty and returned for repair were in fact found to be failure free. For a fuller discussion of test parameters see Irwing(2).

5. RELIABILITY

ARM plans usually address relaibility targets and the methods of relaibility prediction, analysis and demonstration in some detail.

However, even in this aspect there are some common errors which are worthy of discussion.

a. Failure to fully specify target reliability The ARM plan should specifiy the duty cycle environment and failure definition which relate to the reliability target. Failure to do this will make it very difficult for the manufacturer (and the customer) to determine whether the target has been met at the end of the project. Some manufacturers also try to avoid making precise statements of reliability - eg expressing targets as desirable or unquantified aims rather than as essential requirements. Although such practices may be commercially expedient in the short terms, since they effectvely remove the requirement to spend money and effort on reliability, they are unlikely to result in major improvements in reliability and hence improve the manufacturer's long-term standing in the market place.

b. Failure to specify burn-in. Many electronic systems are "burnt-in" to identify and remove manufacturing defects which may otherwise be the source of early-life failures in the field. This testing is often described in the ARM Plan. Comparatively few plans, however, describe the means by which the duration of the burn-in is to be set (eg cease burn-in after 10 hours without failure). This can result in either burn-in being continued for longer than is necessary, at considerable expense to the manufacturer, or burn-in being stopped too early resulting in avoidable early life failures in service. The optimum burn-in duration is dependent upon many production processess and parameters and should be assessed empirically on a continuous basis.

6. TESTING

Errors in ARM Plans concerning testing (eg reliability development testing, reliability demonstration testing, reliability growth testing, maintainability demonstration testing) often impinge significantly on project costs, timescales and availability of prototype equipment. Many ARM plans are inadequate in these areas and have not addressed the project programme, or more relevantly, have not been taken into account in the project programme. Common errors concerning test are:

a. Failure to specify the correct test objective. Testing is the most effective means of identifying design defects during development but is also highly expensive in terms of time, cost and resources. It can therefore be a fundamental part of ARM assurance or improvement. However, many ARM plans consider testing solely in terms of reliability demonstration. Design defects are detected during testing only when they cause failures. Therefore the objective of reliability development testing is to generate the maximum number of design defect-related failures in the minimum length of time. The objective of reliability demonstration testing is to demonstrate equipment reliability by generating few failures in a fixed length of time. Reliability demonstration testing is a very expensive and time consuming activity, and often a shorter test in a more severe environment will be a more efficient identifier of design defects. The correct test objectives must be specified in the ARM plan, or a great deal of money may be wasted measuring ARM achievement when a lesser amount could have been used to improve it.

b. Failure to identify resources required for testing. Effective ARM testing requires time, manpower, representative prototypes and suitable test facilities. It is unlikely that these resources will be available when required unless the ARM plan identifies them at the outset. In several cases this error has resulted in planned tests not taking place or being of limited value (eg only non-standard development samples available for testing). Also, a limited provision of resources in the ARM plan for unforeseen ARM testing in support of development can be very useful (eg to allow extended reliability testing of individual items which have proved unreliable in the main reliability test or to prove the effectiveness of modifications introduced to correct design defects).

c. Failure to specify reporting, analysis and correction procedures. The reliability of a design is improved during development testing only if design defects are identified and removed. The ARM plan should specify the procedures by which failures occuring during testing will be detected, recorded, analysed and corrected. Since testing is expensive, and the test personnel are often under pressure to achieve a certain number of hours testing within a specified period, there may be a tendancy to repair a failure and restart the rest as quickly as possible - ie without fully investigating the failure causes at the time or recording all the information necessary to establish the causes at a later date. In many practical cases, lack of a formal and disciplined approach to failure reporting, analysis and correction has delayed the identification of pattern failures until late in the project, increasing the cost of the corrective action.

7. CONCLUSIONS

This paper has outlined thirteen of the more common errors in ARM Plans. It is not a comprehensive list, and other practitioners will no doubt be able to add to it from their own experience. Nevertheless it should give some indication of the difficulty in preparing a comprehensive and appropriate ARM Plan.

The principal lesson to be learnt is that it is very difficult to correct an error in the ARM plan at a late stage in the project. This is because:

a. ARM activities require a considerable commitment of effort and other resources from Quality and Design departments as well as from ARM specialists. These resources must be planned in from the beginning of a project. The programme should include contingencies for correcting defects revealed by ARM analysis or testing without incurring unacceptable costs or delays late in the project.

b. ARM activities must be correctly phased with respect to other project activities if they are to be effective.

c. ARM requirements, like other design constraints, must be considered at all stages of design if a successful, balanced design is to be achieved. The ARM Plan Programme must be an integral part of, and not an addendum to, the overall project. Indeed in many cases the ARM plan should drive the project programme, rather than vice versa.

Many of the errors described here are errors of omission. In order to assist those responsible for preparing ARM plans, a checklist of items which should be considered for inclusion in ARM Plans is presented in Table 1. Further guidance may be obtained from references (3) and (4).

REFERENCES

(1) Dr B. Branson, Malvern's Program Analysers, RSRE Research Review 1985.

(2) M.H. Irwing, BITE - Past Mistakes, Present Problems and Future Solutions, Reliability 85 Proceedings, Volume 1.

(3) Ministry of Defence, Achievement of Reliability and Maintainability, Part 1: Management Responsibilities and Requirements for R & M Programmes and Plans. Defence Standard 00-40 (Part 1/Issue 1, 17 July, 1981).

(4) British Standards Institute, Reliability of Systems, Equipments and Components, Part 1: Reliability Programme Management. British Standard 5760 1979.

BIOGRAPHICAL SKETCH

Paul Kirrane graduated from Bristol University in 1978 with a Bachelor of Science degree in Aeronautical Engineering. After a period of professional training with the Ministry of Defence, he joined A&AEE Boscombe Down where he was involved in the assessment of the reliability and safety of military helicopter systems. In 1981 he joined Rex, Thompson and Partners of Farnham, Surrey, a firm of consultants specialising in Reliability and Operational Research. He is now a Reliability Consultant with RTP, and has been involved in ARM management and analysis on many major projects in the defence, aerospace and energy industries.

Table 1

Checklist for ARM Plans

Topic	Item No.	Checklist Description	Relevant to	
			R Plan	M Plan
TITLE/ REFERENCE	1	Title	✓	✓
	2	Issue status (and date)	✓	✓
	3	Reference	✓	✓
INTRODUCTION	4	Company identification	✓	✓
	5	Project/Contract identification	✓	✓
	6	Requirement for plan	✓	✓
	7	Scope of plan	✓	✓
ORGANISATION MANAGEMENT and CONTROL	8	Relationship of project ARM personnel within project management structure	✓	✓
	9	Relationship of project ARM personnel with project Quality and Design	✓	✓
	10	Nominated ARM co-ordinator	✓	✓
	11	Statement on ARM Progress Reviews	✓	✓
	12	ARM Design Review Structure	✓	✓
GENERAL REQUIREMENTS	13	Contractor responsibilities	✓	✓
	14	ARM Plans update procedure	✓	✓
	15	Plans address hardware and software (as appropriate)	✓	✓
	16	ARM Targets and Conditions	✓	✓
	17	Current ARM Predictions or Achievement (as appropriate)	✓	✓
	18	Sub-system targets (apportionment)	✓	✓
	19	Statement on redundancy	✓	✓
	20	Eqpt included/excluded from ARM targets	✓	✓
	21	Electronic Component Quality Levels	✓	—
	22	Statement on design safety factor (mechanical)	✓	—
	23	Use of de-rating criteria (electronic)	✓	—
	24	Use of Burn-In and conditions for terminating Burn-In	✓	—
	25	Statement of proposed logistics structure	—	✓
	26	Identification of test equipment requirements and specifications	—	✓
MODELLING AND PREDICTION	27	Develop models and update throughout project	✓	✓
	28	Apply required duty cycles	✓	✓
	29	R, M and A prediction Methods	✓	✓
	30	Identify prediction reports appropriate to the particular Sub-Contractor	✓	✓
	31	Agree to provide supporting data as required	✓	✓
DESIGN ANALYSIS	32	Statement of agreed failure definitions	✓	—
	33	Statement on level of detail for FMECA	✓	—
	34	Use project FMECA worksheet at LRU level	✓	—
	35	FMECA results input to Design Reviews	✓	—
	36	Statement on FMECA method	✓	—
	37	Identify relevant FMECA Reports	✓	—
	38	Statement on Maintainability Analysis method	—	✓

Table 1 (Continued).

Topic	Item No.	Checklist Description	Relevant to	
			R Plan	M Plan
DESIGN ANALYSIS (Contd)	39	Maintainability Analysis input to Design Reviews	–	√
	40	Identify relevant Maintainability Analysis Reports	–	√
	41	Spares Analysis method	–	√
	42	Spares Analysis results input to Design Reviews	–	√
	43	Identify relevant Spares Analysis Reports	–	√
	44	Statement on Critical Items List (CIL)	√	√
	45	Review and progress report on CIL at ARM Progress reviews	√	√
DESIGN EVALUATION	46	Identify items for dedicated R testing	√	–
	47	Statement on R test plans	√	–
	48	Statement on Qualification testing	√	–
	49	Test data input to Design Reviews	√	–
	50	Statement on R Growth Monitor method	√	–
	51	Identify start of R Growth Monitoring	√	–
	52	Growth reported at ARM progress reviews	√	–
	53	Produce Growth report	√	–
	54	Statement on Maintainability Evaluation method	–	√
	55	Identify M Appraisal requirement	–	√
	56	Statement on M Appraisal Plan	–	√
	57	M Evaluation progress at ARM progress reviews	–	√
	58	Produce M Evaluation Report	–	√
	59	Statement on approach to software	√	–
	60	Commitment to collect data	√	–
	61	Identify start of software data collection	√	–
	62	Statement on ARM approach to design change	√	√
DATA COLLECTION and INCIDENT REPORTING	63	Detail Incident Reporting, Analysis and Correction procedure	√	√
	64	Present standard reporting form for project	√	√
	65	Apply standard classification method	√	√
	66	Produce quarterly incident summaries	√	√
	67	Identify phases/eqpts appropriate to incident reporting	√	√
PROGRAMME	68	Relationship with other project activities	√	√

RELIABILITY MODELLING AND ANALYSIS

Reliability Technology — Theory & Applications
J. Møltoft and F. Jensen (Editors)
© Elsevier Science Publishers B.V. (North-Holland), 1986

A PRACTICAL EXAMPLE OF THE MEASURES
NEEDED TO IMPROVE A LOW MTBF RADAR

Robert P.F. LAUDER
Ferranti Defence Systems Ltd., Product Support
Dept.,South Gyle,Edinburgh,Scotland

A brief definition of availability is given showing the equal dependence of reliability and mean down time. A detailed case history is presented of the measures which had to be taken to improve a very low MTBF multimode radar by wholesale redesign of Tx,Rx and power supplies Maintainability and mean down time were improved by designing new items of test gear. The specification of all plugs and sockets were rewritten, and all major cableforms were re-designed and replaced. An operational improvement led to the introduction of a digital computer to replace the existing analogue ones, and allowed the realisation of a many times faster BIT.

1.   INTRODUCTION

The military user of avionics is not primarily interested in reliability per se. What he wants above all is "availability", which can be defined as the probability that a given piece of equipment is ready and working when the user needs it. Numerically it can be described by the equation

$$Ao = \frac{MTBF}{MTBF + MDT} \quad \ldots\ldots\ldots \text{ equation 1}$$

Where MTBF is the mean time between failures in hours, MDT is the mean down time in hours and Ao is the operational availability expressed as a number less than unity. It will be obvious that if MTBF and MDT are equal, the availability can never be greater than 50%. Equation 1 is shown graphically in Fig.1 as a family of curves of MTBF versus Ao with MDT as a parameter.

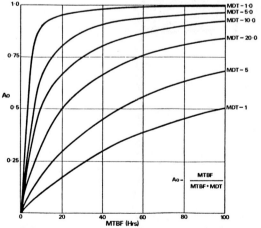

CURVES OF THE RELATIONSHIP BETWEEN Ao, MTBF & MDT

FIG. 1

Reliability of avionic equipment receives massive attention, and countless papers and books have been written on the subject. Maintainability, represented here by MDT, does not receive nearly so much attention, and it is therefore useful to review its component parts.

MDT is the elapsed time between a fault being reported, to the equipment being back in a serviceable state, ready for action. In the case of avionics equipment, the fault has to be diagnosed at the aircraft, and isolated to an LRU (Line Replaceable Unit). This will normally be done by the use of BIT, together with the maintainer's knowledge of the system, because BIT can never be 100% unambiguous. Because time spent at the aircraft on the ground is precious. The maintainer's uncertainty may cause him to withdraw two or even three LRUs to increase the probability of having selected the faulty one. This process will of course increase the No Fault Found (NFF) percentage and add to the cost and time for no good reason, down the repair line. The time to remove the faulty LRU may be insignificant or it may be very considerable. Some cockpit items can only be accessed by removing the pilot's seat. (In extreme cases, some equipments require the removal of the engine for access). Having removed, what is hopefully the correct LRU, a replacement must be obtained from stores and fitted. It may be crucial that the replacement is at the same modification standard as the faulty one, and modification programmes sometimes address the items in store in a different time frame from those in use.

Once the replacement LRU has been fitted, the system must be tested, and this includes those items removed for access, even though they played no part in the original fault.

Much of the time of MDT depends on a fast accurate BIT, good systematic training of the maintainers, and useable technical handbooks which can be carried out to the aircraft, and read under what are often very difficult environmental conditions. All of this is the background to the case history, now to be presented.

2. CASE HISTORY

The air intercept radar which is the subject of this case history is foreign in origin, and was designed in the very difficult period <u>after</u> the invention of the transistor, semiconductor, but <u>before</u> the invention of the integrated circuit. It contained no ICs,LSI or VLSI components. The result was that it contained an excessive number of components. Although the radar was designed over 20 years ago, it was, in its day, a "state of the art" radar. It was multimode, with pulse doppler as its primary mode to give the interceptor a look down, shoot down capability, but with a variety of pulse modes, ground mapping and terrain following sub-modes. The aircraft carries a gun, and a variety of semi-active homing air to air missiles, and an IR homing missile, also air to air. The radar was wholly analogue in design. Security prevents its precise definition but suffice it to say that it belongs to a high performance, air intercept fighter aircraft.

The first problem was to analyse which elements of the radar were responsible for its MTBF being in almost single figures. At first sight, that would seem to be a trivial problem. The Royal Air Force has a central maintenance data computer installed at a station called Swanton Morley in England, and all RAF units send defect data to this station for analysis. As mentioned earlier, this radar was foreign in origin and, as a result, its identifying marks did not conform to normal British practice.

We have to remember that all reliability, or more properly, unreliability data, is first order dependent on a low-ranking military man, filling in a form. In this case, elements of the radar had a drawing number as well as a NATO stock number printed on each item. Different maintainers used different reference numbers on the defect form, describing the same item. When fed to the maintenance data computer, the computer was unable to provide accurate data since it had no cross-reference file. We provided one.

Having arranged matters so that we had reliable defect data to work on, we then set about analysing the defects on an LRU basis, and this is shown in Fig.2 showing only defects and the no fault founds. From this it was clear that certain LRUs were significantly worse than others. The prime candidates for attention were the modulator, the main klystron power amplifier (KPA), the high voltage power supplies, the receiver and the low voltage power supplies.

Starting with the KPA, this was a 1 KW output tube requiring 10 KV cathode voltage, a 7 KW bias supply on the grid, and consequently a 7 KV drive pulse to switch the tube on and off. Since the time of the first design, KPAs in the same family had become available, based on the "shadow grid" principle. In this case there are two grids in the form of metal dishes with holes drilled in them.   The grids are very closely spaced and the pattern of holes accurately aligned.   The one nearer the cathode is

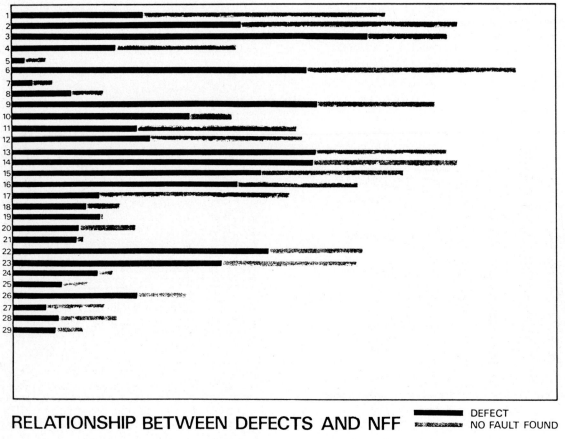

RELATIONSHIP BETWEEN DEFECTS AND NFF

████████ DEFECT
▓▓▓▓▓▓▓▓ NO FAULT FOUND

FIG.2

electrically connected to the cathode and consequently electrons stream out through the
holes like narrow rods. The second active
grid can suppress this electron flow with only
350^V. Thereafter the action of the klystron
is as before.

The use of this KPA allowed many other improvements to be made. Since the modulator only
had to produce drive pulses of 350°, it could
be realised in solid state technology, instead
of the previous one which was vacuum tube
based. The 7 KV bias supply was eliminated
on the high voltage power supply (HVPS). Like
many transmitters, this one suffered from too
many volts in too small a space, and significant improvement was made in reorganising the
routing of high voltage cables.

The front end of the two-channel receiver had
two parametric amplifiers containing varactors
which were pumped at J-band by a vacuum tube
klystron. The parametric amplifiers had a
very large number of interactive tuning
controls, and were very difficult to set up.
The result was that the receiver frequently
flew with sub-standard performance. These
parametric amplifiers, and their klystron
pump tube were discarded and replaced with
two field effect transistor (FET) amplifiers
with no controls, and merely an input and
output socket, and a supply voltage plug. The
FET amplifiers had superior noise performance
over the paramps, and were very flat over the
frequency band. The discarding of the pump
klystron also allowed its associated high
voltage power supply within the modulator
housing to be discarded, thereby reducing the
heat load and increasing the MTBF.

Low voltage power supplies (LVPS) often suffer
from an insidious increase in demand due to
successive modification action. This was the
case here, and the LVPS was redesigned with
foldback current protection which was not
previously available.

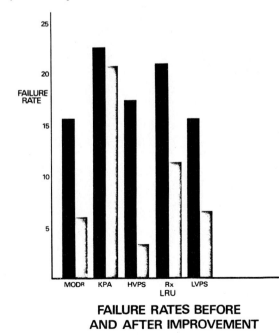

**FAILURE RATES BEFORE
AND AFTER IMPROVEMENT**

FIG. 3

Fig.3 shows a histogram of these specific
changes before and after modification with the
significant improvements to MTBF. Obviously
there were many other areas which we would have
liked to address but there were both financial
and time constraints. This was as far as we
intended to go. However when we came to plan
the implementation of these changes, we examined the condition of main cableforms, and
their associated plugs and sockets.

Most of the plugs were manufactured by the
same company, and were made from phosphor
bronze pins with copper, sometimes silver,
undercoats with gold on top. These were
moulded in Diallyl Phthalate bases. The
condition of all the plugs was very poor indeed with much corrosion and pitting of the
surface of the pins. There were also circum-
ferential scratches on the pins which were
difficult to explain until we visited the
company and saw the manufacturing method. The
raw pins were plated and polished, and then
manually laid up in a moulding tool which was
then injection moulded with the diallyl pytha-
late. The join between the two halves of the
mould produced the inevitable flash which was
difficult to remove, and a metal tool re-
sembling a pencil sharpener was used. This
was causing the scratch marks. It is now
known that gold plated on silver is bad
practice since, in time, silver migrates
through the gold to the outer surface where
silver sulphide products are created causing
blackening and loss of conductivity. All
stocks of plugs with gold over silver were
purged from military stores. Pure gold has
very poor wear properties, and it is usual to
add trace elements to a level of 0.2 to 0.3%
of cobalt, nickel or iron/carbon/potassium to
improve the wear properties. The body material of diallyl phthalate was glass fibre filled
which made it very difficult to remove the
flash. The use of nylon filling allows the
flash to be removed with the fingernail or
much softer tools. The plug specifications
were re-written to reflect all of these
changes.

Turning now to the main cableforms, these had
been in use for 20 years, and the main cable
harness has about 1400 wires which have been
the subject of many update modification
programmes, and contained splices in a variety
of places, some of which were no longer used.
Individual wires often ran to a remote spot
then back to close to the point where it
started. We decided to indulge in wholesale
redesign of all the cable harnesses and to
fit the new type of plugs. This was a major
exercise, but will pay dividends to improved
MTBF.

3.   MAINTAINABILITY IMPROVEMENTS

The above improvements were operational ones
to the radar itself. However, it was freqently
reported that basic parameters such as pick-
up range and lock-on range were poor. With
the co-operation of the RAF, we took two
operational radars, and set up their
receivers with great care. The effect was
instantaneous performance improvement to a
level rarely achieved. However it took many

hours to perform the setting up.

This radar is maintained at squadron workshop level on the "hot bench" principle, i.e. a bench is equipped with a complete radar together with simulated signals from the rest of the aircraft, and a variety of measuring equipment. Setting up the receiver on this bench for a lengthy period prevented its use for any other diagnosis by the substitution method.

We therefore designed and built a free standing item of test gear which would accept the three LRUs forming the pulse doppler receiver, together with a simulation of the other radar signals. This item not only makes the receiver controls much more accessible, but simplifies the setting up, and does not tie up the main bench which can be used for other purposes simultaneously. We also designed another test bench on the same principle to set up the STALO.

While these two items do not affect the radar's MTBF in any way, they contribute to better performance in the air.

For other reasons than this particular attempt to raise the MTBF, it was decided to fit a digital computer in place of the existing three analogue computers. The existing BIT was analogue in nature and was programmed by 105mm photographic film containing about 1000 frames. The frames were read one at a time, and advanced by a Geneva gear mechanism. To run a complete performance profile assuming no faults took about 50 mins. The digital computer brought with it a digital BIT which carried out the same performance profile in about 8 mins. This is another highly significant maintainability improvement.

4. CONCLUSIONS

The purpose of industrial case histories is to learn something from the past which can be applied to the future. There are lessons to be learnt from this particular case which should be borne in mind when designing new radars.

Transmitter design often fails due to inattention to adequate insulation for high voltage. Even pressurised systems will not necessarily be at one atmosphere, and it is naive to assume that insulators will remain clean and dry. Even for relatively modest voltages, the avoidance of sharp points is helpful.

Aim for a receiver with no adjustments. It will probably not be realised, but it is a good aim to have in mind. More than required current capacity available in low voltage power supplies will always be useful.

Be very careful about the specification for plugs and sockets. If a perfect plug and socket existed, its use could probably double the MTBF of almost any system. It is largely responsible for intermittent faults and the low signal paths are particularly vulnerable. It is also very subject to vibration and shock.

In new designs of airborne radars, specify twice as many wires between the nose package and the cockpit as you actually need. They will be needed during the long in-service period and aircraft wiring is very expensive and time consuming to change after the aircraft is built.

The case history of this particular radar is a particularly painful one on which much time and effort was spent. The process is still going on, but the overall MTBF has almost doubled.

The lessons learnt are very simple and obvious, yet we still seem to go on making the same mistakes. Let us hope that it is untrue that the only thing we learn from history is that we learn nothing from history.

R.P.F. Lauder served in the British Army as a Radar Officer from 1943-47. He joined Ferranti in 1947 and spent 15 years designing high power transmitters, 5 years leading a design team for the radar for the Buccaneer aircraft. In 1967 he led a team of 10 engineers to the USA to study a multimode radar design. Since 1986 he has been Engineering Manager of Product Support Department in Ferranti. He is the elected member to the Engineering Assembly in Britain representing electrical engineers in the East of Scotland. He is also an executive director of the Scottish Engineering Training Scheme Ltd company, and a member of the interviewing panel for membership of the IEE. He is married with two sons and a daughter.

Reliability Technology — Theory & Applications
J. Møltoft and F. Jensen (Editors)
© Elsevier Science Publishers B.V. (North-Holland), 1986

RELIABILITY ANALYSIS OF TELEPHONE NETWORK IN SLOVENIA

Franci DERŽANIČ
ZO PTT Slovenije, Ljubljana, Yugoslavia
Alenka HUDOKLIN
University of Maribor, VŠOD Kranj, Yugoslavia

A reliability analysis of the telephone network in Slovenia using fault tree technique is presented. The network was analysed in two ways: in terms of the reliability of its constituent parts, 9 different secondary areas, and as a whole. In both cases the study is based on traffic diagrams which enable the definition of catastrophic failures of the systems of connections in secondary or tertiary areas of the telephone network. The fault trees for the defined failures as top events were constructed and evaluated using available field data.

1.INTRODUCTION

In order to assure an adequate quality of telephone services it is necessary to maintain an appropriate level of the telephone network reliability. It is neither technically nor economically advisable to consider the reliability of a single exchange or a number of ex- changes in a local area as a separate problem. On the contrary, it is neces- sary to consider the reliability of va- rious items in a telephone network as an integral part of the reliability of the national telephone network.

The aim of this paper is to present a study of the dependence of the Sloveni- an telephone network reliability upon the reliability of its constituent parts. The reliability analysis was performed from the subscriber's point of view. Traffic diagrams were chosen as a basis for the definition of failu- res, and a fault tree technique was u- sed for the identification of failure causes.

2.DESCRIPTION OF THE TELEPHONE NETWORK IN SLOVENIA

The telephone network in Slovenia is built up in a usual hierarchical form with the star-shaped backbone network (final choice routes) and the direct routes between certain exchanges. It consists of 9 secondary areas which can be grouped into 3 types according to the different internal structure (see fig.1). The centres in the network are of two types: crossbar and SPC. The ex- changes of the first type are mainly from the family ISKRA-58 which repre- sents a register controlled system de- signed on the principles used in large exchanges made up by selection stages. The SPC exchanges are of the type ISKRA- METACONTA 10 C and METACONTA 10 CN. Transmission paths in the network have either cables or radio relay equipments as transmission media. The transmission mode is mainly of FDM type. Equipments operating on PCM basis are also used. Signalling systems in the network are of DC, one frequency and MFC type.

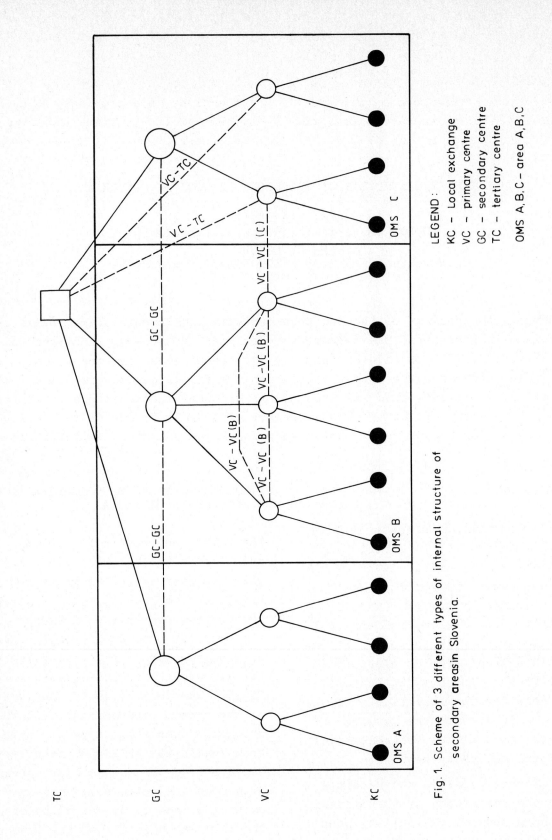

Fig. 1. Scheme of 3 different types of internal structure of secondary areas in Slovenia.

LEGEND:

KC – Local exchange
VC – primary centre
GC – secondary centre
TC – tertiary centre

OMS A,B,C – area A,B,C

3.RELIABILITY OF THE TELEPHONE NETWORK

A basic subject of our reliability investigation of a telephone network is a system of (simultaneous) connections. The term connection in this context means a two way voice channel completed between two points by means of a switching equipment, and capable of transmitting voice and supervising signals. Each exchange in a telephone network is capable of commutating a certain number of simultaneous connections, depending on its technical characteristics. Failures of functional units in the network cause a decrease in the traffic capability of the exchange. Unsuccessful calls due to these failures cause additional trials to set up connections, and thus cause an increase of traffic losses. A certain value of traffic losses could be regarded as a degradation, or even catastrophic, failure of the system of connections through the exchange. Such a failure affects the system of connections in the whole telephone area, depending on the role of the exchange in the area. In the case of secondary centre, a given value of traffic losses represents a catastrophic failure, i.e. a total break down of the system of connections in the whole secondary area. It means that it is impossible to set up connections inside the area nor connections to and from other areas.

3.1 Secondary areas

In our reliability analysis,the system of connections in a secondary area is described in two ways: in terms of its typical elements, and as a whole.

3.1.1 Failures of the elements of the system of connections

Typical elements of the system of all connections in a secondary area are systems of local, trunk and long distance connections. A scheme of trunk connections in a secondary area is shown in fig.2. A traffic flow through such an element depends on failures of different functional units of the telephone network as well as on changes of the incoming and outgoing traffic in different commutation stages. A failure of each element of the system of connections in a secondary area is defined as a certain value of traffic losses. We have chosen the following values: 2%, 5%, 10% and 20%. First three values of traffic losses could be considered as degradation failures, while the third value represents a catastrophic failure of the system of connections.

By using a fault tree technique, we found out possible causes of failures for three typical elements of the system of connections in a chosen secondary area. In this way the functional units of telephone exchanges and transmission paths which could fail and cause different failures of the system of connections were identified. If there are several equivalent units, the number of units which must fail in order to induce a failure of the system of connections was determined. The traffic changes causing these failures were also found out. These changes are due to an interaction of all exchanges participating in trunk or long distance connections.

An example of a failure analysis of trunk connections is given in table 1. The results of this analysis combined with the available field data indicate functional units and points of traffic changes which are critical with respect to the telephone network reliability.

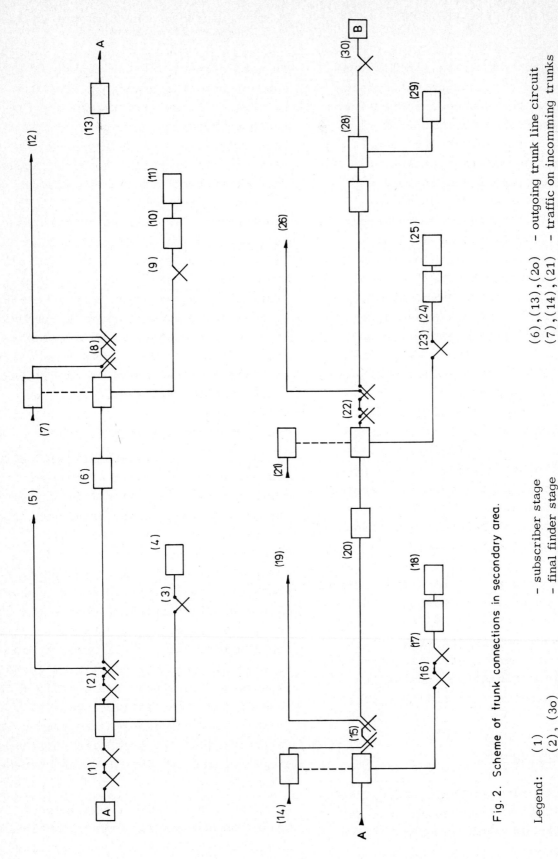

Fig. 2. Scheme of trunk connections in secondary area.

Legend: (1)   – subscriber stage
             (2), (3o) – final finder stage
             (3), (9), (16), (23), (29) – register finder stage
             (4), (1o), (17), (24)   – register
             (5), (12), (19), (26)   – traffic on outgoing trunks

             (6), (13), (2o)   – outgoing trunk line circuit
             (7), (14), (21)   – traffic on incomming trunks
             (8), (15), (22)   – group selector stage
             (11), (18), (25)   – analyser
             (28)   – incoming trunk line circuit

Table 1
Failure analysis of trunk connection in secondary area

Functional units in connect.	Number of units	Traffic capab. Erl.	Number of failed units causing traffic losses exceeding				Change of traffic causing traffic losses exceeding			
			2 %	5 %	1o %	2o %	2 %	5 %	1o %	2o %
(1) - R	2									
(1) - M	2									
(1) - O	8o	57	1o	16	22	31				
(2) - R	2									
(2) - M	4				1	1				
(2) - I	4o	26	5	8	11	16				
(3) - P	4			1	1	2				
(3) - I	4o	19	9	12	16	21				
(4)	6	1	1	2	3	3				
(5)							55	65	96	1oo
(6)	2o	12	1	4	6	8				
(7)							18	31	47	78
(8) - R	6			1	1	2				
(8) - M	12		1	2	3	5				
(8) - I	12o	8o	16	25	34	51				
(9) - P	12									
(9) - I	12o	8o	16	25	34	51				
(1o)	32	16	7	11	13	16				
(11)	2	16	1	1	1	1				
(12)							29	41	53	65
(13)	7o	49	4	1o	17	26				
(14)							2,3	5,8	11,6	23,3
(15) - R	22		3	5	7	9				
(15) - M	44		6	9	13	18				
(15) - I	44o	3oo	58	88	121	171				
(16) - M	2x5		1/2	1/2	1/2	1/2				
(16) - I	4oo	273	52	8o	11o	156				
(17)	114	73	18	27	35	48				
(18)	6	73	1/2	1/2	1/2	1/2				
(19)							1o	15	21	3o
(2o)	7o	49	4	1o	17	26				
(21)							4o	68	1o3	171
(22) - R	6			1	1	2				
(22) - M	12		1	2	3	5				
(22) - I	12o	8o	16	25	34	51				
(23) - P	12									
(23) - I	12o	59	28	49	56	62				
(24)	4o	18	1o	14	18	22				
(25)	2	18	1/2	1/2	1/2	1/2				
(26)							16	24	3o	41
(27)	8	3	1	2	3	4				
(28)	8	3	1	2	3	4				
(29)	4	1		1	1	2				
(3o)	2o	1o	4	6	8	1o				

R - rack, M - marker, O - outlets, I - inlets, P - piece
Numbers in brackets have the same meaning as in fig.2.

3.1.2 Failures of the system of connections as a whole

The system of connections is now defined as a number of simultaneous connections through the secondary centre. Only the catastrophic failure, i.e. the total break down of this system, was considered. On the basis of an experimental evidence we have estimated that the catastrophic failure occurs when a decrease of the traffic capability of the secondary centre exceeds 10%. This value is called the critical traffic change.

The critical traffic change was determined by means of properly constructed traffic diagrams. A traffic diagram allows an evaluation of magnitudes and directions of all traffic flaws in a given telephone area. Besides, an analysis of such a diagram could help in searching for the causes of failures of systems of connections. In fig.3 a scheme of a traffic diagram for a secondary area with 3 primary centres is given. The following quantities must be determined:

N_n - number of subscribers

N_p - number of trunks

y_m - traffic to and from other secondary areas

y_{om} - traffic inside the secondary area

y_v - traffic of the secondary area terminating in a secondary centre

y_p - trunk traffic

$y_{VC,O}$ - outgoing traffic in primary centre

$y_{GC,O}$ - outgoing traffic in secondary centre

$y_{VC,D}$ - incoming traffic in primary centre

$y_{GC,D}$ - incoming traffic in secondary centre

$y_{VC,T}$ - transit traffic in primary centre

$y_{GC,T}$ - transit traffic in secondary centre

The traffic diagrams were constructed for 9 secondary areas. The critical traffic changes and some other critical parameters were determined (see table 2).

The causes of failures of the systems of connections in secondary areas were found using fault tree technique. The fault trees were constructed for all secondary areas. Basic events in these fault trees represent failures of functional units which could be used as test points for an automatic supervision system of the telephone network.

The fault trees were analysed by means of a minimum cut sets method. The resulting minimum cut sets were compared to the available field data on the network reliability, and critical functional units were identified. During supervision and maintenance of the telephone network, special attention should be paid to these critical units. However, we were not able to perform a quantitative evaluation of fault trees because of rather scarce field data.

3.2 Tertiary area

In the same way as for individual secondary areas, a traffic diagram was constructed for the whole network in Slovenia including direct and final choice routes. On the basis of this traffic diagram the critical traffic change, representing a catastrophic failure of the system of all connections in the tertiary area, was determined.

A search for possible causes of the break down of the system of connections in the tertiary area has given the following main causes: a failure of the tertiary centre and failures of direct routes between different centres in the area. However, in our network the occurrence of failures of direct routes independently of the failures of final choice routes is not possible because of

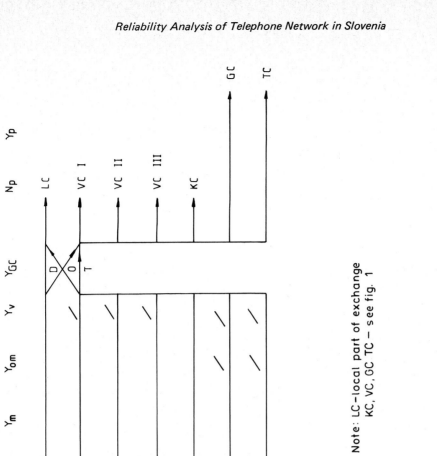

Note: LC–local part of exchange
KC, VC, GC TC – see fig. 1

Fig. 3. Scheme of a traffic diagram

Table 2

Data on structure of secondary centres and critical parameters

Centre or route	N_n	N_p	N_{pm}	y_K (Erl)	N_{Kp}	$N_{K\,reg}$
GC 1	1o4ooo			191,3	23o	33
KC	32ooo					
GC – TC		176/247				
LC (VC) – TC			682/1o15			
GC 2	28o24			77,3	93	18
KC	592o					
GC – TC		37o				
LC (VC) – TC			214			
GC 3	13896			79,8	96	18
KC	o					
GC – TC		348				
LC (VC) – TC			154			
GC 4	8ooo			52,7	75	14
KC	3448					
GC – TC		135/219				
LC (VC) – TC			48/48			
GC 5	8ooo			25,6	39	1o
KC	3948					
GC – TC		166				
LC (VC) – TC			88			
GC 6	11144			38,5	51	12
KC	538o					
GC – TC		165/166				
LC (VC) – TC			1o2			
GC 7	3ooo			16,1	26	8
KC	64o					
GC – TC		64/52				
LC (VC) – TC			29/3o			
GC 8	54oo			28,8	44	1o
KC	23oo					
GC – TC		68/87				
LC (VC) – TC			58/58			
GC 9	54oo			28,7	43	1o
KC	1892					
GC – TC		155				
LC (VC) – TC			48			

where is: N_n – number of subscribers

N_p – number of trunks

N_{pm} – number alternative trunks for traffic to and from another secondary area

y_K – critical value of traffic change

N_{Kp} – number of trunks which cause y_K when simultaneously failed

$N_{K\,reg}$ – number of registers which cause y_K when simultaneously failed

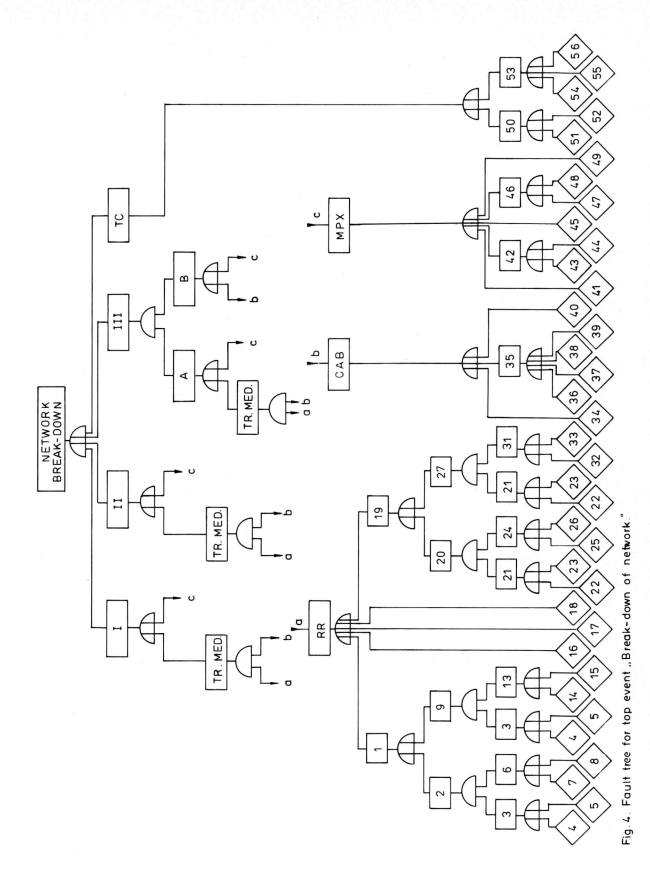

Fig. 4. Fault tree for top event „Break-down of network."

LEGEND to fig.4

I   - main transmission path I
II   - main transmission path II
III - main transmission path III
A   - transmission path A
B   - transmission path B
TR.MED. - transmission media
CAB   - cable
MPX   - multiplex
RR   - radio relay equipment
1 - receiving interruption
2 - failure of receiving unit
3 - failure of receiver I
4 - failure of receiving circuits I
5 - incorrect receiver level I
6 - failure of receiver II
7 - failure of receiving circuits II
8 - incorrect receiver level II
9 - failure of switch
13 - failure of automatic switching
     control
14 - failure of switching circuits
15 - failure of switching power supply
16 - disturbing voltages
17 - failure of antennas
18 - failure of power supply
19 - transmitting interruption
20 - failure of transmitting unit
21 - failure of transmitter I
22 - failure of transmitting circuits I
23 - incorrect transmitting level I
24 - failure of transmitter II
25 - failure of transmitting circuits
     II
26 - incorrect transmitting level II
27 - failure of switch

31 - failure of automatic switching
     control
32 - failure of switching circuits
33 - failure of switching power supply
34 - cable cut off
35 - failure of line equipment
36 - failure of transmitter
37 - failure of receiver
38 - failure of amplifier
39 - failure of remote power supply
40 - change of transmission characte-
     ristics
41 - failure of generator for signal-
     ling frequencies
42 - interruption of speech transmi-
     ssion
43 - failure of generator for frequen-
     cy carriers
44 - failure of transmitting TGM and
     QGM
45 - failure of signalling receivers
46 - interruption of speech receiving
47 - failure of generator for carriers
48 - failure of receiving TGM and QGM
49 - failure of power supply
50 - speech path cut off
51 - power supply cut off
52 - program interruption
53 - program fault
54 - incorrect program flow
55 - incorrect switch of processors
56 - incorrect data

common transmission paths. Besides, certain transmission paths are common for direct and final choice routes of different secondary areas. We shall call such a type of transmission path the main transmission path. A failure of the main transmission path means simultaneous failures of direct and final choice routes of several secondary areas. One must therefore find out critical main transmission paths in the network which could cause, if failed, a catastrophic failure of the system of connections. An analysis of our telephone network has shown that the principal causes of a break down of the system of connections are:

- (i) failure of the tertiary centre TC
- (ii) failure of the main transmission path I
- (iii) failure of the main transmission path II
- (iiii) simultaneous failure of the main transmission paths III A and III B

Fault trees for these events were constructed using our previous results [1], [2], [3]. These 4 fault trees are shown in fig.4 as four branches of a single fault tree for the top event "break down of the system of network connections".

4. CONCLUSION

A fault tree reliability analysis of the telephone network in Slovenia, based on traffic diagrams is presented. It enables the definition of catastrophic failures of the system of connections in terms of traffic losses, as well as the identification of their causes. A comparison of the results of our analysis and the field data exposes critical functional units of the telephone network which require special attention during supervision and maintenance.

REFERENCES

[1] A.Hudoklin and al., Reliability of the telecommunication system - Telephone network, part I : Reliability of the telephone exchange as an element of a telecommunication network, Report RSS, VŠOD Kranj 1980, 166 + 8 pp. (in Slovenian).

[2] A.Hudoklin and al., Reliability of the telecommunication system, part III : Transmission paths, Report RSS, VŠOD Kranj 1981, 116 pp. (in Slovenian).

[3] F.Deržanič, A.Hudoklin, Reliability analysis of telephone exchange in existing network, Reliability in Electrical and Electronic Components and Systems, EUROCON'82, Copenhagen June 14 - 18 1982, Preprints, Part I, 358 - 362.

BIOGRAFIES

Alenka HUDOKLIN graduated and received the M.S. and Ph.D.degrees in physics from the University of Ljubljana, Yugoslavia, in 1958, 1968 and 1969, respectively. From 1958 to 1969 she has worked with the Institute "J.Stefan", Ljubljana, on the investigation of nuclear reactions. In 1969 she joined Iskra-Institute for Automation, Ljubljana, where she was head of Reliability Laboratory. Since 1976, she is professor of stochastic processes and reliability at the University of Maribor.

Franci DERŽANIČ graduated from the University of Ljubljana in electrical engineering in 1972 and received the M.S. degree from University of Maribor in 1981. Upon graduation he joined Iskra-Institute for Automation,Ljubljana, as a research worker in Reliability Lab. Since 1977, he is working with ZO PTT Slovenije,Ljubljana,in Research Department.

Reliability Technology — Theory & Applications
J. Møltoft and F. Jensen (Editors)
© Elsevier Science Publishers B.V. (North-Holland), 1986

"CORRELATION BETWEEN RELIABILITY PREDICTION,
RELIABILITY TESTING METHODS AND FIELD DATA"

K. Stochholm
TERMA ELEKTRONIK AS
Hovmarken 4
8520 Lystrup
Denmark

KEY WORDS: RELIABILITY TESTING. BURN-IN OF SYSTEMS,
RELIABILITY FIELD DATA.

This paper discusses the correlation between reliability figures predicted
on systems and reliability testing.
Different approaches to reliability testing are used on the same system and
based on test data. The methods are compared and discussed.

The database for reliability prediction is MIL- HDBK-217.

Analyses of Weibull plots of Time to First Failure (obtained during Burn-In)
for two systems are discussed and compared to results of reliability testing
as per MIL-STD-781 for the same system.

It is the intention with this paper by means of actual test data to
highlight an useful analytic method for optimizing of the Burn-In effort,
which leaves us with a question mark on the use of the MIL-STD-781 as an
appropriate tool for reliability testing.

For one of the systems the calculated figure is compared to both data
obtained during reliability testing (Burn-In) and data collected during use
at the customer.

1.   INTRODUCTION

The first "case story" from TERMA
Elektronik A/S describes one of the
practical experiments which were
carried out in order to try to verify
the theoretical models for failure
development on, among others,
electronic equipment.
These theoretical aspects together
with practical experiments were the
baseline for a BURN-IN programme,
carried out by the Danish Academy for
Engineers (DIA-E) which resulted in a
book, item 3 in the list of
literature.

This "case story" describes:

- Baseline for selection of specimens
- Planning of experiments
- Reporting
- Results obtained
- Follow up

The following "case story" describes
how advantages can be taken from the
lessons learned during the
experimental phases, and be used as a
tool for reliability testing.

2.   BURN-IN EXPERIMENT

2.1   Theorectical Models (Lit. ref. 3)

The theoretical approach is based on
a Bi-modal distribution of component
failures as per fig. 1

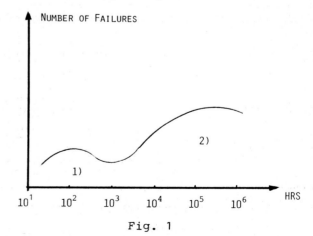

Fig. 1

1) Weak population
2) Strong population

This failure distribution is shown on the Weibull plot below.

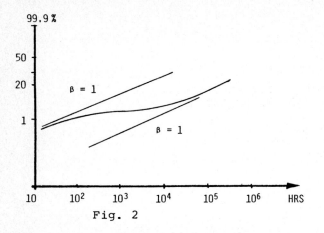

Fig. 2

The first part of the graph is related to a weak population and the last part is the strong population both characterized by the shape paramenter $\beta = 1$.

It was the intention with the experiments to investigate and describe methods for uptimizing the burn-in effect by using this model fig. 2, and in this case also using accelerated test environments.

2.3   Choise of speciments

In order to achieve as much data as possible form the burn-in test, a subject, suitable for an experimental sequence as shown in fig. 3, was sought.

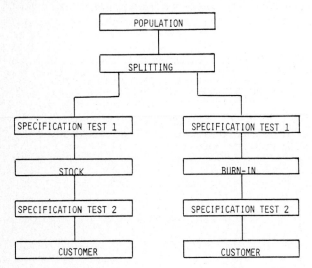

Fig. 3

Requirements to the speciments:

- a reasonable number of units
- easy to handle
- easy to supervise, test, and error-detection
- a predicted MTBF laid down in advance
- a possibility of following up at the customer

The choise fell on the radio RT6 from TERMA's 2061 system, which satisfied our demands on the experiment in this way:

- Number: 72 units at disposal
- Dimension: 28 x 28x 10 cm,
                   app. 5 kg
- Test: < 1 min/unit
- MTBF: calculated app.
                   2300 hrs
- Service Environment: Temp. -40°C
                   to +65°,
                   vibration
                   shock, bumps
- Customer :   Hærens Materielkommando
                   Army Material command
                   (HMAK)

2.4   Planning of Experiment

The experiment itself was defined by DIA-E and was carried out according to the following specifications:

- 72 pcs. RT6 exposed to A-test (specification test 1). Hereafter the 72 pcs. RT6 are divided into two groups.

- Group 1:   30 pcs. undergo burn-in at TERMA.

- Group 2:   42 pcs. are delivered to customer without burn-in.

- Customer reports on all 72 pcs. RT6 when/if failures occur. Way of report and time is agreed upon with the customer.

- Testplan for RT6 burn-in:

  Duration: 10 days and nights
  Voltage: 15 V DC (nominal
                   12 V DC)
  Temperatur: 65°C
  Functional test: Transmission for 2
                   min. and receiving
                   18 min. by turns
                   at the frequency
                   31.500 MHz
                   (continouos
                   service).
                   During
                   transmission the
                   output P-out is
                   recorded. During
                   receiving the
                   reciever
                   sensitivity
                   (signal/noise
                   conditions) is
                   recorded.
  Supervising: The above
                   mentioned
                   recordings will be
                   made as follows:

   0 - 8   hours: recording every
                  hour
   8 - 24   -   : recording every
                  2nd hour
  24 - 48   -   : recording every

6th hour
2 - 10 days : recording every
24th hour

At failure: Time of failures is
          recorded. The failures
          are found and reported
          on Failure Analysis
          Report
          a) Description of
              failure.
          b) Failure mechanism
              and cause.
          c) Solution/repair of
              failures.
          Repair is to be carried
          out as soon as
          possible. Having
          finished the repair,
          the units are
          reinstalled in the
          functional test.

After burn-in an A-test is carried
out (specification test 2) before
the units are delivered to the
customer.

2.5   The Results from the Experiment

(Extract from the DIA-E's company
report)

While testing (during the burn-in
period) the following number of
failures were recorded:

0 failure on 22 pcs. = 0 failure
1 failures on 6 -   = 6 failures
2 failures on 1 -   = 2 failures
3 failures on 1 -   = 3 failures

   30 pcs. with 11 failures
    8 pcs. with at least 1 failure

Weibull-Analysis

Here only times to first failure are
used. The times are arranged and
given rank values according to the
formula:

$$r = \frac{i - 0.3}{n + 0.4}$$

i = number of failure in order of
    sequence.
n = number of units being tested.

Number of Failure value	Time	Rank
1	1	2.3
2	4	5.6
3	14	8.8
4	14	12.2
5	42	15.5
6	48	18.8
7	120	22.0
8	216	25.3

The result is plotted in fig. 4.

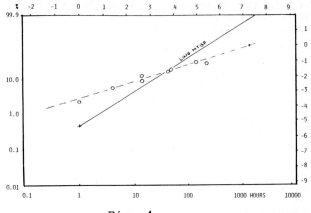

Fig. 4.

The curve has such a slight curvature
that you are not immediately able to
say anything about the completion of
the burn-in. If you, however,
tentatively draw a straight line
through the 8 points, you will get a
line with $\beta \approx 0.5$ and η = 1400 hours.

As previously mentioned the MTBF was
predicted to be 2300 hours and it
should be assumed that this MTBF
applies to a rather different
environment.
If the burn-in environment is assumed
to decrease MTBF with a 10 factor to
230 hrs, you get "line MTBF"
in fig. 4.

The assumption of the 10-factor is,
among other things, based upon the
fact that an increase of the
temperature from 25°C to 65°C
will increase the failure rate app.
16 times according to the Law of
Arrhenius.

2.6   The Bayes'-Estimate

As the last failure is on the right
hand side of "line MTBF", a new line
with the slope β = 1 through the last
failure is drawn, which consequently
enables us to estimate the "strong"
radios' parameters at
$\beta_2 = 1, \eta_2 = 750$ hours.

As to the weak devices, they are
estimated from fig. 4: $\beta_1 = 0.7$
and η_1 = 20 hours. Afterwards a
Bayes'-estimate is made with the
following result:

Time of Failure	S1
1	0.97
4	0.97
14	0.95
14	0.95
42	0.83
48	0.78
120	0.10
216	0.00
$\hat{p}$	18 %
$t_{50\%}$	75 hours

From the S1 column the probability of
failure due to a weak unit can be
seen from the parameters given.

$\hat{p}$ is the Bayes'-estimate, in
percentage, of weak devices in the
population, and $t_{50\%}$ is the time
where the probability of the failure
being due to a weak or strong device
is fifty-fifty (50%).

From the S1 column it can be seen
that the failures from 1 to 48 (48
included) are likely to be weak .
These 6 points are therefore plotted
alone, assuming that they represent
all the weak ones:

Time of Failure (h)	Rank (%)
1	10.9
4	26.6
14	42.2
14	57.8
42	73.4
48	89.1

See fig. 5

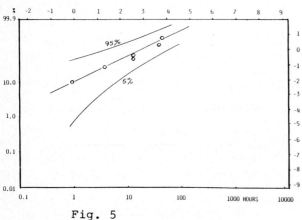

Fig. 5

Furthermore 5% and 95 % rank values
are marked in fig. 5. The 6 plotted
points follow a straight line with
the parameters
$\beta_1 = 0.75$ and $\eta_1 = 20$ hrs. These
parameters correspond so well with
the first estimate that no further
Bayes'-estimates are made.

In addition it can be seen from fig.
5 that 99% of the weak units can be
removed by a burn-in, having a
duration of 160 hrs.

In order to investigate the strong
devices, a cut-off of the first 48
hrs is made which leaves us with only
30 - 6 = 24 units.

First Time of Failure	New Time of failure	Rank (%)
120	72	2.9
216	168 hours	7.0

The result is plotted in fig. 6.

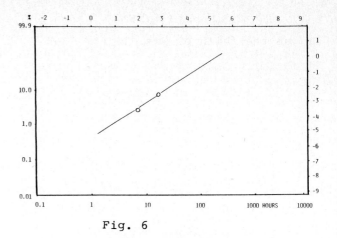

Fig. 6

Even though it can be dangerous to
make estimations on the basis of only
2 points, a line with slope $\beta = 1$ has
been drawn.
This line gives a $\widehat{MTBF} = \hat{\eta}_2 = 2200$
hrs, which corresponds very well with
the predicted MTBF of 2300 hrs.

2.7   Follow up with the Customer

One of the advantages of choosing the
RT6 equipment was the fact that HMAK,
as customer, agreed to the marking of
the equipment in connection with a
one-year follow up period.
Each transceiver was marked by means
of cutting ridges in the handles, and
a manila label listing the action to
be taken in case of failures, in
order to get quick response back to
ourselves.

We have had the opportunity of
following the 72 RT6 radios with the
army, and during this period we have
received the following feed back:

Order No. of Failure	Time	Failure
	MAR 80	Start of the follow up
9	DEC 80	Noise
10	DEC 80	Broken wire
11	JAN 81	Push Botton
12	MAR 81	Transistor short circuit

Failure Order No.		Time	Burn-in/ No Burn-in
No			
9	9 months	= 1350 hrs	X
10	9 -	= 1350 -	X
11	10 -	= 1500 -	X
12	12 -	= 1800 -	X

Conditions: 1 month = 25 days
             1 day = 6 hours service

Out of the four transceivers reported
having failed, three, namely Nos. 10,
11, and 12, had been burn-in tested.

2.8 Burn-in Environment

It was the failures Nos. 10 and 11, which initiated the analysis mentioned below, with the purpose of roughly dividing the failures into mechanical failures and component failures, which must be regarded as independent of a mechanical environment.

Order No. of Failure	Mechanical Failures	"Purely" Component Failures
1		X
2	X	
3	X	
4		X
5	X	
6		X
7	X	
8		X
9		X
10	X	
11	X	
12		X
Total	6	6

Out of a total of 12 failures half turned out to be mechanical failures, and as it can be seen from the responses from the customer not all were detected in the chosen burn-in environment.
One should therefore consider if a mechanical environment for this very equipment also should be used during burn-in.

3. RELIABILITY TESTING

Reliability testing is a controlled method used to demonstrate that the reliability is within the specified requirements. The set of parameters specified here is:

- Test duration each equipment
- Test environment
- The number of failures allowed during the test period.

MIL-STD-781 describes in principle the above method and consists of a set of different plans each with different boundaries depending on the decision risk of the test to be carried out.

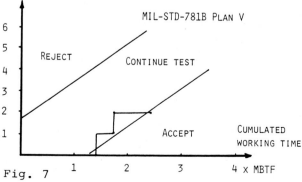

Fig. 7

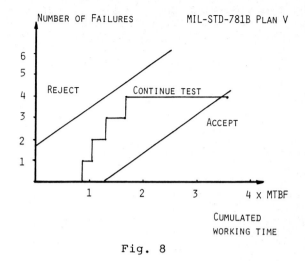

Fig. 8

The graphs, fig. 7 and 8, are examples where the customer has accepted the equipment based on MIL-STD-781.

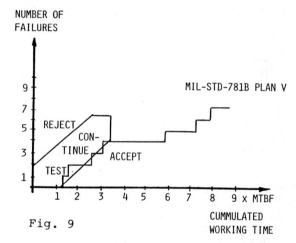

Fig. 9

Fig. 9 is a plot of failures found during lot acceptance testing of 38 radar equipmemnts. The graph shows that according to the failures plotted, the equipment would have been accepted after approx. 2.3 x MTBF (accumulated test hrs) and only two failures have occured.
In order to make further investigations on the validity of the above method the test was continued, and for the equipments failing during test the Time To First Failure (TTFF) was measured.

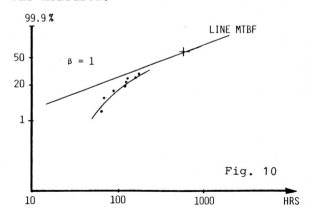

Fig. 10

Fig. 10 shows the Weibull plot of Time To First Failure. 8 failures were found during the test and the percentages (ordinate axis) is calculated in rank values as follows:

$$R = \frac{i - 0.3}{n + 0.4} \times 100;$$

where i is number of failure in the order of sequence and n the total number of equipments under test.
The line MTBF is drawn through the point of predicted MTBF and provided $\beta = 1$, i.e. (constant failure rate). In order to explain the interpretation of the graph an ideal curve is shown in fig. 2.

Comparing the two graphs, fig. 2 and fig 10, we can see that:

a) due to the slope we have probably still to do with early failures, one maybe two more, before the strong population takes over,

b) the graph is approaching the line MTBF from below, which means that the actual MTBF is greater than predicted.

Using the MIL-STD-781, fig. 9, method as the requirement for accepting the equipment, it is seen that in this case there is still "early failures" left.

The advantage by using the Weibull analyses as the basis for acceptance of the equipment is that you can screen all "early failures" or predict the remaining numbers. In the present example the epuipment, fig. 9, was accepted after two failures only, but six more were found later. These failures would have occurred relatively early in the equipment life period and would have been very expensive to repair on the user location. From the user point of view the availability of the equipment is increased.

The above illustrates that the MIL-STD-781 testing method is doubtful when you have early failures in the equipment.
When starting a life test without previous burn-in the number of early failures could leave you with a rejected lot of equipment which is only lacking a screening procedure. From the above example the Weibull plot tells both producer and customer the advantages of carrying on with the life testing.

4.   <u>CONCLUSION</u>

The burn-in experiment has supplied us with a well supporting background to plan and decide on a burn-in

sequence. Moreover, it has pointed out a good relation between predicted MTBF and measured MTBF.
In this case the subsequent "feed back" from the customer has caused considerations concerning the test environment, which will be used as basis for future burn-in tests.

Reliability testing, as shown by the examples using the Weibull plots, tells most of the past and future failure distributions. Therefore, with reference to the examples both customer and producer will benefit from having a Weibull plot and analysis specified as basis for lot acceptance rather than a simple life test.

LIST OF LITERATURE

1. MIL-HDBK-217C: Reliability Prediction of Electronic Equipment

2. MIL-STD-718B: Reliability Tests: Exponential Distribution

3. Jensen, Finn and Pedersen, Niels Erik, Burn-in: An Engineering approach to the Design and analysis of Burn-in Procedures, John Wiley & Sons, (1982).

BIOGRAPHY

KURT STOCHHOLM, born in 1941 in AARHUS, DENMARK. ENGINEER B.Sc. 1968, Aarhus Technical Institute of Electrical Engineering.

Joined TERMA Elektronik AS since 1969 as Quality Assurance Engineer. Participated in the ESRO IV Housekeeping Project and as from 1971 full time P.A. manager for various space programmes.

As from 1st of March 1985 he is appointed as project manager for EURECA-SGF Power S/S and EGSE.

PUBLICATIONS

"Safety Assessment as a Part of a Company Product Assurance programme" EOQC Copenhagen 1976

"Safety and reliability Interferences" Sintom 11, Borgholm, Sweden, 1976

"Reliability Assessment/Evaluation of Results" DFK, Copenhagen 1979

"From Space Methodology to Consumer
Products"
ESA International Coloquium
Palais de l'Europe, Strasbourg
(France) April 1980

"Failure Mode Analysis & Fault Tree
Analysis tool for Optimizing Design"
DFK, Aarhus 1982

"Considerations on how to specify
Reliability Requirements"
EUROCON '82, Copenhagen 1982

**RELIABILITY ANALYSIS
OF OPTICAL SYSTEMS**

Reliability Technology — Theory & Applications
J. Møltoft and F. Jensen (Editors)
© Elsevier Science Publishers B.V. (North-Holland), 1986

RELIABILITY DEMONSTRATION FOR THE UK-BELGIUM
FIBRE OPTIC SUBMARINE COMMUNICATIONS SYSTEM

ROBERT H MURPHY
STC PLC Submarine Systems Limited,
Christchurch Way, Greenwich, LONDON UK

At the commencement of the reliability growth programme for the first international fibre optic submarine system, very little was known about the predominant failure mechanisms of the main components. It was soon realised that traditional techniques for reliability assurance, such as very high temperature accelerated testing and burn-in, would be inapplicable in some cases and more generally the scale of reliability demonstration exercises based purely on time to catastrophic failure testing would be prohibitively costly in both time and resources. This paper reviews the innovative techniques that were employed to overcome these problems and presents some of the results in detail on the major components.

1. INTRODUCTION

Prior to the development phase for the UK-Belgium fibre-optic digital communication system, a typical analogue submarine system took some years to implement if a major new active component was required to be incorporated. As an example, the NG 45MHz Frequency Division Multiplex (FDM) System, which demanded the fabrication and reliability assurance of the British Telecom 40 Series Transistor family[1], began its development phase in 1972 and entered service in 1977.

Analogue system reliability growth programmes were extremely challenging at the time, by virtue of the need to extend the frontiers of knowledge of UHF transistor performance, failure prediction and analysis techniques, packaging and mounting methods. The programmes revolved around the now traditional approach of testing very large numbers of the critical active devices to destruction under accelerated test conditions. The well known Arrhenius equation[2] was applied to extrapolate the results of three or more very high temperature overstress tests to the system operational environment, and a projected 25 year cumulative hazard figure (eg. 1 in 500 or 0.2%) so obtained.
The standard 'time-to-failure' descriptor employed in this exercise was the Log-Normal Cumulative distribution of failures (Cdf), but ultra-severe screening (burn-in, etc) had to be employed prior to life testing to ensure an acceptable degree of fit. The Log-Normal Cdf was an adequate representation of wear-out failures only and could not be readily incorporated in a mixed distribution describing the entire 'bath-tub' reliability characteristic for complex electronic devices. In retrospect, fairly arbitrary methods were employed to predict the absence of early-life and random failures (collectively, 'rogue' failures) in the working environment, but it must be stressed that history has proved that these methods were extremely effective. For example, the non-occurrence of a failure of any description in 40 Series transistors to-date in over 7,500 repeater-years of NG system operating experience, when combined with prior operational and shadow life test experience at the transistor manufacturer's laboratories, has confirmed a 90% Upper Confidence Limit (UCL) for the average failure rate to-date of better than 5 FITs (25 year Cumulative Hazard, 0.11%) for these transistors.

Whereas it had long been recognised that rogue failures would stand out in accelerated test regimes, by failing to fit the log-normal Cdf describing the wear-out portion of the bath tub curve, it was not (and still is not, in some circles) appreciated that they obey different laws with regard to the activation energies/acceleration factors involved. Analogue system reliability assurance largely circumvented this problem (albeit at extremely high cost in terms of the numbers of destructive samples involved in the tests) by employing such severe screening techniques on essentially homogeneous, dedicated facility devices that no early life failures were encountered in the final overstress certification procedure.
If the results did not fit log-normal Cdf's similar to those obtained on previous devices, entire batches were rejected and the lengthy qualification phase restarted after suitable corrective action.

In 1980, when fibre-optic submarine systems were first seriously considered[3][4] as suitable candidates for a vastly enhanced undersea network capable of interfacing with, and extending, the new Integrated Service Digital Networks (ISDN's) being proposed for terrestial communications throughout the World, several major challenges were already apparent:

- The necessary electro-optic and electronic components (laser diodes, receive modules, very high speed integrated circuits, etc) had not yet been developed into even moderately reliable devices, and certainly had no prior history of the kind demanded of submarine system active devices.

- Although analogue systems had taken a quarter of a century to reach the degree of perfection that enabled specifications requiring up to 25 years fault-free operation to be accepted with a high confidence of achievement, fibre-optic systems would demand these standards from the outset.

- Timescales for implementation would be extremely contracted; 5-7 years being completely untenable for development and reliability growth programmes in the accelerating pace of deployment of fibre-optic technology world-wide.

- Traditional techniques for submarine system reliability assurance would need drastic revision, partly because of these timescale constraints, and partly because some of the components (notably the laser diodes) could not even be operated under highly accelerated (high temperature) test conditions, let alone subjected to rigorous screening and overstress test regimes for long periods of time.

Less than two years later (April 1982), STC Submarine Systems Limited accepted a commitment to develop and supply the first international fibre-optic digital submarine system. This was to be a 280 Mbit/s, 3 fibre-pair link with 3 undersea repeaters between the UK and Belgium. It was specified to have an MTTF for the entire submerged plant of greater than 10 years and a probability of greater than 70% that there would be fewer than 3 failures in 25 years, based on 90% UCL component reliability predictions for this 'design life'.

THE UK-BELGIUM SUBMARINE SYSTEM

The System[5] comprises 112km of cable as a package for 6 single mode 1.3um optical fibres and the power feed arrangement for 3 submerged repeaters. Each repeater contains 6 one-way regenerator modules for the standard receiving, re-timing and regeneration (3R) functions at a line data rate of 280 Mbit/s.

Hence there are 3 fibre pairs each capable of handling the equivalent of 4000 telephone channels between the UK and the continent of Europe.

A regenerator module is shown schematically in Fig.1 which also indicates the component count.

RELIABILITY ASSURANCE AND PREDICTION TECHNIQUES

With regard to the following description of the overall reliability assurance methodology and prediction techniques that were employed for the UK-Belgium system, it is important to recognise the particular relevance of the three phases of the classical 'bath-tub' failure rate descriptor in the context of optical submarine systems.

Starting in the middle with the random failure rate portion, it has long been realised that, since the undersea environment is particularly benign from the constant temperature/ultra low mechanical and environmental stress viewpoints, random failures due to transient operating stresses exceeding the inherent 'strength' parameters of well constructed/highly screened components, simply do not occur at a significant level. Hence the prime tool of reliability engineers in other disciplines, viz. the FIT analysis technique, is totally inapplicable, since it is based on the 'a priori' assumption of constant failure rate regimes.

The 'wear-out failure' portion of the 'bath-tub' is potentially applicable to all components, when design lives exceeding 25 years are considered, and for many active devices (eg. laser diodes), 'wear-out' usually means a drift with time of one or more major device parameters in the direction of a known system tolerance limit, with the implication that this limit will eventually be exceeded. However, it also encompasses the more traditional concept of catastrophic failure due to the degradation of the inherent

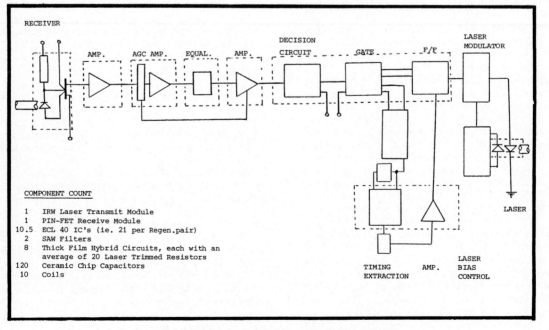

FIG.1 REGENERATOR SCHEMATIC & COMPONENT COUNT

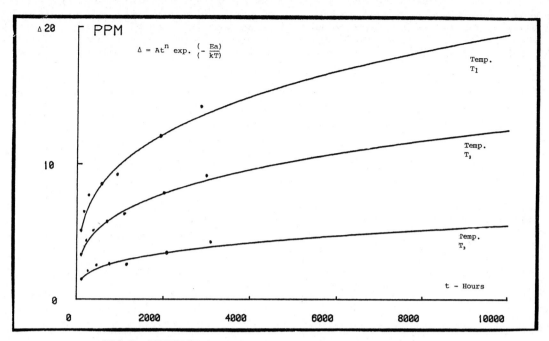

FIG.2 DEGRADATION ANALYSIS & PREDICTION TECHNIQUE

'strength' of the component until the latter reaches a point where it can be overcome by a continually operating stress in the system environment.

The 'early-life failure' portion is hopefully at a very low residual level for a fully screened product, but must be specifically confirmed to be so, and accurately identified, on all major components. Any higher than expected residue of early life failures is usually catastrophic in the context of virtually un-repairable systems or systems where repair is extremely costly in terms of both the physical demands and the lost revenue aspects. The nature and extent of both the early-life and wear-out failure portions must also be thoroughly evaluated from the viewpoint of planning the screening techniques in the first place, because there is no point in devising an ultra-severe and efficient screen for a particular component if this leaves very little of the wear-out life for system operation.

Hence the extensive pre-qualification reliability growth programmes on all major components for the UK-Belgium system followed the general philosophy[6] of identifying relevant failure modes, redesigning active element and packaging configurations where necessary, repeating accelerated life tests and developing suitable screen limits by isolating early-life failure occurrences with respect to main wear-out failure distributions.

COMPONENT CATAGORISATION

Based on previous experience with analogue system components, some of whicn were carried over into optical systems, it was found convenient to divide the components into 3 catagories according to whether they were known to be:-

A. Susceptible to both wear-out and early life failures.

B. Predominantly susceptible to wear-out failures.
C. Predominantly susceptible to early-life failures.

Catagory (A), which included laser diode transmit modules, PIN-FET receive modules, ECL40 Bipolar integrated Circuits and silicon zener diodes, comprised those active devices that were entirely new to the submarine systems field, or were known to have early-life and wear-out failure distributions that were likely to interact at the ultra-low failure level required during the system design life. Hence this catagory needed careful attention from a screen development viewpoint.

Although Catagory (B) components, which included thick film hybrid circuit resistor elements and Surface Acoustic Wave (SAW) filters, would of course also be susceptible to rogue failures, it was known that very few early-life catastrophic failures had ever been experienced on well designed, fabricated and screened passive devices of this type. Hence rogue failures in this catagory could safely be ignored in the context of the component numbers to be deployed in the system.

Similarly Catagory (C) components, which included capacitors, coils, thick film crossover elements, fibre and electrical joints, etc, were known to have wear-out lives that occurred on 'geological' timescales, whereas the level of deployment implied that rogue failure would be by far the more important consideration.

WEAR-OUT LIFE PREDICTION FROM PARAMETER DEGRADATION

At the outset of the UK-Belgium development programme, it was known that the semiconductor laser diodes fabricated to date, and those feasible within the foreseeable future, were limited by maximum operating temperatures and

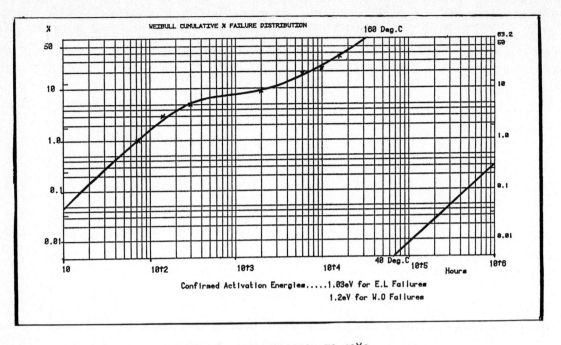

FIG.3   PIN DIODE Cdf & TRANSLATION TO 40°C

relatively low activation energies of predominant failure mechanisms to genuine acceleration factors for overstress testing to catastrophic failure in the range 1.4 to 16.

This implied that several years, or even decades, of accelerated testing on the traditional attributes basis would have been necessary before confidence could be firmly established in the wear-out life of the product.

Fortunately an alternative method[7] (Fig 2) was available and ultimately proved to be very successful in resolving this problem. The method, which was already being employed for thick film resistor and SAW filter wear-out life prediction, involved the very accurate measurement of the degradation profiles of major parameters under mildly accelerated conditions, and the fitting of these measurement points to a known degradation law

which incorporated a temperature acceleration factor. This curve fitting, which was carried out by the 'least-squares' method with calculable upper confidence limits on the spread of residuals (differences between measured and predicted levels at each time interval) was preferably applied to the means and standard deviations of the parameters for entire batches of components, but could also be applied on an individual device basis. Examples of the above are threshold current and power output for constant monitor current on laser diodes, resistance values on laser-trimmed thick film resistors and centre frequency on SAW filters.

As it turned out, the temperature acceleration factor in the prediction law for the laser diodes was even lower than expected and made a negligible contribution (except as a safety factor) to the increased confidence in the 25

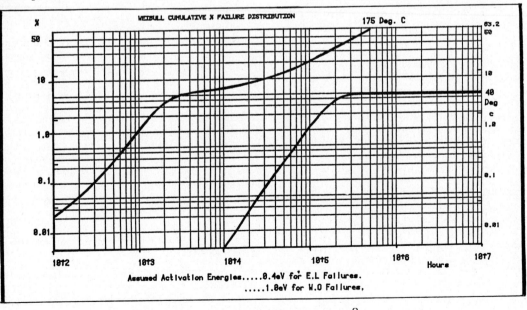

FIG.4   ZENER DIODE Cdf & TRANSLATION TO 40°C

year life predictions. Hence it was ignored on an individual device basis, and although the life testing was carried out at 50°C, the extrapolation was assumed valid (and conservative) for the 35°C maximum operating case temperature in the system environment.

The work on the complete laser module package was supported by up to 23,000 hours of continuous testing at 50°C on submounted laser diode chips under typical operating conditions (5mW CW power output) and this showed a continuous decrease in the residuals (within measurement accuracy) throughout the entire period, hence validating the approach to within a logarithmic decade of the system design life.

The thick film resistor and SAW filter evaluations extended to 30,000 hours and 8,000 hours respectively, and proved extremely useful in setting the burn-in limits for the allowable parametric drift. For example, detailed analysis of the high temperature burn-in results for thick film resistors showed that the allowable drift for mid-range values during the 500 hour burn-in period could be as high as 0.71% before the total screen plus operating life drift would exceed the system tolerance limit set at 1% over 25 years. However, selection for less than 0.2% drift during burn-in was demonstrated to be well within the process capability for overglazed and laser-trimmed parts within the resistor paste series chosen. Hence this figure was applied, with the net result that the total compliment of resistors could be safely considered ineffective (25 year cumulative hazard less than 0.01%) with regard to their contribution to the overall system reliability prediction.

USE OF THE MIXED WEIBULL Cdf

The second major catagory A device, viz. the PIN-FET receive module, comprises an InGaAs P-Intrinsic-N photodetector diode in series with a GaAs Field Effect Transistor to form an extremely sensitive and wideband integrating front end to the regenerator module. For the purpose of preliminary reliability evaluation and screen development, it was decided to assess separately the PIN and the FET in small hermetic packages, which (although unrepresentative of the final sub-mount configuration in a single module) would give an indication of the likely time-to-failure characteristics of these components.
To this end, over 16,000 hours of overstress testing at temperatures up to 160°C were accumulated on the PIN, and over 8000 hours at temperatures up to 200°C on the FET.
These temperature-time profiles were more than adequate to provide a detailed description of the cumulative hazard functions for both devices at the relatively high temperatures employed, and served to resolve the activation energies for the predominant failure mechanisms relevant to both early-life and wear-out failures.

The Cdf for the PIN diode [see Fig.(3)] turned out to be of particular interest, as it fitted neither the simple Log-Normal nor the simple

Weibull characteristic within the limits of any of the standard tests for 'goodness of fit'. The mixed Weibull distribution[8] was the obvious candidate for the resolution of this difficulty, and computer programmes were rapidly developed for exploring the implications of this. In retrospect, it proved surprisingly easy to handle this type of time-to-failure distribution, and to resolve the early-life and wear-out portions as separate entities, each with its own acceleration factor. Early-life failures had previously been assumed to conform to a decreasing failure rate pattern (Weibull shape parameter <1), but this was found to be not the case. In conjunction with the wear-out distribution, a complete picture of the 'bath-tub' for this component was built up and the implications for the necessary temperature-time profile of the burn-in screen became immediately obvious.

However, there was yet another hurdle to face. This concerned the fact that the screen could not be made 100% efficient in removing early-life failures within the constraints of the allowable temperature-time profiles of the exposed (sub-mounted) PIN diode before assembly into the final package, or during burn-in on the latter after assembly. Taking full advantage of the descriptive value of the mixed Weibull distribution, this problem was resolved by striking a fine balance between the severity of the two screens and the allowable residual hazard due to early-life failures in the system environment. Since a high confidence existed by now on the conservative acceleration factors that could be separately applied to derive the early-life and wear-out failure levels at the operational temperature in the system, it was relatively straightforward to calculate an acceptable 25 year cumulative hazard figure from the results for fully screened product subjected to these overstress tests.

The FET reliability evaluation proceeded along similar lines, but fortunately there was not a significant early-life failure contribution in this case, and the Weibull distribution for the wear-out portion was merely combined with that for the PIN diode to obtain a composite characteristic for the complete module. This was later confirmed, as far as possible within the overstress constraints imposed by the assembly procedures for the receive module, by subjecting the latter to long term operational life tests.

Following the success of this exercise, the mixed Weibull distribution was applied to other devices in the Group A catagory, and proved to be of immediate value. The zener diodes were found to have an early-life failure characteristic as shown in Fig.(4) on high temperature (200°C) overstress tests, which, although at an extremely low level, was found to be far more significant than the wear-out failure distribution. The latter was projected (from previous experi- ence of the activation energy involved) to occur well beyond the system design life at the normal operating temperature. Although acceptable within the constraints of the planned

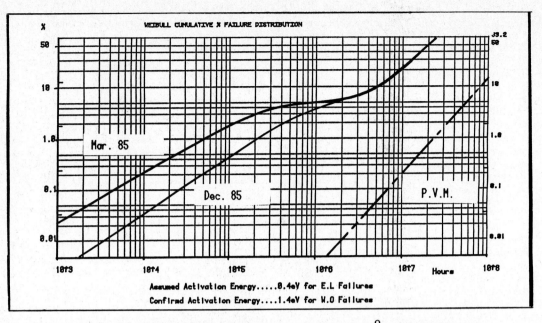

FIG.5 90% UCL UPPER BOUNDARIES FOR ECL 40 IC 40°C PREDICTION Cdf.

cumulative hazard for the UK-Belgium system, it was shown that future long-haul systems could be placed in jeopardy by this early-life failure characteristic, and steps were taken immediately to redesign the screening procedures for zeners on the evidence provided. The challenge for the ECL 40 ICs was of a slightly different nature, but the Mixed Weibull distribution proved to be equally effective. In this case, a vast amount of prior reliability data had been accumulated on 40 Series transistors, on which the technology for the ICs was based (even to the extent of employing exactly the same diffusion profiles, metallisation system and design rules where applicable). Hence the time domain for the likely wear-out failure characteristic and the appropriate activation energy for the predominant failure mechanism were already identified and well understood. A comprehensive step and over-stress test programme on a large number of Process Validation Module (PVM) transistors (integrated with every IC chip and in the final chip carrier style device package) rapidly confirmed this. Data on overstress tests for the complete range of circuit functions involved was potentially rather limited, because of the difficulty of obtaining meaningful operation at high temperatures and the life test jig/fixture complexity required. Nevertheless, tests were carried out at 145°C for 11,000 hours on two circuit types, and in conjunction with the 96 hour, 160°C burn-in and 1000 hour, 60°C operating life tests on over 660 devices covering the complete family with zero failures. The mixed Weibull distribution was then employed to place an upper boundary on the Cdf for the ICs [see Fig.(5)]. The very low (pessimistic) activation energy of 0.4eV was assumed for the early-life failure portion, and the previously evaluated figure (1.4eV) was applied to the wear-out characteristic to obtain an acceptable combined distribution for system life prediction purposes.

In the course of this work it was discovered that for all components, the simple or mixed Weibull Cdf's provided an extremely cost-

COMPONENT	QUANTITY @ TEMP.DEG.C	LIFE TEST HOURS (SPEC.CONDS)	TECHNIQUES EMPLOYED IN ANALYSIS				90% UCL 25 YEAR HAZARD	
			ATTRIBUTES TEST	DEGRADATION PREDICTION	SIMPLE WEIBULL	MIXED WEIBULL	PER COMPONENT	PER REG.
LASER MOD.	88 @ 50 36 @ 150 5 @ 2 5 @ 25 87 @ 50	22KHrs(Chip) 14KHrs(Chip) 4KHrs 4KHrs(Stg) 4-16KHrs		✓		✓	0.53%	0.53%
PIN FET MOD.	200 @ 100-160 200 @ 100-200 50 @ 90	16KHrs(PIN) 8KHrs(FET) 4KHrs(MOD)	✓			✓	0.07%	0.07%
ECL 40 ICs	400 @ 60 890(Matrix) 20 @ 145 660 @ 160	1KHrs 5KHrs(PVM) 16KHrs 96Hrs(Burn-in)	✓			✓	1.09%	10.91%
ZENER DIODE	48 @ 60 50 @ 175 51 @ 250	2KHrs 2KHrs 2KHrs	✓			✓	5.80%	2.90%
SAW FILTERS	40 @ 85 46 @ 40	12KHrs 1-2KHrs		✓	✓		<0.01%	Ineff.
CAPACITORS T/F RESISTS X-OVERS,etc	3000(Matrix) 150 @ 100-200 10 Substr.(Mat.)	30KHrs	✓ ✓ ✓	✓	✓ ✓ ✓		<0.01%	0.64%

TABLE I   FORMAL LIFE TEST SUMMARY

effective solution for estimating hazards or placing upper boundaries on hazards for system prediction purposes. They also formed better representations than the previously revered log-normal distributions and, since the mathematics involved are rather straight-forward in comparison, it is intended to adopt them for most future component evaluations, if they remain appropriate.

COMPONENT LIFE TEST/QUALIFICATION PROGRAMMES AND RESULTS

Table 1 provides a comprehensive summary of the component life test programmes completed for the UK-Belgium system, the reliability assessment and prediction techniques employed and the outcome of these programmes in terms of the 90% UCL 25-year cumulative hazard figures derived.

In most cases the life tests themselves were conducted by the component suppliers after the completion of the engineering evaluation and screen development phases, which were extensively monitored and audited by STC Submarine Systems Limited component engineers. The entire manufacturing and test sequences were developed in close collaboration with these engineers, and specifications mutually agreed on an iterative basis. When finalised, the manufacturing and test sequences were subjected to comprehensive engineering and QA audits by STC and British Telecommunications plc as the approval authorities.

As may be seen from the table, the life tests ran for over 2 years in some cases. Many lessons were learned about electro-optic component handling/testing procedures/calibration and repeatability aspects of 'state-of-the-art' electro-optic measurement techniques/electro-static damage precautions and data reduction on a large scale. The results of the life tests were continuously analysed and updated for both internal and external (system customer) reviews. This analysis was largely carried out by STC with, of course, regular feedback to the component manufacturers, and feed-forward from them on device technology/failure analysis/prediction techniques and anomalous results.

Comprehensive mechanical and environmental qualification approval exercises were carried out on all components by the manufacturers, in parallel with the life test exercises, and the results obtained fully supported the effectiveness of the design procedures and corrective actions taken during the engineering development and evaluation phases. Of particular relevance to the laser and PIN-FET receive modules, hermeticity and internal package atmosphere analyses were performed on all assembly batches, and specifically moisture, oxygen and hydrogen contaminations were tightly controlled to the 'few hundred ppm' levels. This was accomplished by the use of relatively severe pre-sealing bake-out procedures, which in turn had necessitated package design with epoxy-free internal assembly techniques, all metal fibre glands and halar coated fibre tails able to withstand the high temperature-time profiles involved.

The thick film hybrid circuit qualification exercise deserves special mention since it followed a 'capability approval' format rather than the evaluation of the specific circuits used in the system.

After the manufacturer's agreement of the design rules and processing parameters, based on the prior work in a comprehensive proto-typing and techniques development laboratory at STC, six Process Validation Module Circuits (PVMCs) were designed and manufactured in quantities ranging up to 150 of each. These PVMC's embodied the worst case aspects of all of the design rules. For example, one set encompassed over 2m of conductor tracks at minimum allowed width and spacing and resistor patterns of the minimum and maximum allowed aspect ratios, all maximally laser trimmed. Another set had over 900 track cross-over points with critical dimensions, and a third set used all of the component mounting techniques (with specially wire-bonded internal configurations for the chip-carriers) to detect short- and open- circuits that might be caused by the severe handling during mounting. The last of the PVMCs was a typical circuit employed in the system, and as such was fully testable at the line bit-rate before and after complete mechanical, environmental and endurance testing at levels far higher than those to which the actual system components would be subjected. There were no failures and few anomalous results in the entire capability approval exercise, and the anomalous results were all explainable by extraneous problems such as faulty test jig construction or verified test equipment error. Tests were, of course, repeated, after appropriate corrective action in these cases.

SYSTEM PREDICTION FROM THE LIFE TEST RESULTS

The Cdf curves for the various components depicted the actual test results from the (normally) accelerated test regimes conducted in the relevant reliability growth programmes, as modified by the acceleration factors that had been evaluated in the course of these accelerated tests. To obtain a combined cumulative hazard prediction for each regenerator on a more appropriate timescale (see Fig.6) it was necessary to:

- Translate the relevant Weibull scale parameters from 'hours' to 'years' in each case and check that the 25-year cumulative hazard parameters did, in fact, agree with the previously calculated values.

- Apply a 90% UCL to each curve, because otherwise sample destructive test results could not be said to truly reflect the appropriate reliability predictions for related product in system deployment.

Having accomplished these translations, the combined cumulative hazard plot gave a graphical insight into the relative contributions of the various components, and the following comments are appropriate:

- By far the largest contribution to the early-life hazard was the curve for the ICs.

This was mainly due to lack of data at the time this curve needed to be finalised, and has since shown a rapid reduction as the long-haul (TAT-8) system burn-in and operating life test results became available.

- Where curvature of the individual component Cdf plots occurred beyond 25 years but prior to 100 years, this was ignored as being irrelevant to the UK-Belgium design life predictions. Hence the combined Cdf curve, and the following system probability plot (fig 7) are distinctly pessimistic in this region.

- In most cases, as expected, the wear-out portions of the individual Cdf plots were ineffective, and only the early life failure predictions contributed to the 25 year cumulative hazard per regenerator. It was necessary to establish their location and parameters to prove this, but having done so, they may now be safely ignored for future short-haul systems.

Fig.7 illustrates the probability curves for C=0,1,2,3 failures versus time for the 18 regenerators of the UK-Belgium system, as derived from the cumulative hazard plot for Fig.6. This shows that there is a 72.9% probability of 3 or fewer failures in the 25-year design life. The MTTF is calculated from the integral of the C=0 curve on Fig.7 to be 11 years. These figures comfortably exceed the system reliability specification, but even so, should be confirmed as demonstrably pessimistic as the reliability growth programmes continue for the medium and long-haul systems now under consideration.

FIG.6 UK-BELGIUM COMPONENT HAZARDS PER REGENERATOR

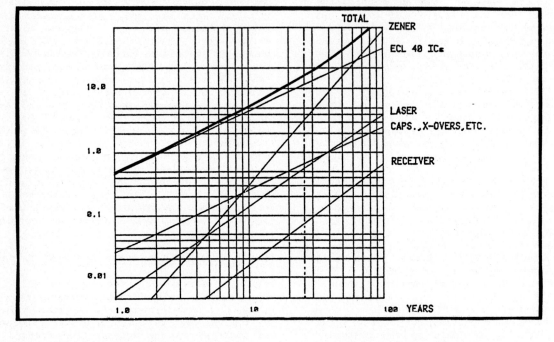

FIG.7 UK-BELGIUM PROBABILITIES OF 0,1,2 & 3 FAILURES VERSUS SYSTEM LIFE

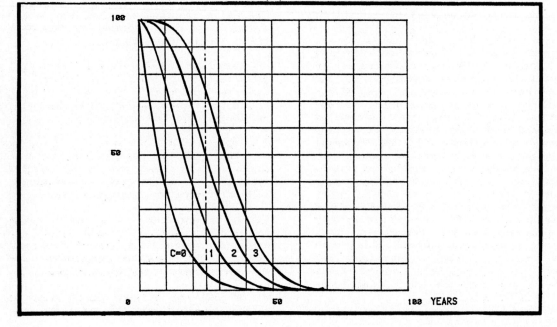

REFERENCES

[1] D W Brown, A J Tew, R J D Scarborough: <u>Submarine System Transistors, Type 40 Testing for Reliability</u>. IEE Conference Publication No.183 - Submarine Telecommunications Systems, 1980 Feb.

[2] J M Groocock: <u>Accelerated Life Testing and Overstress Testing of Transistors</u>. Electronics Reliability and Microminiaturisation. 1963, May.

[3] P E Radley, A W Horsley: <u>Technology Trends for Future Optical Submerged Systems</u>. IEE Conference Publication No.183 - Submarine Telecommunications Systems, 1980, Feb.

[4] P Runge: <u>A High Capacity Optical-Fibre Undersea Communication System</u>. VI ECOC, York, 1980, Sept.

[5] J P Burvenich, R L Smith, A J Jeal: <u>The First European Optical Highway; UK-Belgium No.5 System</u>. Sub-Optic 86, Paris, 1986 Feb.

[6] R H Murphy: <u>Reliability Growth Programme for Undersea Communications Systems</u>. IEEE Transactions on Reliability, Vol.R-32, 1983, Aug.

[7] R H Murphy: <u>The Application of Degradation Models to Laser Diode Wear-Out Life Prediction</u>. IX ECOC, 1983, Sept.

[8] J H K Kao: <u>A Summary of Some New Techniques on Failure Analysis</u>. 6th National Symposium on Reliability and Quality Control. 1960, Jan.

AUTHOR

Robert H Murphy; STC Submarine Systems Limited Christchurch Way, Greenwich, London SE10 0AG ENGLAND.

Robert H Murphy was born in London, England in 1935. He received his BSc(Hons) degree in Electronics Engineering from Queen Mary College, University of London, in 1958. From 1964 to 1973 he was engineering manager at the European transistor and integrated circuit manufacturing facility of Transitron Electronic Corp., and was responsible for all aspects of semiconductor device technology from design and development to quality control and reliability assurance. He joined the Submarine Systems Division of STC PLC in 1973 and is project manager for Component and Hybrid Circuit Development. In addition to overall responsibility for electronic and electro-optic device development, specification and system deployment, this position involves management of reliability growth and prediction programmes for new systems; specifically those applying advanced technology to fibre-optic communications.

PLENUM SESSION

Reliability Technology — Theory & Applications
J. Møltoft and F. Jensen (Editors)
Elsevier Science Publishers B.V. (North-Holland), 1986

RELIABILITY MODELS FOR REPAIRABLE SYSTEMS

Harold E. ASCHER

Naval Research Laboratory
Washington, D.C., USA

Probabilistic modeling of repairable systems is presented, followed by a discussion of the shortcomings of such models. Considerable emphasis, therefore, is placed on the need for more, and better performed, statistical analyses of repairable systems failure data. There are fundamental differences between the analyses of repairable and non-repairable systems data. As stressed by O'Connor [12], however, these essential differences have usually been overlooked, or incorrectly addressed, in reliability texts and papers. Great emphasis will be placed on explaining why basic-and inherently simple-concepts have so often been misunderstood.

1. INTRODUCTION

The Provisional Programme flyer for this Conference introduced the Advanced Professional Seminars in the following terms:

> The Advanced Professional Seminars will allow reliability engineers to discuss new information or new theories in reliability engineering. In many ways traditional reliability engineering has reached a point where a new, sometimes unconventional and controversial look, is called for. For instance: Is there really a period of constant hazard rate? What are the trade-offs between component and equipment screening? How do we best deal with the analysis of repairable systems?

This paper will address the questions, "Is there really a period of constant hazard rate?" and "How do we best deal with the analysis of repairable systems?" Approaches for answering these questions, particularly the latter one, will be presented in terms of *standard* techniques which have been developed over a period exceeding two centuries. The Programme statement that these techniques are "unconventional and controversial" in the reliability field, is nevertheless, *accurate*. Considerable emphasis will be placed, therefore, on the reasons why many basic-and often simple-concepts have been either ignored or poorly understood/interpreted by the reliability community.

Although this paper is primarily concerned with repairable systems, it begins with a very brief discussion of models for non-repairable items, in order to establish some fundamental concepts and terminology. Some basic models for repairable systems are introduced, followed by a discussion of major omissions from existing-and probably future-models. A very brief examination of statistical techniques for repairable systems is then presented. This is followed by two examples which show how the concepts, terminology and notation of the reliability field are inadequate to *describe* two simple situations, much less provide the framework for their resolution.

After the above background material is covered, the two questions, Is there a period of constant hazard rate? and How do we best deal with the analysis of repairable systems? will be addressed. It will be shown that the former question, as phrased, does not pinpoint whether nonrepairable or repairable systems are being considered. Morever, the resolution of this point has a very important bearing on how to proceed with answering the question. The latter question will be investigated from two viewpoints: 1) use of the many existing techniques and 2) presentation of techniques and results in books and papers in such a manner as to make it clear that repairable systems are being treated.

Space limitations preclude the presentation of a detailed survey of the modeling of repairable systems here. Many important concepts are discussed only briefly or even ignored. Others are presented in simplified form because of the space constraint, as well as for clarity of exposition. The reader is referred to [1] where almost all of the following material is covered in much greater detail.

2. MODELS FOR NONREPAIRABLE ITEMS

In this section, some basic notions for nonrepairable items, henceforth called parts, will be presented.

2.1 Force of mortality (FOM) and probability density function (PDF) of a part

If X is defined as the random variable (RV), time to failure of a part, then $F_X(x) \equiv Pr\{X \leqslant x\} \equiv$ distribution function of X. Assuming F to be absolutely continuous, then $h_X(x) \equiv \dfrac{f_X(x)}{1 - F_X(x)} \equiv$ FOM of X, where $f_X(x) \equiv F'_X(x)$ and $1 - F_X(x) \equiv R_X(x) \equiv Pr$ {part survives past time x}. The FOM is best interpreted by means of the following equation:

$h_X(x)\,dx = Pr\{x < X \leqslant x + dx \mid X > x\}$ i.e., $h_X(x)\,dx$ is the conditional probability that a part, from the population with distribution F, put into service at $x = 0$ and known to have survived until x, fails in $(x, x + dx]$.

In the terminology of Elandt-Johnson and Johnson [2, pp. 12-14] $h_X(x)$ is a *relative* rate since at time x, the derivative of $F_X(x)$ is divided by the probability of surviving past time x. In the terminology of Elandt-Johnson and Johnson [2, pp. 12-14 and p. 51] the probability density function (PDF) $f_X(x) \equiv F'_X(x)$ is an *absolute* rate, since it is defined as a derivative with no normalization.

It has often been claimed that the FOM is a conditional PDF. If this were true, the demarcation between the FOM as a relative rate and the PDF as an absolute rate would be, at best, blurred. The fact that the FOM is not a density, conditional or otherwise, can be seen from the following well known relationship,

$$R_X(x) = e^{-\int_0^x h_X(y)\,dy}$$

The requirement that $R_X(\infty) = 0$ implies that

$$\lim_{x \to \infty} \int_0^x h_X(y)\,dy = \infty, \text{ rather than } 1.$$

2.2 Mean time to failure

The mean time to failure (MTTF) of a population of parts with failure law F and PDF f is

$$E[X] \equiv \int_0^\infty x \, f_X(x) \, dx \equiv \mu.$$

2.3 Part wearout

For the purposes of this paper, we will say that a part wears out if it is a member of a population with failure law F such that $h_X(x)$ is a strictly increasing function of x. Then, in terms of the previously given interpretation of $h_X(x)\,dx$, a part wears out if the conditional probability of failure in an interval of fixed length dx, $(x, x + dx]$, increases as the part's age x increases.

3. MODELS FOR REPAIRABLE SYSTEMS

3.1 Introduction

We will begin by defining, "repairable system," following which the natural model for such a system, the "stochastic point process" will be introduced. Several point process models will be discussed and examples of their suitability/unsuitability for modeling repairable systems will be given. The section concludes with a brief examination of the shortcomings of the probabilistic models presented here and elsewhere in the reliability literature.

3.2 Definition of repairable system

Any system which, after failing to perform at least one of its required functions, can be restored to performing all of its required functions by any method, other than replacement of the entire system, is a *repairable* system.

Comment 1: The above definition is worded to include the possibility that no parts are replaced. For example, the system might be repaired by an adjustment or by a well directed kick.

Comment 2: Some systems have redundant paths which are repaired. If such a system is discarded as soon as it fails to perform at least one of its required functions satisfactorily, it is *not* considered to be a repairable system. That is, such a system is a nonrepairable system with redundant, repairable *sub*-systems. Henceforth, when we refer to a system, we will mean a repairable system.

Comment 3: It is empirically obvious that most systems are designed to be repaired, rather than discarded, after failure. Hence, the concepts and techniques surveyed in this paper pertain to most systems.

3.3 Stochastic point processes

A stochastic point process is a mathematical model for a physical phenomenon characterized by highly localized events distributed randomly in a continuum. In the present application, the continuum is time and the highly localized events are failures which are assumed to occur at instants within the time continuum. We will assume that repair times, which usually are short compared to up times, are negligible; i.e., repair times will be ignored throughout the paper. Fig. 1 shows a portion of a sample path of a stochastic point process, representing the successive failures of a single system.

3.3.1 Arrival and interarrival times and rate of occurrence of failures

T_i, $i = 1,2,3,\ldots$, measures the total time from 0, a convenient fixed origin, to the ith failure and is called the *arrival time* to the ith failure. T_i is a random variable (RV). The real variable t measures the total time since the startup of the process,

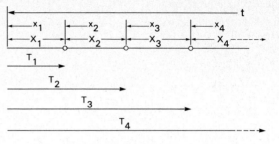

Fig. 1 System arrival and interarrival times

i.e., time from the fixed origin. We will use the term "global" time for t.

X_i, $i = 1,2,3,\ldots$, is the *interarrival time* between the $(i-1)$st and ith failures, where $X_0 \equiv 0$. X_i is an RV. Since the origin for X_i is the arrival time of the $i - 1$st failure, we will say that the X_i's are *chronologically ordered*. The real variable x_i measures the time elapsed since the most recent failure, i.e., $x_i = t - T_{i-1} \mid T_{i-1} = t_{i-1}$ where $T_0 \equiv 0$. We will use the term "local" time for x_i. Note that the x_i's are analogous to x for part models but t has no direct analogy in part modeling. Clearly, $T_k \equiv X_1 + X_2 + \ldots + X_k$. Then the RV, $N(t)$, is defined as the maximum value of k for which $T_k \leqslant t$, i.e., $N(t)$ is the number of failures which occur during $(0, t]$. $\{N(t), t \geqslant 0\}$ is the integer valued *counting process* which includes both the number of failures in $(0, t]$, $N(t)$, and the instants T_1, $T_2\ldots$, at which they occur. The expected value of $N(t)$ is denoted $V(t)$, i.e., $V(t) \equiv E[N(t)]$. We will assume that $V(t)$ is absolutely continuous. Its derivative, $v(t) \equiv V'(t)$, is the time rate of change of an expected number of failures and therefore, will be termed the rate of occurrence of failures (ROCOF). The ROCOF is obviously an *absolute* rate in the sense of [2, p. 12]. ROCOF must not be confused with "failure rate" which is usually defined to be equivalent to the *relative* rate, force of mortality $(h_X(x))$ *but in practice, is also widely used in the ROCOF sense*. The ROCOF has the following simple interpretation: $v(t)dt$ is the probability that a failure, not necessarily the first, occurs in $(t, t + dt]$. This contrasts sharply with the interpretation of $h_X(x)\,dx$ which is the *conditional* probability of first and only failure in $(x, x + dx]$, given survival to x. In the latter case, the condition is essential since it is meaningless to consider the probability of failure of a part after time x, if it has already failed and been discarded before that time. Because $v(t)$ is defined as the derivative of an expected number of failures, $\int_0^{t_0} v(t)dt = E[N(t_0)]$, the expected number of failures in $(0, t_0]$.

We can define a force of mortality, $h_{X_i}(x_i)$, of each of the interarrival times, X_i, $i = 1,2,\ldots$, in a manner completely analogous to the FOM of parts, i.e.,

$$h_{X_i}(x_i) \equiv \frac{F'_{X_i}(x_i)}{1 - F_{X_i}(x_i)}.$$

We emphasize that the FOM is a property of each RV (interarrival time) rather than a sequence of RV'S (stochastic point process). Note that x_i is measured from T_{i-1}; this is necessary since $h_{X_i}(x_i)dx_i = Pr\{$failure in $(x_i, x_i + dx_i]\mid$ survival to $x_i\}$. Naturally, there is considerable connection between the properties of a stochastic point process and its constituent interarrival times. However, as we will show in section 5.2, information about some properties of interarrival times, e.g., knowing that each has strictly increasing FOM, is *not* very useful without additional knowledge about the underlying point process.

3.3.2 Independent increments

A counting process $\{N(t), t \geqslant 0\}$ is said to have independent increments if for all $0 < t_1 < \ldots < t_k$, $k = 2, 3, \ldots$,

$N(t_1) - N(0), \ldots, N(t_k) - N(t_{k-1})$ are independent random variables. In words, if a counting process has independent increments, then the number of failures in an interval is not influenced by the number of failures which occurred in any strictly earlier interval (i.e., with no overlap).

3.3.3 Repairable system deterioration (or improvement)

For the purposes of this paper, we will say that a system is deteriorating if its successive interarrival times are tending to become smaller. This corresponds -roughly- to strictly increasing ROCOF. The correspondence is only approximate, since the properties of early interarrival times, especially X_1, may make the ROCOF nonmonotonic, even if there is a tendency towards more frequent failures. We will say that a system is improving if its interarrival times are tending to become larger, corresponding, again roughly, to strictly decreasing ROCOF. Some of the subtleties in establishing rigorous definitions for deterioration/improvement are discussed in [1, pp. 25-26, 37-46, 169-171].

We note that what is called improvement above, is roughly synonymous with "reliability growth." As discussed in [1, p. 100], however, a distinction is sometimes made between these terms, so we will consistently use the term, improvement.

3.4 Homogeneous Poisson process (*HPP*)

The most straightforward way to define the *HPP* is as a nonterminating sequence of independent and *identically* exponentially distributed X_i's. We stress the necessity of the condition that the X_i's must be identically distributed, i.e., have the same mean, since this requirement has often been overlooked. There are several equivalent definitions of the *HPP*. In order to show how the nonhomogeneous Poisson process (*NHPP*) is a direct generalization of the *HPP*, we also give the following definition:

The counting process $\{N(t), t \geqslant 0\}$ is said to be an *HPP* if

(a) $N(0) = 0$

(b) $\{N(t), t \geqslant 0\}$ has independent increments

(c) The number of events (in our context, failures) in any interval of length $t_2 - t_1$ has a Poisson distribution with mean $\rho(t_2 - t_1)$. That is, for all $t_2 > t_1 \geqslant 0$,

$$Pr\{N(t_2) - N(t_1) = j\} = \frac{e^{-\rho(t_2 - t_1)}\{\rho(t_2 - t_1)\}^j}{j!},$$

for $j \geqslant 0$. From condition (c) it follows that

$$E\{N(t_2 - t_1)\} = \rho(t_2 - t_1)$$

where the constant, ρ, is the rate of occurrence of failures (ROCOF).

The *HPP* is the appropriate model for a socket, where parts from the same exponentially distributed population (i.e., with the same mean) are inserted one after the other, as the previous part fails. It also is assumed that the total stress exerted on the successive parts does not change with cumulative operating time, t.

3.5 Renewal process

A renewal process is defined as a nonterminating sequence of independent, identically distributed (IID) nonnegative random variables, $X_1, X_2, \ldots$, which with probability one are not all zero. Hence, it is a direct generalization of an *HPP*.

An example where a renewal process is an appropriate model is a socket in which light bulbs are operated one after the other under the following conditions: 1) the bulbs are from the same population and 2) the stress, say voltage and operating temperature, does not change with increasing t. We will

assume that instead of constant FOM which would result in an *HPP*, each bulb has strictly increasing FOM, i.e., each bulb wears out.

A renewal process, in general, is **not** a good model for a repairable system. The fact that most repairs involve the replacement of only a small proportion of a system's constituent parts, makes it implausible that such repairs renew the system to its original condition. This assertion will be bolstered by means of the following example.

Assume that you are on "Honest" John's used car lot. He takes you over to a tired looking old vehicle. When you ask, "How old is this 'hunk of junk'?" he replies, "It wouldn't start two days ago so we charged the battery. It starts all right now, so the car is two days old." Clearly, this answer is absurd, but if the car were modeled by a renewal process, it would be two days old!

3.6 Superimposed renewal process (SRP)

A *superimposed renewal process* (SRP) is developed as follows. Assume that *n* renewal processes are operating independently of each other, see fig. 2. Then the process formed by the union of all events, as shown on the bottom line of fig. 2, is known as an SRP. In general, the SRP will *not* be a renewal process. In fact, Çinlar [3] shows that if the superposition of two independent renewal processes is a renewal process then all three processes must be *HPP's*. However, it has been shown, e.g., by Drenick [4], that the superposition of an infinite number of independent renewal processes, all of which have been operating for an infinitely long period, becomes an *HPP*.

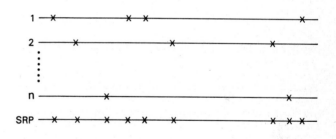

Fig. 2 Superposition of renewal processes

The SRP is sometimes an appropriate model for a repairable system. As discussed in section 3.5, a renewal process is a good first order model for a socket. Then, if all parts in the system are in series and they operate, fail and are replaced independently of each other, the system's failures are modeled by an SRP. This result has often been used to justify an *HPP* as the almost universal model for a system. This is not the case, however. A paper [5] titled, "The curse of the exponential distribution in reliability," shows that many systems do not operate long enough to reach the steady state *HPP* model. This point will be discussed further in section 6 in connection with the question, "Is there a period of constant ROCOF?" That section will also stress that a renewal process is **not** necessarily the appropriate model even for a socket.

If "Honest" John's car were modeled by a renewal process it would be two days old. If it were modeled by an *HPP*, it would have no effective age at all!

3.7 Nonhomogeneous Poisson process (NHPP)

The NHPP differs from the HPP only in that the rate of occurrence varies with time rather than being a constant. That is, conditions (a) and (b) of section 3.4 are retained and condition (c) is modified to be:

(c') The number of failures in any interval (t_1, t_2) has a Poisson distribution with mean $\int_{t_1}^{t_2} \rho(t)\,dt$. That is, for all $t_2 > t_1 \geqslant 0$

$$Pr\{N(t_2) - N(t_1) = j\} = \frac{e^{-\int_{t_1}^{t_2} \rho(t)\,dt} \left\{ \int_{t_1}^{t_2} \rho(t)\,dt \right\}^j}{j!}$$

for $j \geqslant 0$. From (c') it follows that

$$E[N(t_2) - N(t_1)] = \int_{t_1}^{t_2} \rho(t)\,dt,$$

where $\rho(t)$ is the time variant ROCOF. When $\rho(t)$ is an increasing (decreasing) function of t, the NHPP can be used to model a deteriorating (improving) system.

The minor change in the definition leads to a major difference between the HPP and NHPP models. Under the latter model the X_i's are *neither* independent *nor* identically distributed. That is, not only are they not exponentially distributed, neither are they independent samples from the Weibull distribution nor any other distribution. Because of condition (b), the NHPP retains the independent increments property. This is a very useful feature, because it makes the NHPP a very tractable model. Moreover, this property is often a good approximation for a repairable system. This will be shown by returning, for one last time, to the used car lot scenario.

The first thing a prospective purchaser wants to know about a used car is its total age, measured in terms of odometer reading and/or its year of manufacture. In other words, to a first approximation, the failure/repair history of a complex system like a car does not affect its future reliability. The NHPP, therefore, is the car's first order model.

Some authors have gone to great lengths to emphasize that the NHPP will seldom, if ever, be an exact model for a repairable system. They are correct—there will be *some* effect due to previously performed repairs, so the independent increments property will not hold exactly. (Since the HPP also has the independent increments property, it never will be an *exact* model either.) Nevertheless, the NHPP is a good first order model for an improving/deteriorating repairable system. Moreover, parameter estimation based on limited data, is relatively straightforward and, in some cases, simpler than when dealing with parts. In addition, the superposition of two or more independent NHPP's (without fixed discontinuities in their ROCOF's) is known to be an NHPP, so data from two or more system copies can be pooled when data are scarce. Overall, then, the NHPP is an exceptionally useful model for repairable systems analysis. A few reliability books discuss this model. At least one other book discusses it, while making it appear that distribution functions are being treated, rather than NHPP's. Most reliability books *do not mention* the NHPP!

3.8 Shortcomings of probabilistic models

We have presented a very brief overview of the probabilistic modeling of repairable systems. The treatment was concise, so concise, in fact, that some important models have been totally ignored. This was partly due to space limitations, but also because all existing probabilistic models ignore the effects of many important factors on reliability. Chapter 4 of [1, pp. 63-69] lists 18 such factors, which are seldom included in models. That reference also presents some evidence indicating the practical importance of some of these factors. Here, we will concentrate on the difficulties in realistically modeling *repairable* systems and on the effects of what might be the most important overlooked factor, on/off cycling.

Harris [6] discusses the problems inherent to predicting the reliability of even nonrepairable systems. Once repair is considered, the effects of corrective (and preventive) maintenance

are so complex that they cannot be modeled in any precise manner. Furthermore, the failure of a part can damage associated parts in a myriad of ways. Existing probabilistic models are useful for predicting the distribution of time to *first* system failure. Nevertheless, they become progressively more inaccurate as factors which they ignore exert ever increasing effects on reliability.

Frequent on/off cycling may have a major effect on system reliability. Kujawski and Rypka [7] found that copies of frequently cycled systems had ROCOF's several times larger than nominally identical copies which were seldom turned off. Except for relays and switches, MIL-HDBK-217 ignores the effect of on/off cycling on part reliability. Most systems do not have enough relays or switches for these types of parts to have a major influence on their reliability. For such systems, the implications of MIL-HDBK-217 would be to shut them off whenever possible, even for very brief periods. This would minimize operating time and hence, supposedly maximize reliability over any given calendar time period. According to [7], and common sense, such a policy would be much more likely to degrade reliability than to improve it.

Most of the reliability literature has concentrated on nonrepairable items. When repairable systems have been addressed, the probabilistic modeling of such systems has received most of the emphasis. In our opinion, such probabilistic modeling *cannot* be relied on as the primary source for ascertaining the reliability of repairable systems. Even if all 18 real world factors discussed in [1] could be accounted for probabilistically, the problem would remain of determining how important they were in the real world, i.e., the estimation of relevant parameters would still be necessary. This leads us to our next section, on statistical analysis of interarrival times.

4. STATISTICAL ANALYSIS

4.1 Introduction

Two aspects of data analysis will be very briefly covered here. The first approach, Section 4.2, is applicable when there are a relatively large number of failures of a single system copy. The methods of Section 4.3 are particularly useful when most copies have few failures but a large number of copies are available for analysis.

4.2 The point process approach

Fig. 3 outlines the basic approach for analyzing the interarrival times of a repairable system. Testing for trend means testing for improvement or deterioration. For example, application of an appropriate trend test [1, p. 79] provides statistically significant evidence that the "happy" system of fig. 4 is improving and the "sad" system is deteriorating. NHPP's with decreasing and increasing ROCOF's are fitted to the "happy" and "sad" system data, respectively, in [1, pp. 83-84]. The "noncommittal" system does not show trend, either by "eyeball" analysis or by formal application of the [1, p. 79] trend test. Hence, an NHPP model would *not* be fitted to it. [1, pp. 95-96] shows that fitting a Weibull distribution to the noncommittal system data is *much more complicated* than fitting NHPP's to the "happy" and "sad" systems.

Testing for independence of interarrival times is covered in [1, pp. 88-89] and [8]. The branching Poisson process is discussed in [1, pp. 36-37, 52, 89-91].

If no trend or dependency among successive interarrival times, in their original chronological order, has been detected, *then* it is appropriate to ignore the chronological order and to reorder by magnitude. This is shown symbolically on fig. 3 as $X_i \rightarrow X_{(i)}$. If the data for the "happy", "sad" and "noncommittal" systems are reordered by magnitude, all three data sets become:

$$15, 27, 32, 43, 51, 65, 177.$$

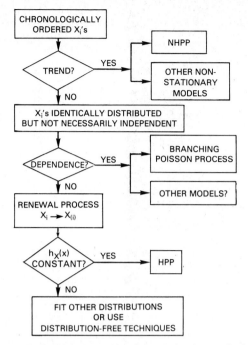

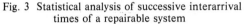

Fig. 3 Statistical analysis of successive interarrival times of a repairable system

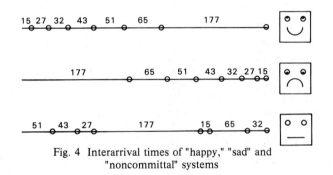

Fig. 4 Interarrival times of "happy," "sad" and "noncommittal" systems

In other words, exactly the same results will be obtained for all three data sets and these results will be WRONG for both the happy and sad systems! There is no real inconsistency - data which show trend should not be reordered by magnitude in the first place. Nevertheless, most reliability texts ignore testing for trend or dependence and reorder data by magnitude as the standard operating procedure.

4.3 Regression techniques

Point process techniques often cannot be applied or can be applied only under unverifiable assumptions. In many cases there are too few failures on any one system, at least during the limited observation intervals usually encountered, to apply such techniques. Even if there are a number of copies of a system available for analysis, data can be pooled only under the assumption that each system is modeled by an HPP or NHPP. It is only under such assumptions that the probabilistic law for the superposition of a finite number of point processes, observed for finite intervals, is known. Even in cases where the HPP/NHPP hypothesis is plausible, it cannot be tested against the available data, in the small sample size case being considered. Therefore, even if the HPP or NHPP model is assumed for each system copy, it still may be beneficial to have an alternative analysis.

In many situations, even though any individual system copy has not failed often, there are a large number of similar copies available for analysis. In such a case, there usually are known differences which may have a marked effect on reliability.

The same type of system may be operated on different platforms, in different positions on a given platform (e.g., a multi-engine plane), with different stresses, operators, maintenance men, etc. In addition, different copies often will have different configurations of installed design fixes. Reference [9] extends Cox's [10] "proportional hazards" model to the analysis of recurrent events such as the failures of an ensemble of repairable systems. Some papers have been published which emphasize the pitfalls encountered when using this technique. As with any other powerful method, it is not difficult to go astray when applying [9] to system reliability problems. The potential advantages, however, provide ample justification for its use. For example, the techniques in [9] can be used to evaluate the effects of differing patterns of on/off cycling, among different system copies, in the presence of other sources of heterogeneity. In summary, [9] can be used to *exploit* heterogeneity, i.e., to use the differences among system copies to help pinpoint the causes of the copies' varying reliability.

5. SHORTCOMINGS OF BASIC CONCEPTS

5.1 Introduction

In Sections 3 and 4, we have emphasized that some basic concepts and techniques for repairable systems have received too little attention. In this section we will show that equally basic concepts cannot even be *described* in consistent terms, when virtually standard terminology and notation are used. These examples will provide particularly emphatic evidence for the vicious circle of fig. 5. The "subtle" concepts of the figure will not be discussed here; they are addressed in [1, pp. 32-33, 162-168].

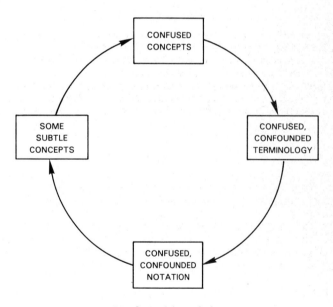

Fig. 5 A vicious circle

5.2 Simultaneous wearout *and* improvement

Consider a situation where light bulbs are being operated in a single socket. That is, when the first bulb fails, it is immediately replaced with the second bulb; when the second bulb fails the third bulb is inserted at once, etc. Each light bulb is know to wear out, i.e., it is from a population modeled by a wearout distribution, as defined in Section 2.3. In words, the longer a bulb survives, the more likely it will fail in say, the next hour. How it has been determined that each light bulb wears out is immaterial here; perhaps this has been verified from data on times to failure of bulbs or it can simply be assumed that we are dealing with wearing light bulbs.

First we will assume that each bulb's time to failure is a sample from the same wearout distribution as all the others and that the failure times are independent of each other. In this case, the pattern of failures in the socket is modeled by a renewal process, as was discussed in section 3.5.

Now we will assume that, even though each bulb wears out, it is from a better population than the preceeding bulb in the socket. Specifically, we will assume that the ith bulb has a mean time to failure, $\mu_i = i$, for $i = 1,2,3,\ldots$. We now have a situation where the interarrival times are tending to become larger, in spite of the fact that each bulb is wearing out. Figure 6 depicts the first four failures of a typical sample path for the "happy" socket.

Fig. 6 Interarrival times of a "happy" socket

We have just presented a simply stated and physically plausible situation. The only way in which the situation is not realistic is that one would not keep inserting better and better parts in a socket *indefinitely*. (Some of the complications which result when a stochastic process terminates after a finite number of events have occurred - in contrast to the process being observed for a finite observation period - are discussed in the second example below.) If we tried to state what is happening, in the standard terminology and notation of the reliability field, we would be led to such inconsistencies as those presented in the following statements. We stress that the terms in double quotes should not be used at all and that the terms in single quotes should not be used in the senses indicated. Justification for these rules is provided by the inconsistencies which result when the rules are not followed.

1) the "failure rate" is increasing (since each bulb wears out) and *simultaneously* the "failure rate" is decreasing (since the number of failures per unit time in the socket is tending to become smaller),

2) wearout is occurring (since each bulb, by assumption is wearing out) but the opposite of 'wearout' is also occurring (since a tendency for successive interarrival times to become smaller - rather than larger - is *also* almost universally called 'wearout'), and

3) as operating time, 't', increases we have increasing "failure rate"/wearout, and *simultaneously* as operating time, t, increases, we have decreasing "failure rate"/improvement.

Statement 3) uses the same symbol, t, for two different time scales, as depicted in fig. 7. This notation is, unfortunately, commonly used in practice. It would not be used if fig. 1 were drawn; in practice, fig. 1 virtually never appears in papers, or even in texts.

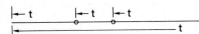
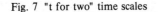

Fig. 7 "t for two" time scales

To clarify the recommended terminology and notation - and therefore, to also help clarify the underlying concepts - statements 1) 2) and 3) are repeated below, using appropriate, distinguishing terms:

1) each bulb's FOM is increasing but the socket's ROCOF is decreasing,

2) each bulb is wearing out but the pattern of failures in the socket is showing improvement,

3) as a bulb's age increases (i.e., as local time, x, increases) we have increasing FOM/wearout and simultaneously as the total operating time in the socket increases (i.e., as global time, t, increases) we also have decreasing ROCOF/improvement.

A variant of the above example is one where worse and worse wearing bulbs are put into the socket. At first glance, the inconsistency between simultaneously increasing and decreasing "failure rates" *seems* to disappear since both FOM and ROCOF are increasing in this situation. It should be clear, however, from the first example that this variant is an insidious one: both "failure rates" are increasing but they are still DIFFERENT "failure rates"! Lest our position about usage of the term "failure rate" not be completely clear, we will explicitly state it; "failure rate" should not be used at all, since in practice, it has proven *impossible* to use it for only FOM or for only ROCOF!

A repairable system is composed of many parts, in many sockets, which are never all in series. These parts interact with each other in very complicated ways. Moreover, a system's reliability is also influenced by many external factors such as quantity - and quality - of preventive maintenance and the number of on/off cycles it has experienced, cf. section 3.8. We have just demonstrated that the standard concepts, terminology and notation in the reliability field are inadequate to describe the situation where a sequence of increasingly reliable but wearing parts are put into one socket. It is hardly startling, therefore, that the framework for modeling repairable systems reliability, in other than the simplest terms, is virtually nonexistent.

In spite of the inconsistencies encountered when standard terminology is used to describe what is happening in the successive-light-bulbs-in-a-socket case just presented, it is still possible to understand the physical situation. However, this is because the situation is so simple that even confused terminology and notation cannot completely confound it. When it is recalled, however, that almost any repairable system will be many orders of magnitude more complex than a single socket, the need for better nomenclature and notation becomes apparent.

5.3 Will the **real** "failure rate" please stand up!

Section 5.2 stressed that the *only* way that the example of a nonterminating sequence of better, but wearing, light bulbs in a socket is unrealistic, is that the process will not continue forever. That is, in practice, only a finite number of replacements will be made. In this section, the properties of a finite sequence of IID exponential interarrival times will be used to help explain the fundamental difference between the ROCOF of an *HPP* and the *only*-numerically-equal-FOM of each of the *HPP's* interarrival times.

We will consider the properties of a sequence of IID exponentials which terminates after M failures have occurred, M known. We will call this process a *terminating HPP*. Note that this process differs from an *HPP* since the *HPP* consists of a *non*terminating sequence of IID exponentials. We will demonstrate an important distinction between the constant FOM of each interarrival time and the *strictly decreasing* ROCOF of the process, whenever $M < \infty$. In one sense, the terminating *HPP* is an unrealistic model since the number of failures a system will experience, before being discarded, will seldom be known a priori. On the other hand, it *is* known a priori that the number of failures will be finite. For example, there will never be an infinite number of light bulbs available to be put into a socket.

It is shown in [1, pp. 61-62] that for a sequence of M IID exponentials, the ROCOF $v(t) \equiv \dfrac{d}{dt} E[N(t)]$ is

$$v(t) = \sum_{j=1}^{M} f^{(j)}(t)$$

$$= \gamma e^{-\gamma t}\left[1 + \gamma t + \dots + \frac{(\gamma t)^{M-1}}{(M-1)!}\right],$$

where $f^{(j)}(t)$ is the j-fold convolution of the exponential distribution's PDF. In the special case where $M = 1, v(t)$ reduces to the time variant exponential PDF $\gamma e^{-\gamma t}$, rather than to the FOM, γ. In other words, the special case of the ROCOF when only one failure can occur, is the PDF, *NOT* the FOM. In the special case $M = \infty$ (i.e., an *HPP*), $v(t) = \gamma e^{-\gamma t} e^{\gamma t} = \gamma$. Since $\int_{0}^{\infty} v(t)dt = M$, it is apparent that $v(t)$ must be time variant for $M < \infty$. That is $\int_{0}^{\infty} v(t)\, dt < \infty$ implies that $\lim_{t \to \infty} v(t) = 0$. It is shown in [1, pp. 61-62] that $v(t)$ monotonically decreases from γ to 0, as t increases from 0 to ∞, whenever $M < \infty$.

We have just demonstrated that the ROCOF of the process of IID exponentials is *not even numerically equal* to the FOM of each of these exponential distributions, for any $M < \infty$, down to and including $M = 1$. How could a constant FOM be equivalent to-i.e., identically equal to - the ROCOF for $M = \infty$, when the ROCOF is time variant for any $M < \infty$? The special - and physically impossible - case of $M = \infty$ just happens to be virtually the only assumed alternative to $M = 1$. This makes it appear that it is the only possible alternative to $M = 1$, thus perpetuating the myth that the FOM of the exponential distribution and the ROCOF of the *HPP* are interchangeable functions. Of course, the fact that both are usually referred to as constant "failure rate," also exacerbates the confusion, see fig. 5. If we were to use "failure rate" for both FOM and ROCOF-as, unfortunately, is standard practice - we would have to say that the "failure rate" of each interarrival time is constant, but the "failure rate" of the process is strictly decreasing, unless M happens to be infinite.

The NHPP with power law ROCOF,

$$\rho(t) = \lambda \beta t^{\beta-1}$$

has been widely used in reliability growth modeling. This NHPP is often confused with a Weibull *distribution*, for many reasons, not the least of which is that the Weibull distribution's FOM is of the same functional form

$$h_X(x) = \lambda \beta x^{\beta-1}.$$

The discussion about how numerical equality between the *HPP's* ROCOF and the exponential distribution's FOM does not imply equivalence between these functions, carries over to the lack of equivalence between the power law NHPP's ROCOF and the Weibull distribution's FOM. We will add the following example, moreover, to show that the *same* mathematical function can be interpreted in *drastically* different ways, depending on the physical situation it is modeling. The power law function $\lambda \beta t^{\beta-1}$, in the special case where $\beta = 3$ and $\lambda = g/6$, represents the *distance* traveled in t seconds, by an initially stationary freely falling body in earth's gravity field. Obviously, nobody is going to confuse this interpretation of the same function, with its dimensions of length, with either ROCOF or FOM. Many reliability workers, however, believe that ROCOF and FOM are equivalent, partly because they can be represented by the same mathematical function. The fact that all ROCOF's and FOM's have the same dimensions, reciprocal time, makes it *particularly important* to use different names for them, as well as to avoid the use of "t *for two*," *see figs. 5 and 7.*

6. ADDRESSING BASIC ISSUES

6.1 Introduction

We now have established the background to address the questions posed at the beginning of the paper. Please note the use of the term "address" rather than "answer." The questions being considered do not have-and probably never will have-definitive answers. The second question, especially, is posed in such comprehensive terms that this section should be considered to provide the summary and conclusions of the paper.

6.2 Is there a period of constant hazard rate?

The term "hazard rate" is usually defined to be equivalent to what we have called FOM, $h_X(x)$. In some cases, however, e.g. in [11], "hazard rate" is used in the sense of ROCOF. We will address the question, "Is there a period of constant ROCOF?" It should be clear from the discussion in section 5, that the question, "Is there a period of constant FOM?" would have to be approached from a different perspective, [see 1, p. 177].

There are two widely believed-but conflicting-rationales for constant ROCOF's of repairable systems. Many reliability personnel postulate the existence of a bathtub curve for repairable systems, see the lower plot of fig. (8). The middle, more or less constant, portion of this curve is the constant ROCOF region. Many others justify constant ROCOF from the results of the asymptotic theorems concerning the superimposed renewal process model discussed in section 3.6. In this case, however, the ROCOF settles down to a constant value as operating time, t, increases without bound, see fig. 9. What is not usually recognized, is that these two justifications for constant ROCOF conflict with each other. As shown in fig. 9, just where the asymptotic theorems indicate that the ROCOF is stabilizing, the bathtub curve shows it to be increasing rapidly. A possible reconciliation of this apparent anomaly is presented in fig. 10. For example, the popular conception of an automobile's reliability is that its ROCOF is on an upward swing after about 100,000-150,000 kilometers. If repairs were always made, regardless of how many were needed, the car would eventually be composed of parts with a randomized mix of ages and it is plausible that its ROCOF would eventually settle down to a constant value. In practice, of course, many cars are scrapped precisely because they are on the upward swing of a bathtub curve, so that the asymptotic therorems do not become applicable to them.

An infinite time span must pass before the asymptotic theorems become a fully applicable justification for constant ROCOF. In practice, a finite time would suffice. However, [5] shows that this finite time can be of the order of *centuries*. In addition, the superposition of renewal processes basis for these theorems is never totally appropriate. Real systems do not have all their parts in series so system failures are not merely the union of all part failures in the system's constituent sockets. Moreover, a renewal process is not necessarily appropriate for modeling a socket. It is often claimed that, at least for electronic systems, see [12, pp.176-177], once initially defective parts are culled out, their replacements last indefinitely. In other words, with high probability, if a part fails quickly, it is not representative of a typical part to be placed in the same socket. Hence, electronic systems may have decreasing ROCOF's for much or all of their operational lives.

The asymptotic theorems provide a questionable basis for justifying constant ROCOF. The repairable system bathtub curve may be applicable in some cases but it conflicts with the asymptotic theorems. In our opinion, there is *no* theoretical, or a priori empirical, basis for settling the question, "Is there a period of constant ROCOF?", much less the question of *when* it occurs, given its existence. Instead, the question must be approached from the opposite viewpoint, analysis of data. Moreover, there may be two or more periods of constant ROCOF, see fig. 10.

A constant ROCOF corresponds, at least roughly, to the absence of a trend. Therefore, the techniques referred to in section 4.2 and covered in [1, pp. 73-83] should be applied to help determine when a constant ROCOF exists. Techniques

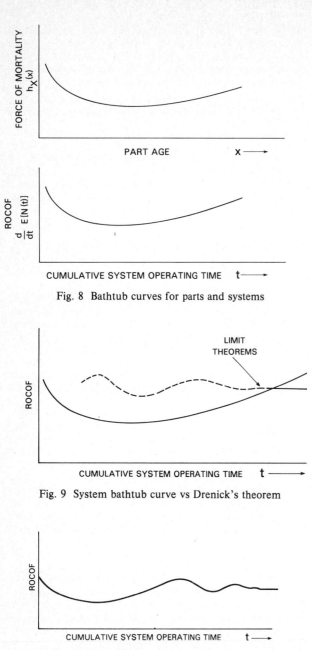

Fig. 8 Bathtub curves for parts and systems

Fig. 9 System bathtub curve vs Drenick's theorem

Fig. 10 Reconciliation between bathtub curve
and Drenick's theorem

involving change points [1, pp. 177-178] can be used to estimate when a constant ROCOF period begins and/or ends. Note that constant ROCOF does *not* imply that an HPP is the appropriate model. In fact, even a renewal process is not a necessary condition for constant ROCOF to hold [1, p. 22]. Conversely, a sequence of exponential interarrival times does *not* necessarily have the constant ROCOF property-the means of the successive distributions may be changing.

We note that there has been considerable confusion over the existence of **two** bathtub curves, one for parts and one for systems, see fig. 8. Since the curves *look* the same, the same mathematical function can be used to describe them. As emphasized in section 5.3, however, the mathematical function is interpreted differently in the two cases. [1, pp. 136-139] discusses the different interpretations of these curves. The main reasons that they are confused, of course, is that 1) they have the same shape, 2) in most cases both ordinates are called "failure rate" and 3) both time scales are denoted t, see figs. 5 and 7.

6.3 How do we best deal with the analysis of repairable systems?

Since most systems are designed to be repairable, this question is practically equivalent to asking how to do reliability analysis. Thus, it is so vast that it can be addressed only very briefly here.

At the outset, it should be emphasized that far more probabilistic modeling has been done than has been even alluded to in this paper. Nevertheless, in our opinion, there are major drawbacks to how much can be accomplished with the probabilistic approach to repairable systems. Section 3.8 briefly points out major shortcomings of existing probabilistic models; this topic is covered in much greater detail in [1, Chapter 4]. An example is given in [1, p. 54] where an enormously oversimplified problem becomes mathematically intractable when it is made *slightly* more realistic. Even to the extent that more realistic problems can be formulated and solved mathematically, the question remains, how will their parameters be estimated? In other words, statistical analysis of repairable systems failure data is essential for improved understanding of such systems.

We will summarize our opinion of probabilistic modeling in the following terms. Such models are useful in developing system architectures, i.e. in determining how much redundancy to use and specifying its allocation; this statement is equivalent to saying that modeling is useful in comparing alternative designs. Models also are useful in providing "ballpark" estimates of system reliability. MIL-HDBK-217 implicitly acknowledges the approximate nature of reliability prediction since many baseline "failure rates" (actually FOM's) and π factors are estimated to only two or even one significant figure. In practice, many predictions go into excruciating detail about some modeling aspects-while totally ignoring many other important factors-and then present results to five or six or even more significant figures. This approach has rightfully earned the epithet, "Numbers Game"-but, unfortunately, it is still almost standard practice. In an editorial entitled, "Predicting the Unpredictable," O'Connor [13] has pointed out that the actual reliability level a system achieves is largely dependent on the effort-or lack of effort-of the people who design, assemble, operate and maintain the system. No amount of mathematical machination is going to change that!

A great need exists for more statistical analyses to help determine what really causes system, i.e., repairable system, failures. There is also a great need for performing these analyses correctly. The preponderance of the reliability literature treats nonrepairable items and/or makes the a priori IID assumption. The overemphasis on the IID assumption is not unique to the reliability field, but it is particularly ironic that it is usually invoked in reliability applications-one often (usually?, always?) hopes for "happy" systems rather than noncommittal ones! The best known set of failure data in the reliability literature is the air conditioner data originally presented in [14]. These data, in actuality, are the interarrival times of 13 repairable systems. Nevertheless, in most references, they are treated, a priori, as if they were the times to failure of 13 batches of nonrepairable items. The correct interpretation of these data differs *markedly* from the interpretation which would apply *if* the same numbers were the times to failure of parts, see [1, p. 151]. Morever, even eyeball inspection of the chronologically ordered data [1, pp. 144-145] reveals important features which, nevertheless, almost always have been ignored. As [15] puts it, "...let's face it, statisticians are so used to IID assumptions that it is hard to break the habit." If this is true of statisticians, it is not surprising that practitioners have been so thoroughly convinced-maybe "hoodwinked" would be a more accurate term-that *any* given set of data *must* be a sample from a single distribution. Not only is it important to analyze repairable systems failure data, it is also important to analyze it correctly. This last statement is, of course, a truism. Judging from the track record of the reliability field, however, it is a truism which has seldom been true in practice.

Finally, it is important to pose analyses of repairable systems reliability in terms which (1) make it clear that repairable systems are being considered and (2) present the results unambiguously. Because of some of the subtle concepts briefly alluded to in fig. (5), see [1, pp. 32-33], it is technically feasible to discuss the NHPP as if one were dealing with a distribution function. This insidious approach actually has been adopted in some papers! Recall that the prevalent concepts, terminology and notation are inadequate to *describe* the situation where better but wearing parts are successively put into a socket. Even more treacherously, the prevalent framework appears-INCORRECTLY-to be adequate to describe the situation where worse and wearing parts are put into the socket. Hence, it becomes clear that an improved framework for probabilistic modeling of repairable systems, as well as for performing, and presenting the results of, repairable systems data analyses is *essential*.

REFERENCES

[1]. H.E. Ascher and H. Feingold, *Repairable systems reliability: modeling, inference, misconceptions and their causes*, Marcel Dekker, New York and Basel, 1984.

[2]. R.C. Elandt-Johnson and N.L. Johnson, *Survival models and data analysis*, John Wiley, New York, 1980.

[3]. E. Çinlar, *Introduction to stochastic processes*, Prentice-Hall, Englewood Cliffs, 1975.

[4]. R.F. Drenick, The failure law of complex equipment, *Journal of the Society of Industrial Applications of Mathematics*, vol. 8, 1960, 680-690.

[5]. S.B. Blumenthal, J.A. Greenwood and L.H. Herbach, The curse of the exponential distribution in reliability, *Proceedings of twenty-third conference on the design of experiments in army research, development and testing*, ARO report 78-2, 1978, 457-471.

[6]. L.N. Harris, The rationale of reliability prediction, *Quality and reliability engineering international*, vol. 1, 1985, 77-83.

[7]. G.J. Kujawski and E.A. Rypka, Effects of "on-off" cycling on equipment reliability, Proceedings of Annual Reliability and Maintainability Symposium, IEEE-77CH 1308-6R, 1978, 225-230.

[8]. A. Bendell and L.A. Walls, Exploring reliability data, *Quality and reliability engineering international*, vol. 1, 1985, 37-51.

[9]. R.L. Prentice, B.J. Williams and A.V. Peterson, On the regression analysis of multivariate failure time data, *Biometrika*, vol. 68, 1981, 373-379.

[10]. D.R. Cox, Regression models and life tables (with discussion), *Journal of the Royal Statistical Society*, Series B, vol. 34, 1972, 187-220.

[11]. G. Härtler, Graphical Weibull analysis of repairable systems, *Quality and reliability engineering international*, vol. 1, 1985, 23-26.

[12]. P.D. O'Connor, *Practical reliability engineering, second edition*, Wiley, New York, 1985.

[13]. P.D. O'Connor, Predicting the unpredictable, *Quality and reliability engineering international*, vol. 1, 1985, p. 69.

[14]. F. Proschan, Theoretical explanation of observed decreasing failure rate, *Technometrics*, vol. 5, 375-383.

[15]. R.G. Easterling, Review of [1], *Technometrics*, vol. 27, 439-440.

BIOGRAPHY

Harold Ascher is an Operations Research Analyst at the Naval Research Laboratory, Washington, D. C. He has participated in reliability programs on the POSEIDON and TRIDENT missiles and has performed reliability growth analyses on marine gas turbines such as those installed in SPRUANCE class destroyers. He was a member of a Tri-Service Committee for the preparation of MIL-HDBK-189 on reliability growth management. Mr. Ascher received his BS in Physics from City College of New York in 1956 and his MS in Operations Research from New York University in 1970. He is the author of over 25 reliability papers and is a member of the American Statistical Association, IEEE, Society of Reliability Engineers and Sigma Xi.

SYSTEM ANALYSIS

Reliability Technology — Theory & Applications
J. Møltoft and F. Jensen (Editors)
© Elsevier Science Publishers B.V. (North-Holland), 1986

CONTRIBUTION OF SAFETY ANALYSIS METHODS
IN THE FRAMEWORK OF SAFETY AUDITS

Jean-Paul JEANNETTE
Socotec Industrie
1, avenue du Parc
78180 Montigny-le-Bretonneux
France

Nikolaos LIMNIOS
Université de Technologie
B.P. 233
60206 Compiègne cédex
France

ABSTRACT:
In France, industrials risks prevention is regulated by law and involves auditing the most
dangerous plants. Though this approach is not too satisfactory in somes cases, it enables
to shed light on some inadequacies. This article shows the necessity to get additional
information for the audit: the safety analysis. Besides it appears that these two approaches
are not competing with one another.

-1-INDUSTRY IN FRANCE: Legal dispositions

The urge to enforce through regulations prevention of industrial risks and nuisances has been strong in our country for a long time. Actually, the first dispositions date back to 1810. Since then, because of industrial evolution and technological progress, the original text has been retreated lots of times.

Today, it is the Bill of July 19, 1976, so-called "Installations classified by the environment protection" -Installations classées pour la protection de l'environnement-, that sets the legal and mandatory framework for industrial plant installation and operation.

The application framework of this bill is defined in its first article:
"The shops, plants, storages, construction areas quarries, and all installations that can show dangers or drawbacks either for safety, or for neighborhood facilities, or for agriculture or wild life and environment protection, or site and monument preservation are regulated by law".

The authorization to operate is granted after a rather complicated process, of which the major steps are the following:
  -public investigation (collecting ideas and advice from the endangered population),
  -impact study (evaluation of consequences on natural environment),
  -health and safety documents (protection of workers),
  -hazards study.

This last point must-list the dangers (fire, explosion, noise, radioactivity...) that an installation or a plant can show in case of accident,
                   -justify the proper measures to decrease the occuring probability and their effects,
                   -list and describe accidents likely to happen.

-2- RISK PREVENTION IN EUROPE

Following the accident of July 10, 1976 in Seveso (Italy - dioxine leakage), governments of the EEC countries have jointly come up with a text entitled "Notice of the european council of June 24,1982 about the risk of major accident of some industrial activities" (Directive du conseil du 24 juin 82 concernant les risques d'accident majeur de certaines activités industrielles) currently called "Notice SEVESO" (Directive SEVESO).

The basic principles that are developed are the following:
  -industries dealing with hazardous substances must take the necessary steps to prevent accidents that could lead to serious consequences for the people and the environment and to limit their effects,

  -public authorities must be informed of these risks by appropriate procedures and must have a control on the industrial activities. They must be informed in case of major accidents and be able to take all necessary steps.

  -necessity of information of the workers and the public,

  -collaboration between member countries,

  -harmonization of regulations enforced in industries within each member countries.

In France, it is through the law on classified installations (installations classées pour la protection de l'environnement) that this bill is enforced.

-3-SAFETY AUDITS

When one considers the general principle of no retro-activity in the French law, article 18 of the September 21, 1977 notice is the exception enabling the Commissary of the Republic (Commissaire de la République) to come up anytime with an additional notice to

enforce any new complementary requirements to
an already existing.

Since today's law is relatively recent and
that numerous installations authorized to
operat in the past have been modified to such
an extent that they are far from their original description,the Administration has set up
a procedure:the safety audit, of which the request is confirmed by state notice.

The point of audit is to go through a thorough examination of the installation from
the conformity control standpoint,enforcement
of bills and regulations in use,standards...

    -Examinations of reports from control bo-
dies: .electric appliances
         .safety means
         .devices under pressure
         .measures of air pollutions
         .measures of water pollutions
    -Visit and examination of installation
operating conditions,as far as design is concerned,states of equipment,operation habits,
instructions,incidents or accidents that occured in the past.
    -Analysis of risks for environment.

-4-DEFICIENCY OF APPROACH:

This first approach that already enables to
spot with much accuracy the actual risks of
the industrial installation shows,though,seve-
ral deficiencies.
It is very difficult in such an investigation
to accurately analyze the industrial process,
operating conditions,influence of various subparts between them and on environment and vi-
ce-versa.
Moreover,the methodology is very much based
on knowledge of incidents or accidents happened in similar installations,on comparative
studies between processes when operating conditions can change from one another: age,tech-
nology,annual production,personnel training
and qualification,specific environment...
As a result,in this framework,it appeared to
us necessary to call in a complementary technology enabling a deeper analysis allowing to
back up our opinions with tangible facts.This
complement is the Safety Study.

-5-SAFETY STUDY:

Safety Study is a well-known methodology of
which the findings in electronics,nuclear or
aeronautic areas are proven.This is, though, a
technique requiring the reunion of multi-
discipline teams carrying out long and expen-
sives researches.
Hence,it is necessary to set up a strategy
with which we can get in return the accuracy
of analysis of safety method for systems,but
that,on the other hand,enables to decrease its
drawbacks.
This strategy is based on two aspects:
    -a traditionnal approach such as the one
previously presented
    -an approach,the safety analysis type.

-6- TRADITIONAL APPROACH:

This one carries the different study steps as
defined in paragraph 3,but is tackle in a more
systematic way.
The system on which the audit is carried out
breaks down in functional subparts,each of
them being the object of an overall FMECA
through joint knowledge of industry people
and experts.
Criticality analysis has not been forgotten
since it enables a future classification of
events by a gravity scale in the risk domain
such as fires,blasts,pollution...

-7-SAFETY ANALYSIS APPROACH:

Safety analysis is only carried out for
one or several subparts or situation of which
criticality is maximum
The approach used is classical:
    .functional break-down
    .preliminary analysis of risks
    .FMECA at the components level
    .Setting of the fault tree and qualitative
and quantitative analysis

-8-EXAMPLE 1 : "Oil extraction unit" case

8-1 OPERATION PRINCIPLE (see exhibit 1)
Oil grains,after mashing,are treated under a
solvent flow.This solvent gets filled with oil
The following distillation enables recupera-
tion of oil and solvent recycling.

8-2 SAFETY AUDIT
Audit was on study of safety as far as fire
and explosion were concerned.Traditional approach has enable to spot the major aspect of
the cooling system of installations that in
case of failure led to a pressure increase
with explosion risks.

8-3 ANALYSIS OF THE COOLING CIRCUIT
Operation : see exhibit
Fault tree (see exhibits)

-9-EXAMPLE 2: atomization tower case

9-1 OPERATION
Liquid to be deshydrated is sent to a turbine
at the tower top where it is pulverized in
small drops.When getting down,the product
loses its humidity by contact with hot air.

9-2 SAFETY AUDIT
Audit was on study of safety as far as fire,
explosion,operation and storage conditions
risks were concerned.

    9-3 ANALYSIS OF THE RISK OF POWDER MILK
              IGNITION IN TOWER
Operation:see exhibit
Fault tree : see exhibit

-10- CONCLUSION :

Regulation provides means and methods to
check at the component or equipment level,
thus enabling increase of each safety parame-
ter(reliability,safety).
But it cannot take into account between equipment interaction.

In fact, in example 1, electric supply failure of air-cooling devices and no-shut down of the high pressure valve leads to a critical situation.

In case nimber 2, failure of temperature sensors, together with the gaz manometer and vibrating devices leads directly to a high risk of ignition.

The fault tree technique , still not widespread in food industry, happened to be particularly didactic to shed light on accident scenarios.

It is getting known as a new means of equipment design checking, optimization of safeties, simulation of accidents and analyses of procedures of conducts.

This type of study has also become a means of motivation for a better follow up in maintenance time: MTBF, MTTR and failures causes recording.

REFERENCES

1) Loi n° 76.663 du 19 juillet 1976
   Installations classées pour la protection
   de l'environnement
   Brochures du Journal Officiel de la
   République Française n°1001-tomes 1,2 et 3.

2) Directive du Conseil du 24 juin 1982
   concernat les risques d'accidents
   majeurs de certaines activités industrielles
   Journal Officiel des Communautés Euro-
   péennes (82/501/CEE)

3) Sûreté des installations classées -
   Activités industrielles et agricoles
   Ministère de l'Environnement Industriel
   Septembre 1984

4) Les audits de sûreté dans les installations
   classées
   Document interne SOCOTEC - 20 décembre 1984

5) LIMNIOS N., JEANNETTE J.P. -Le logiciel
   ARBRE: Traitement automatique des arbres
   de défaillance
   3ième séminaire européen sur la sécurité des
   systèmes-Cannes France 19, 20, 21 septembre 1984

6) JEANNETTE J.P. - Les méthodes d'évaluation
   des risques
   Les risques industriels et les sapeurs-
   pompiers
   C.N.I.P.C.I. Club Chaptal - Paris France
   2 décembre 1985.

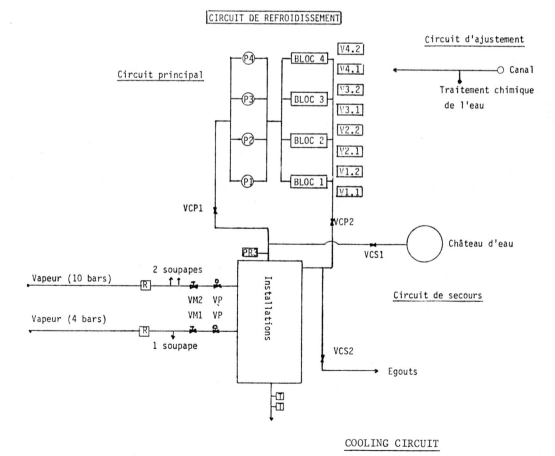

COOLING CIRCUIT

FAULT TREE OF COOLING CIRCUIT

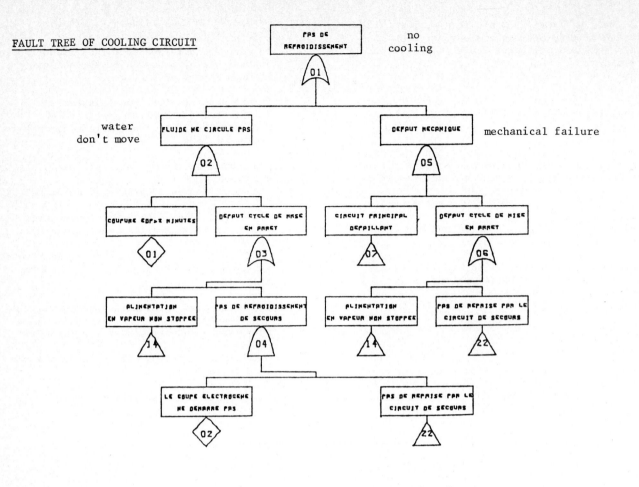

water
don't move

mechanical failure

no
cooling

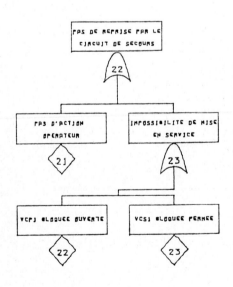

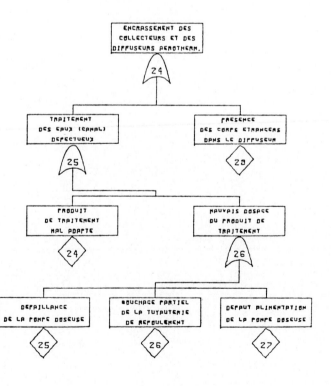

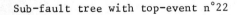

Sub-fault tree with top-event n°22

Sub-fault tree with top-event n°24

COOLING CIRCUIT
COOLING CIRCUIT

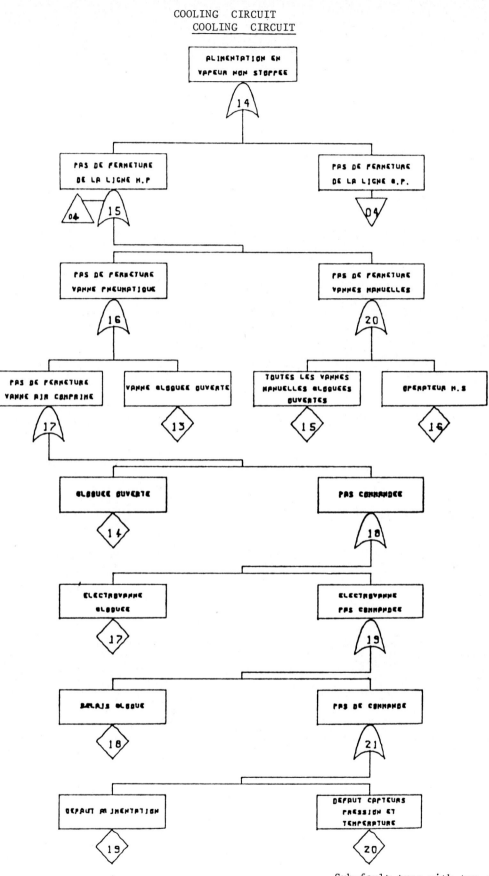

Sub-fault tree with top event n°14

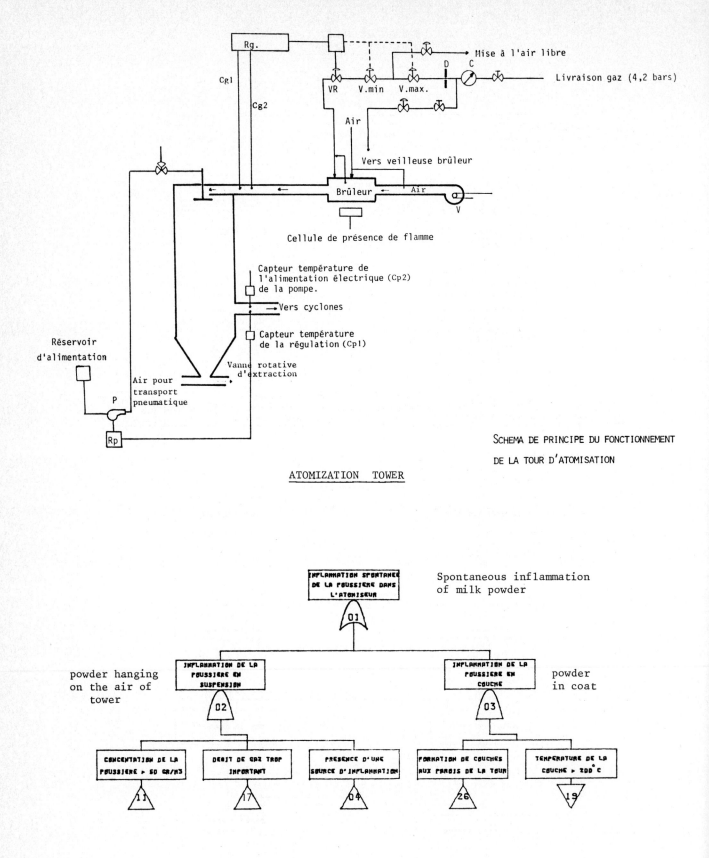

ATOMIZATION TOWER

Scʜᴇᴍᴀ ᴅᴇ ᴘʀɪɴᴄɪᴘᴇ ᴅᴜ ꜰᴏɴᴄᴛɪᴏɴɴᴇᴍᴇɴᴛ
ᴅᴇ ʟᴀ ᴛᴏᴜʀ ᴅ'ᴀᴛᴏᴍɪsᴀᴛɪᴏɴ

Spontaneous inflammation
of milk powder

powder hanging
on the air of
tower

powder
in coat

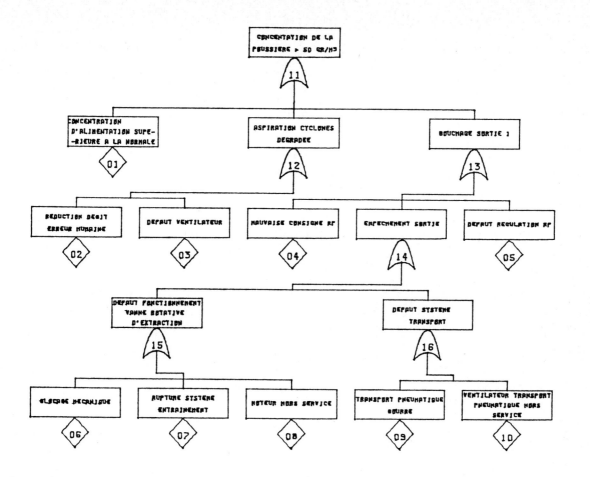

ATOMIZATION TOWER: sub fault tree with top-event n°11

PRESSION DE LIVRAISON TROP IMPORTANTE — 11

DEFAUT TRANSMISSION COMMANDE — 25

PRESSION DU GAZ A L'ENTREE TROP IMPORTANTE

RUPTURE DU DIAPHRAGME — 12

21

DEBIT D'ARRIVEE DE GAZ TROP IMPORTANT

20

DEFAUT CAPTEUR TEMPERATURE — 13

DEFAUT MOTEUR — 26

PAS DE FERMETURE DE LA LIGNE — 22

DEFAUT REGULATEUR — 14

IMPOSSIBILITE DE FERMETURE DE LA VANNE VR — 24 / 24

BLOCAGE MECANIQUE — 27

PAS DE REGULATION PAR LA BOUCLE DE REGULATION 1 — 23

DEBIT DE GAZ TROP IMPORTANT — 19

COMBUSTION EXCESSIVE — 18

TEMPERATURE D'AIR > 190 ARRIVEE D'AIR TROP CHAUD — 17

OUVERTURE EXCESSIVE DE LA LIGNE D'ALIMENTATION — 22

V-MIN OUVERTE — 24

PAS DE FERMETURE DE SECURITE — 25

V-MIN OUVERTE — 24

PAS D'ALIMENTATION ELECTRIQUE — 28

ATOMIZATION TOWER

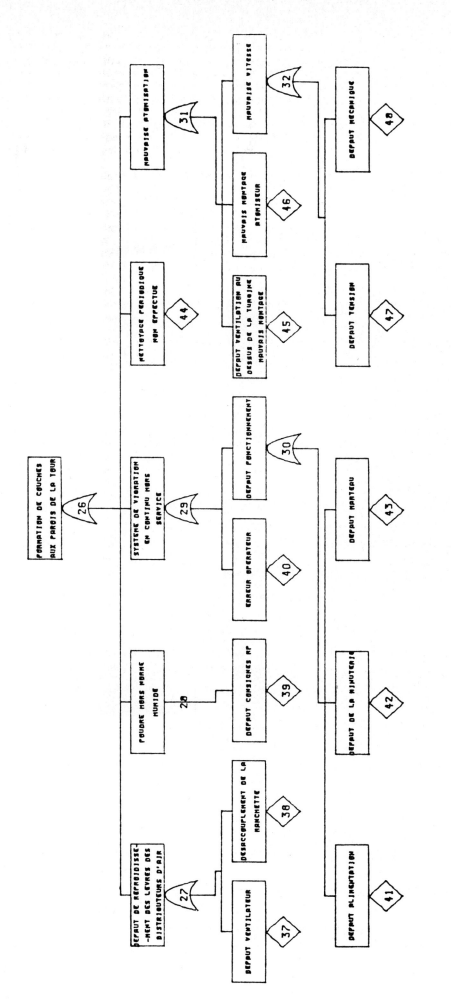

ATOMIZATION TOWER

Reliability Technology — Theory & Applications
J. Møltoft and F. Jensen (Editors)
© Elsevier Science Publishers B.V. (North-Holland), 1986

IMPROVEMENT OF SAFETY OF SEQUENTIAL SYSTEMS

BY ON-LINE SYNTACTIC ANALYSIS METHOD

Jean DEFRENNE Laboratoire d'automatique
Université des Sciences et Techniques de Lille

The safety of production automatic system is linked to the reliability of the components of
the process plant (mechanisms, sensors, actuators...). The observation of reports coming from
the process plant to the control computer, allows us to show the faults in these elements. In
this paper, setting of a checking system based on a syntactic analysis of this information
flow is proposed. Different grammars requiring different levels in the studying of the process
plant are proposed. The checking system allows a detection of errors and a location of failures, which is most useful for any intelligent reaction of the control computer.

1 FAILURE EFFECT UPON PRODUCTION AUTOMATIC SYSTEMS

1.1 Components of automatic system

The different components of an automatic system are:
- the process plant (mechanisms, sensors, actuators...);
- the control computer;
- the supply devices;
- the driving interfaces.

All these components are linked by mechanic, electric or fluidic connections.

Any failure on one of these components disturbs the good running of the machine. Some other reasons of disturbances are machine-user's faults and design errors.

The control computer sends "orders" to the process plant which returns "reports" by means of sensors.

1.2 Failure effect upon safety

The improvement of safety is often an application of redondant techniques. But, if the continuous running is not necessary for safety, this method may have serious consequences from the point of view of reliability. As the electronic component reliability is good, the control computer failures are not the main cause of disturbances. Some programmable controler makers say that only 5% of running disturbances are caused by control-computer failures.

Our suggestion aims at the setting of an error electronic detection device which realises an on-line testing of process plant. This testing device gives some necessary information to come a decision (running-off strategy, configuration...) after a failure recognition.

2 DIAGNOSTICS BY SYNTACTIC ANALYSIS OF REPORTS

2.1 Trajectory pattern

Sensors permit observation of some measurements. For each mesasurement, only a finite set of values is observed. We call elementary trajectory of a measurement all the points of evolution space obtained from a point, without modifying the report observed.

We assign to each elementary trajectory a conventional sign s. All the signs used for the elementary trajectories of the measurement j form an alphabet Sj. Every evolution of this jth measurement gives a sequence of the components of Sj.

In the observation space, each component s of Sj is coded by an elementary report r which is a component of an alphabet Rj. We added to Sj a conventionnal element noted s_o. This component is the arbitrary elementary trajectory which is obtained at the starting point observation.

The sequence of reports can be obtained from s_o by application of production rules such as:

$$- s_1 \longrightarrow r_1 s_2 \quad \text{or}$$
$$- s_1 \longrightarrow r_1$$

with $s_1, s_2 \in Sj$ and $r_1 \in Rj$.

The grammar $Gj = (Rj, Sj, s_o, P)$ where P is a set of similar production rules is a regular grammar.

This grammar can be modeled by a finite-state automaton $Aj = (Rj, Qj, qo, qf, Dj)$ where

Qj is a set of nodes; each component s of Sj gives a node q of Qj;

$qo \in Qj$ is the initial node associated to s_o;

$qf \in Qj$ is a final node which is obtained by an abnormal report; every transition to qf corresponds to rule such as $s_1 \to r_1$;

Dj is a set of arrows corresponding to the set P of rules.

Fig 1 shows an application of this method.

2.2 Test and limitaion included in the model.

The finite state automaton is implanted in the testing device. Every report modification generates a transition in the model which is synchronous with the process plant evolution. Every transition to the final state qf corresponds to an abnormal transition in the failure recognition.

Often, with discrete sensor setting, more than one elementary trajectory have the same report. Thus the automaton is a non deterministic automaton. Figure 1shows this case. The transformation in a deterministic automaton is not an acceptable operation for test. Indeed, in this transformation we obtain final nodes which are not absorbant nodes. So the test procedure is ambiguous.

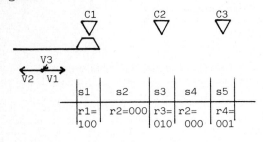

$S=\{s1,s2,s3,s4,s5\}$
$R=\{000,001,010,....,111\}$

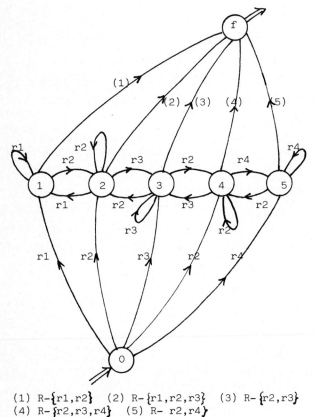

(1) $R-\{r1,r2\}$ (2) $R-\{r1,r2,r3\}$ (3) $R-\{r2,r3\}$
(4) $R-\{r2,r3,r4\}$ (5) $R- r2,r4\}$

Process plant and his model

- figure 1 -

3 INTRODUCING SEMANTIC ELEMENTS IN THE MODELING

3.1 Programmed grammar

We call programmed grammar a grammar such as
G= (R, S, s0, V, P)

where V is a set of predicats assigned to each production rule of P.
The predicat value is fixed according to a programmed evaluation.

We design a weighted automaton
Ap= (X, Q, q0, T, V, D)   where

X is an alphabet;
Q is a set of nodes;
$q0 \in Q$ is the initial node;
$T \subset Q$ is a set of final nodes
V is a set of predicats assigned to arrows of D;
D is a set of arrows.

At any time, we can define the weight pi gives to the node $qi \in Q$.
Let αik be an arrow with qi as origin and qk as extremity. Let Iik be a weight interval. A predicat $V(\alpha ik)$ assigned to the αik arrow is

pi IN Iik   (or pi $\in$ Iik).

By convention, for a set of values of weight at any moment, only arrows such as αik for which predicat $V(\alpha ik)$ is true, are present.

3.2 Process plant modeling by weighted automata

Suppose that a control vector defining the evolution of measurement exist. We suppose this evolution occurs again within some tolerance interval if the same control vector value is applied from the same state of process plant.

Let Cj be the control vector for the jth measurement. Two hypotheses are proposed.

i/ all control vectors such as Cj, are independent.

ii/ the Cj value is constant during an evolution inside an elementary trajectory. A control vector modification is always linked to a transition between an elementary trajectory and the next one.

This second hypothesis is verified by the correlation between orders and reports which is established by the process controller.

Weight meaning: The weight pi assigned to qi is the duration kept in the correspondent state.

Associated interval Iik:This interval is an average duration of keeping in qi state within a tolerance value.

This duration is modified by the control vector applied, and by the fact that a modifying of the evolution orientation was commanded or not since its coming in the qi state, by the controller. Figure 2 shows two different evolutions from qi state with the same control vector value v.

$I^2_{\alpha ik,v}$ is obtained without orientation change of evolution;

$I^1_{\alpha ik,v}$ is obtained with an orientation change. This average duration is certainly lower than $I^2_{\alpha ik,v}$.

Determination of weight pi
  - for every transition αik from qi to qk $\neq$ qi, the weight pi is reset to 0;
  - for every transition to qi, the weight pi is incremented; this is true for arrow αii from qi to qi as well.

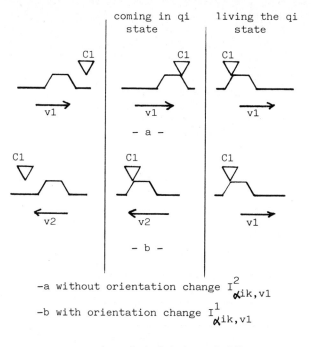

coming in qi   living the qi
state   state

-a-

-b-

-a without orientation change $I^2_{\alpha ik,v1}$

-b with orientation change $I^1_{\alpha ik,v1}$

Associated interval Iik

- figure 2 -

Testing method: the previous method is available, but the automata are evolutive at any time according to the present control vector value applied.

4 PERFORMANCE EVALUATION

We define a failing automaton by introducing failure in the model.
Let d be an application such as:
$$d: r1 \longmapsto r2 \quad \text{with } r1, r2 \in R,$$
which is caused by a failure (on a sensor for example).
In a weighted automaton, we can replace any arrow such as (qi,r1,qk) by arrow (qi,r2,qj) if this second arrow is initially present in the weighted automaton. We obtain a failing automaton Ad.
Figure 3 shows transformation of figure 1 automaton into failing automaton if the failure is blocking up C1 sensor.

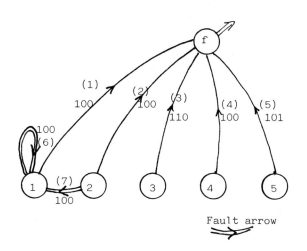

Fault arrow

Failing automaton

- figure 3 -

If qj is not the final state qf, we obtain a fault arrow. A fault arrow transition cause a location loss for the testing device. Figure 3 shows arrows 6 and 7 are fault arrows.

Recovery rate

Let P*i be probability to be in qi state in a normal running. Recovery rate is the sum of probabilities assigned to nodes qi which are origin for arrows such as (qi,r,qf) in failing automaton.

Detection rate

Detection rate is the sum of probabilities assigned to each node which is starting point for a way to final state qf in failing automaton.

Table 1 shows recovery rate for Ad automaton (figur 3). Δt is total duration in correspondent state. T is duration of observation from starting point moment.

5 SETTING OF THE TEST DEVICE

5.1 Architecture

Some architectures mono or multiprocessing are available. But the inpout/output device duplication is to be avoided in order to maintain the reliability. Any architectures such as networks are an interesting approach (figure 4).

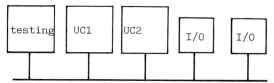

network bus

network with testing device

- figure 4 -

5.2 Modeling and data-base

For the setting of the model, we have defined a descriptive language. The process plant is decomposed into two parts.
The first part generates the controls from the orders. It is a coupling of sequential cells (flip-flop,...). These cells are grouped in a library.
The second part includes the automata modeling the trajectories according to the controls issued from the other part.
The data-base is automatically created in relation with the descriptive file. The elements for the necessary starting sequence are automatically added. This data-base is used by the on-line test monitor. The durations used in predicats are obtained in a learning phase at the starting point of the process.

6 CONCLUSION

Some restrictive hypotheses have been necessary. So our testing method is available for discrete applications. In the respect of these hypotheses, we have show that we can obtain a

good recovery rate of the errors. With this
testing method, we can find the location of
failure causing an error observed.
This system allows a detection of errors and
location of failures, which is most useful for
any intelligent reaction of control computer.

REFERENCES

1 M.CORAZZA <u>Techniques mathématiques de la
  fiabilité prévisionnelle</u> Sup-Aero editor 1975

2 J. DEFRENNE Modélisation de la partie opéra-
  tive; impact sur la sécurité et la maintenance des systèmes à évolution séquentielle
  <u>Thèse de Docteur</u> es-sciences Université des
  Sciences et Techniques de Lille.

3 K.S.FU Syntactic methods in pattern recognition <u>Mathematics in science and engineering</u>
  vol 112 Academic Press 1974

4 M.F.Mc INALLY Machine diagnostics improved
  with programmable controllers <u>Control Engi-
  neering</u> juilet 1982

5 J.M. TOULOTTE, J. DEFRENNE, R. HACHEMANI
  Etude et réalisation d'un automate à sécurité intégrée <u>Contrat ADI 83/351 Report</u>

arrows	report	probabilities			extremity
		V1	V2	V3	
1	100	O	X	X	
2	100	$P*2$	$P*2(1-\Delta t/t2$	$P*2$	
3	110	$P*3$	$P*3$	$P*3$	qf
4	100	$P*4$	$P*4$	$P*4$	
5	101	$P*5$	$P*5$	$P*5$	
6	100	$P*1$	X	X	$\neq$qf
7	100	X	$\Delta t/t2 \cdot P*2$	X	

X no purpose

recovery rate = $1-P*1$ for V1
recovery rate = $1-P*1- \Delta t/T$ for V2
recovery rate = $1-P*1$ for V3

recovery rate for blocking up C1 sensor

        - table 1 -

Reliability Technology — Theory & Applications
J. Møltoft and F. Jensen (Editors)
© Elsevier Science Publishers B.V. (North-Holland), 1986

STATISTICAL MODELLING OF ACCIDENT CONSEQUENCE PROCESSES

L. Olivi
Commission of the European Communities
Joint Research Centre - Ispra Establishment
21020 Ispra (Va) - Italy

A modelling method of accident consequence processes within the scope of the system identification by statistical inference is presented. The prior knowledge and expert information with the data of experience are formally dealt with in the same context: the information space. The model identification and the parameter estimation is provided by following a Bayesian approach: the Zellner approach. Performances of the method are the classification of the variables according to their importance in the physical process and the probabilistic assessment on the consequence of the accident.

1. INTRODUCTION

Technological development and complexity of the plants are closely related. The management of possible consequences of abnormal behaviour is vital from both safety and economic points of view. It requires the knowledge of the physical processes which could be triggered off by some abnormal state of the plant and their evolution. Generally, the only information available is based on design technical data, engineering experience and data from simulation scenarios. The task of the analyst is an adequate modelling of the accident consequence processes addressed to the decision making. That can be done by an optimal exploration and interpretation of the information basis. Indispensable requirements are:

- choice of the information pieces according to optimality criteria;
- modelling coherently with some theoretically well-founded background;
- criteria of adequacy available;
- modelling flexible to the increase of information quantity.

This lecture presents a procedure of analysis developed at JRC Ispra within the scope of the system identification by the techniques of statistical inference [1,2,3]. The main steps of the procedure are:

- identification of an enlarged experiment space (the information space) on the universe of information (i.e. subjective judgement, previous experience, simulation scenarios and technical data);
- definition of the techniques of information detection in the information space, that is the observation campaign;
- definition of the model identification procedures in such a space by the Bayes approach;
- disaggregation of the system under study, according to a subjective judgement, in subsystems grouping variables with different effects on the output, i.e. negligible, uncertain and important;
- analysis of the subsystem producing the uncertain effect by an observation campaign on its variable subset plus an important variable taken as a reference;
- rank ordering with respect to the reference variable and identification of the variables to be joined to the important ones;
- definitive disaggregation of the system in two subsystems grouping the important variables (model variables) and the unimportant ones (latent variables);
- modelling and experimental design on the model variable space; simultaneous random design on the latent variable space;
- fitting of the model; residual error analysis;
- uncertainty propagation through the identified model by Monte Carlo simulation.

2. THE SPACE OF THE SYSTEM, THE SPACE OF THE EXPERIMENTS, THE SPACE OF THE INFORMATION

The behaviour of a system is, in general, function of a large set of variables and physical parameters. In modelling applications, its identification is performed on a reduced subset of variables, considered as the most important in producing the particular effect of interest. A preliminary approach to system identification is, then, a sensible definition of the space of the experiments, i.e. the space of the variables where the system is observed.

As far as "natural" systems are concerned, classical statistics makes use of techniques for an optimal choice of the important variables, being considered as a random disturbance that cannot be controlled or without an important effect on the output of the system. In this case the choice of the experiment space is quite obvious and it is suggested from the context.

In general, a system can be defined in a space, the space of the system S_s, where all the effects are taken into account and the system behaviour can be fully represented by a function of all its independent variables
$x^* \equiv (x_1^*, \ldots, x_s^*)'$

$$y = f(\underline{x}^*) \tag{1}$$

In this picture the space of experiments S_E can be identified as the subspace of the m variables which most effectively determine the response of the system.

An observation in this space provides a response corresponding to the value of $f(\underline{x}^*)$ for well defined values (designed values) of the m variables of the experiment space and for uncontrolled values of the remaining s-m not observed variables (latent variables). If for sake of simplicity one refers to a linear modelling, the system behaviour can be easily represented by the standard multiple linear regression model.

$$\underline{Y}_E = X_E \underline{\beta}_E + \underline{u}_E \qquad (2)$$

where X_E is the n×m non-stochastic design matrix of the m observed variables, $\underline{Y}_E$ is the n×1 vector of the observations on $\underline{y}$, and β_E is the m×1 unknown parameter vector, while $\underline{u}_E$ is the n×1 error vector which can be thought of as drawn from a multivariate normal distribution $N(0,\sigma^2 I_n)$. It represents the random contribution in the experiment space S_E of the residual effect of $\underline{Y}_E$ of the latent variables $\underline{x}_L \in S_L = S_S \cap CS_E$. This effect is due to the contribution of all the $\underline{x}_L$ singularly, their powers, their interactions and their interactions with the experiment space variables. When these interactions become not negligible, the previous assumptions on the error have to be reviewed because of the correlations among the errors.

From the experimenter point of view, the experiment space is the space he can access to or the space he decides to investigate, while S_S is the space where all the information is contained but not always accessible.

The latent variables space contains two groups of variables. The ones, either observable or unobservable which globally can be considered definitely unimportant according to previous experience and founded knowledge. They will be indicated with $\underline{x}_U \in S_U$. The other ones, indicated with $\underline{x}_R \in S_R$, which singularly need a procedure of identification in order to be rejected from the model.

$$S_L = S_R \cup S_U$$

Now the information space can be defined, i.e. the space where all the available information either the "actual" or the "potential" is contained

$$S_I = S_R + S_E$$

The "actual" information is that one already available before the experiment, it condenses all that one can know about the system from the previous experience and expert knowledge. The "potential" information is that one carried out by replicating either direct observations on the system or indirect observations such as those obtained by a simulation program. The information space is, therefore, a generalization of the usual experiment space, once it is established how to perform the observations to extract and homogeneously quantify either the

"actual" or the "potential" information. In such case, in fact, the modelling procedures are unified and pieces of information coming from different sources can be represented in the same formal context. Hence, a linear model of (1) can still be represented by

$$\underline{Y} = X\underline{\beta} + \underline{u} \qquad (3)$$

where X is the n×k non-stochastic design matrix of $\underline{x} \in S_I$, β the k×1 coefficient vector, while $\underline{u}$ is the error due to the noise contribution of only the globally unimportant variables $\underline{x}_U \in S_U = S_S \cap CS_I$, assumed as drawn from a multivariate normal distribution $N(0,\sigma^2 I_n)$.

3. THE OBSERVATION CAMPAIGN IN THE INFORMATION SPACE

A campaign of observation in the information space is a sampling on y as it is generated by the model (3). Since, in general, several methods of observation can be used, the quality of the observation received depends on the particular sampling applied.

Refering again to "natural" systems, when the experimenter is able to decide that all $\underline{x}_R$ can be rejected, the representation of the model (3) coincides with (2), because the $\underline{x}_R$ variables do not need singularly to be identified as uneffective on the system response. The modelling becomes more complex when the variables to be introduced into the model need to be assessed according to specific criteria, as it occurs in most of the technological studies. As an example, one can take the study of an accident consequence process in a nuclear power plant.

In this case, the analyst has to perform preliminary experiments in the information space in order to correctly identify the definitive experiment space. The "actual" information is provided by the acquired experience on the problem, subjective expertise and technical design data. The "potential" information is carried out by observation of data, i.e. the data observed directly in the plant during the accident and by simulation programs.

With reference to the model (3), it can be envisaged that three possible samplings can be generated

$$\underline{Y}_o = X\beta + \underline{u}_o \qquad (4)$$

where $\underline{Y}_o$ represents a "conceptual" sampling made on the "actual" information sources, and $\underline{u}_o$ the associated random error,

$$\underline{Y} = X\underline{\beta} + \underline{u} \qquad (5)$$

where $\underline{Y}$ represents a sampling made by a simulation program in the information space for exploring the "potential" information, and $\underline{u}$ the error associated with the indirect observations,

$$\underline{Y}_D = X\underline{\beta} + \underline{u}_D \qquad (6)$$

where $\underline{Y}_D$ is the direct observation of the accident variable y in the points of design matrix on the plant to be studied, and $\underline{u}_D$ the error associated with the direct observations.

There is a qualitative difference among the data obtained; the conceptual sampling provides values assessed by subjective assessment, they are considered as random variables as well as those obtained by measurement, while the degree of precision of the assessment of an expert plays the same role as the precision of a measure.

The simulation program can be considered as an instrument of measure to investigate in the information space and, in principle, one can attribute a degree of precision to the data received according to subjective criteria or on the basis of sample tests.

The direct observations, which are not feasible in practice, unless some pilot experiment in laboratory, provide the same quality of data as that obtained in a standard experiment on a "natural" system. The "actual" information can be formalized as informative prior distribution, the "potential" information can be estimated as a normal likelihood function of the sample observation. Finally, the posterior probability density function of $\underline{\beta}$ vector obtained by applying Bayes theorem, will condensate all the information available.

4. THE ASSESSMENT PROCEDURE

The topic developed in the previous chapters implies that the formalization of the a priori knowledge and the expert judgement in terms of informative prior distributions is feasible. This important point has been dealt with by Zellner [4] (1983), who provided a very effective method to obtain a satisfactory representation of the a priori knowledge by informative prior distributions formalized by him as reference informative priors (RIP's) because they are able to indicate a standard of reference. The same approach has been exploited by Calvi Parisetti and Olivi [5] (1985) for the identification of important parameters in complex technological systems.

In the following, the procedure of assessment of the unknown model parameter vector $\underline{\beta}$ by Zellner approach will be briefly reminded. It will clearly appear that both the "actual" and "potential" information can be fully exploited, the posterial mean $\underline{\beta}_p$ will provide an assessment of the parameter vector $\underline{\beta}$ and the covariance matrix $V_p(\underline{\beta})$ a measure of the precision of such assessment.

Let us refer to the model (3) in the information space and consider the conceptual sampling (4) assuming $\underline{u}_o$ drawn from $N(0, \sigma_o^2 I_n)$. $\underline{Y}_o$ is generated by the same design matrix, but σ_o^2 is allowed to be different from σ^2, let be $\sigma^2 = g\sigma_o^2$, $0 < g < \infty$. With a diffuse prior for $\underline{\beta}$ and σ, $p(\underline{\beta}, \sigma) \propto 1/\sigma$; the posterior probability density function based on the sampling (4) is:

$$p(\underline{\beta}, \sigma | \underline{Y}_o, X) \propto \sigma^{-(n+1)} \exp\{-g(\underline{Y}_o - X\underline{\beta})'(\underline{Y}_o - X\underline{\beta})/2\sigma^2\}$$

$$\propto \sigma^{-(n+1)} \exp\{-g[\nu s_o^2 + (\underline{\beta} - \hat{\underline{\beta}}_o)'X'X(\underline{\beta} - \hat{\underline{\beta}}_o)]/2\sigma^2\}$$
(7)

$$\nu = n-k$$

$$\hat{\underline{\beta}}_o = (X'X)^{-1}X'\underline{Y}_o$$

$$\nu s_o^2 = (\underline{Y}_o - X\hat{\underline{\beta}}_o)'(\underline{Y}_o - X\hat{\underline{\beta}}_o)$$

It can be seen that the posterior pdf is expressed as a function of $\underline{\beta}_o$ and s_o^2, which, in principle, could be estimated after a direct sampling on the system.

Let us assume to have performed this "conceptual" sample on the system as it can be generated by (4), where the direct observations have been substituted by the expert elicitation. In this experiment, the model does not change whilst the quality of the information received does change. In particular, we suppose that anticipated values for β and σ^2 are available. Zellner, in his approach, invokes the rational expectations hypothesis (Muth, 1961) [6] and then takes:

$$\underline{\beta}_a = E(\underline{\beta} | \underline{Y}_o, X) = \hat{\underline{\beta}}_o$$
(8)

and

$$\sigma_a^2 = E(\sigma^2 | \underline{Y}_o, X) = \nu g s_o^2 / \nu - 2$$

where $\underline{\beta}_a$ and σ_a^2 are the anticipated values for $\underline{\beta}$ and σ^2, and $E(\underline{\beta} | \underline{Y}_o, X)$ and $E(\sigma^2 | \underline{Y}_o, X)$ are the means associated with (7). Putting

$$g s_o^2 = \frac{\nu-2}{\nu} \sigma_a^2 \equiv \bar{\sigma}_a^2$$

the joint RIP for $\underline{\beta}$ and σ is

$$p(\underline{\beta}, \sigma | \underline{Y}_o, X, \bar{\sigma}_a, \underline{\beta}_a) \propto \sigma^{-(\nu+1)} \exp\left[-\frac{\nu\bar{\sigma}_a^2}{2\sigma^2}\right] \sigma^{-k}$$

$$\exp\{-g(\underline{\beta} - \underline{\beta}_a)'X'X(\underline{\beta} - \underline{\beta}_a)/2\sigma^2\}$$
(9)

with marginal prior pdf's

$$p(\sigma | \nu, \bar{\sigma}_a) \propto \sigma^{-(\nu+1)} \exp(-\nu\bar{\sigma}_a^2/2\sigma^2)$$
(10)

and

$$p(\underline{\beta} | \underline{\beta}_a, g, \nu) \propto \{\nu\bar{\sigma}_a^2 + g(\underline{\beta} - \underline{\beta}_a)'X'X(\underline{\beta} - \underline{\beta}_a)\}^{-(\nu+k)/2}$$
(11)

The joint prior in (9) is in the inverted-gamma-normal form, the marginal prior for σ in an inverted gamma and that for $\underline{\beta}$ is a multivariate Student-t.

The mean and the covariance matrix can now be derived from (11):

$$E(\underline{\beta}) = \underline{\beta}_a$$

$$E\left[(\underline{\beta}-\underline{\beta}_a)(\underline{\beta}-\underline{\beta}_a)'\right] = (X'X)^{-1}\nu\bar{\sigma}_a^2/g(\nu-2) =$$

$$\frac{\sigma_a^2}{g}(X'X)^{-1} \qquad (12)$$

and from (6)

$$E(\sigma^2) = \frac{\nu\bar{\sigma}_a^2}{\nu-2} = \sigma_a^2$$

The expression in (9) is the joint RIP for $\underline{\beta}$ and σ. Therefore, if $\underline{\beta}_a$ and σ_a are given by the previous knowledge and by acceptance of the rational expectations hypothesis, one can deal with RIP's as the prior pdf's and obtain posterior pdf's based on a normal likelihood function for the sample observation as it is generated by (5) and (6). In particular, the posterior mean, $\underline{\beta}_p$ and the covariance matrix $V_p(\underline{\beta})$ are:

$$\underline{\beta}_p = E(\underline{\beta}) = (\hat{\underline{\beta}} + g\underline{\beta}_a)/(1+g)$$

and

$$V_p(\underline{\beta}) = E\left[(\underline{\beta}-\underline{\beta}_p)(\underline{\beta}-\underline{\beta}_p)'\right] = \frac{\nu_1 s_1^2}{(\nu_1-2)(1+g)}(X'X)^{-1}$$

$$(13)$$

where

$$\nu_1 = 2n-k$$

$$\nu_1 s_1^2 = (\nu-2)\sigma_a^2 + (\underline{Y}-X\hat{\underline{\beta}})'(\underline{Y}-X\hat{\underline{\beta}}) +$$

$$+ g(\underline{\beta}_a-\underline{\beta}_p)'X'X(\underline{\beta}_a-\underline{\beta}_p) +$$

$$+ (1+g)(\hat{\underline{\beta}}-\underline{\beta}_p)'X'X(\hat{\underline{\beta}}-\underline{\beta}_p)$$

Equations (13) enlighten the meaning and the role of g as the degree of precision of the assessment of the experts and as a measure of their contribution to the posterior pdf's.

5. THE IDENTIFICATION OF THE EXPERIMENT SPACE

The ultimate goal of the presented methodology is the modelling of an accident consequence process, i.e. the modelling of a given physical consequence as a function of the significant variables which single out the accident state of the system, namely the accident space, as the space which has to be explored in order to obtain the "potential" information concerning the consequences. In this context the accident space coincides with the experiment space.

The previous development indicates the procedure to be followed:

i) Identification of the information space, i.e. identification of all the variables which singularly need to be assessed as significant or not significant. In the specific case they can belong either to S_E or to S_R.

ii) With reference to the model (3), the Zellner assessment procedure can be applied and the RIP's for β can be used to calculate the mean and the covariance matrix as in form (12), which depends on g.

If g is extremely large, in principle one receives directly from the expert elicitation the prior information needed to decide on the screening of the variables. That is because the experts certainly know which variables to introduce in the model and which to exclude. By decreasing the g value, the boundary between the two classes of variables, the ones to be certainly included and the ones to be certainly excluded, becomes more and more uncertain. The insufficient degree of precision makes the decision on the attribution to either class of a finite subset of variables unfeasible. Such a subset will be indicated as the subset of the "fluctuating" variables.

iii) As a consequence, an assessment on the importance of the "fluctuating" variables becomes indispensable and, therefore, an experiment of direct sampling to get the "potential" information needed has to be carried out. In principle, one can design an experiment on the whole set of the important and "fluctuating" variables and, then, assess the posterior pdf's. Such a way is often unfeasible because of the very large number of observations needed. In addition, the already gained information on the important variables is not optimally exploited. It is reasonable to think that the parameters of the physical processes occurring during the system evaluation do not drastically change, in the sense that the major effects contributing to the output conserve their relative importance within the range of the variables observed. Therefore, it seems sensible to observe the variation of the output only for the limited subset of the "fluctuating" variables, being fixed at their working values the important ones and randomized in their ranges the globally unimportant ones ($\underline{x}_U \in S_U$). In addition, a term of reference for the importance has to be introduced in order to apply a unique classification criterion. For this reason the experiment space for such preliminary experiment is set up by all the "fluctuating" variables plus a "reference" variable chosen as one of intermediate importance among the important variables.

iv) The analysis is performed as follows:

- The variables chosen to be definitely included in the model, except one chosen as a "reference" variable, are fixed at their nominal values (usual working values according to the system under study).

- The excluded variables ($\underline{x}_U \in S_U$) take random values, sampled in their range.

- The "fluctuating" variables plus the one chosen as a "reference" are designed according to a simple design (e.g. a factorial one) in order to identify the principal factors.

- A submodel of the form (3) is identified.

And by the results:

- The multiple correlation coefficient is an index of soundness for the a priori exclusion of the $\underline{x}_U$ variables.

- Taking as a reference the posterior pdf of the regression coefficient for the "reference" variable, the posterior pdf's of the parameters can be compared, i.e. the probability that each of the "fluctuating" variables is more or less important than the "reference" variable can be assessed.

This choice procedure can be tested on the final model by analysing the random error due to the randomization of the whole latent variables ($\underline{x}_L \in S_L$) subset and by possible replicates in significant points by applying statistical tests.

The accident variables can now be definitely established. They are: the ones included in the model exploiting the "actual" information and the features of RIP's for β to single out the "certainly" important variables according to expert elicitation, and those added to them after the previous procedure of choice among the "fluctuating" variables. Henceforth, their space will be considered as the ordinary experiment space S_E where all the "potential" information is contained.

6. MODELLING AND EXPERIMENTAL DESIGN

The identification of the experiment space S_E and of the latent variable space S_L enables the accident-consequence modelling to be performed according to the ordinary approach of statistical inference. The observation campaign can be carried out in S_E according to the most appropriate experimental design for the model, whilst the S_L variables are randomized according to their probability density function.

In order to limit the number of observations, sequential design [1,7] can be taken into account in order to aggregate the most effective model terms according to their rank ordering. One can envisage, for instance, a first subdesign as the central composit design (CCD) for a second-order polynomial model without interactions, and identify - in this submodel - the rank ordering of the principal factors. The aggregations in interactions of the most effective model terms can, then, be decided. Other parallel subdesigns, e.g. fractional factorial designs (FFD), for interaction term inclusion, can be experimented. In this way, either the model (5) or (6) can be applied to exploit the "potential" information contained in S_E. The adequacy of the model can be verified by applying the usual criteria of statistical inference, being the noise coming from the randomization of S_L the only contribution to the random error. Residual error analysis and possible replicates in some crucial point can be exploited for the application of statistical tests.

However, the ultimate judgement of the model's adequacy is always dependent on the problem solving context, the statistical features and other figures of merit remain an indication for the decision of the analyst. In particular, the identification of an adequate

response surface requires that its behaviour fits to a possible expected reality, within the problem to be studied [7]. Therefore, the response of the model has to be critically analysed and judged, taking into account both the limit of the information sources exploited, e.g. the number of the observations in S_E, and the engineering expertise.

7. UNCERTAINTY PROPAGATION

Ultimate goal of a safety study is the assessment of the probability that a given consequence is produced in the accident studied. The probabilistic nature of the consequence is due to the uncertain knowledge about the accident state of the system; in general the accident variables are identified as random variables in order to represent in an accessible way, from a formal point of view, the information about their state.

The identified model of the accident consequence process provides a formal relationship between the accident space (the previously explored experiment space S_E) and the consequence under study (the onput of the model, e.g. (5) or (6)). In the limits of the adequacy of the modelling, one can use the model form in order to evaluate how the uncertainties on the input variables contribute to the uncertainty on the output. The Monte Carlo simulation technique is the natural tool to be used.

8. CONCLUSION

The aim of this work was to provide a systematic rigorous approach to the safety analysis of complex technological systems. A major effort has, therefore, been devoted to the foundation of the method rather than to the specific procedures already otherwise presented [1]. Either the inferential performance or the limits of the application of the method have been enlightened. A case example, recently published in the literature [7] has clearly shown how to explore the "actual" information by a very simplified procedure in order to assess the preliminary choice of the accident variables and, then, the practicality and the effectiveness of the method. This approach is generally known as the Response Surface Methodology (RSM) [1] and it is of current use in the most advanced reliability studies in the nuclear safety background, but it starts to be extended to other fields [8] as a tool of sensitivity and uncertainty analysis.

ACKNOWLEDGEMENTS

The present work could profit of an indispensable contribution given to the presented methodology by Prof. C. Calvi Parisetti [5]. The author is much indebted to her because of the considerable effort provided in the common research.

REFERENCES

1   L. Olivi ed., RSM Handbook for Nuclear
    Reactor Safety Analysis, EUR 9600 EN (1984).
2   L. Olivi, F. Brunelli, P.C. Cacciabue,
    P. Parisi, Response surface methodology
    for sensitivity and uncertainty analysis:
    performances and perspectives. Proc. of
    ANS/ENS Int. Top. Meet. on Probabilistic
    Safety Methods and Applications, S. Fran-
    cisco, California, 24/2 - 1/3, 1985.
3   L. Olivi, The universe of information as
    experiment space in uncertainty analysis
    for accident consequence models of complex
    systems. Workshop on Methods for Assessing
    the Off-site Radiological Consequences of
    Nuclear Accidents, CEC, Luxembourg, April
    15-19, 1985.
4   A. Zellner, Application of Bayesian ana-
    lysis in econometrics. The Statistician
    32 (1983).
5   C. Calvi Parisetti and L. Olivi, Identifi-
    cation of the important parameters of
    complex systems by a Bayesian approach.
    Proc. of the 5th EuReDatA Conference,
    Heidelberg, April 9-11, 1986.
6   J.F. Muth, Rational expectations and the
    theory of price movements. Econometrica
    29, 315-35 (1961).
7   P.C. Cacciabue, L. Olivi, P. Parisi and
    F. Brunelli, Assessing the uncertainties
    in reactor accident consequences by means
    of RSM: the case of an ATWS in a PWR.
    Nuclear Engineering and Design 88 (1985)
    179-194.

BIOGRAPHY

L. Olivi is Doctor in Physics. In such discipline he worked up to 1973, when the JRC programs required a major effort on simulation
and modelling of chemical and physical process
interesting either environmental systems or
nuclear safety assessment.

He devoted his major interest to modelling of
systems in population dynamic problems and
statistical modelling of accident consequence
processes in nuclear safety, producing Conference and Review papers.

Since 1980 he received the charge to develop
and adapt Response Surface Methodology techniques for nuclear safety purposes.

The Response Surface Methodology Handbook,
edited in 1984, summarizes all the results obtained at JRC Ispra as well as from external
collaborations.

Reliability Technology — Theory & Applications
J. Møltoft and F. Jensen (Editors)
Elsevier Science Publishers B.V. (North-Holland), 1986

THE USE OF TIME SERIES ANALYSIS IN THE RELIABILITY FIELD

Elena Madiedo
Metro-Dade Transportation Administration
Miami, Florida, USA

This paper illustrates by an existing application the building of a time series model and the use of such model in the reliability field.

This paper describes how the data is collected, and explains the iterative stages in the building of a time series model from the data. The objective of creating this model is to assess statistically the impact of a unit modification program on a certain failure condition. Emphasis is placed in the iterative stages to build a time series model by illustrating it with an existing application.

1. INTRODUCTION

Due to the excessive number of failures of the propulsion unit of the vehicles used in the rapid transit system, a unit modification program was implemented to reduce the existing number of failures.

A time series model was built to test the null hypothesis that the unit modification program caused an impact in the average daily number of failures occurrence.

This paper is set out in two sections:

The first section provides an explanation of Box-Jenkins approach to time series analysis, and the iterative stages used to build a time series model, [1] and [2].

The second section illustrates the building of a time series model from failure data occurring prior to the implementation of the unit modification program, and how this model is used to assess statistically the impact after implementing this program.

2. SECTION I

In this section the concepts of Box-Jenkins approach to time series analysis, [1] and [2], are presented.

A time series is a set of observations generated at equally spaced intervals of time. A stochastic process is a sequence of observations in that each observation has a finite number of outcomes with given probability. Thus, a time series of N observations is considered as a sample from an infinite population of such time series that has been generated by the stochastic process.

Time series analysis refers to the iterative stages used to build a model of the stochastic process that generated the time series. The time series models are built around three stochastic process components: The AutoRegressive, Integrated, and Moving Average components. These models are called ARIMA models, and they are denoted as ARIMA (p,d,q) models, where p,d, and q represent the autoregressive, integrated and moving average stochastic process components. ARIMA seasonal models are denoted by ARIMA (p,d,q) $(P,D,Q)_S$, where P,D, and Q are analogous to p,d, and q, and S indicates the length of the seasonal cycle.

The time series models used by Box-Jenkins are based on the assumption that a time series Y_t, Y_{t-1},...., in which successive values are highly dependent, is considered as generated from a series of independent random variables a_t, a_{t-1}, ..., which have mean zero and a constant variance σ_a^2. The sequence a_t is a white noise process or stochastic process of the time series.

The time series models for describing time series may be either stationary or nonstationary. A stationary time series is represented by an ARIMA (0,0,0) model or white noise process that assumes that the stochastic process remains in equilibrium about a constant mean level. A nonstationary time series is represented by an ARIMA (0,d,0) model, implying that a time series is a white noise process after being differenced d times. While stationarity is a necessary condition of an ARIMA model, however, it is not a sufficient condition, the time series must also have a stationary variance. The series can be made stationary in both level and variance by transformation and differencing prior analysis.

Autoregressive processes of order p are modeled by using p lagged observations of the stochastic process and a random variable a_t to predict the current observation at time t.

Stationary and nonstationary autoregressive processes are represented by ARIMA (p,0,0), and ARIMA (p,d,0) models respectively. These models contain $p+2$ unknown parameters μ, Φ_1,..., Φ_p, σ_a^2 that are estimated from the data. The parameter σ_a^2 is the variance of the white noise process a_t, u is the mean in which the stochastic process varies, and the Φs are the autoregressive parameters.

Moving average processes are characterized by a finite persistence of a random variable that persists for no more than q observations, and then it disappears. Stationary and nonstationary moving average processes are represented by ARIMA (0,0,q), and ARIMA (0,d,q) models respectively. The model contains $q+2$ unknown parameters μ, Θ_1,..., Θ_q, σ_a^2 that are estimated from the data. The Θs are the moving average parameters.

The Box-Jenkins interative stages used to build an ARIMA model are:

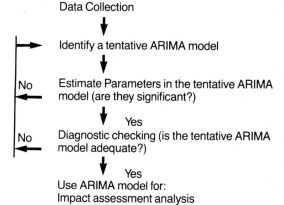

Data Collection

Identify a tentative ARIMA model

No — Estimate Parameters in the tentative ARIMA model (are they significant?)

Yes

No — Diagnostic checking (is the tentative ARIMA model adequate?)

Yes

Use ARIMA model for:
Impact assessment analysis

2.1 DATA COLLECTION

An essential requirement in using time series analysis is that observations must occur at equally spaced intervals of time, and because the ARIMA models are built empirically from the data, relatively long time series are required. At least 50 observations or more should be used.

2.2 IDENTIFICATION

A preliminary identification of the ARIMA model components is accomplished by an examination of the autocorrelation function (ACF), and the partial autocorrelation function (PACF) In this stage, the patterns of the ACF, and the PACF estimated from the time series are examined to identify a tentative ARIMA model.

For a time series process, Y_t, the ACF at lag k is a measure of correlation between Y_t and Y_{t-k}. The estimated auto-correlation ACF (k) is calculated for $k = 0,1,2,..,L$, where L is not larger than N/4, and N is the number of observations.

The PACF at lag k is a measure of correlation between time series observations k units apart after the correlation at intermediate lags has been controlled.

For ARIMA (0,0,0,) model or white noise process, the estimated ACF is zero for all lags. The ACF of a nonstationary process has a positive value for ACF (1) and the successive lags of the ACF die out slowly to zero. Autoregressive processes are characterized by spiking PACFs, and ACFs that decay exponentially. An autoregressive process of order p has p nonzero spikes in the first p lags of its PACFs. All successive lags of the PACF are zero. Conversely, moving average processes are characterized by spiking ACFs and PACFs that appear to decay exponentially. A moving average process of order q have q nonzero spikes in the first q lags of its ACFs. All successive lags of the ACFs are zero. Mixed autoregressive and moving average processes have both decaying ACFs and PACFs. Finally, for time series with a seasonal cycle S, the patterns of spiking and decaying are analogous to the regular ARIMA processes, except that they occur at seasonal lags.

2.3 ESTIMATION

In this stage, the parameters of the tentative model are estimated. The estimated parameters must be statistically significant and must lie within the bounds of stationarity and invertibility for autoregressive and moving average parameters respectively. For ARIMA (1,0,0) (1,0,0)$_S$ and ARIMA (2,0,0) (2,0,0)$_S$ models the bounds of stationarity for autoregressive parameters are: $-1 < \Phi_1$, $\Phi_S < 1$, $-1 < \Phi_2$, $\Phi_{2S} < 1$, $\Phi_1 + \Phi_2 < 1$, $\Phi_2 - \Phi_1 < 1, \Phi_S + \Phi_{2S} < 1$, $\Phi_{2S} - \Phi_S < 1$. For ARIMA (0,0,1) (0,0,1)$_S$ and Arima (0,0,2) (0,0,2)$_S$ models the bounds of invertibility for moving average parameters are: $-1 < \Theta_1$, $\Theta_S < 1$, $-1 < \Theta_2$, $\Theta_{2S} < 1$, $\Theta_1 + \Theta_2 < 1$, $\Theta_S + \Theta_{2S} < 1$, $\Theta_2 - \Theta_1 < 1$, and $\Theta_{2S} - \Theta_S < 1$. If the estimated parameters do not satisfy either criteria, the tentative model must be rejected, and another model must be identified, and its parameters estimated.

2.4 DIAGNOSTIC

To test if the tentative model is adequate, the estimated ACF for its residuals must satisfy two criteria:

(i) The ACF for the model residuals at the first two lags must have no statistically significant values, that is, ACF (1) = ACF (2) = 0, and for seasonal models, ACF (S) = ACF (2S) = 0.

(ii) The ACF for the model residuals must be distributed as a white noise process. The Q statistic is used to test whether the ACF for the entire residuals of the model is a white noise process. The Q statistic is distributed approximately chi-square with the degrees of freedom (DF) = k-p-q-P-Q. If the Q statistic for the residual ACF is statistically significant, the model residuals are not white noise, and the tentative model must be rejected.

If the model is rejected, another model must be identified, its parameters estimated, and its residuals diagnosed. The iterative stages continues until a statistically adequate model has been built for the time series.

2.5 IMPACT ASSESSMENT ANALYSIS

Impact assessment is defined, [2], as a test of the null hypothesis that a postulated event caused a change in the stochastic process measured as a time series. Because the change agent is an event, it is represented in the impact assessment model as a dummy intervention variable or step function such that:

$$I_t = \begin{cases} 0 & \text{prior to} - \text{Preintervention} \\ & \text{the event} \\ 1 & \text{thereafter} - \text{Postintervention} \end{cases}$$

The impact assessment model is written as $Y_t = F(I_t) + N_t$, where N_t denotes the noise component, an ARIMA model, and $F(I_t)$ denotes the intervention component. The intervention component itself describes the deterministic relationship between an event represented by the variable I_t, and the time series. The noise component describes the stochastic behavior of the time series around the $Y_t = F(I_t)$ relationship.

Because the impact of the event in the time series may be so large that it can distort the ACF and PACF, the impact assessment analysis begins with an ARIMA model for the preintervention series only. This ARIMA model describes the stochastic behavior of the time series process, and it refers to the noise component of the impact assessment model. After a statistically adequate ARIMA model has been built for the preintervention series, an intervention component is added. The parameters of the impact assessment model, both noise and intervention components, are then estimated. If the estimated parameters of the noise component are not statistically significant or lie outside the bounds of stationarity-invertibility for autoregressive and moving average respectively, then another noise component must be identified and its parameters estimated.

After a statistically adequate noise component has been built, and the estimated parameters of the impact assessment model are statistically significant and acceptable, the impact assessment model must be diagnosed, that is, the ACF of the model residuals must be distributed as a white noise process. If the model is not statistically adequate, it must be rejected, and the iterative stages continues until a statistically adequate impact assessment model is built.

3. SECTION II

Figure 1 shows the propulsion unit failures occurring in time intervals of one day. On the 102nd observation the unit modification program was implemented. Then through 99 days the data was collected for analysis impact.

We want to test the null hypothesis that the unit modification program caused an impact in the average daily number of failures of the propulsion unit.

To build an impact assessment model, an ARIMA model must be built first for the preintervention time series, the first 101 observations. This model is identified as the noise component of the impact assessment model.

The time series analysis is accomplished by the use of the

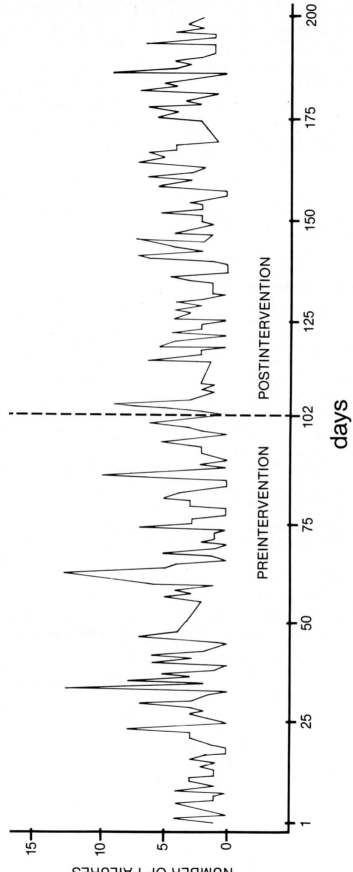

Figure 1. Propulsion unit failures by day

PACK software [3].

The ACF and PACF for the preintervention time series are shown in figures 2 and 3. Confidence intervals of ± 2 errors (SE) have been plotted around the ACFs and the PACFs. The estimated ACF (k) and PACF (k) within the confidence intervals are not significantly different than zero. By examining the ACF and PACF we identify an ARIMA $(1,0,0)$ $(2,0,0)_7$ model. The ACF decays exponentially at lag 1, lag 7, and lag 14, and the PACF spikes at lag 1, lag 7, and lag 14.

An analysis following the iterative stages previously described indicate that the ARIMA $(1,0,0)$ $(2,0,0)_7$ model is adequate. The parameters $\Phi = .24$, $\Phi_7 = .20$, and $\Phi_{14} = .16$ are statistically significant and lie within the bounds of stationarity for autoregressive parameters. The average daily number of failures is 3.22. The residuals show no spikes at lag 1 and lag 2 or at the seasonal lags. The $Q = 18.11$ statistic with 21 DF is not statistically significant.

The significant spikes at lag 7, and lag 14 imply that failures occur more frequently on every seventh day. The probable cause of this is due to a delay in the failure reporting process.

The intervention variable is represented in the impact assessment model as:

$$I_t = \begin{cases} 0 \text{ for the first 101 observations} \\ 1 \text{ for the 102nd and subsequent observations} \end{cases}$$

A detailed analysis following the interative stages described in section I indicates that the impact assessment model is adequate. Parameter estimates are statistically significant and lies within the bounds of stationarity for autoregressive parameters. The residual has no significant spikes at all, and the Q statistic is not statistically significant.

The impact assessment analysis indicates that after the implementation of the unit modification program the average daily number of failures dropped to 2.1. Thus, there is evidence to support the hypothesis that this program had an impact in the average daily number of failures of the propulsion unit.

4. CONCLUSION

This paper illustrated by an existing application the building of a time series model and the use of such model in the reliability field to assess statistically the impact of a unit modification program on a certain failure condition.

The model presented in this paper was able to verify the quantitative achievement resulting from the unit modification program by identifying a 34% reduction in the average daily number of failures occurrence.

REFERENCES:

1 – Box, G.E.p and G.M. Jenkins, Time Series Analysis: Forecasting and Control (Holden-Day, San Francisco, California, USA, 1976).

2 – McCleary, R. and R.A. Hay, Jr., Applied Time Series Analysis (Sage Publications, Beverly Hills, California, USA, 1980).

3 – Pack, D.J., A computer program for the analysis of time series models using the Box-Jenkins philosophy (Ohio State University, Columbus, Ohio, USA, 1977).

Biography

Mailing Address:

Elena Madiedo
Metro-Dade Transportation Administration
System Safety and Assurance Division
111 N.W 1st Street, Suite 910
Miami, Florida, USA 33130

Elena Madiedo is a Reliability Engineer with the Metro-Dade Transportation Administration in Miami, Florida, USA. She is responsible for developing and managing the computerized reliability and availability system for Miami's rapid transit system. She received her BS and MS degrees in mathematics from Florida Atlantic University, majoring in probability, statistics and operations research. Her main professional interests are: applied probability and statistics, designing and developing software, and simulation.

Figure 2. Autocorrelation function for the noise component.

Figure 3. Partial autocorrelation function for the noise component.

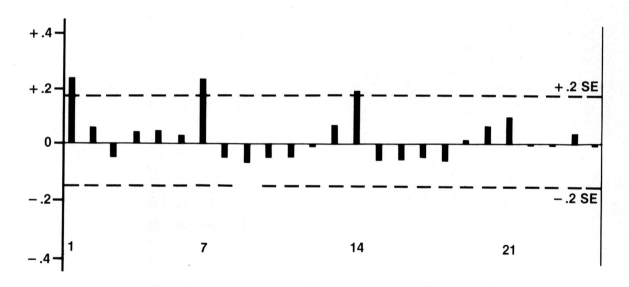

Reliability Technology — Theory & Applications
J. Møltoft and F. Jensen (Editors)
© Elsevier Science Publishers B.V. (North-Holland), 1986

215

INTEGRATED VOTER FOR TRIPLE MODULAR REDUNDANCY

Tapio PULLI
Technical Research Centre of Finland
Electrical Engineering Laboratory, Espoo, Finland

In the paper an idea of integrating the voter of TMR (Triple Modular Redundancy) into the modules is described. Its operation and characteristics are considered and the advantages and also the disadvantages are presented. Some comparisons are made with the known TMR systems.

1. INTRODUCTION

In known, classical TMR systems there is a separate voter outside the modules. In fig. 1 there is a logical realization of the voter for one binary signal, and also a hardware realization with 6 transistors and 10 resistors {1}. The realization is quite simple, but there is a need for a big voter circuit, e.g. in the case of voting on 16 binary signals, the voter must be a 66 pin circuit in minimum. If the fault detection is desired, some more hardware and pins are needed.

In this paper a different approach is described. The voter logics are integrated into the triplicated modules. The modules are entirely identical. The respective output signals of each module are wired together to compose the system output and there is no need for extra pins in the modules, so far. However, a control bus is needed, via which some information is changed between the modules; accoreding to that information the modules can conclude if some of them is faulty.

In the following the operation and characteristics of the system are described. It is compared also with known TMR-structures and the advantages and disadvantages are discussed.

2. TMR WITH INTEGRATED VOTER

In fig. 2a we have a TMR system in which the respective output signals of every module are wired together to compose the output of the system. The output buffers of the modules must be either of three-state or open-collector type. The modules are also connected by a control bus. Each module includes the following blocks: the actual module logics (e.g. microprocessors, memories, A/D converters), the output buffers and the voter logics, fig. 2b. In the voter logics there are: the comparators, the majority indicators and the status registers, fig. 2c. Of course there will be also system inputs, either separate or bidirectional with the system outputs, although they are not drawn in the fig. 2.

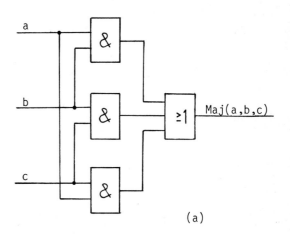

(a)

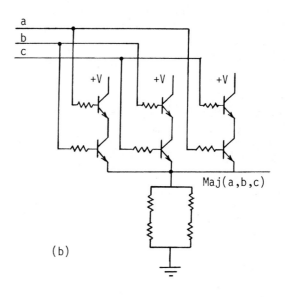

(b)

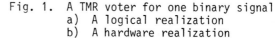

Fig. 1. A TMR voter for one binary signal
a) A logical realization
b) A hardware realization

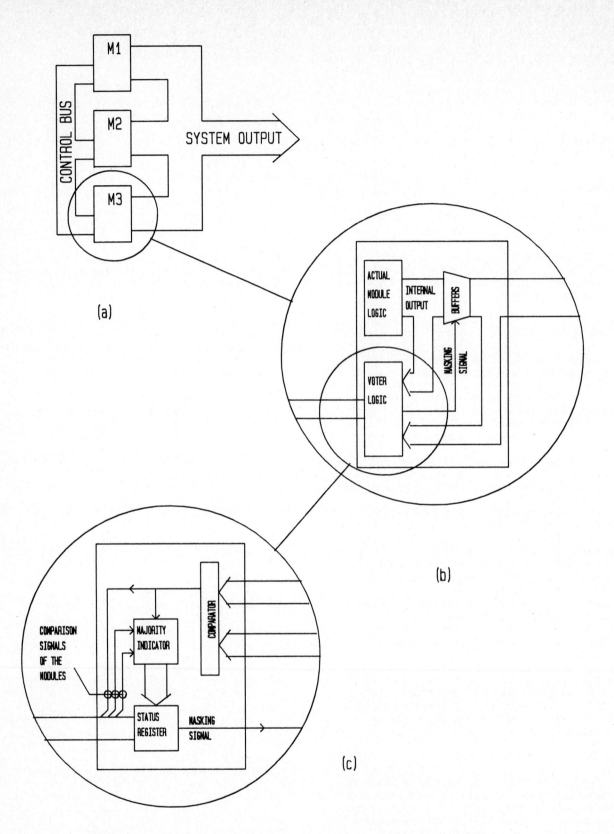

Fig. 2 a) A TMR system without voter logic outside the modules,
           the outputs of the modules are wired together
        b) The block diagram of one module
        c) The block diagram of the voter logic

2.1 Principle of operation

During normal operation only the outputs of one module are active while the outputs of the other two are masked (i.e. the outputs are either in high-impedance state or, for oc-outputs, the outputs are high). All the actual module logics are performing the same task simultaneously and the comparison of the system output and the internal output of each module takes place in the voter logics (comparators) continously. As long as there is no fault in the system, the comparisons will indicate a match and no corrective actions are needed. In the case of a fault in one of the module logics there will be two possibilities:

1) the fault occurs in the module logic, whose outputs are active, or
2) the fault is in one of the modules with masked outputs

In the first case the fault will temporarily appear also in the system output and the comparison in the voter logics of the other two modules will give an indication of a mismatch. Via the control bus the majority indicator of each module will get the information of the comparison from the other two modules and according to that the majority indicator and status register of the faulty module will mask its output buffers. At the same time the adjacent module is informed (via the control bus) of the masking and the output buffers of this module are made active releasing the correct data to the system output. Now the two nonfaulty modules agree with the system output and the system task can continue. With a proper hardware design the corrective action can take place during one clock cycle, so there will be no delays in the processing. In the second case, when the fault is in one of the module logics with masked outputs, no fault will appear in the system output. In that case the two nonfaulty modules, i.e. the majority, agree all the time with the system output, and no corrective actions are needed.

In the both described fault cases a fault detection signal can be aqcuired from the control bus, for instance for error logging purposes. The detection of a faulty module can be based on the comparison signals in the control bus. The fault detection option can be used to facilitate the repair in repairable systems.

The faults can take place also in the voter logics and in the output buffers. Most of the faults in the voter logics will be non-critical, i.e. they have no immediate effect on the system operation. Unfortunately, most of the voter failures are also latent. The output buffers are a special case in some respects, because part of the failures can be tolerated, namely the stuck-at-one faults (if not shorted to the supply voltage) and stuck-at-high-impedance faults. In those cases the other modules can control the faulty signal line. About a catastrophic failure in the output (e.g. stuck-at-zero), a detection signal can be formed in the voter or outside

the modules based on the comparison agreement signals.

2.2 The hardware realization

A hardware realization for the voter logics has been designed and it was done with 8 and/or-gates, 2 inverters, 2 flip-flops and n exclusive or-gates (n=number of the voted signals). Depending on the module logics also latches may be needed in the output buffers. The control bus should be seven bits (irrespective of the number of the voted signals), including the comparison agreement signals for the modules (3), the fault detection signal for the outputs (1), reset signal and two module status signals. Assuming that the voters are integrated into the ICs, there will be 21 extra pins in the system, seven in each module. The described solution will be pc-board effective if the number of voted signals is > 5.

In principle it is also possible to put the buffers and the voter logics alone into separate ICs and use them as buffers in electronic pc-board modules. Three such modules can simply be plugged into the mother board bus (in which there is also the control bus) to make up the TMR structure.

Other majority voter structures, like 3-out-of-5 etc., can be constructed with the described principle. The voter logic hardware will in that case be a little more complicated and the reliability characteristics will be in some respects worse than for the TMR.

3. RELIABILITY CHARACTERISTICS

The described system is a TMR system, in which the voting is performed on the whole output vector, not bit by bit. This means that all double failures taking place at the same time in two separate modules, will cause also a system fault. In classical TMR some simultaneous double faults can compensate each other, when the voting is performed bit by bit. However, this kind of compensating faults are probably quite rare in such module logics for which the described system will be useful. Some faults will not produce an incorrect output for all inputs, and, as these are also regarded as compensating faults {2}, the described system will tolerate all such faults in two module logics. Also the fault detection option will detect all simultaneous non-similar double and triple failures in the module logics.

For short mission times and for repairable systems the single point failures will be dominating from the reliability point of wiev. In the described system there will be single points primarely in the outputs, but also in the voter logics some failures will be critical as single. In the module logics there will be no single points. As was mentioned earlier, the stuck-at one faults and the stuck at high-impedance faults in the outputs will not be absolutely critical, at least the system will behave fault tolerantly despite those faults. Some correct data may be lost in some cases but the fault detection can be used to prevent the faulty data to spred into the rest of the system.

In order to compare the single points of the classical system and the described system we should consider one output signal line of both systems. In the classical system with fault detection there will be two gates connected to one output line; one output of a gate (the output buffer) and one input of a gate (the feedback to the fault detection logic). The hardware realization in fig. 1b is fault tolerant in principle, in practice, however, there will be single points, e.g. wirebonds if the voter output is also an output of an IC. In the described system there will be six gates connected to one output line; three outputs (the output buffer of each module) and three inputs (the input to the comparator in each module). However two of the three buffers are most of the time inactive and they can be assumed to be more reliable than an active buffer. Also, as was mentioned earlier, part of the failures will not be critical. By taking into account all the mentioned aspects, a failure rate estimate for the single points could be calculated, but it has not been made more accurately so far in this study. Most obviously the failure rate of the single points will be less in the classical system, but probably the difference will not be very big. One should also here remember the fault detection, which will detect most of the single point failures, too.

4. DISCUSSION

In the paper the possibility of realizing the TMR without any voter logic outside the modules was considered. With the described solution it is possible to construct such a system and the need for additional pins in each module is seven regardless of the number of the voted signals. In the case of integrating the voters into ICs, the described solution will be pc-board effective compared with the classical system, if the voting is performed on more than 5 signals. The voter logics can be realized with quite a little logic in the modules. An important feature for the repairable systems is that the fault detection is very easy to get from the system, either for module logic failures or for system output failures.

For short mission times and repairable systems the single point failures will be dominating in the TMR. Estimating the failure rate of the single points in the described system is not straightforward, and it has not yet been performed so far. The voting is performed on the whole output vector, and therefore the system cannot correct simultaneous compensating faults but the system is able to tolerate all nonsimultaneous compensating faults in two modules. All compensating faults are also detected by the fault detection logic.

In principle it is possible to integrate the whole triplicated system in one circuit, the result would be a fault tolerant IC. Also other majority voting structures like 3-out-of-5 etc. can be constructed with the described principle.

REFERENCES

{1} Pat. U.S. 4375683. Fault tolerant computational system and voter circuit. August Systems, Tigard, Oreg. (John H. Wensley). Appl. 205935, 12.11.1980. 9 p.

{2} York & al., Compensating faults in Triple Modular Redundancy. Proc. of the 15 Ann. Int. Conf. on Fault-Tolerant Computing. Ann Arbor, Michigan, 19-21 June 1985. IEEE Computer Society Press. p. 226-231.

Mr. Tapio Pulli was born in Teuva, Finland in 1952. He received his MSc. in digital systems engineering from the Helsinki University of Technology in 1978. In 1979-1982 he was working as a quality control engineer at the Terminal Systems Department of Nokia Electronics and since 1982 he has been working at Technical Research Centre of Finland as a research scientist. His main field is the reliability of electronic components and systems.

RELIABILITY MANAGEMENT

Reliability Technology — Theory & Applications
J. Møltoft and F. Jensen (Editors)
© Elsevier Science Publishers B.V. (North-Holland), 1986

RELIABILITY ENGINEERING IN RISK MANAGEMENT

Oddbjørn Jonstad and Svein A. Øvergaard
Nor-Risk as.
Asker, Norway

This paper discusses the relationships between Risk Management and Reliability Engineering, and the problems of definition and identity common to two new professions having developed in parallel during the later decades.

The meaning of words

Words like "reliable" and "risk" are used both colloquially and as technical terms, and this may lead to confusion. Allow me therefore to dwell for a while on the concepts of Risk Management and Reliability Engineering. There is a close connection between them, but I also feel that they share problems of definition and identity.

Risk Management

The term Risk Management was coined some 25 years ago in the US. To many, Risk Management only denotes a somewhat extended insurance company service, comprising a more thorough scrutiny of possible losses, their causes and consequences. The aim will then be to obtain a better balance between insurance and actual loss prevention.

Insurance, as you know, does not prevent losses from occurring and will not diminish total risk. Insurance is a way of levelling out the financial costs of expectable, irregularly occurring losses on the members of a pool of insurance takers, willing to pay annual premiums roughly corresponding to their proportional shares of the total loss estimated for the pool. Pool estimates can be made more precisely than individual ones, and will reflect the greater stability of pooled losses from year to year.

A satisfactory explanation of Risk Management cannot start with insurance. Insurance is one, but only one of the useful tools of a risk manager.

Risk Management forms part of the more general management function. It will suffice to talk about the management of an industrial enterprise, shortly called the company. Necessary translation to management of other types of organisation will be immediate.

The management is responsible for the resources of the company. It may be conservative, aiming primarily at the preservation of resources at a level deemed satisfactory, the minimum usually being a level sufficient to ensure the continued existence of the company. Minimizing losses may then take high priority. An aggressive management will aim more at expanding the resources,perhaps by short term maximizing of profits to be taken out of the company, or perhaps only with a greater weight on developing future earning potential than on protecting static wealth. It should be noted that the "resources" we talk about may have an intrinsic, static "good to have" value, but even more important may be their potential as means for acquiring future gains.

Let me insert here that Risk Management is not only for the conservative gereral manager. An aggressively managed company will usually be more vulnerable, and accordingly should be even more careful about keeping the risks taken under close observation.

An aggressive management may be willing to accept fairly high losses if this is considered necessary to exploit possibilties for important gains. Losses that are known, necessary consequences of a profitable activity we simply call costs.Some losses we may hope to avoid, but cannot always escape since we lack sufficient information to ensure that we do so. The production and marketing of a product for which the demand is uncertain may lead to such losses. Some authors call the possibility of loosing by a decision made voluntarily, but on insufficient information to predict the outcome, a "dynamic risk". The word may also denote an estimate of the maximum possible loss. Management, aggressive or conservative, will have to accept dynamic risk. Every decision can be said to comprise an element of dynamic risk, the loss being the difference between the optimal and the actual outcome.

To include into Risk Management all management having to consider dynamic risk leads to unreasonable consequences. The risk manager then would have to be a jack of all trades, able to supplant or to advise any kind of manager in his own profession. Some delimitation of the types of risk considered in Risk Management is necessary. To avoid misunderstanding, I want to point out that the expert Risk Manager may be a good adviser also in evaluation of dynamic risks. But he will then meet the question of whether he is acting in his own professional sphere or trespassing into others. One reason not to be too restrictive is his understanding of loss propagation effects. Such effects of a loss materialising out of the dynamic risk pertaining to a managerial decision, may be very similar the propagation effects of losses caused by such events as must be the prime concern and the particular responsiblity of the Risk Manager.

Those are the events that we know may occur, but not that they will occur, not when, nor exactly under what conditions. Events that are potentially harmful, and that we wish to avoid, but for the occurrence of which we may at most have statistical information. If probabilities can be calculated or satisfactory empirical statistics are available, an expected value for the losses in a certain future period may be calculated as the quantitative expression of the risk. This will be the model situation of the Risk Manager. In practical situations he will very often have to content himself with looser "guesstimates" based on practical experience and good common sense.

The rational goal of Risk Management will be to keep the risk within sensible, acceptable limits, mostly by searching for ways to reduce risks by measures influencing the probability or frequency of loss-causing events, or such that will reduce the extent of the damage that may ensue out of such events, if and when they occur. In his analysis and in his advice to general management the risk manager will have to weigh the costs of possible risk reducing actions against their effectiveness. Absolute risk minimizing is inconsistent with even the most conservative management principles, and sound advice from a risk manager may well include dropping risk reducing precautions if the perhaps marginally increased risk is worth taking. Some types of risk may be given a particular weight, e.g. those concerning events

that will "kill" the company, even if their probability is so low that they do not contribute heavily to the statistical expectation of losses. Risk management should be rational, but rationality does not mean exclusion of psychological factors, and certainly not holding a superstitious belief in the possibility of always expressing risk relevantly as the product of probability and consequence. This world is not a linear one even if that would have been mathematically convenient.

I have wanted to stress here that the essence of Risk Management is the analysis of unwanted, non-predictable events, their effects on a certain object or system, and the finding of suitable, economic ways to control the risk, i.e. the expected volume of losses ensuing from such events. In the literature, this type of risk is called "static risk", not because nothing can be done to change it, but to differentiate between it and the "dynamic" type where the causing event is a voluntary act and where no stochastical event need enter the picture at all.

Stated in this way, I think the close connection between Risk Management and Reliability Engineering becomes very visible.

Reliability Engineering
Like the word "risk", "reliable" is used with several meanings and often very vague ones. The result is that a term like "Reliability Engineering" may be misunderstood so as to comprise any engineering aiming at the production of objects that are reliable in any one of the possible senses of that word. The unreasonable consequence of this again would be that all good engineering is reliability engineering.

We have to sort out a certain type of engineering problems forming the realm of the reliability engineer. In many problems, all relevant information is of an absolute and not of a statisical kind, no stochastical events being necessary to consider, no probabilities entering the picture. Our factual information may be incomplete, and our theoretical understanding not sufficient to warrant a fully satisfactory solution, and this may result in bad or even unsafe constructions. The remedy is to obtain more of the relevant information and to utilise what is available in a better way. But only when statistical considerations are important can the problem be regarded as belonging to Reliability Engineering.

One single, isolated simple object may be good or bad, faulty og correctly made, in accordance or not with given specifications. You may have to rely on it for a certain usage, and if it turns out good it may be called reliable in the sense of everyday language. Whether it is reliable in the sense "has the qualities stated for it" cannot be answered by any particular method belonging to reliability engineering, but rather by specialists in some other branch of engineering and in the quality control particular to that branch. Only when simple objects are seen as samples out of a lot of similar objects, with somewhat varying attributes due to events in the production process, or to variations in the materials used, over which we can only have statistical information, can the "reliability" of the objects be regarded as a technical term, approaching the essential meaning of the word in reliability engineering. But there is no need to introduce "reliability engineering" as another name for the engineering disciplines aiming at securing high and homogeneous quality for mass production of isolated objects.

The crucial problems of reliability engineering must remain the design, construction and evaluation, possibly also maintenance, of systems of a certain degree of complexity, and necessarily consisting of components for which certain significant properties are known only within statistical terms. Rational goals will then be to obtain desirable system properties such as reliability, availability, maintainability, etc., each of these words having been given a fairly precise meaning, expressing various aspects of the ability of the system to function as intended and when intended.

The systems considered in practice will usually be open ones, influenced by factors external to the system. Such factors may comprise stochastical events or not, but in any case the internal existence of component properties known only in a statistical sense allow for random effects within the system. If the unwanted consequences of such effects can be assessed, e.g. in economic terms, expected losses and so also the risks involved can be estimated on the basis of reliability figures.

But the typical problem of reliability engineering will still be one where the system may be regarded mainly as a closed one, and where only to a limited degree external influences need be taken into regard as "operating conditions", varying within specified limits and according to some distribution known at least in principle.

Reliability Engineering in Risk Management

To exemplify, I found it natural to talk about a company as the object of the risk manager and of his risk analysis. The company is the object that may be damaged and sustain losses. The company may be vulnerable to strain materialising out of threats directed against it. The company is the object carrying risk.

A company may be a very complicated object. Complicated objects we often find it convenient to regard as "systems", thereby indicating that they can be decomposed into subsystems, and recursively so until the decomposition ends up with components for which no further decomposition is possible or convenient.

Some subsystems of a company very often will be purely physical, engineered ones, important to the company in the sense that significant losses arise if they do not function as and when intended. The reliability of such subsystems, the availability of their intended functioning, are of consequence to the risk carried by the company. This is trivial, and evidently a reliability analysis will in such cases form part of a wider risk analysis task. A very important part if the company is, say, a manufacturing one, basing its whole existence on complicated technical systems consisting of components for which significant attributes can be known only statistically. Constructive reliability engineering will be among the means considered for reducing the risk carried by the company.

To attack the complete risk of a company from a reliability engineering wiewpoint will generally not be feasible. Very important components of the company are human beings in unique and critical positions. Certainly they may behave erratically, and to a certain extent we can have statistical information about human behaviour appearing random to an external observer. When many persons function under fairly standardised conditions as components in the company system, they can possibly be taken into account in the same way as technical components. But mostly the human beings functioning within the company will have to be regarded as individuals, good or bad, but not characterized sufficiently by their belonging to a population for which some statistics are available. The operator or the executive in a critical position holds a too wide and varied potential for acting than can be coped with by the methods of reliability engineering.

The risk manager, being well aware of
the importance of internal malfunctioning as a threat to the company,
will more than the reliability engineer
have to study external factors that may
cause loss, damage or even destruction
of the the company. Natural extra-
ordinary events like storms and flood,
or attack by terorists, or the company
being cut off from ordinary supply
sources because of international
dispute, do not enter into reliability
engineering in any reasonable way.
Variations within certain limits in the
"environmental conditions" of a system
may be taken into consideration in a
reliability study, but to really
include any major part of the external
factors considered by the risk managers
would necessarily lead to unwieldy
models.

It would, however, be very interesting
to try to extend the range of systems
considered by reliablity engineering to
those comprising humans as components
performing in a ways that do not
strongly emphasize their individual and
particularly human qualities - their
being thinking, feeling and volitional
beings. Very important then would be to
set proper limits to such endeavours.
And also to secure that no misunderstanding arises about "these engineers
without compassion trying to treat
people as if they were just mechanical
or electronic components". These are
questions pertaining to the more
general problem of treating the design
and maintenance of organisations, i.e.
of very complex systems of people and
their various physical and informa-
tional tools, as an engineering task.
The technification also of the
administrative branch of organisations
by data technology, and the replacement
of so many traditionally human function
by programmed routines executed
electronically, justifies such an
outlook.

To conclude these observations, I think
we should not overestimate the possibilities for making direct use of
reliability engineering methods in the
more general problems of risk management. Human error and human misunderstanding very often is the cause of
such losses as the risk managers try to
avoid. Personnel is certainly a major
"risk factor", and the unreliability of
man a major headache for the risk
manager. But few of those problems lend
themselves to solution by the mathematically and statistically based
methods that are the distinguishing
tools of the reliability engineer.

Similarity in approach
Let me rather turn to some similarities
i the working methods of the risk
manager and the reliability engineer.

Both will for their object typically
have a fairly complicated system. The
risk manager will ask what types of
damage or loss may appear, what events
may possibly cause such unwanted
effects and under which conditions, and
what should be changed to reduce the
expectancy of loss, if this is considered unsatisfactory. The reliability
engineer will ask in what ways his
system may not function, what may cause
dysfunctioning and what system effects
may result out of component failures.
Formally their problems are quite
similar. They both face the question of
whether to use a" top down" approach,
starting with an possible inital
causing event and following its
consequences down through several
system levels, or a "bottom up"
approach starting with a possible,
undesirable ultimate situation and
chasing its causes upwards through
several levels. Or perhaps to choose
some obviously critical situation from
which both to deduce further consequences and to search for preceeding
events necessary for such a situation
to arise. The risk manager as well as
the reliability engineer will find that
there is no single approach eminently
commendable in all cases, and usually
no unique way of describing a complicated system, which will ensure that
the model considered does reflect the
actual system in the best possible way
for their purposes.

The risk manager probably will be faced
with more comprehensive, more complicated and more unmanageable systems
than are left to the reliability
engineer to cope with. This may lead
him to give up too soon in the tasks of
quantifying, doing serious statistical
studies etc. He may be tempted to fall
back too easily on common sense in
cases where scientific methods might
support or negate his conclusions, and
to regard his profession as belonging
more to the arts than to science. The
reliability engineer, concentrating on
purely technical systems, will see more
possibilities for carrying through
mathematical and statistical reasoning,
and may then fall into the trap of
trusting his model and his calculations
too far. But he should also be in a
better position to see the true
limitation of scientifically based
methods, particularly when system
components and external effects appear
for which sufficient data are not of
the quality making them suitable as a
basis for calulation of reliability
figures that can really be relied upon
as a basis for crucial decisions.

Concluding remarks
Reliability Engineering and Risk
Management both have emerged as
professional disciplines during the

last few decades. As I said, they share
a problem of identity and definition. I
think they would both gain by not
trying to extend their domains in ways
that will make it difficult for both
insiders and outsiders to understand
what is the real content of the
professed professional expertise. The
two professions should coexist in
fertile coooperation.

Reliability engineers, concentrating
primarily on the fairly well defined
technical systems, should develop even
better tools for coping with the
problems of propagation of fault
conditions throughout a complicated
system. Since so often we lack
sufficient knowledge about probabilty
distributions, and since calculation of
propagation effects based on
statistical methods may be very
cumbersome, it might be well worth
exploring more thoroughly the
possibilities offered by fuzzy set
approaches.

The risk manager for his analysis will
have to lean on the reliability
engineer. He may be able to support the
reliability engineer in studies of
techno-social systems where the human
factors appear in ways that can safely
allow their inclusion in a statistical
model.

THE AUTHORS:
Oddbjørn Jonstad (42) graduated from
the Technical University of Norway,
Trondheim, and has held positions as
Safety Manager (Norw. Water Resources
and Electricity Board) and as Section
Manager, offshore engineering (Kvaerner
Engineering A.S). Presently Managing
Director, Nor-Risk as.
Svein A. Øvergaard (60), M.Sc. (U. of
Oslo) has held positions as Research
Officer (Norw.Defence Staff, Norw.
Defence Research Est.),lecturer (U. of
Oslo; Techn U. of Norway), Director
(Blindern-Kjeller Computer Inst.),
Board Chairman (Norw.Computing Centre),
Deputy Director General (Ministry of
Consumer Affairs and Govt.
Administration. Presently Director,
Nor-Risk as. Former Chairman of the
Norw.Soc. for Information Processing
and the Norw. Data Processing Society.

Reliability Technology — Theory & Applications
J. Møltoft and F. Jensen (Editors)
© Elsevier Science Publishers B.V. (North-Holland), 1986

PROCESS ORIENTED RELIABILITY SIMULATION OF SATELLITE COMMUNICATION SYSTEMS.

Bjarne E. Helvik and Norvald Stol

ELAB
N-7034 Trondheim-NTH
NORWAY

The paper summarizes problems assosiated with the reliability evaluation and replenishment of the space segment of a communication satellite system. It gives a general model of and a methodology for the simulation of a system during its mission time. The space segment consists of a number of satellites with potentially different structures. Emphasis is put on modelling the operating organizations' effort towards an optimal resource utilization during the mission. Parts of a system evaluation made by the implemented simulator are included as an example.

1. INTRODUCTION

The space segment of a satellite communication system is supporting a number of communication channels, each at a separate frequency band, during a period called the system's mission time. This space segment consists of many different satellites with their separate internal substructures and lifecycle, plus a control subsystem optimizing the use of system resources with respect to system service.

As in most systems, the basic reliability engineering challenge is to provide highly available service during the mission time at minimum cost. Special for this kind of system, however, is the stepwise cost function. One less satellite launched represents a considerable saving. The degrees of freedom we have, in addition to the choice of components and launching service, are the redundancy structure of the satellites, when to order long lead items, and when to launch the satellites. From the reliability evaluation point of view, these systems pose a number of interesting problems:

* we are interested in the reliability of a logical rather than a physical structure, namely the communication channels. These may have different resource demands in terms of bandwidth and power, and may be assigned different priorities.

* the satellites and their subsystems form a highly dynamic redundancy structure, i.e. the physical and logical structures of the system change during the mission time. For instance, the launching of a new satellite changes the physical structure of the system, while the failure of a satellite item changes the logical structure as it may cause a complete reallocation of the physical resources to the different channels, in order to provide service to the channels with the highest priority.

* the evaluation must account for the operating organization's optimal use of the instantly available resources to provide the expected service. For instance, it maximizes the utilization of the satellites with the shortest remaining life and the utilization of many partly defect satellites in order to provide a complete service.

* the evaluation must also take into account that the operating organization may decide to order new long lead items or to launch a new satellite on the basis of the current or predicted situation. For instance, a new satellite is launched if a high priority channel has no spare or a satellite approaches its maximum life.

Simulation is necessary in order to perform a realistic reliability evaluation of such satellite systems. The most important reasons for this are the complex optimization and decisions, outlined above. They cause very strong unit interdependencies, an untreatably large state space. In addition, the satellite life cycle has non-Markovian properties. Because of the highly dynamic redundancy structure of the system, it is cumbersome to base a simulator on rather static decription methods like fault-trees or reliability block schemes. We have chosen to use a process oriented approach, where each type of subsystem, e.g. satellite, is modelled by a process class. These process classes model, in a parameterized way, the behaviour of the different types of subsystems through their life cycles. Behaviour, in this context, refers to both the item failure pro-

cesses and the relevant operational
behaviour like for instance satellite
item delivery times, launching delay,
and time required for in orbit testing.
The specific subsystems are generated
from the appropriate class by the
choice of parameters. The different
subsystems/processes cooperate (interact) by sending signals to each other.

This modelling approach is described in
the next chapter. The third chapter
gives some implementation details, and
the final chapter contains a brief example.

2. MODEL AND METHODOLOGY

The first section of this chapter
describes the basic model of a satellite system. In the next sections, we
outline how the different parts of the
system are modelled by processes. Each
of these processes describes a section
of the course-of-life or the whole
life, of the corresponding part of the
system. In the final section we discuss
how these processes cooperate.

2.1 Basic Model.

The satellite system is divided into
three main parts or process classes in
our model :

* The satellites and their units
  represents the physical resources
  needed to obtain the desired service
  from the satellite system.

* The channels represents the service
  itself. A given channel is at a
  given time based upon one particular
  combination of physical resources in
  one satellite. Some of these resources are shared with other channels,
  while others are exclusive. The
  channels are abstract entities.

* The control structure represents all
  management aspects of the satellite
  system, especially aimed at obtaining a good resource utilization and
  a high availability of the channels.
  This includes switching of satellite
  resources to support channels, and
  the ordering and launching of satellites.

This division is intuitively the most
natural one. Each part represents a
main element of the satellite system.
While each of the parts has a rather
complex internal structure and behaviour, only a small amount of information has to be exchanged between these
main parts in order to make them work
together. This makes the modelling (and
realization) of one part rather independent of the other two.

A more thorough description of the main

parts of the model are given in the
subsequent sections.

2.2 Satellite model.

The satellites are constituted by four
types of units.

Platform:
  The basic part of the satellite, and
  the part that all other units of the
  same satellite are dependent on,
  i.e. the mechanical framework, power
  supply etc. We use a truncated
  Weibull distribution to model the
  platform's life-time, i.e time to
  failure.

Module:
  A part common to a number of
  transponders (varying from one up to
  all). If this unit fails, all transponders connected to it will be
  useless. This unit may for instance
  be used to model antennas or divided
  power supply. The life-time of the
  module is modelled by a Weibull
  distribution.

Transponder:
  Those parts of the satellite which
  are exclusively allocated to receiving and transmitting one channel
  except for the TWT, see next paragraph. We use a negative exponential
  distribution to model the life-time
  of the transponder. The failure rate
  depends on whether the transponder
  is active (in use) or passive
  (spare). The transponder may also
  fail, with a given probability, the
  first time it is powered.

TWT (Travelling Wave Tube):
  This is an important and vulnerable
  component of the transponder.
  Because this component is more
  vulnerable than the other parts of
  the transponder, there are usually a
  number of spare TWTs, reachable from
  all or a number of the transponders.
  Hence it is modelled as a separate
  unit. A negative exponential distribution is used to model the lifetime
  of the TWTs. The failure rate
  depends on whether the TWT is active
  or passive, and we have a
  probability of initial failure, as
  for the transponder.

It is assumed that each unit fails
independently of the others. The life
time distributions listed above may of
course be altered if desired. The
reliability structure of the satellites
are indicated by the quasi reliability
block scheme in figure 1. These will be
explained in some more detail in the
next section.

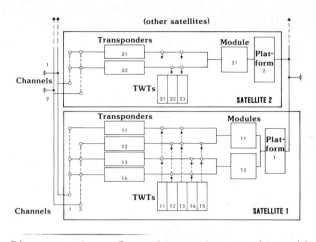

Figure 1. *Example of a (quasi) reliability block scheme for a set of satellites.*

A satellite is modelled as having four main phases of life, as shown in figure 2. It is modelled by processes, describing a part of or the whole of its life. Initially one process is enough to describe the satellite's life.

The pre-orbit life of a satellite is modelled by two phases. The interpretations of these are arbitrary. For simplicity, however, let us use the following ones. We start with the ordering of parts and the assembling of these into a satellite. This is the ordering phase. The second phase, the launch phase, is the time from a launch is initiated and until the actual launching takes place. The launch phase can not start before the ordering phase is completed. Both these phases are modelled by empirical distributions. This means that an arbitrary distribution may be given for the duration of each of these phases, for each satellite.

The launch of a satellite may be successfull or unsuccessfull with an individual probability. If it is unsuccessfull, the satellite process is terminated, otherwise it continues with an in-orbit-test phase. The phase is deterministic, though it may vary from one satellite to another. After the testing is finished, the satellite enters its final phase, normal operation. At this point the satellite process is divided into subprocesses. The platform and modules are modelled as processes, while transponders and TWTs are modelled as passive resources, for efficiency reasons. Hence, we restrict ourselves to the following two kinds of processes:

- the satellite (platform) itself continues as one process. It contains information about the transponders and TWTs. This includes the intact vs. defect status, the current resource utilization, and the usability matrices, described in

the next section. The maintenance of this status is partly performed by other processes.

the modules. One process for each module. These processes' prime task is to assure that the failure of the module is reported to the control structure, and that the proper updating of resource status is performed, i.e. the corresponding transponders marked as useless.

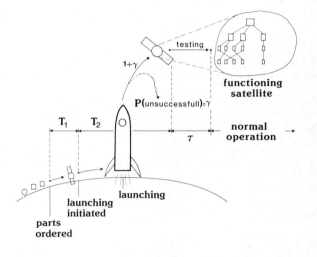
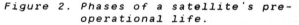

Figure 2. *Phases of a satellite's pre-operational life.*

In addition, the satellite process also logs a number of event times. These are to be used to obtain satellite specific statistics, for instance, the ordering and launching time distributions.

The description above concerns one general satellite. Each satellite used in the satellite system is described by its own set of parameters. Hence we may use many different satellite types in the system. During a mission we will, most of the time, have a number of satellites in orbit, probably partly defect, and one or more in the preorbit phases. In the simulator, there will be a set of processes for each satellite used. These, and their corresponding resources, are managed by the control structure process.

2.3 Channel model.

The channels are a set of transmission paths. Each channel occupies a frequency band, received by a satellite and retransmitted towards the earth within a new frequency band, with a specified power. These channels represent the service given by the satellite system. A channel can not be associated with any specific physical components permanently, but is, at a given time, the result of a combination of such components. The quasi reliability block scheme shown in figure 1, illustrates the dependencies between the channels and the satellite resources for an example system. One specific TWT may be switched to only one transponder at a time. One specific channel is intact (operative) if it is assigned one usable transponder and one usable TWT, where the module and platform supporting the transponder have not yet failed. Usable in this context means both that the resource may be utilized by the channel, directly or indirectly, and that the resource is intact. In this example we have (at least) two satellites in orbit, sharing the support of two channels. One specific channel is, of course, supported by one specific satellite at one specific time.

Table 1 shows the usability matrices for the example system in figure 1. These matrices show which TWTs may be utilized by which transponders, and which transponders may be utilized by which channels. The matrixes list theoretically possible combinations. Whether a given channel actually can utilize a transponder or TWT at a given time is dependent on the intact/defect status of the resource, on the priority of the channel, (since there may be another channel with a higher priority that has occupied the resource), and on whether the utilization of the resource (transponder) is allowable within the power restrictions of the satellite.

Table 1.
The usability matrixes of the satellites in figure 1

SATELLITE 1:

Matrix A.

Transp. Channel	11	12	13	14
1	X		X	X
2		X	X	X

Matrix B.

TWT Transp.	11	12	13	14	15
11	X		X		
12		X	X		X
13	X			X	X
14		X		X	

SATELLITE 2:

Matrix A.

Transp. Channel	21	22
1	X	X
2	X	X

Matrix B.

TWT Transp.	21	22	23
21	X		X
22		X	X

The allocation of resources to channels is done by the control structure according to the priority of the channels. The priority is a significant parameter of a channel. In fact there are two kinds of priorities. Ordinary priority, giving the order in which the channels are to be allocated resources, and a high/low priority bound indicating whether the amount of resources available for a channel shall influence

the ordering and launching of new satellites, cf. the next section. An example of channel priorities is shown in table 2.

Table 2.
An example of channel priorities.

Priority	Channel	Type	Group
5	1 2 3	Broadcast	Cause the launch of a new satellite, if the spare requirement is not full-filled
4	6 7	Communication	
3	4 5	Broadcast	
2	8 9	Communication	Does not influence the launch procedure.
1	10	Spare	

The channels constitute a process class. One specific channel is modelled as a process from this class, parameterized with its priority and power requirement. This process may be in one of two states, active or passive. Active means that the channel is supported by the system of satellites, passive that it is not. Figure 3 shows one possible event sequence of a channel process.

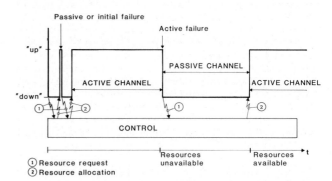

① Resource request
② Resource allocation

Figure 3. Active and passive periods of a channel, and its interaction with the control process.

Whenever a channel becomes passive, i.e. by initialization or after a failure, it signals this to the control structure. The control structure tries to allocate resources to the channel, to make it active.

The channel process also handles the failure event processes of the transponder and the TWT allocated to the channel. This is convenient since these satellite units were modelled as resources, not as processes. Passive failures of the resources, i.e. failures when they are not allocated to a channel, are "drawn" immediately before they are put into operation. The channel process also updates the status information concerning these resources when they have failed.

Another important task of the channel process is to log it's own active and passive periods in order to obtain the performance of the channel and the priority level which it belongs to, e.g. the availability of the channel and its down time distribution.

2.4 Control structure model.

In many aspects, the control structure models the organization operating the technical part of the satellite system. At every instant during the mission time the objective of this organization is to keep as many channels operative as possible, given the channel priorities and the resource constraints of the system, i.e. to maximize the instantaneous availability of the channels. This must be done in an optimal way, e.g. use the resources in the satellites with the shortest remaining life ought to be used first and the resources with the highest applicability ought to be "saved" as long as possible. The long term objective of the operating organization is to provide sufficiently available channels at a minimum cost, i.e. to order and launch as few satellites as possible. The means to achieve these objectives are:

- the switching of transponders and TWTs to channels, and

- the ordering and launching of new satellites.

This is illustrated in figure 4. The exact optimal choices of switch settings, ordering and launching instants are rather complex problems. Within a simulator it demands too much computer time. Hence we have developed some heuristic algorithms. How these operate within the simulator is summarized below. They will be described in more details in chapter 3.

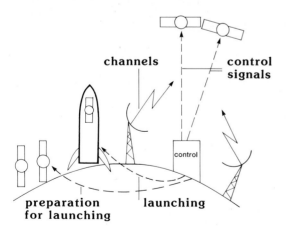

Figure 4. The primary functions of the control structure.

The control process manages the system. At every event in the system (an event may for instance be a failure) the control structure process is started. It checks the status of the channels, and tries to (re)allocate resources if any of the channels are unsupported. Resources may be withdrawn from a channel with a lower priority. The process tries to support as many of the channels as possible, starting with the channels of highest priority. Two different optimization algorithms are used to achieve this. The simplest is used when all channels can be supported easily and no reallocation is necessary, while the other performs a total reallocation. This last algorithm solves a rather complex combinatorical problem.

The other main tasks of the control structure process are to order long lead items for future satellites and to initiate the assembly and launch of new satellites. The schedule may be preprogrammed, i.e. predecided before the mission time starts, or it may be decided during the mission according to a number of redundancy criteria. The three criteria that may be used (in arbitrary combination) are shown in figure 5. We are currently working on a new ordering and launching procedure based on the predicted future reliability of the highpriority channels.

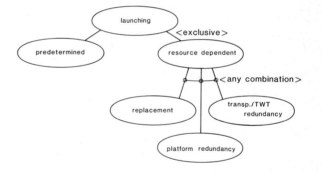

Figure 5. Outline of the possible factors influencing the ordering and launching decisions.

In figure 5 'Replacement' means that a new satellite is ordered and launched in time to replace one that is soon reaching its maximum life time. 'Platform redundancy' means that we always shall have at least n satellites in orbit able to support all high-priority channels (n is usually equal to 2). This is to prevent loss of service in case of an early platform failure. 'Transp./TWT redundancy' means that we demand a minimum number of spare resource-combinations for each of the high-priority channels.

2.5 Process Interworking.

The interworking of the processes, described in the previous sections, is most easily explained by an example. The processes cooperate by sending signals. These signals interrupt the receiving process, and are handled in a manner similar to failure events. This is illustrated by the signal sequence diagram in figure 6. The diagram must be interpreted qualitatively as the "time"- axis only gives the sequence of events and not the actual time-span between them. The magnitudes of inter-event times vary greatly among the different event types, and therefore the actual values can not be given

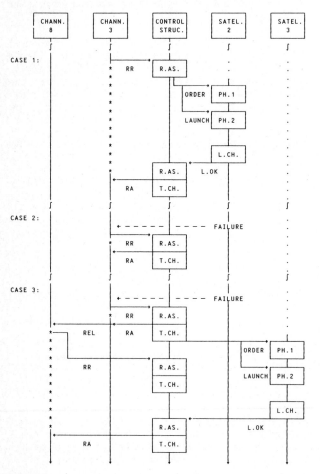

LEGEND:

RR   : Resource Request
RA   : Resource Allocation
REL   : RELease Resource
ORDER   : Start Phase 1 ("ORDERing")
LAUNCH : Start Phase 2 ("LAUNCHing")
L.OK   : Launch OK
PH.1   : PHase 1
PH.2   : PHase 2
R.AS.   : Resource ASsignment
T.CH.   : Total resource situation CHeck
L.CH.   : Launch successfull CHeck

Channel state:

*
* :passive
* (down)

| :active
| (up)

∫ :"time-jump"

Figure 6. Signal sequence diagram showing the cooperation between the processes in the simulator. Note that the diagram is not complete, and the time axis is not drawn to scale.

Three different incidences are shown in the diagram. The first one shows the ordering and launching of a new satellite when no resources are available for the "high priority" channel 3. The lower priority channel 8 is in this case assigned resources which can not be used by channel 3, e.g. channel 3 requires higher power, which is not available. After the successful launch of this new satellite the overall situation is checked and found satisfactory.

The second case shows a failure in a resource (TWT or transponder), followed by a resource request and a resource allocation. The overall situation after the last failure is checked and is still satisfactory.

The last case shows another resource failure, followed by a resource request. Satellite 2 is now transmitting high priority channels (not shown) previously transmitted by satellite 1. This removes the power restriction of satellite 1 and resources initially assigned to channel 8 may now be used by channel 3. Hence, when the higher priority channel 3 experiences a failure, resources are released from channel 8 and assigned to channel 3. The overall situation after this failure is not satisfactory, and a new satellite is ordered. Note that this is not necessarily due to lack of resources available to channel 8.

Channel 8 may have too low priority to influence on this decision. The reason may as well be that there are too few spare resource combinations for the high priority channels. After the successfull launch channel 8 is assigned resources and the overall situation is again satisfactory.

This concludes the model and methodology chapter.

3. REALIZATION

The first section of this chapter gives a cursory description of the heuristic algorithms used by the control structure process in the assignment of resources to channels. The strategy for ordering and launching of new satellites is also briefly outlined. The implementation of the satellite system reliability simulator is summarized in the final section.

3.1 Resource allocation algorithms.

The long term objective of the control structure (operating organization) is to provide sufficiently available channels at a minimum cost. This means that we want to use our resources (TWTs,transponders, satellites, e.t.c.) as efficiently as possible, and at the same time ensure that the availabilty requirements of the channels are met. As mentioned in the section dealing with the control structure, algorithms are implemented in the simulator in order to meet this objective. The main elements of our heuristic approach may be summarized as follows :

- Use resources located at the older satellite first, because this satellite has a shorter remaining life than the newer ones.

- Allocate resources according to channel priorities.

- Put resource combinations into operation in such a way that the ones which may be utilized by few channels are put into operation before those who may be utilized by more channels. A resource combination is one transponder and one TWT.

- Delay the ordering and launching of new satellites as long as possible without violating the redundancy requirements.

The first three of these are incorporated in two different switching algorithms, while the last one is handled by an algorithm which performs a number of status checks, and decides upon ordering or launching from a set of criteria.

When a resource fails, a simple switching algorithm seeks an idle resource combination to replace the one that failed. If this simple and fast algorithm does not succeed, a more complex algorithm is utilized. This algorithm may, if necessary, make a total rearrangement of the resource allocation. The algorithm starts with all the resources still assumed to be intact marked free. As mentioned, the channels are divided into priority groups, and the algorithm starts with allocating resources to the channels with highest priority. The algorithm starts with the oldest satellite, proceeds with the second oldest and so on. The first resource combination marked assigned is the one that may be utilized by fewest channels, leaving as many resource combinations as possible to the other channels. This number of possible combinations is determined by transforming the usability matrixes to binary matrixes, and by matrix operations on these. If we use "A" for the binary channel-transponder usability matrix, and "B" for the transponder-TWT matrix, the product

$$C = A \times B$$

will give us a channel-TWT matrix, where the elements tell us via how many transponders one given channel can "reach" one given TWT. If we compute

$$C \times (1\ 1\ ..\ 1)^T$$

where the number of ones equals the number of TWTs, we get the total number of possibilities for each channel. By using expressions of this kind, we can find the "least-usable" resource combination at any time. In this context, the transponders represent a kind of bottleneck because they "tie together" channels and TWTs. After trying to allocate resources to all channels of a given priority, i.e. the resources are marked assigned to channels, we proceed to the next (lower) priority level. After each resource allocation the matrixes and the power account of the corresponding satellite are of course updated. After allocating, or trying to allocate resources all channels, the algorithm checks the markings made by the algorithm against the existing allocations, and executes the necessary reallocations.

The possible criteria for the ordering and launching of new satellites, in the current version of the simulator, are shown in figure 5. These are described in section 2.4, dealing with the control structure. To apply the resource dependent criteria, a number of status checks are necessary. The 'replacement' criteria use an internal timer for each satellite. This timer alarms the control structure when a replacement satellite ought to be ordered and launched. A replacment satellite may however already be in pre-launch phase, due to another criteria. These alarms do also result in a change in the satellites operational status, with respect to the other criteria. The 'platform redundancy' criteria require a check that all the channels in the high priority group can be carried by at least n satellites (n usually 2). The 'Transponder/TWT redundancy' criteria require a check that all the channels in the high priority group have a sufficient number, i.e. at least m, of spare resource combinations available. The resources assigned to channels in the low priority group are, in this context, regarded as spares. Resource of satellites approaching their maximum life are ignored. This check is carried out by matrix operations, in a way similar to the algorithm described above. The final order and launch decisions are based on the outcome of these status checks and the number of satellites in the two pre-launching phases. Because an arbitrary combination of the three criteria is allowed, the resulting logical expressions become rather complex, since we also must avoid ordering and launching too many satellites. Further details are given in <HELV 85a>.

3.2 Implementation.

The SIMULA programming language is used to implement the simulator <BIRT 79a>. The SIMULA based DEMOS (Discrete Event Modelling On Simula) - package, which contains most of the elementary building blocks necessary for discrete event simulation, is used <BIRT 79b>. This package contains primitives as queues, resources, semaphores, variate generators etc., and a number of statistical- and reporting aids. We found it, however, necessary to make some extensions. A number of variate generators had to be made, in order to give the satellite elements their appropriate life time distributions, e.g. the truncated Weibull distribution of the platformes. Statistics of events at different instants during the system's mission time are of special interest. For instance statistics showing the ordering- and launching times of new satellites and statistics showing the availability of the channels during the mission time. Neither of these could be registrated nor reported in a convenient way by using the existing DEMOS primitives. The necessary aids were made and included. These kinds of statistics are shown in the next chapter.

Each process class in our model is implemented by using the ENTITY CLASS - primitive in DEMOS. This primitive allows us to describe the course-of-life of the processes at a high level, cf. <HELV 85b>.

The size of the different parts of the simulation software is as follows : DEMOS - about 2650 lines of program code, extension of DEMOS - about 1050 lines, and the simulator - about 1800 lines. The simulator part includes the procedures for resource switching and ordering and launching of new satellites.

4. AN EXAMPLE

This example briefly describes a subset of a satellite system, and some typical results from the simulator are presented and discussed.

The launching scenario used is as follows :

- Satellite 1 is successfully launched at time zero.

- Satellite 2 is launched after 2 years, with prob. 0.9 of a successfull launch. Satellite 1 is utilized in a new way, confer the description below.

- The launching procedure of satellite 3 starts when it is needed to meet the system's redundancy requirements. The satellite is laun-

ched between one and one and a half years (uniformly distributed) after the launch is initiated. However, not earlier than 3 months after satellite 2.

- The need for a 4'th satellite is also registrated, but this satellite is never put into operation.

- The system is designed for a useful mission period between the second and the twelfth year after the launch of the first satellite.

Satellites 2 and 3 are identical and able to support all three high priority channels and some of the lower priority channels. During its first two years satellite 1 supports two high priority channels and two with lower priorities. Later it supports the three high priority channels.

The high priority channels have dedicated transponder-TWT assemblies in satellite 1. In the subsequent satellites each high priority channel may be transmitted by four transponders according to a switching scheme. Some of these may olso be used by other high priority channels. Transponders which are not used by the high priority channels, may be used by the lower priority channels as long as the power restriction of the satellite is not exceeded. Each transponder has its dedicated TWT in these satellites. Modules are regarded as fault free units. In table 3, the matrix shows which channels may utilize which transponders in satellites 2 and 3. The maximum life of satellite 1 is seven years, and of the other satellites ten years. Table 4 gives some other important reliability parameters of the system. Successors are ordered and launched by the control structure, cf. figures 4 and 5. We demand that each high priority channel has at least one set of spare resources. If not, a new satellite is ordered and launched.

Table 3
Usability matrix for the channel transponder combinations in satellites two and three.

Frequency band	Transponder 1 2 3 4 5 6 7 8 9 10 11 12	Channel number
F1	X X	5
F2	X X	6
F3	X X X X	1
F4	X X X X	3
F5	X X X X	4
F6	X X X X	2
F7	X X	7
F8	X X	8
FA	X X X X	9
FB	X X X X	11
FC	X X X X	12
FD	X X	10

Table 4
Some reliability parameters in the example system.

Table 4
Some reliability parameters in the example system.

	SATELLITE	1	2 and 3
Transponder			
- probability of initial failure		0	0.03
- failure intensities (constant)		*	
. when active		6500 FITS	6500 FITS
. when passive		650 FITS	650 FITS
Plattform			
(Truncated Weibull distribution.)			
- max. useful life		7 years	10 years
- form parameter		1.88	1.88
- scale parameter		$6.8 \ 10^{-6} \ h^{-1}$	$4.8 \ 10^{-6} \ h^{-1}$
Probability of launch failure		0	0.1
Switching delays (i.e. min. down times)			
- active failure		3 hours	
- passive and initial failure		3 hours	
- provoked		5 min.	

$*1 \ FITS = 10^{-9} h^{-1}$

To keep this example simple, we focus on channel 1, which has the highest possible priority, and may be supported by all the satellites, of course provided that they are in orbit and have the necessary resource s intact i.e. at least one usable transponder and the platform. We consentrate on the two basic quanteties:

- the (un)availability of channel 1, which is obtained by logging the "up" and "down" events of this channel, cf. figure 3, during approximately 1600 repetitions of the mission period. The result is shown in figure 7.

- The cumulative distributions of the launching times of the satellites in our scenario are shown in figure 8. The launch of satellite 3 (and 4) is determined by the control structure from the resource situation at any instant.

The simulation repeats the mission period a number of times and the stocastic events are registrated.

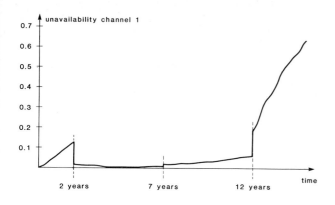

Figure 7. The unavailability of channel 1 during the satellite system's mission period.

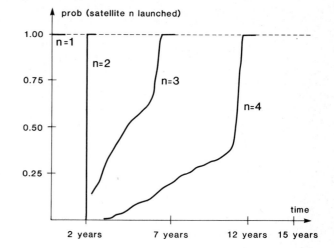

Figure 8. The cumulative distributions of the satellites' launching times.

Note that the launching times of satellite 4 in figure 8 give the times that a new satellite would have been launched, if available. In our example we use only three satellites.

From figures 7 and 8 we see that :

0 - 2 years : Only satellite 1 may be operative. Channel 1 can use only one specific transponder on satellite 1. If this transponder (or the platform) fails, the channel is "down" until a new satellite is successfully launched. The probability of the channel being "down" immediately before satellite 2 is launched (after two years) is approx. 0.13.

2 - 7 years : After two years satellite 2 is launched with a success probability of 0.90. Hence the unavailability drops to 0.014. If the launch is unsuccessfull, satellite 3 is ordered immediately and launched as soon as possible. Satellite 3 may also be ordered earlier if satellite 1 has failed during its first 2 years. The earliest possible launch of satellite 3 is at time 2 years and 3 months. We see from figure 8, that the probability of the earliest possible launch of satellite 3 is approx. 0.155. Hence the unavailability is decreasing until it reaches a minimum of 0.004 after 3.5 years. The unavailability is then slowly increasing to 0.008 after seven years, mainly because of failing transponders resulting in down times because of spare switching (3 hours) and resource exhaustion. Satellite 1 has a maximum life of seven years, and is, since we demand at least two satellites in space, replaced by satellite 3, if this satellite is not already launched. This results in a sharp increase (to 1.0) in the launching probability of satellite 3 immediately before seven years have elapsed, cf. figure 8. We see, however, that in approximately two thirds of the missions satellite 3 is needed in order

to meet the redundancy requirements before satellite 1 needs replacement.

7 - 12 years : Satellites 2 and 3 may be operative. The unavailability of channel 1 is now increasing more rapidly, from 0.02 after seven years to 0.063 after twelve years. Remember that, in addition to units failing in orbit, the launch of satellite 2 or of satellite 3 may have been unsuccessfull. Hence, in order to meet the redundancy requirements, a next satellite (number 4) ought to be launched before 7 years in approx. 20% of the cases, and before 12 years in approx. 40% of the cases.

12 - 15 years : Satellite 2 has surely failed because of its maximum life time. Hence only satellite 3 may be operational during this time span, probably with some failed transponders. This can be seen clearly from the unavailability curve for channel 1. The unavailability is rapidly increasing from approx. 0.185 after twelve years to approx. 0.635 after fifteen years.

5. CONCLUSIONS

Our experiences with the application of process oriented simulation for reliability evaluation have been very satisfactory. The highly dynamic redundancy structure of a satellite system is easily handled. The development and execution of the rather complex optimization and decision procedures turned, in fact, out to be more demanding than the development and execution of the reliability simulator itself. The incorporation of these procedures in the simulator was, however, straightforeward. Note also that the concept of process classes allows us to develop generic reliability models of satellites and channels, where each specimen, also within the same simulation, may be different. Hence, the process approach allows an easy parameterization of a wide range of systems.

In general, we suggest that process oriented simulation is considered for reliability evaluation when simulation is necessary and when some units (or subsystems) have a complex inner life, when there are functional interdependencies between units affecting their reliability behaviour, when the reliability structure of the system is dynamic, or when generic models of units or subsystems is advantageous.

ACKNOWLEDGEMENT

This simulator was developed under contract with the Norwegian Telecommunication Administration. The views expressed, however,are those of the authors. We want to thank H. Fjøsne and P.M. Bakken for useful discussions during the modelling phase of the project.

REFERENCES.

<BIRT 79a> Birtwistle G., Dahl O.-J., Myhrhaug B., Nygaard K.: "SIMULA BEGIN". Studentlitteratur, Lund, Sweden, 1979.

<BIRT 79b> Birtwistle G.: "DEMOS reference manual", University of Bradford, 1979.

<HELV 85a> Helvik B.E., Stol N.: "Model and Simulators for reliability Evulation of Satellite Systems". ELAB report STF A85050, 1985. (In Norwegian)

<HELV 85b> Helvik B.E., Stol N.: "Simulators for Reliability Evaluation of Satellite Systems. Program documentation". ELAB report STF44 85049, 1985. (In Norwegian.)

Bjarne E. Helvik received the Siv. Ing. and the Doctor Technicae degrees from the Norwegian Institute of Technology in 1975 and 1982, respectivly. He is with ELAB (Electronics Research Laboratory), Trondheim, Norway, as a Senior Scientist. His research interests are within the areas of traffic and reliability evaluation of telecommunication systems. He is currently involved in reliability analysis of distributed, fault tolerant communication control systems, in reliabilty simulation of satellite systems, in the design and evaluation of fault tolerant off-shore control systems, and in the development of a reliability planning course.

Norvald Stol received the Siv. Ing. degree from the Norwegian Institute of Technology in 1983. He is with ELAB (Electronics Research Laboratory), Trondheim, Norway, as a Scientist. His research interests are within the areas of traffic and reliability evaluation of telecommunication systems. He is currently involved in traffic simulation and evaluation of the Nordic Public Data Network, in reliability simulation of satellite systems, and in a data security project. He is a voting member of the Association for Computing Machinery (ACM).

SOFTWARE RELIABILITY

Reliability Technology — Theory & Applications
J. Møltoft and F. Jensen (Editors)
© Elsevier Science Publishers B.V. (North-Holland), 1986

TRANSIENT FAULT EFFECTS IN MICROPROCESSOR CONTROLLERS

Janusz SOSNOWSKI
Institute of Computer Science,
Warsaw Technical University,
Warszawa 00-665, Poland

This paper deals with transient faults in microprocessor controllers. We analyse controller upsets caused by transient faults on circuit and system levels. The main point of this analysis is investigation of program flow disturbancies (e.g. deadlock, false loops) due to errors appearing in various parts of the controller. The occurence of long duration failures is estimated for some microprocessors and single-chip microcomputers. Finaly, we approach the problem of designing controllers which are capable to overcome these failures. Error avoidance and error removal techniques are taken into account.

1. INTRODUCTION

Microcomputers' decreasing size and cost create a rising interest in low-cost controllers used in automobiles, appliances, process control,electrical and mechanical devices, slave modules of complexed systems etc. Quite often these controllers operate in noise conditions resulting from supply line transients, electromagnetic pulses, electrostatic discharges, radio frequency interference etc [8]. This noise is the source of random transient faults (physical level) in the controller. In practice transient faults may result also from timing coincidences and incorrect states of controlled objects.

Short duration transient faults may lead to circuit errors (logical level) resulting in system failures. We distinguish three classes of errors and failures: transient (short duration), steady state (long duration) and fatal (circuit or system damage). Steady state errors or failures can be removed by resetting procedures. Fatal errors and failures are avoided by careful design and will be not discussed. Fig. 1 illustrates system responses to transient faults. Dashed lines refer to systems with redundancy in hardware or software.

Steady state and sometimes transient failures are undesirable in controllers. Hence arises the problem of designing controllers which are proof against transient faults. We consider this in relevance to the analysis of transient fault effects.

Investigating transient fault upsets in various controller elements we will find possible errors and failures. For many elements this analysis is quite simple. For example combinational circuits (ALU, ROM, PLA) respond to transient faults with transient errors. Transient faults in memory elements (flip-flops, regis-

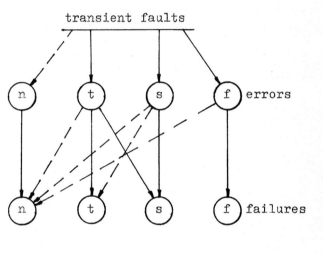

transient faults

n-no   s-steady state
t-transient   f-fatal

Fig. 1 Controller responses to
transient faults

ters, RAM's) result in steady-state errors: false setting or resetting of bits, cells etc. These errors influence basic controller functions: data processing, data storage, data transfer and control. Disturbancies of the first three functions are well known and various approaches allow to tolerate them [1]. In the paper we study disturbancies of control function. Moreover, we are mainly interested in steady-state failures (transient failures in many applications are not critical [5]). Meaningful estimation of occurrence of these failures is presented.

2. CONTROL FLOW DISTURBANCIES

Specification of control flow in microprocessor controllers is not an easy task due to the fact that controllers are composed of sequential circuits (loosely or tightly coupled). Moreover, the complexity of these circuits increases accordingly to development of technology. So apart from simple machines we have processors, coprocessors, interrupt and DMA controllers, programable protocol controllers etc. The operation of simple sequential machines is preferably described using state transition graphs. These graphs may be inordinately large in the case of sophisticated machines. Nevertheless, the structures of microprocessor family circuits allow their operation to be modelled by a set of acceptably sized graphs. Sometimes they can be represented with nested graphs i.e. graphs relevant to the entire system where nodes represent a part of the system or one of its functions. In turn, these nodes are defined by graphs etc.

Analysis of control flow disturbancies in microprocessor controllers can be based on examination how transient faults contribute to errors in controller elements and how these errors create system failures. This analysis can be accomplished using state transition graphs. In the paper we are concerned with steady state failures which mostly result from steady state errors. For the purpose of the analysis we distinguish three classes of sequential machines: single mode, multimode and programmed.

Single mode machines realize one specified function and can be modelled with single state transition graph. Examples of these machines are: synchronizing and timing logic, processor sequencers, bus arbiters. Some of them realize state graphs in which the next state depends upon the previous state only. In such machines transient faults result in transient errors (temporary disturbance of a state sequence). If the next state is a function of the previous one and the state of some inputs, steady state errors may appear. It is illustrated in fig. 2 which shows simplified state transition graph of 8085 microprocessor sequencer [10]. It can be easily checked that spurious transitions (due to transient faults) within this graph mostly result in transient errors (their contribution to system failures will be discussed later). However, there is one state (WH) which creates the danger of a steady state error. State WH is set by HLT instruction and can be left only with interrupt signal IN (or Reset). In the case when interrupts are not armed a false transition into WH state leads to deadlock.

In general machine states which cannot be easily abandoned (e.g. due to rarely encountered conditions will be called trap states. Another example of the trap state is given in fig. 3. The presented graph is relevant to the sequencer of DMA controller (8257 [10]). False transition to state SO when input signal DRQ_n is not active results in deadlock due to the fact that signal HLDA is delivered by cooperating microprocessor in the case when signal HRQ was issued. Signal HRQ is produced by a flip-flop which can be set by DRQ_n signal if the sequencer is in state SI. In state SO there is no possibility to set this flip-flop.

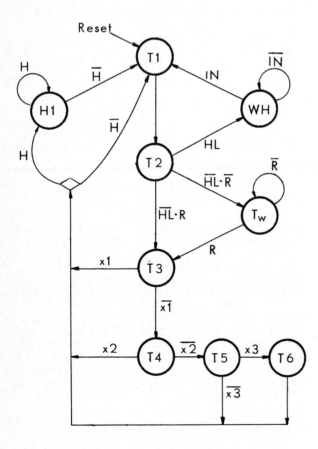

Fig. 2   A simplified state transition graph of 8085 CPU

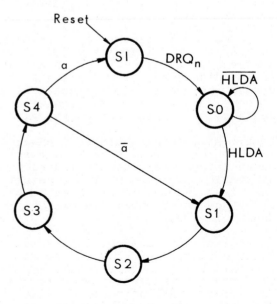

Fig. 3   DMA operation state graph

It is worth noting that sequential machines may incorporate not used states (resulting from binary state coding) also called parasitic states. In fault free environment these states are never entered. Transient fault may set the machine into a parasitic state. In well designed machines this state is easily abandoned and oughtn't to cause deadlocks.

Multimode machines allow to realize one function which is choosen from a set of specified functions. Function selection can be accomplished by hard or soft programming of the machine. Hard programming involves making specified circuit connections. Soft programming requires issuing specified programming commands. We deal with hard programmed machines as with single mode machines described by selected state graphs. Soft programmed machines we treat as sophisticated machines described by a complex graph comprising state graphs representing all specified functions. An example is universal synchronous asynchronous receiver transmitter (USART 8251) which can realize various communication formats. These formats are programmed by setting the circuit into so called "Idle" state and then issuing control words [10].

Analysing transient fault effects in soft programmed machines we have to take into account spurious state transitions within state subgraphs representing all specified functions and false transitions between subgraphs including invalid (parasitic subgraphs). This is illustrated in fig. 4. In most cases false transitions between subgraphs are as critical as trap states (trap subgraphs). For example in USART circuit false transition from asynchronous into synchronous transmission mode is a steady state error resulting in critical controller failure [7]. Parasitic subgraphs are not encountered in well designed machines.

Programmed machines are universal and can realize unlimited number of functions. In fact these functions are accomplished by a set of sequential machines (processor sequencer, interrupt controller etc) executing a specified program. Errors in these machines result in program execution failures. For example transient errors in the processor sequencer lead to skipping (or adding) instruction cycles and finaly to instruction misinterpretation, wrong fetch of the next instruction etc. Errors in program counter or address calculation circuits transfer program execution to a spurious point (e.g. within data memory). Most steady state errors lead to steady state program failures. Transient errors can be analysed in a similar way as in non-programmed machines (using program flow charts). However in this case we have to take into account a lot of parasitic trap states and subgraphs (e.g. false loops) resulting from the above mentioned disturbancies in program execution. We deal with this problem in the sequel.

In controllers composed of several cooperating machines a transient fault in one machine may result in steady state error of the other one due to state locking or appearance of not allowed input sequences. These errors can be analysed in a similar way as transmission protocols.

3. PROGRAM EXECUTION UPSETS

In general controllers receive data from an input, process it and deliver final results to outputs. The particular processing to be done is specified by the program. The program is stored in the program area. Inputs for computations and intermediate results are held in the data area.

The data area can be partitioned into register file, data memory and I/O subareas. Data memory subarea is represented mostly by read-write memories (RAM's). However, in many applications there are used tables of constants (e.g. look-up tables) which are stored in ROM's. I/O subarea is represented by I/O ports and registers. Sometimes it is merged with data memory subarea.

In many microprocessors the program and data memory areas are implemented within one memory address space (e.g. 8085, Z80, 8086, 8096). In this case a transient fault may lead to execution of a false program comprising instructions fetched from data memory. Such situation can be eliminated in microprocessors with separate program and data memory address

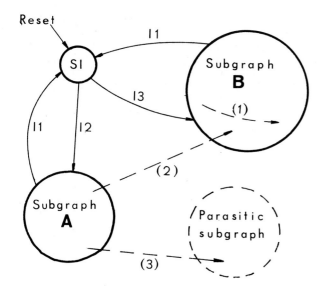

Fig. 4   Transient fault effects in a complex circuit (1,2,3 – false transitions)

spaces (e.g. MCS48,MCS51 single-chip microcomputer families). Nevertheless, these microprocessors allow to store constants in program memory and load them into working registers. Investigating upsets in program execution due to transient faults we will start with the analysis of instruction sets.

3.1 Statistics of instruction codes

Most program upsets can be considered as spurious transitions of program execution into data memory or accidental jump into any point within the program (e.g. second byte of a multibyte instruction). Analysing these upsets it is helpful to have some information on instruction formats.

In table 1 we give the statistics of non branch instructions for some microprocessors. Table entries represent percentage of non branch instructions of specified length L (in bytes) within the whole instruction set. Conditional branches we treat as non branch instructions with probability 0.5.

In general we can distinguish four classes of branch instructions: jumps,calls, returns, restarts and software interrupts. They transfer control from current value of the program counter (PC) to a target address. An important feature of branch instructions is the way of specification of target addresses. Target addresses are taken directly from instructions and indirectly from appropriate registers or memory locations. They are defined in an absolute or relative way. Relative specification defines the target address as an offset (or displacement) from the current state of the program counter, within the current memory page, memory bank or segment [10,11]. Some branch instructions may specify fixed target addresses e.g. restart and software interrupt instructions.

In table 2 we present the probabilities of branch instruction execution for various classes of these instructions (p_l - long range, p_s - short range, p_f - fixed address, $p=p_l+p_s+p_f$). These probabilities are calculated as the ratio of the number of branch instructions to the number of all possible opcodes. In brackets we give the range of target address specification. Similarly as in table 1 we take into account conditional branches with probability 0.5.

The presented figures must be treated as some sort of approximation due to the fact that interpretation of not specified opcodes was not known. Moreover, we assume that instruction opcodes are defined in the first byte. In some processors there are several instructions which spill the opcode over the next byte.

Table 1
Statistics of non branch instructions

L	Microprocessor					
	8085	Z80	8086	8048	8051	8096
1	74.7	66.5	38.0	74.0	53.1	3.8
2	8.1	13.7	38.2	17.3	26.5	12.8
3	7.2	8.0	8.3		6.6	37.6
4			4.0			19.7
5						8.9
6						1.7

Table 2
Probabilities of branch instructions

Proc.	p_l	p_s	p_f	p
8085	.064 (2)	–	.032	.096
Z80	.064 (2)	.016 (6)	.032	.112
8086	.024 (3)	.024 (8)	.016	.104
		.04 (6)		
8048	.008 (1)	.008 (5)	–	.076
		.06 (4)		
8051	.016 (2)	.064 (5)	–	.140
		.06 (6)		
8096	.016 (2)	.064 (6)	.008	.150
		.064 (7)		

Address ranges: (1) - 4k, (2) - 64k, (3) - 2M (absolute); (4) - $\pm$256, (5) - $\pm$2k (within the page); (6) - $\pm$128, (7) - $\pm$1k, (8) - $\pm$32k (PC relative).

3.2 False loops within data memory

Let us study the occurrence of a false loop resulting from program bump into data memory space. Creation of a loop requires at least one branch instruction. An example is given in fig. 5. A false transition (W) to address A1 initiates execution of a sequence of non branch instructions. At point B1 a branch instruction is executed which transfers the program control to point C1 within the previously executed sequence or to point D1 preceeding entry point A1. The sequence of codes between D1 and A1 points does not incorporate branch instruction codes. A sequence of non branch instructions ended with branch instruction we will call the branch segment (e.g. segment A1-B1 in fig. 5). A sequence of non branch instructions leading to the branch segment we will call branch segment extension (e.g. segment D1-A1 in fig. 5). In general false loops may be created as a sequence of branch segments (segment i branches to segment i+1 etc.) with the last branch segment transferring the program control into any branch segment of this sequence (including the last one) or to its extension.

In [7] we have analysed creation of false loops in the system comprising P loca-

tions of program ROM and D locations ($P+D=2^n$) of data RAM. Moreover, we have assumed:
1) all data memory states are equally probable and mutually independent,
2) branch instruction may point to any location within the whole memory space (absolute addressing)
3) execution of instructions within the loop doesn't cancel this loop
4) non branch instructions are single byte
5) branch instructions are unconditional
6) branching to the valid program stops creation of a false loop.

The probability (p_D) that a transition into data memory space leads to a false loop is estimated as:

$$p_D \simeq \sum_{i=1}^{\infty} p_{Di} \quad ; \quad p_{Di} = \left(\frac{D}{2^n}\right)^{i-1} \frac{(2S-1)i}{2^n}$$

where $S=1/p$ expresses expected value of the number of instructions within a branch segment, p is the probability of a branch instruction code, p_{Di} is relevant to the probability of the loop with i branch segments.

The first factor of p_{Di} accounts for the probability that a transition into data memory space leads to execution of a sequence of i branch segments. The second factor is relevant to the probability that the branch instruction in the last branch segment transfers the program control to any point within executed branch segments (there are on average Si such points) or to any point within the extensions of these segments (there are on average $(S-1)i$ such points).

The presented model can be easily extended for systems with conditional and unconditional branch instructions. It is only required to take conditional branches with factor 1/2 in calculation of p (compare table 2).

Next we will take into account branch instructions with fixed target addresses. It is reasonable to assume that the branch segment ended with such instruction stops creation of a false loop. Usually the above mentioned addresses point to well defined program procedures. Denoting by ϵ the ratio of the number of branch instructions with

not fixed target addresses to the number of all branch instructions we have

$$p_{Di} = \left(\frac{D}{2^n}\epsilon\right)^{i-1} \frac{(2S-1)i}{2^n}\epsilon$$

$$p_D \simeq \epsilon \frac{(2S-1)}{2^n}\left(1-\epsilon\frac{D}{2^n}\right)^{-2}$$

Coefficient ϵ can be found from table 2 as $\epsilon = (p-p_f)/p$.

The given model is relevant to 8085 microprocessor. From table 2 we get $\epsilon=0.67$, $S=10.4$, $n=16$. Hence, assuming $D/2^n=1/2$ we obtain $p_D\simeq1/2000$.

Taking into account the existence of multibyte non branch instructions we can introduce coefficient α denoting mean length of non branch instructions (which can be found from table 1). Now the upper bound of the probability p_D can be found by replacing parameter S (in expression for p_D) by αS. For 8085 processor we have $\alpha=1.15$.

The probability of false loops grows if between program memory (lower addresses) and data memory spaces there is unused memory space such that only non branch instruction codes (e.g. NOP) can be fetched from this space. In this case all branches to unused memory space (U locations) lead to the beginning of data memory space creating bigger chance for looping. A similar effect is possible if a part of data memory is realized as ROM (containing constants). In [7] we have estimated the probability p_{UD} of a false loop within data and unused memory spaces (due to a transition into data memory space). Assuming U>>S and including coefficient ϵ we have

$$p_{UD} \simeq \epsilon \frac{U}{2^n}\left(1-\epsilon\frac{D}{2^n}\right)^{-2}$$

Taking $U=D=P=2^n/3$ we get $p_{UD}\simeq0.14$ for 8085 microprocessor

From Table 2 it can be easily seen that microprocessors may comprise branch instructions with limited target address ranges. These instructions increase significantly the probability of false loop creation, especially in the case when the address range is less than data memory space. For this model we will derive only a rough estimation of the probability of false loop creation (in the case when transition into data memory occured). We take into account three classes of branch instructions: with fixed, short range and long range addresses. Long range addressing covers the whole memory space, short range addressing is relevant to a part of memory space. The upper bound of the above mentioned probability can be estimated as

$$p_{UB} \simeq \frac{p_s}{p}\sum_{i=0}^{\infty}\left(p_1\frac{D}{2^n}\right)^i = \frac{p_s}{p}\frac{1}{(1-p_1D/2^n)}$$

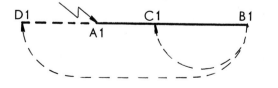

Fig. 5   Creation of a loop with one branch instruction

where p_s, p_1 are the probabilities of
execution of short range and long range
branch instructions, respectively; p,D
were specified previously.

Calculating the upper bound we assume
that any branch segment (within data
memory) ended with a short range branch
instruction leads to a false loop. The
contribution of long range branch instructions to loop creation is considered as negligible. Nevertheless, we take
into account branch segments within
data memory which are activated by long
range branch instructions (term $p_1 D/2^n$).

We find the lower bound of the estimated probability taking into account
false loops which are given in fig. 6.
A transient fault transfers program
control to point A1 (beginning of a
branch segment in data memory space).
At the end of the first branch segment
(B1) a branch instruction is executed.
If the target address of this instruction is fixed or long range specifying
a point in program memory, we assume
that false loop will be not created
(exit point LF). A short range branch
instruction may close the loop in point
C1 (within the first branch segment) or
D1 (within the branch segment extension)
or it may transfer program control to
the next branch segment (point A2) etc.
We take into account only loops which
are closed within the last branch segment or its extension.

The probability of the loop closing
within a branch segment or its extension
can be estimated as

$$p_{BS} \simeq \frac{p_s}{p} \frac{(2S-1)}{R}$$

where R is the number of memory locations accessible by short range branch
instructions, 2S-1 refers to the number
of instructions in the branch segment
and its extension (it was discussed
previously, $1/p < S < \alpha/p$).

The probability p_{LB} of creation of a
false loop responding to the structure
given in fig. 6 is:

$$p_{LB} \simeq p_{BS} \sum_{i=0}^{\infty} \left(\frac{p'_s}{p} - p_{BS} \right)^i =$$

$$= \frac{p_{BS}}{1 + p_{BS} - p'_s/p}$$

where $p'_s = p_s + p_1 D/2^n$

The i-th term of the series is relevant
to the situation in which a sequence of
i+1 branch segments (within data memory)
was executed and the branch instruction
of the last segment transferred the
control to any point within this segment
or its extension. The probability of
branching from one segment to another re-

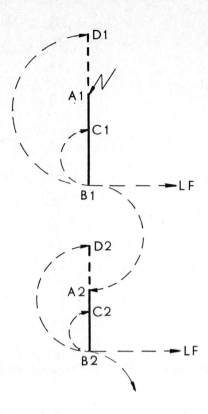

Fig. 6   The model of simple
         loop structures

lates to p'_s. Here we assume that branch
instructions with short range addresses
lead always to data memory (if fetched
from there). It is obvious that there
are possible other false loops (e.g.
closing within more than one segment).
Hence, the probability p_{DS} of a false
loop for the analysed model is

$$p_{LB} < p_{DS} < p_{UB}$$

As an illustration we present p_{DS} estimation for Z80, 8086 and 8096 microprocessors. Taking data from table 2
and assuming $D/2^n = 1/2$ we get:

  Z80 : $0.01 < p_{DS} < 0.14$

  8086: $0.07 < p_{DS} < 0.61$

  8096: $0.17 < p_{DS} < 0.85$

Calculating p_{BS} for 8086 and 8096 we
take R=256 and p_s equal to the probability of branch instructions with specification 6 in table 2. These instructions mostly contribute to loop creation
due to the shortest address range. Finding p_{LB} and p_{UB} we take as p_s the probability of all short range instructions.
The probability p_{DS} for 8086 is closer
to the upper bound due to segmented memory structure (especially in the case
of disjoint program and data memory
segments).

3.3 False loops within program memory

Consider the problem of false loop creation within program memory area for
processors with physically separated
address spaces of program and data memories (e.g. 8048, 8051). If there are

no constant parameters within the program memory a false loop may be created due to a spurious program transition into a point which is not the first byte of an instruction (so called internal byte). Loop structures are similar to those described in previous section. The only difference is relevant to branch segments. Now these segments comprise a sequence of false instructions. Moreover, they are not so easily created as in data memory. It is due to the fact that fetching of a valid program instruction stops activation of a branch segment (the control is returned to the program). Examples of branch segment structures are given in fig. 7. First bytes (defining opcodes) of valid program instructions are denoted by OP. False instructions creating the branch segment are specified as FI_{1-3} (FI_3 is a branch instruction).

Assume that branch instructions may point any of 2^n program memory locations and that all states of internal bytes are equally probable. The probability of false loop creation due to a spurious transition within program memory can be estimated as:

$$p_P \simeq \sum_{i=1}^{\infty} (Ip_B)^i \frac{(2S'-1)i}{2^n} = \frac{Ip_B(2S'-1)}{2^n(1-Ip_B)^2}$$

where I is the probability of branching into internal byte (ratio of the number of internal bytes to 2^n), S' is the expected value of the branch segment length, p_B is the probability of creation of a branch segment.

The i-th term of p_P is the probability of loop activation in a sequence of i branch segments. The branch instruction of the last segment transfers the control to any point within the segments of the sequence including their extensions.

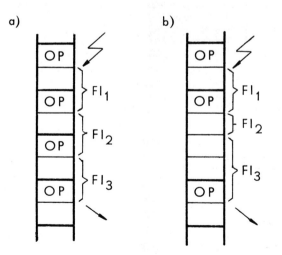

a)   b)

Fig. 7   Branch segment creation
within the valid program

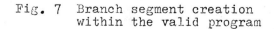

As an illustration we have estimated the probability p_P for 8048 processor. Taking into account low probability of two-byte non branch instructions (see table 1) we can assume $S'=1$, $p_B=1/p$. Moreover, we take into account branch instructions with addressing range 256 (within the page). Hence, we get $p_P \simeq 4 \cdot 10^{-4}$. For a sample of programs we found I=0.3, so $p_P \simeq 10^{-4}$.

3.4 Other deadlocks

Beyond false loops there are possible other deadlocks in program execution. A spurious transition from one point of the program into another may result in transient or steady state failure. We can analyse this in a similar way as errors in sequential machines (using graphs or flowcharts). Nevertheless, it is reasonable to consider separately hazards resulting from spurious execution of some critical instructions (which create the danger of deadlock).

Here we give examples of such instructions for some microprocessors:
8085 : HLT (set halt state), DIS (disable interrupt), SIM (set interrupt mask);
Z80 : HLT, DIS, $IM0 \neq 2$ (set interrupt mode);
8086 : HLT, WAIT (enter wait state), DIS, load segment register;
8048 : DIS, SELRB, SELMB (select register or memory bank), STRT, STOP (start or stop timer);
8051 : IDL, PD (set idle or power down mode), load function register.
Critical instructions may set the processor into a risky state or change program execution conditions. Risky states are potential trap states e.g. halt, wait, idle states when interrupts are disabled or disarmed.

Any program (or its part) is executed in a specified environment which is defined by states of some flip-flops of the processor etc. An illustration of this are: interrupt enable and interrupt mode flags, memory bank flags, function registers, segment registers, I/O circuit states etc. [10,11]. Some instructions may change this environment and cause steady state failures. For example wrong setting of memory bank or segment register can transfer the control beyond memory area of the valid program.

Let us assume that a transient fault caused spurious transition in the program execution. The probability of fetching a critical instruction in this situation can be estimated as:

$$p_c \simeq 1-(1-p_i)^N$$

where p_i is the probability of the critical instruction code, N is the number of false instructions fetched in consequence of the false transition (e.g. instructions fetched from data memory).

In the case of single byte critical instructions (e.g. HLT) the probability p_c may achieve significant values. The number N is relevant to the number and length of erronously executed branch segments within data memory or false instruction segments in program memory. For the analysed microprocessors expected value of the branch segment length in data memory is of the range 7÷15. Hence, the number N ranging up to 100 is quite probable. For single byte instructions we have p_i=1/256 so, for N=30 we get p_c=0.11 and for N=100 p_c=0.7.

Two byte instructions which specify a critical condition in the second byte (e.g. address of critical function register) are less risky. Nevertheless, we must take them into account in the case of large data memory spaces. The probability p_i for these instructions can be estimated as $1/2^{16}$. Hence, for N=100 we get p_c=0.0016 and for N=1000 p_c=0.015.

4. TRANSIENT FAULT TOLERANCE IN MICRO-PROCESSOR CONTROLLERS

Taking into account the results of the analysis presented in the previous sections we can state that transient faults may create serious problems in microprocessor controller operation (for example steady state failures). Hence arises the necessity of designing controllers which are insensitive to transient faults. This can be achieved using two techniques: fault-avoidance and fault tolerance [1].

Fault avoidance involves various techniques (e.g. shielding, noise reduction) which allow to eliminate transient faults. However, even one follows the best guidelines for designing for a noisy environment, it's always possible for a transient fault to occur. So, we must refer to fault tolerance. This approach incorporates techniques which ensure that acceptable system service is maintained by coping with the faults which remain despite the use of fault prevention. In the sequel we deal with fault tolerance in microprocessor controllers.

In general we can distinguish two methodes of achieving fault tolerance: error avoidance and error removal with error recovery. These methodes ought to ensure failure free operation of the controller. Here it is worth noting that in many controllers transient failures are not critical. For example controllers cooperating with electromechanical devices. In this case the inertial characteristics of attached physical equipment may absorb temporary aberrations so long as subsequent control signals are valid. In fact an erroneous

signal is compensated by subsequent signals. Similar situation arises in data acquisition systems [7]. Hence, we will restrict our considerations to controllers with no steady state failures. It is very difficult to eliminate transient failures so, practically the controlled objects ought to cope with these failures.

Some steady state failures can be eliminated or their occurence will be reduced with error avoidance techniques. These techniques base on error analysis including more sophisticated risk assessment methods [4]. Here we give only a short illustration of this approach. Designing sequential circuits we ought to avoid trap states. For example the trap state of the DMA controller discussed in section 2 (fig. 3) can be eliminated by introducing redundant setting of the flip-flop HRQ if the controller is in state SO. Many critical situations will be avoided if soft programmed multimode machines are not used. There are also other possibilities of risky upsets' reduction. First it is recommended to ensure that references to unused memory locations are not critical (compare section 3.2). In order to achieve this we fill unused memory areas with NOP instructions followed occasionally by jumps to an error recovery routine (e.g. reset instruction).

Using single-chip microcomputers with physically separated program and data memory address spaces we avoid false loops within data RAM memory, which in other cases are quite probable. It is also possible to eliminate false loops within program memory.

Analysing internal bytes of instructions within the valid program we can find those which comprise opcodes of a branch instruction. These false instructions may create false loops in the case of some spurious program transitions. Checking target addresses of these instructions we can easily verify if false loops are possible. It is helpful to use for this purpose a special analysing program. If we detect a false loop we can cancel it by inserting NOP instructions into the program (they will disturb target addresses of branch instructions creating the loop). This approach is effective thanks to the low probability of false loops in program memory.

In the case of lookup tables stored in program memory the probability of a false loop or a critical state increases. This can be reduced or eliminated using coded values in tables (which do not incorporate opcodes of critical instructions). Moreover, partitioning the tables and separating the parts with initialization sequences we ensure a speedy recovery. An initialization sequence may comprise k single byte NOP instructions followed by a reset instruction where k is maximal length of microprocessor instructions (e.g. k=3 for 8051).

Steady state errors which cannot be eliminated by error avoidance techniques must be removed by error recovery techniques. The most important aspect of these techniques is the provision of an effective means of transforming an erroneous state of the controller into a well defined and error free state [1,2,9]. Realization of this approach requires two things: error detection and recovery. Both of these require some sort of redundancy, spatial or temporal. An example of spatial redundancy is excessive coding. Temporal redundancy can be accomplished by providing the program with checking sequences etc.

Fault detection is implemented by means of testing. Taking into account that controllers comprise several logical modules we have to introduce local and global testing procedures. Local testing is performed on logical module level. In practice it is realized in intelligent modules comprising microprocessors. Global testing allows to find errors which cannot be covered by local testing. For example errors in non intelligent modules. Multimode sequential machines (e.g. USART) require external verification if they operate in the preprogrammed mode etc. It can be accomplished by reading status registers, output signal checking etc. Moreover error detection can be simplified if intermodule testing is involved.

Effective testing requires software and hardware redundancy. In most microprocessor controllers it is no need for using sophisticated testing hardware e.g. processor monitor [3]. Usually it is sufficient to equip the controller with a watchdog timer [6]. However, this circuit must be free of steady state errors. Moreover, it is helpful to use some program assertions which may reveal an error by detecting an illegal use of program data, range checking [7], data checksums, checks on invariant relationships between variables etc. These techniques are quite efficient in cyclic controllers. In controllers which provide some services on request (e.g. voting machine [8]) it is possible to repeat calculations several times and compare results. It is important to be aware that the controller environment may provide information which is helpful in error detection. For example a floppy disc controller can use some checking data delivered by disc drive.

The fault detection methodes give the foundation for the next step - the implementation of controller recovery. It is assummed that whenever an error is detected, either by software or hardware checking, the control passes to an error handler. The error handler must determine of the extent of error effects and then initiate appropriate error recovery measures. In most controllers it will be sufficient to reset the whole controller (global recovery) or its part (local recovery).

The global recovery usually is done with reset signal (e.g. issued by watchdog timer). It can be realized also by direct branching from the testing procedure into reset handling program. Local recovery refers to a part of the controller. Hence, it can be realized by issuing initialization commands to the module which is in an erroneous state. Here we must be aware of the fact that this module may be in a spurious state and its initialization may require additional synchronizing commands (e.g. in 8251 circuit [10]).

Taking into account various critical situations (trap states, deadlocks) which were described previously, it can be seen that some of them can be avoided by preventive error removal procedures. In soft programmed multimode machines we can occasionally reprogram them, no matter if an error occurred or no. However, there is one drawback in this approach, we must be sure that reprogramming procedure doesn't interfere with normal operation of the machine. There is no problem if the circuit allows to set operation mode flags at any moment (e.g. 8255 programmable peripheral interface [10]). Some microprocessor circuits can be reprogrammed after setting them in an idle state (e.g. 8251). In other circuits every reprogramming procedure starts the operation from the beginning (e.g. 8254 programmable interval timer [10]). In both of the these cases normal circuit operation may be disturbed by reprogramming procedures.

It is worth noting that in microprocessors we can reinitialize preventively quite a lot of status flip-flops and control registers e.g. interrupt enable and mode flip-flops. Moreover, it is reasonable to reinitialize (periodically) control variables in data memory (e.g. data buffer pointers). This simplifies testing and speeds up recovery. For example it is very difficult to check if a DMA controller realizes rotating or fixed priority. In this case a potential priority error can be easily removed by preventive circuit reprogramming.

5. CONCLUSIONS

In the paper we deal with transient fault effects in microprocessor controllers. It is reasonable to assume that these faults are more probable than fatal (persistant) faults. Hence, in many applications it is acceptable to neglect fatal faults, but transient faults must be taken into account.

A systematic analysis of transient
fault effects in microprocessor controllers has been presented. Estimating
the probabilities of various deadlock
situations (for some off-the-shelf microprocessors) in response to a transient fault, we have found that the
deadlock probability may exceed 0.1.
Hence, it arises the necessity of fault
tolerance techniques in microprocessor
controllers.

Most controllers seem able to operate
acceptably despite momentary degradation of service. So, transient failures
may be not critical. Steady state failures practically disallow the controller to perform correctly its function.
It has been shown that introducing some
redundancy in software and adding simple
hardware measures one can easily eliminate the occurrence of steady state
failures in microprocessor controllers.

REFERENCES

[1] A.Avizienis, Fault tolerance, the
    survival attribute of digital sys-
    tems, Proceedings of IEEE, vol.66,
    no.10, Oct.1978, 1109-1125.
[2] T.Anderson, J.C.Knight, A framework
    for software fault tolerance in
    real time systems, IEEE Transactions
    on Software Engineering, vol. SE-9,
    no. 3, March 1983, 355-364.
[3] J.B.Eifert, J.P.Shen, Processor
    monitoring using asynchronous sig-
    natured instruction streams,
    Proceedings of the FTCS-14, 1984.
[4] M.O.Fryer, Risk assessment of
    computer controlled systems, IEEE
    Transactions on Software Engineering
    vol. SE-11, no.1, Jan.1985, 125-129.
[5] R.E.Glaser, C.M.Mason, Transient
    upsets in microprocessor control-
    lers, Proceedings of the FTCS-11, 1981
    1981.
[6] F.Novak, Watchdog timers in system
    dynamic checking, Proceedings of
    the 10th Euromicro Symposium, 1984.
[7] J.Sosnowski, Transient fault tole-
    rance in a data acquisition system,
    Microprocessing and Microprogram-
    ing, 16 (1985), 255-260.
[8] J.Sosnowski, Simulation of transient
    fault effects in microprocessor
    systems, Advances in Modelling &
    Simulation, vol.6, no.1, 7-17.
[9] I.S.Upadhyaya, K.K.Saluja, A hard-
    ware supported general rollback
    technique, Proceedings of the 14th
    FTCS, 1984, 409-414.
[10] Microsystem Components Handbook,
     Intel Corp., 1984.
[11] Microcontroller Handbook, Intel
     Corp., 1984.
[12] Microcomputer Z80 Data Book,
     Mostek Corp., 1982.

Janusz Sosnowski received the M.Sc.deg-
ree in electronics in 1969 and the Ph.D.
degree in computer science in 1976,
both from Warsaw Technical University,
Warszawa, Poland. Since 1969 he has
been working in the Institute of Computer Science, Warsaw Technical University. At present he is an Assistant Professor. He is also a Vice Director of
the Institute. Dr. Sosnowski is the author and coauthor of aproximately 40
professional publications including
three books. His current research interests include fault tolerant computing, local area networks and machine
architecture. He is engaged in a project
of a local area network for industrial
purposes.

Reliability Technology — Theory & Applications
J. Møltoft and F. Jensen (Editors)
Elsevier Science Publishers B.V. (North-Holland), 1986

CODING FOR MEMORIES WITH DEFECTS

Dr.ir. Han Vinck and Guy Kerpen
University of technology, Eindhoven,
the Netherlands.

We discuss the application of coding for yield and performance improvement of memories with defect-(hard) and soft errors. We consider memories with a word and a string organization. In a word organized memory, coding gives a positive efficiency in a product process in the area where uncoded memories have zero yield. This result is noticable for relative small, but practical, word lengths. In a string organization we use convolutional coding for "bursty" defect correction.

In the last part we deal with the Mean Time Before Failure for coded and uncoded memory systems.

1. INTRODUCTION

One of the important parts of a computing system is the memory. There is a tendency in producing large memory systems on one chip. This is possible due to improvements in process technology and clever circuit design that lead to high packing densities. Obviously, packing density has its limits. A high packing density may cause defects in memory cells. We distinguish between 0-defects and 1-defects, i.e. between defective cells that always produce a "0" or a "1", respectively, when being read. We assume that the writer is able to determine the location and value of a defect. Suppose that a memory cell is defective with probability p. Then the probability of an error free chip goes to zero as the number of cells increases. If N denotes the number of memory elements of size n, then the chip error probability is given by

p(chip error)=1-prob (all memory elements correct) $\qquad$ =$1-((1-p)^n)^N$

The yield of a production process is the fraction of good chips. In the sequel we discuss the application of coding techniques for the improvement of yield and performance. These techniques can be incorporated in the chip design or build in on a separate circuit. As an illustration of the power of coding we use a 64k-4 bit word memory chip.

2. CODING

2.1 Defects as random errors

In this chapter we discuss the application of coding to protect 4 bits of information against defects (hard) and/or random (soft) errors. We first study the performance of the (4,7) one error correcting Hamming code. This code stores 4 information bits together with 3 parity digits. Hence, the word length is extended to contain 7 memory cells. The input/output relation is given in Fig. 1, where we assume modulo-2 arithmetic.

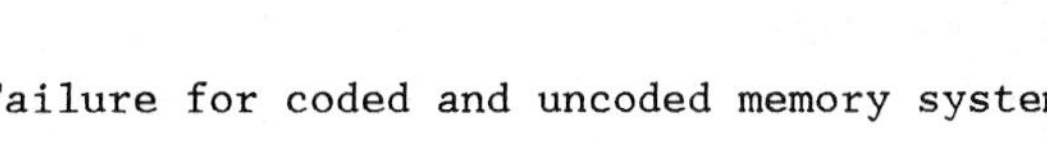

$$(i_1,i_2,i_3,i_4) \longrightarrow \boxed{\begin{array}{c}\text{Hamming}\\\text{encoder}\end{array}} \longrightarrow (i_1,i_2,i_3,i_4,P_1,P_2,P_3)$$

$$(i_1,i_2,i_3,i_4) \begin{bmatrix} 1 & & & & 1 & 1 & 1 \\ & 1 & & & 0 & 1 & 1 \\ & & 1 & & 1 & 0 & 1 \\ & & & 1 & 1 & 1 & 0 \end{bmatrix} = (i_1,i_2,i_3,i_4,P_1,P_2,P_3)$$

Fig. 1. (4,7) Hamming encoding.

The error correcting capability of the code is determined by the minimum distance between any two codewords. The distance between two codewords is the number of components in which the two codewords differ. By inspection one can verify that the minimum distance of the above code is 3. A possible decoder implementation is to make a comparison between a word read from memory and one of the 16 possible stored words. There are faster decoding algorithms

available. The decoder decodes the word that is
closest in distance to the word that is read.
If one single error is made, then the decoder
is able to give the correct word as output. For,
the distance between the received word and the
correct word is one, whereas the distance to
any other possible word is at least two. If two
or more errors occur, then the decoder is mislea-
ded, and gives the wrong word as output. If we
apply this (4,7) Hamming code to our defect problem, then a chip is regarded as bad if there
is a word with 2 or more defects. Hence, the
efficiency of the production process is given
by
$R(4,7)=\frac{4}{7}((1-p)^7+7p(1-p)^6)^{64k}$,
where the factor $(\frac{4}{7})$ stems from the fact that
only $(\frac{4}{7})$th of the total surface is used to store
the 4 bits of information. The performance of
this scheme can be found in Fig. 2. The advantage of the code is that it can correct 1 defect
(hard) or one so called soft error that may
occur later. Characteristic for the above method
is that errors are corrected after a word is
read from memory. We will show that performance
can be improved if we take into account the
knowledge of defects in a word before storing.

2.2 Defects in a word organized memory

Suppose that it is possible, for instance by
testing, to determine the location and value
of a defective cell within a word. Then, according to Kuznetsov and Tsybakov {1} it is possible to store messages at an efficiency of
$R=k/n\leq1-p$ (p is fraction of defects). We will
show that this is indeed possible in certain
cases. We first explain an optimal strategy
if there is only one defective cell per word of
5 cells. As mentioned before, 4 cells can be
used to store information. Hence, per word we
must be able to store one out of 2**4 messages.
These messages are numbered from 0 to 2**4-1=15.
The strategy is as follows:
- A message m is encoded by the codeword
  $\underline{x}=(0,m_1,m_2,m_3,m_4)$.
- From $\underline{x}$ we form
  $\underline{\bar{x}}=(1,\bar{m}_1,\bar{m}_2,\bar{m}_3,\bar{m}_4,)$,
  where $\bar{m}$ is the inverse of m.
- If m is to be stored in the memory, then we

first look whether $\underline{x}$ can be stored error free
(i.e. defective cells must match with $\underline{x}$). If not,
then the vector $\underline{\bar{x}}$ must match, as each symbol is
inverted.
The encoding complexity is proportional to a
write/read cycle plus an additional write instruction if $\underline{x}$ does not fit. Note that this strategy
can be extended to arbitrary length. The efficiency remains optimal.
The efficiency of the above strategy translated
into production efficiency is given by

$$R(4,5)=\frac{4}{5}((1-p)^5+5p(1-p)^4)^{64k},$$

and can be found in fig. 2. The above example is
a specific case of additive coding. The general
additive coding scheme can be found in {1}. We
proceed by looking at the problem of two defects
per word. The codeword $\underline{c}$ to be stored is called
defect compatible if it can be stored without
any changes, i.e. the components of a codeword
agree with the value of the 2 defects. The codeword $\underline{c}$ itself depends on the defects and the
message m. We first construct a codematrix for
which any pair of defects is present in some row.
The construction is as follows:
1) take all binary (2a-1) tuples of weight (a)
   as columns of the codematrix.
2) add the all zero row to the matrix.
3) select $\lceil log2a \rceil$ columns such that all rows of
   the $2a\lceil log2a \rceil$ submatrix are different. ($\lceil \ \rceil$
   is round off upwards).
ad 1). If we compare two arbitrary columns, then
it contains at least the combinations (01), (10)
and (11) in one of the rows. This can be concluded from the fact that a column of length (2a-1)
has more than halve ones, and hence there must
be some overlap between two specific columns.
Together with 2) we see that any pair of defects
is present in some row.
ad 3). First take $2**\lceil log2a \rceil$ different rows of
length $\lceil log2a \rceil$. The columns all have an equal
number of zero's and ones. By deleting complementary pairs of rows, except for the all zero
and all ones row, the equal weight property
remains valid. We may delete until (2a) rows
are left, with the property that all columns
have (a) ones. As this is part of the matrix
constructions, we place this submatrix in front

of the original matrix by column permutations.
Note that each row is uniquelly specified by
the first $\lceil \log 2a \rceil$ digits. This property plays
an important role in the sequel. To be more
specific we consider the case where a=3. The
code matrix has 6 rows and 10 columns. For our
64k-4bit system we shorten each row to length
7, i.e.

$$C = \begin{bmatrix} 0 & 0 & 0 & 0 & 0 & 0 & 0 \\ 0 & 0 & 1 & 0 & 1 & 1 & 0 \\ 0 & 1 & 0 & 1 & 0 & 1 & 1 \\ 1 & 1 & 0 & 0 & 1 & 0 & 1 \\ 1 & 0 & 1 & 1 & 0 & 0 & 1 \\ 1 & 1 & 1 & 1 & 1 & 1 & 0 \end{bmatrix}$$

Note that any defect of multiplicity 2 in a
vector of length 7 can be found in a row of C.
This property will be used in the encoding of
m. Secondly, the first 3 digits uniquelly specify each row of C. The message m is represented by the vector.

$$\underline{x}=(0,0,0,m_1,m_2,m_3,m_4).$$

Hence, $\underline{x}$ selects one out of 16 messages and
the efficiency of the code is 4/7. Now, suppose
that $\underline{x}$ is not defect compatible in two of its
components. Then we look for a rowvector $\underline{c}(\underline{x},d)$,
such that

$$\underline{c}=\underline{x} \oplus \underline{c}(\underline{x},d)$$

is defect compatible. The sign $\oplus$ means component
wise modulo-2 addition. The constructed vector
$\underline{c}$ is stored instead of $\underline{x}$. Decoding is done as
follows. The vector $\underline{x}$ has 3 all zero initial
components. Hence, the decoder (reader) knows
which row of C is used in order to make $\underline{x}$ defect
compatible. This row is added (mod-2) to $\underline{c}$ and
the last 4 components specify m again. For,
$(\underline{x} \oplus \underline{c}(\underline{x},d)) \oplus \underline{c}(\underline{x},d)=\underline{x}$. Suppose that we would like
to store the message (1101) as $\underline{x}$=(0001101) and
the memory has a stuck-at-1 defect in the third
position and a stuck-at-0 defect in the fourth
position. The encoder adds modulo-2 the vector
$\underline{c}(\underline{x},d)$=(1011001) to $\underline{x}$ and stores $\underline{c}$=(1010100).
The decoder sees as three initial components
(101), and adds the vector (1011001) to $\underline{c}$, which
results in a decoded vector $\underline{x}$=(0001101) and
message (1101). The production efficiency of
the 2-defect method is given by

$$R(4,7)=4/7((1-p)^7+7p(1-p)^6+\binom{7}{2}p^2(1-p)^5)64k.$$

and can be found in Fig. 2. The above coding
strategy can be generalized. The problem is to
minimize the initial number of zero's for a given codeword length and defect multiplicity. It
can be shown that there exists a coding matrix
C that facilitates utilization of the fraction
(1-p) of non-defective memory cells. If we do
not know the position of the defects in advance,
then a fraction p/2 read errors will occur. From
classical information theory it then follows
that the amount of information that can be stored
error free is less than 1-h(p/2) bits per memory
cell, where h(.) is known as the binary entropy
function. Note that when p=1/2, we can store at
most 1-h(1/4)=.18872 bits per memory cell. The
remaining fraction of non-defective memory is
necessary to inform the reader about the location of the defects. If we use a generalized
additive coding scheme, we can store (1-p)=1-½=.5
bits per memory cell, an improvement of a factor
3.

2.3 Defects and random errors

We will now give a simple example of a 1-defect
and 1-random error correcting code. The defects
are known before writing, the random errors not.
Observe that in Fig. 1, the information input
vector (1111) gives rise to a codevector (1111111).
Then, observe that the 1-defect correcting strategy is equivalent to adding the vector (1111111)
to $\underline{x}$ if it is not defect compatible. Combination
of the two is possible as follows. Use only the
components m_1,m_2 and m_3 to encode the information.
The component m_4 is normally equal to zero, but
is set automatically to one if the all ones codeword is added modulo-2 to the encoded information.
Suppose, for example we would like to store the
message (110) and the memory has a stuck-at-1
defect in the third position. The encoder encodes
this information as (0011011). If in time a random error occurs, then by normal minimum distance
decoding this error can be corrected, for (0011011)
is a regular codeword. Not that this coding scheme
is also able to correct two defects as the second
defect can be treated as a random error. Performance of this scheme is determined by the proba-

bility that more than 2 defects occur in a word.

$$R(3,7)=(3/7)*((1-p)^7+7p(1-p)^6+\binom{7}{2}p^2(1-p)^5)^{64k},$$

where the factor 3/7 follows from the information rate. In Fig. 2 we combine the calculations
for all previous schemes. It can be observed
that in the coded case there is still an acceptable yield. The efficiency is reduced by a
factor equal to the code rate. Hence, for
high defect probabilities, where we use long
codeword lengths, there is a positive yield but
low efficiency.

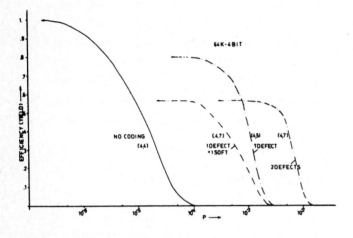

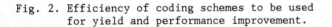

Fig. 2. Efficiency of coding schemes to be used
for yield and performance improvement.

2.4 Bursts of defects

Suppose that we have the availability of a memory that is used to write large data blocks,
such as digitized television pictures. If we do
not want to break up the information stream into small blocks, some coding problems arise.
The complexity of finding codes with long block
lengths for a fixed fraction of defects grows
exponentially. Therefore, we have to use other
methods that are less complex. Another problem
arises when defects occur in bursts. We give a
method that corrects bursts of defects in long
data streams. The principle of additive coding
is that information words are made defect compatible by adding a codeword to it. This codeword depends on the information and the defects.
We explain additive coding for information

sequences with Fig. 3, where I is a binary information sequence and z a binary sequence that depends on I and the defects of the memory.

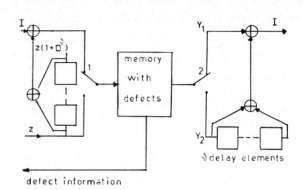

Fig. 3. Defect compatible scheme for sequences.

All additions are modulo-2 component wise.
Switch 1 is used to place both memory input
streams on odd and even numbered subsequent locations. The read switch separates the memory data
again into two output streams. If synchronized,
input and output are the same. At the decoder
one has the sequences $(y_1,y_2)=(I+z(1+D^{\nu}),z)$,
where zD^{ν} is a ν time instants delayed version
of z. If we form the sequence $y_1+y_2(1+D^{\nu})$, as
indicated, then the original information sequence is found back. Note that this is valid for
any sequence z. The question arises whether
the sequence z can be used to create a defect
compatible sequence. The answer follows from
the circuit of Fig. 4.

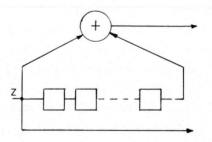

Fig. 4. Circuit that generates $(z(1+D^{\nu}),z)$.

This circuit accepts each time unit one digit
from the sequence z, and shifts the content of
the delay elements one place to the right. The
input/output relation can be written formally
as $zG=z(1+D^{\nu},1)$. It is easy to see that we can

form any output of ν pairs or length 2ν, by proper selection of the input and the contents of the delay elements. If after such a period we refill the delay elements with ν new digits, determined by the desired output, we are again able to construct any output sequence of ν pairs. Consequently, as the decoder of Fig. 3 is insensitive to the output of the circuit of Fig. 4, we can use this output to make information sequences defect compatible. With the circuit of Fig. 4 we are able to correct bursts of ν defect pairs if the burstperiod is preceded by ν defect free pairs. Secondly, observe that by inverting the input, we also invert the output. By proper input selection we even make the burst length two digits longer. One in the beginning, and one at the end. the possibilities are summarized in Fig. 5.

$$\text{time}$$

$$\begin{array}{ccccccccccccc} x & x & x & x & | & 0 & 1 & 1 & 0 & | & x & x & x & x & \text{even} \\ x & x & x & x & | & 0 & 0 & 0 & 1 & | & x & x & x & x & \text{odd} \end{array}$$
$$\;\;\;\text{guard}\quad\text{burst}\quad\text{guard}$$

a) Burst of ν pairs of defects

$$\begin{array}{cccccccccc} 1 & 0 & x & 1 & x & 0 & 0 & 1 & x & 0 & \text{even} \\ x & x & 0 & x & 1 & x & x & x & 1 & x & \text{odd} \end{array}$$

b) Only single defects per pair

$$\begin{array}{ccccccccccccc} x & x & x & x & | & x & | & 1 & 0 & 1 & 1 & | & 1 & | & x & x & x & x & | & \text{even} \\ x & x & x & x & | & 1 & | & 0 & 1 & 0 & 1 & | & x & | & x & x & x & x & | & \text{odd} \end{array}$$
$$\quad\;\;\nu\qquad\quad\;\;\nu\qquad\qquad\;\nu$$

c) Expension of burstlength with 2 digits

Fig. 5. Possibilities for defect compatible scheme of Fig. 3.

The information rate of the scheme is $\frac{1}{2}$, whereas the defect fraction is $\frac{1}{2}$. It can be extended to efficiency $R=k/n$, and defect fraction $p=1-R$, see {2}. If in the above scheme $(1+D^\nu)$ is replaced by $G(D)$, then the above scheme can be used to match random defects. The performance is determined by the freedistance of the convolutional code $(G(D),1)$.

3. MEAN TIME BEFORE FAILURE (MTBF)

In the previous section we consider memories that can be tested directly <u>after</u> production. We now consider the situation where the error free chips are used in a computing system. Errors that occur during operation are supposed to be hard errors, i.e. they remain in the system. The system tests a memory word for defects only when being used, see Fig. 6.

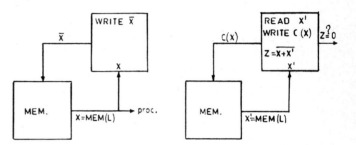

Fig. 6. Two step testing procedure.

Suppose that a word $\underline{x}=\text{MEM}(L)$ is read from memory location L. Directly after reading the inverse $\bar{\underline{x}}$ to $\underline{x}$ is restored at location L, while the word $\underline{x}$ is normally processed. Subsequently the word $\underline{x}'=\text{MEM}(L)$ is read and if there are no errors then $z=\underline{x}+\underline{x}'$ equals $\underline{0}$. If $z\neq\underline{0}$, an alarm is given and processing stops. To create the original situation as before testing, $\underline{x}$ is again written in the memory at location L. This testing procedure can be done in parallel, independent of the processing of $\underline{x}$.

There are several different coding applications;
- error detecting codes can be used to replace the testing cycle.
- error correcting codes can be used to detect and correct errors after reading. Processing stops whenever the number of errors in a word exceeds the error correcting capability of the code in use. It can be expected that this increases the mean time before failure.
- combined defect matching-error correcting codes can be used to correct errors and thus locates position and value of new defects. The restored codeword is chosen such that it matches the defect pattern. This can be done until the defect

matching capability of the code is reached. Here we assume that new errors can be corrected.

The disadvantage of the above methods is the decoding/encoding delay, and the need for redundant chip surface.

We proceed with a general definition of the MTBF and apply this to different memory configurations. We assume that after each unit of time a specific memory location is selected at random and tested for errors. The system is said to be in failure if a selected word is found to be in error. Thus, the Mean Time Before Failure (MTBF) can be defined as

$$MTBF \overset{\Delta}{=} 1 \cdot \bar{P}_1 + 2P_1 \bar{P}_{2/1} + 3P_1 P_{2/1} \bar{P}_{3/2} + \dots , \qquad (1)$$

where P_i is the probability that a word is correct up to time i, and $\bar{P}_{i/i-1}$ is the probability that a word is in error given that it was in order at the previous time instants. The probability that a memory <u>cell</u> at time i is in error given that it was in order at time (i-1), is p. We distinguish between uncoded and coded systems.

3.1 <u>Uncoded memory systems</u>

- <u>Single word</u>: We first consider the case where we have only one word consisting out of k cells. In this case the MTBF is given by

$$MTBF = 1 + q^k + q^{2k} + \dots$$
$$= \frac{1}{1-q^k}$$
$$\cong \frac{1}{kp} ,$$

where q^{ik} is the probability that a word is found correct at time i.

- <u>N words</u>: In this case it is difficult to give a general expression. We give an upperbound for the MTBF. First observe that the mean time before there is at least one memory word in error is given by

$$f_1 = 1 + (q^k)^N + (q^k)^{2N} + \dots$$
$$= \frac{1}{1-(q^k)^N} .$$

Then, as all words are selected with probability 1/N, it takes on the average N units of time before a specific word is selected. Hence, as

an upperbound

$$MTBF(N) \leq \frac{1}{1-(q^k)^N} + N . \qquad (2)$$

Note that for small values of p, $f_1 \approx \frac{1}{Nkp}$.

For small values of N the MTBF can be calculated with a Markov state approach. As an example, for N=2 there are three states 1) both words are good, 2) one word is stuck, but not yet used, 3) stuck state.

Here

$$MTBF(2) = \frac{2+q^k-2q^{2k}}{(1-q^{2k})(2-q^k)} \approx \frac{1}{2kp} .$$

- <u>Infinitely many words</u>: In this case (1) reduces to

$$MTBF(\infty) = 1 + P_1 + P_1 P_2 + P_1 P_2 P_3 + \dots , \qquad (3)$$

where P_i is the probability that a selected word is correct at time i. Due to the fact that the memory has infinitely many words, one can show that $P_{i/i-1} = P_i$. Thus, for words consisting out of k cells, $P_i = (q^k)^i$, and

$$MTBF(\infty) = 1 + q^k + q^{3k} + q^{6k} + \dots$$
$$= \sum_{i=1}^{\infty} (q^k)^{\frac{i(i-1)}{2}} . \qquad (4)$$

This can be approximated with Θ functions [3] as

$$MTBF(\infty) = \frac{\sqrt{2\pi}}{2\sqrt{1-q^k}} . \qquad (5)$$

In Fig. 7, we plotted for k=1, the functions $\frac{1}{Np}$ for N=4,64 and 1024, respectively, together

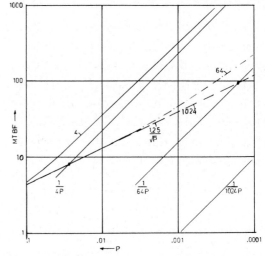

Fig. 7. MTBF(N), for N=4,64 and 1024; MTBF(∞) and simulations for N=4 and 64.

with (5). The curves cross for $p=\frac{1}{(1.25N)^2}$. In the same figure are simulation results for memories of size 4,64 and 1024, respectively. For our 64k-4bit chip, the crosspoint is found at $p_c=40.2^{-40}$. Hence, using (2) and (5) and p_c, we are able to give an estimate of the MTBF of large memory systems.

3.2 Coded memory systems

The question arises whether the application of coding can improve the MTBF. We again consider 3 different configurations.

- Single word: Words of k digits long are extended to n digits by an error correcting code. If a codeword of length n is able to correct t errors, then the probability that a word has less than (t+1) errors at time i is

$$P_i=\sum_{j=0}^{t}\binom{n}{j}(1-q_i)^j q_i^{(n-j)}. \qquad (6)$$

Using (1), the MTBF for a single coded word is

$$MTBF_c(1)=1+P_1+P_2+P_3+\ldots, \qquad (7)$$

where we used the fact that $P_i \cdot P_{i+1/i}=P_{i+1}$. We can use (6) and (7) to calculate the $MTBF_c(1)$ in the coded situation. In general, if we use a rate k/n code with distance d, the MTBF is

$$MTBF_c(k/n) > \frac{\lfloor\frac{d}{2}\rfloor+1}{np},$$

whereas in the uncoded case the

$$MTBF_c > \frac{1}{kp}, \quad k\geq 2.$$

An approximation follows from the fact that the first term in the $MTBF_c(k/n)$ is determined by the probability that $(\lfloor\frac{d}{2}\rfloor+1)$ single errors occur in reading cycles that are on the average $\frac{1}{np}$ cycles apart. All other terms are of an higher order and thus vanish faster when p goes to zero.

Suppose that after a failure the memory is repaired. Then, one is interested in a long time between two failure events. Therefore, we define the time gain factor as

$$\eta \triangleq \frac{MTBF(k/n)}{MTBF(1)} = (\lfloor\frac{d}{2}\rfloor+1)(k/n)$$

If on the other hand chip surface is costly, then one is interested in the average amount of chip surface needed to realize a time T. The chip surface gain is defined as

$$\gamma \triangleq \frac{\frac{kT}{MTBF(1)}}{\frac{nT}{MTBF(k/n)}} \cong (\lfloor\frac{d}{2}\rfloor+1)(k/n)^2$$

For the well known class of 1-error correcting Hamming codes, $(\lfloor\frac{d}{2}\rfloor+1)=2$, $n=2^m-1$ and $k=n-m$. Here, η and $\gamma>1$ for $m\geq3$ and 4, respectively. The maximum gain we can expect for this class of codes is 2 for large values of m. It is easy to see that coding should be applied carefully, for the gain factor can be less than 1 for certain values k and n.

-N words: Here we use the same strategy as used in the previous section. We calculate the mean time before the memory system has at least one error. Using the fact that $P_i P_{i+1/i}=P_{i+1}$, (1) reduces to

$$MTBF_c(N)\geq 1+P_1^N+P_2^N+\ldots, \qquad (8)$$

where the probabilities P_i are given in (6). Again, for small values of N the MTBF can be calculated exactly by using a Markov state approach. However, this method is getting very complex for interesting values of N. For N=2 and a 1-error correcting code we have 3 additional states in the Markov state diagram. These states are 1) one word with a defect, 2) both words with a defect, 3) one word with a defect, the other is stuck but not read. Here

$$MTBF_c(2)\approx\frac{1.25}{np}.$$

For the (4,7) Hamming code the time gain factor is

$$\eta=\frac{1.15}{7} \cdot \frac{4}{7} \cdot 8$$

whereas

$$\gamma=\frac{10}{7} * \frac{4}{7} = \frac{40}{49} \ !$$

- Infinitely many words: Here we substitute (6) in (3). In Fig. 8 we plot the simulation results for the $MTBF_c$, when n=7, N=4, t=0,1,2 and 3, together with MTBF$_c$(N) and MTBF$_c$(∞).

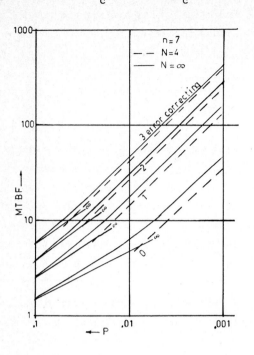

Fig. 8. Simulation results for the MTBF$_c$, when
n=7, N=4; and t=0,1,2 and 3, together with
MTBF$_c$(4) and MTBF$_c$(∞).

CONCLUSIONS

We show that coding can be applied in memory
systems. Although we use a simple model, it
follows that yield and performance can be improved by using relatively short codes. Furthermore, we do calculations on the Mean Time
Before Failure of coded and uncoded memory
systems.

{1} A.V.Kuznetsov and B.S.Tsybakov, Coding
     for memories with defective cells, Proble-
     my peredachi informatsii, Vol.10, no.2,
     pp.52-60, 1974.

{2} A.J.Vinck, Convolutional codes and defects,
     6th Symposium on Information Theory in
     the Benelux, pp.55-61, May 1985.

{3} H.Dym and H.P.McKean, Fourier series and
     integrals, Academic Press New York and
     London 1972, page 52 and 225.

Han Vinck was born in Breda, the Netherlands,
on May 15, 1949. He received the M.Sc. and the
Ph.D. degree in electrical engineering both from
the Technical University of Eindhoven, in 1974
and 1980, respectively. Since 1980 he is associate professor at the same university. In 1981
and 1986 he was a Visiting Scientist at the
Communications division of the German Space
Agency, DFVLR, Oberpfaffenhofen, Germany. His
research interest include information theory
and coding. He is a member of the Board of the
Dutch Electronic and Radio Engineering society,
where he is responsible for the scientific
program.

Guy Kerpen was born in Venlo, the Netherlands,
on May 4, 1959. He is currently working towards
the M.Sc. degree in electrical engineering at
the Technical University of Eindhoven.

MAINTAINABILITY

Reliability Technology — Theory & Applications
J. Møltoft and F. Jensen (Editors)
© Elsevier Science Publishers B.V. (North-Holland), 1986

THEORY OF MAINTAINED SYSTEMS: THE STATE OF THE ART

C.A. CLAROTTI* and F. SPIZZICHINO**
*ENEA TIB-ISP CRE Casaccia, S.P. Anguillarese 301, 00100 Rome, Italy
**Dept of Matematics "G. Castelnuovo", University of Rome "La Sapienza", 00100 Rome, Italy

Different methods for dealing with maintained reliability systems are surveyed. The role of exponentiality in the theory of maintained systems is clarified. Component aging is proven to be immaterial (i.e. it does not affect calculations) if the systems of interest are reliable.

1. INTRODUCTION

In seventies many approaches have been proposed for calculating the reliability of maintained systems, namely: the Markov approach, the fault-tree technique, the device of stages, the method of supplementary variables. These approaches have been understood by reliability analysts as having nothing in common but the aim of estimating the reliability of systems whose components are maintained. In the sequel it will be shown how all the above mentioned approaches can be framed into a unique view if the theory of maintained systems is regarded to as a part of the theory of the Markov-processes; doing this not only leads to some kind of unification of maintained-system-reliability-models but also permits us to establish to what extent assuming that system-components are exponential yields accurate system-reliability-estimates.

2. THE ROLE OF EXPONENTIALITY IN MAINTAINED SYSTEM-THEORY

It is commonly believed that:

(i) The minimal-cut-set reliability-bound for maintained systems is a trivial consequence of the inclusion-exclusion principle.

(ii) Exponentiality is needed only if system reliability is calculated via the Markov-approach. Markov chain theory and fault-tree-analysis have nothing in common.

(iii) Since exponentiality means lack of aging it suits only to electronic components.

All the statements above are false; in this section we will prove (i) and (ii) to be false.

2.1 Commenting on (i)

Arguments in {1}, {2} can be summarized as follows:

As far as system-reliability calculation is of concern the stochastic process representing the system evolution must stop at time of first system failure, we are then entitled to consider only the events which are well defined on that censored process.

Denote by

$P_r\{\ \}$: the probability of the event in brackets

T_s : the system failure time

$R_s(t)$: system-reliability at time t

T_j : the failure time of the min-cut j

$R_j(t)$: the reliability of the min-cut j

N : the number of system min-cuts.

For supporting the min-cut-reliability bound

$$1 - R_s(t) \leqslant \sum_{j=1}^{N} (1 - R_j(t)) \qquad (2.1.1)$$

the inclusion-exclusion principle is invoked together with the decomposition

$$\{T_s < t\} = \bigcup_{j=1}^{N} \{T_j < t\} \qquad (2.1.2)$$

but the events on the right-hand-side of (2.1.2) are not well defined on each trajectory of the process; for instance the event $\{T_j < t\}$ is not well defined on all those trajectories which end at a time $t_1 < t$ due to the failure of a min-cut $i \neq j$. (What to do in the time interval (t_1, t)? In (t_1, t) the repair of failed components which are shared by i and j must be undertaken or not?).

As opposite the events

$$\{T_j > t\}, \ j = 1,..., N \qquad (2.1.3)$$

do have a precise definition on each trajectory of the process which stops as soon as the system fails. If the latter failed at time s then an index r exists such that $T_r = s$; no matter what the value of s is ($s \gtreqless t$) it will be unambiguou sly $T_j = +\infty$ for any $j \neq r$, we are then able to say whether or not the event $\{T_j > t\}$ occurs for any j and for any process-stopping-time.

In view of the above, the event $\{T_s > t\}$ complementary to (2.1.2)

$$\{T_s > t\} = \bigcap_{j=1}^{N} \{T_j > t\} \qquad (2.1.4)$$

must be the starting point for proving that (2.1.1) holds in the case of maintained systems. (2.1.1) is implied if it results:

$$R_s(t) = P_r \left\{ \bigcap_{j=1}^{N} \{T_j > t\} \right\} \geqslant \prod_{j=1}^{N} P_r \{T_j > t\} =$$

$$= \prod_{j=1}^{N} R_j(t) \qquad (2.1.5)$$

Of course (2.1.5) is not trivially true, for it to hold some conditions are needed which are discussed below.

2.2 Commenting on (ii)

The performance process $X_i(t)$ of the generic component i of a system is defined according to (2.2.1)

$$X_i(t) = \begin{cases} 1 \text{ if component i is up at t} \\ \\ 0 \text{ otherwise} \end{cases} \qquad (2.2.1)$$

For (2.1.5) to hold for a coherent maintained system {3} it suffices that:

the $X_i(t)$'s are time-associated and independent of one-another.

The $X_i(t)$'s are independent if system-components fail independently and are separately maintained (i.e. the system operates under an unrestricted repair-policy).

Readers interested in definitions and theory of time-associated stochastic processes are referred to {2}, {3}, {4}, {5}, here we want just to remark that the only sufficient condition of practical use for proving that a process is associated in time is the following:

a stochastic process X(t) is time associated if it is Markov and it results

$$P_r \{X(t) > x | X(s) = y\} \text{ increasing in y for any}$$
$$x, s < t.$$

If component i is maintained its performance process $X_i(t)$ is Markov if, and practically only if, the component-life and the component-repair time are exponential. In principle $X_i(t)$ is Markov also if the transition-rates between component-states are assumed to depend upon the calendar-time (time since the component started operating), but as a matter of fact the latter

hypothesis is even more unsatisfactory than the memory-less model (as to the dispute on and the comparison between the good-as-new-model and the bad-as-old-model see {1}, {6}, {7}).

In the case of maintained systems, previous argumentations show that those which theoretically are sufficient conditions, practically become necessary and sufficient conditions for the minimal-cut-reliability-bound to hold; exponentiality of system components is then needed in order to estimate the reliability of complex-repairable-systems via the min-cut-parallel-system-reliabilities.

Two further remarks are useful for clarifying the role of exponentiality and of the related Markov chain-theory in the frame of fault-tree-analysis of maintained systems.

Remark 2.2.1: Markov-chain-theory {8} makes the following handy estimate available for $R_j(t)$ (the reliability of min-cut j of the maintained system at hand)

$$R_j(t) \geqslant \exp(-\Lambda_j t)$$

$$\Lambda_j = \frac{Q_j \sum_{r \in j} \mu_r}{1 - Q_j} \qquad (2.2.2)$$

where

μ_r is the repair rate of component r

Q_j is the asymptotic unavailability of min-cut j

$\sum_{r \in j}$ means: summation over components belonging to min-cut j

Should (2.2.2) not be available $R_j(t)$ ought to be calculated by means of the time consuming Markov approach.

Remark 2.2.2: The Keilson-Uniformization-Principle {9}, another important result of the Markov-chain-theory, reconducts the problem of proving that a continuous-time-Markov-process is time-associated to the problem of proving that a discrete-time-Markov-chain is such. This is a relevant aid in stating whether or not the min-cut-reliability-bound applies in the case of a given maintained system {10}.

3. ADEQUACY OF THE EXPONENTIAL MODEL

Since assuming that system-components are exponential is compulsory in estimating maintained system-reliability, showing that the exponential model is adequate is a matter of primary importance. In order to do that the relationship must be enlightened between the device of stages and the method of supplementary variables.

The common acceptation of the device of stages is

as follows:

The pdf's below

$$f(t) = \sum_{i=1}^{N} w_i \; \lambda_i \; e^{-\lambda_i t} \qquad (3.1)$$

$$\sum_{i=1}^{N} w_i = 1$$

$$g(t) = \frac{\lambda^n t^{n-1}}{(n-1)!} e^{-\lambda t} \qquad (3.2)$$

can be represented as arising from an n-stage-Markov-process, the stages being "in parallel" for the pdf (3.1) and "in series" for the pdf (3.2). Then if a system of independent components with failure time and repair time distributions such as (3.1) and/or (3.2) is considered, the system can be transformed into an exponential one by introducing suitable dummy states where the system spends exponentially distributed random times; the device of stages at first sight seems to be based on heuristic grounds. In {11} and {12} it is shown how the device of stages is related to the rigourous theory of non-exponential maintained systems (the $X_i(t)$'s will be assumed to be independent throughout the discussion).

Define the stochastic processes

$$Z_i(t) \equiv (X_i(t), U_i(t), V_i(t))$$

$$U_i(t) \begin{cases} \text{time since component i last began ope-} \\ \text{rating if } X_i(t) = 1 \\ \\ 0 \text{ if } X_i(t) = 0 \end{cases} \qquad (3.3)$$

$$V_i(t) \begin{cases} 0 \text{ if } X_i(t) = 1 \\ \\ \text{time since component i last failed if} \\ X(t) = 0 \end{cases}$$

If component lives and component-repair-times have general distributions the $X_i(t)$'s are no more Markov processes but the $Z_i(t)$'s are if both the component-failure-epochs and the component-repair epochs are renewal-epochs for the $X_i(t)$'s. In {11} and {12}, by making use of Markov process theory, equations are derived for the probabilities

$$P_{\underline{x}}(\underline{u},\underline{v},t)d\underline{u}\,d\underline{v} = P_r \{\underline{X}(t) = \underline{x},\; \underline{U}(t) \in (\underline{u},\underline{u}+d\underline{u}),$$
$$\underline{V}(t) \in (\underline{v},\underline{v}+d\underline{v})\} \qquad (3.4)$$

$$\overline{F}_{\underline{x}}(\underline{u},\underline{v},t) = P_r \{T_F > t | \underline{Z}(0) = \underline{x},\underline{u},\underline{v}\}$$

(these are the only probabilities which are meaningful from a rigorous point of view, probabilities introduced by the device of stages are formal); where

T_F is time to first system failure (the system is supposed to have n components)
$\underline{X}(t) \equiv (X_1(t),\ldots, X_n(t))$
$\underline{U}(t) \equiv (U_1(t),\ldots, U_n(t))$
$\underline{V}(t) \equiv (V_1(t),\ldots, V_n(t))$
$\underline{Z}(t) \equiv (\underline{X}(t), \underline{U}(t), \underline{V}(t))$

It is also shown in {12} that if system components start fresh and their lives and their repair-times have distributions such as (3.1) and (3.2), the device of stages is the quickest computational tool for numerically obtaining the reliability and the availability of a non-exponential system starting from the equations that probabilities (3.4) satisfy.

Having validated the device of stages, the latter can be used for stating in which cases assuming that system-components are exponential leads to accurate results. Arguments which will follow are a simplification of the treatment in {13}; the authors hope to have kept the subtle essence of {13} in this more engineering-oriented reasoning.

Assume that system components fail independently and are separately maintained. Let system-component-sojourn-times (both in the up-state and in the down-state) be non-exponential. Any non-exponential distribution can be approximated as close as needed by suitable combinations of distributions having rational Laplace transforms (i.e. distributions such as (3.1) and (3.2)). The desired degree of approximation is obtained by equating a sufficient number of moments of the two distributions (the given distribution and the approximating one).

Under the hypoyheses above the reliability behaviour of the maintained system is represented to close approximation by the device of stages. The latter ensures that:

(i) the derivative of the expected number of system failures approaches an asymptotic value as $t \to \infty$,

(ii) if component-mean-repair-times are much shorter than component-mean-lives (as reasonable for "reliable" systems), then the onset time to the asymptote is at a rate equal to the inverse of the largest component-mean-repair-time.

If the system at hand is "reliable", then it also possesses the properties:

a. the expected number of system failures at time t $E[N(t)]$ has almost the same value as system unreliability $\overline{R}_s(t)$ at time t,

b. the largest component-mean-repair-time is much smaller than the system mission time t_M.

(ii) and b. entail that

$$\frac{dE[N(t)]}{dt}$$

is almost constant in $(0, t_M)$ so that we finally have:

$$\overline{R}_s(t_M) \cong E[N_s(t_M)] = \int_0^{t_M} \frac{dE[N(x)]}{dx}\ dx \cong$$

$$\cong t_M \left.\frac{dE[N(t)]}{dt}\right|_{t=+\infty} \qquad (3.5)$$

Recalling that

$$\left.\frac{dE[N(t)]}{dt}\right|_{t=+\infty}$$

depends only on the mean failure times and the mean repair times of the components and not on their distributions, from (3.5) it results that system reliability has to a good extent the same value as it would have had, had system components been exponential.

To conclude this section we want to remark that equations describing a non-exponential maintained system can be used only for the sake of theoretical investigation, they are not at all suitable as computational tools.

4. CONCLUSIONS

The theory of Markov-processes has been shown to play a central role in the theory of maintained systems. Markov processes more general than continuous time chains must be defined in order to handle non-exponential maintained reliability-systems. The latter can be described by equations suitable only for the theoretical investigation and not for the sake of computation (related numerical calculations are too much involved); as a consequence, numerical estimates of the reliability of maintained systems become available only if system-components are assumed to be exponential. Exponentiality has been proven to be adequate as long as the systems of interest are reliable, maintenance is of good quality (components after repair are good as new) and the maintenance policy is such that components do not queue up for the repair; components must fail independently of one another.

REFERENCES

{ 1} Clarotti, C.A., Limitations of Minimal Cut Set Approach in Evaluating Reliability of Systems with Repairable Components, IEEE Transactions on Reliability, Vol. 30, no 4 (1981) 335-338.

{ 2} Spizzichino, F., Time Associated Random Pro

cesses and Minimal Cut Reliability Bounds for Coherent Maintained Systems, in Serra, A. and Barlow, R.E. (eds.), International School of Physics E. Fermi, Proceedings of Coure XCIV. "Theory of Reliability" (North-Holland, Amsterdam, 1986).

{ 3} Esary, J.D. and Proschan, F., A Reliability Bound for Systems of Maintained Interdependent Components, Jrnl American Statistical Association, Vol. 65 (1970) 329-338.

{ 4} Barlow, R.E. and Proschan F., Theory of Maintained Systems: Distribution of Time to First System Failure, Mathematics of Operations Research, Vol. 1 (1976) 32-42.

{ 5} Barlow, R.E. and Proschan, F., Statistical Theory of Reliability and Life Testing (Holt, Rinehart and Winston, New York, 1975).

{ 6} Balaban, H.S. and Singpurwalla, N.D., Stochastic Properties of a Sequence of Inter-Failure Times under Minimal Repair and under Revival, Dept. of Operations Research, George Washington Univ. (1982).

{ 7} Parry, G.W., Regeneration Diagrams UKAEA Report SRD-R 143, January 1979.

{ 8} Brown, M., The First Passage Time Distribution for a Parallel Exponential System with Repair, in Barlow, R.E., Fussel, J.B. and Singpurwalla (eds.), Reliability and Fault-Tree-Analysis (SIAM Philadelphia, 1975).

{ 9} Keilson, J., Markov Chain Models-Rarity and Exponentiality (Springer-Verlag, New York, 1979).

{10} Clarotti, C.A., Aspetti probabilistici nell'integrazione Markov-Fault-Tree, ENEA-NIRA report NWBA 1ET4B85001, ENEA Casaccia Rome Italy, September 1985.

{11} Clarotti, C.A., Finzi-Vita, S. and Spizzichino, F., Non-Exponential Maintained Reliability Systems: how to Handle them? In Proceedings of 4th International Conference on Reliability and Maintainability (Perros-Guirec, France 1984).

{12} Clarotti, C.A., The Method of Supplementary Variables and Related Issues in Reliability Theory, in Serra A. and Barlow, R.E. (eds.), International School of Physics E. Fermi, Procedings of Course XCIV "Theory of Reliability" (North-Holland, Amsterdam, 1986).

{13} Keilson, J., Robustness and Exponentiality in Redundant Reparaible Systems, School of Management Science, University of Rochester (December 1984).

C.A. Clarotti is Head of Lab. Analisi di Affida
bilità of ENEA (National Committee for Nuclear
Energy of Italy). He is project leader, on ENEA
side, of the research program in the reliability field jointly carried out by ENEA and the
University of Rome "La Sapienza". His interests
lie in the theory of stochastic processes and
their applications in reliability.

F. Spizzichino is Associate Professor of Mathematical Statistics at the University of Rome
"La Sapienza". He is advisor for reliability
problems of several Italian industries and go-
vernamental organizations. He is doing research
in Bayesian statistics with special emphasis on
the theory of stochastic filtering.

Reliability Technology — Theory & Applications
J. Møltoft and F. Jensen (Editors)
Elsevier Science Publishers B.V. (North-Holland), 1986

CONTINUOUS TIME MAINTENANCE MODELS
FOR COMPLEX MULTI-STATE SYSTEMS

Kasra HAZEGHI
Institute for Operations Research
Swiss Federal Institute of Technology
Zurich, Switzerland

Abstract: This paper introduces some models for preventive maintenance of multi-state systems. The deterioration of the state is assumed to be governed by a markov chain in continuous time. Maintenance actions are preventive and emergency renewals of the system as well as its inspection for the purpose of state identification if the latter is not monitored continuously. Markov-renewal programming, in one case combined with the method of steepest descent, is applied to determine the optimal maintenance policy each time.

Zusammenfassung: Es werden einige Unterhaltsmodelle für komplexe Mehrzustands-Systeme vorgestellt. Die Zustandsentwicklung wird jeweils durch eine zeitkontinuierliche Markov-Kette beschrieben. Unterhaltsaktionen sind Präventiv- resp. Noterneuerungen des Systems sowie dessen Inspektion zwecks Zustandsermittlung, falls der Zustand nicht kontinuierlich beobachtet werden kann. Markov-Erneuerungsprogrammierung, in einem Falle kombiniert mit der Methode des steilsten Abstieges, wird zur Ermittlung der optimalen Unterhaltspolitik eingesetzt.

1. INTRODUCTION

The concept of age renewal, i.e. preventive replacement of a device at a prescribed age, arises quite naturally when considering maintenance optimization problems. For a device with life time distribution function $F(x)$ and renewal costs C_P for each preventive replacement respectively C_E for each emergency replacement the expected maintenance costs per unit time K amount to

$$K = \frac{C_P \bar{F}(T) + C_E F(T)}{\int_0^T \bar{F}(x)dx} \qquad (1)$$

Herein T denotes the prescribed age for execution of preventive renewals and $\bar{F}(x) = 1-F(x)$ is the so called survival function.

In real world applications we have usually aging devices (that means devices with increasing failure rate) and at the same time $C_E > C_P$, in which case K has a unique minimum at T^* as shown in Fig. 1 (see for example [1]).

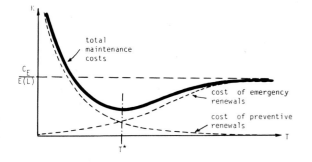

Fig. 1: Expected maintenance costs per unit time as function of renewal age T. ($E(L)$ is the expected life time of the device, without preventive renewals.)

Preventive maintenance based on age alone is of course meaningful only for single components or devices being like, in the sense that they have only one functioning state. For multi-state systems and devices with build in redundancy the actual state should be taken into account when considering preventive maintenance. Let us look at such a system with $N+1$ distinguishable states $0,1,2,...N$ and suppose the state deterioration is governed by a markov chain in continuous time. Most systems can be brought into this class by suitable choice of state space. Of special importance however are systems consisting of components with exponentially distributed life times. Because of the memoryless property of exponential distribution the markovian state deterioration is inherent to such systems. Despite of constant failure rate of their components these systems undergo aging as a result of diminishing redundancy. To this kind of systems we shall confine ourselves in the sequel.

Let $\Lambda = (\lambda_{i,j})$, $0 \le i, j \le N$ be the matrix of transition rates (infinitesimal generator) of the state deterioration in the absence of any maintenance activity. The entries of Λ are obtained in a rather straight forward manner from the failure rates of the system components. In the absence of selfregeneration, which will be assumed throughout this paper, no state is ever revisited after having been left. It is therefore always possible to number the states in such a way, that Λ becomes an upper diagonal matrix. 0 indicates then to the completely new system and N to the failed system, the latter being the only persistent state.

From Λ, and because of its triangular form, it is easy to obtain the transition probability functions $p_{ij}(t)$, $t \ge 0$, $0 \le i \le j \le N$ by solving the Kolmogorov's (forward) differential equation system

$$\begin{cases} \dfrac{dp_{ij}(t)}{dt} = \sum_{k=i}^{j} p_{ik}(t)\lambda_{kj} & 0 \le i \le j \le N, \quad (2) \\ p_{ij}(0) = \delta_{ij} \end{cases}$$

which can be solved recursively (see [13]). In particular for each $i\epsilon\{0,1,...N-1\}$ the function $p_{iN}(t)$ represents the life time distribution function of the system from state i and $1-p_{ii}(t)$ the distribution function of the sojourn time in that state. The latter is exponentially distributed with parameter $-\lambda_{ii}$, so that the mean sojourn time in state i is $1/(-\lambda_{ii})$ (see for example [7]).

The cost of preventive replacement of a multi-state system may depend on its effective state at time of replacement. So we have possibly N-1 different quantities C_{Pi}, $1 \le i \le N-1$, giving the replacement (or overhaul) costs for each functioning state as well as the generally much higher cost C_E for replacement of the failed system. Each replacement is assumed to bring the system immediately to state 0, i.e. to completely new state. The objective of maintenance optimization is to find a practicable maintenance policy, which keeps the expected average maintenance costs as low as possible.

2. SYSTEMS WITH NO CONTINUOUS MONITORING OF THE STATE

In the most simple case, namely the case of systems with completely ordered states, the maintenance problem is solved rather easily. Such systems are characterized by the fact that the degree of deterioration increases monotonically with the state and an example of this kind of a system is the N-unit parallel system, whose state is the number of failed units. With the rather obvious assumption $C_{Pi} \le C_{Pj}$, $1 \le i \le j \le N-1$ the optimal maintenance policy for this kind of systems is expected to be of control limit type, i.e. the system is to be replaced as soon as

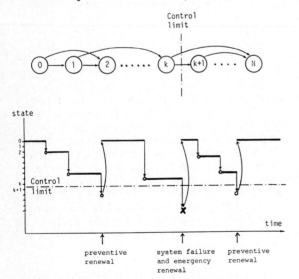

Fig. 2: Illustration of control limit maintenance policy for a system with ordered states

its state reaches or passes a critical level (see Fig. 2 and for some sufficient conditions for this case [11]). The determination of the critical level is done by at most N evaluations of the maintenance costs.

For most systems, specially those with complex redundant structure, it is not possible or at least not easy to rank the states. Instead of giving only a control limit, a partitioning of states into critical and non-critical ones is then necessary to define a maintenance policy (see Fig. 3).

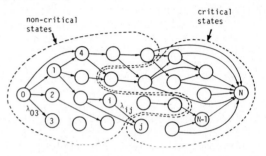

Fig. 3: State diagram for a system whose states are not (completely) ordered

Preventive replacements have to be carried out as soon as a critical state is reached. In worst case 2^{N-1} different maintenance policies compete with each other (state 0 is non-critical and state N critical by definition). Determination of the best one is done preferably by means of linear programming or policy iteration algorithm (dynamic programming), since the problem can be viewed as a markovian decision process (see [8], [14]). There are two possible actions in each state $i=1,2,...N-1$, namely "carrying out a preventive replacement" or "doing nothing". Each maintenance policy may thus be characterized by a boolean vector $\underline{\rho} = (\rho_0,\rho_1,...\rho_N)$, whose i.th entry ρ_i takes the values 0 or 1 depending on whether the state i is to be considered as critical or non-critical ($\rho_0 = 1$, $\rho_N = 0$ by definition). To perform the optimization one needs the transition probability matrix of the imbedded markov chain, the mean sojourn times and the expected one-transition costs as function of policy vector $\underline{\rho}$ (see [15]). The transition probability matrix $\phi(\rho) = (\varphi_{ij}(\rho_i))$ of the embedded markov chain (system state immediately after transitions) is readily seen to have the entries

$$\varphi_{ij}(\rho_i) = \begin{cases} \dfrac{\lambda_{ij}}{-\lambda_{ii}}, & \text{if } \rho_i = 1 \text{ and } 0 \le i < j \le N \\ \dfrac{\lambda_{0j}}{-\lambda_{00}}, & \text{if } \rho_i = 0 \text{ and } 1 \le i, j \le N \quad (3) \\ 0, & \text{otherwise} \end{cases}$$

The mean sojourn time in each state $i\epsilon\{0,1,...N\}$ is

$$\mu_i(\rho_i) = \begin{cases} \dfrac{1}{-\lambda_{ii}}, & \text{if } \rho_i = 1 \\ \dfrac{1}{-\lambda_{00}}, & \text{if } \rho_i = 0 \end{cases} \quad (4)$$

and the corresponding one-transition (immediate) cost is

$$R_i(\rho_i) = \begin{cases} 0 & \text{, if } \rho_i = 1 \\ C_{pi}, & \text{if } \rho_i = 0, \ i \neq N \\ C_E & \text{, if } i = N \end{cases} \qquad (5)$$

Following is a brief outline of the policy iteration algorithm as it is used for the above problem.

Using (3), (4) and (5) one could principally evaluate the average maintenance costs K for any given policy vector $\underline{\rho}$ from

$$K = \frac{\sum\limits_{i=0}^{N} \pi_i R_i(\rho_i)}{\sum\limits_{i=0}^{N} \pi_i \mu_i(\rho_i)}, \qquad (6)$$

where $\underline{\pi} = (\pi_0, \pi_1, \ldots \pi_N)$ is the stationary distribution of the embedded markov chain and satisfies

$$\begin{cases} \sum\limits_{i=0}^{N} \pi_i \varphi_{ij}(\rho_i) = \pi_j, \ 0 \leq j \leq N \\ \sum\limits_{i=0}^{N} \pi_i = 1 \end{cases} \qquad (7)$$

The maintenance costs K may equally be obtained from the linear system of equations

$$V_i = R_i(\rho_i) - K \cdot \mu_i(\rho_i) + \sum\limits_{j=0}^{N} \varphi_{ij}(\rho_i) V_j, \ 0 \leq i \leq N \quad (8)$$

which in addition to K also delivers the so called relative values $V_0, V_1, \ldots V_N$, the latters however only up to an additiv constant. Furthermore, if for another maintenance policy $\underline{\rho}' = (\rho_0', \rho_1', \ldots \rho_N')$ one has

$$V_i \leq R_i(\rho_i') - K \cdot \mu_i(\rho_i') + \sum\limits_{j=0}^{N} \varphi_{ij}(\rho_i') V_j, \ 0 \leq i \leq N \quad (9)$$

where K and V_i, i=0,1,...N are the solution of (8) and corresponds to $\underline{\rho} = (\rho_0, \rho_1, \ldots \rho_N)$, then it is easily seen that $\underline{\rho}'$ is better than (or at least as good as) $\underline{\rho}$ in the sense that it gives rise to less (or at most same) maintenance costs (see [15]). On the basis of this property the policy iteration algorithm is settled up which determines the optimal maintenance policy $\underline{\rho}^*$ after having gone through a sequence of value determinations and policy improvements (see Fig. 4).

3. SYSTEMS WITH NO CONTINUOUS MONITORING OF THE STATE

In many situations continuous monitoring of system state is not practicable. Usually an inspection is then necessary to determine the actual system state at any given time and this obviously gives rise to some extra cost C_I per inspection (assumed as fixed). Beside defining the critical and non-critical states each main-

tenance policy has now to specify also a time schedule for inspections.

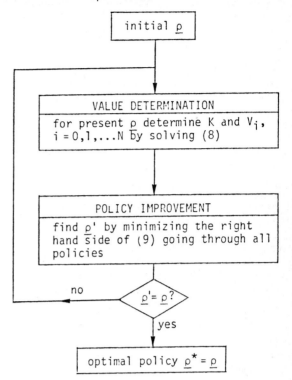

Fig. 4: Policy iteration algorithm for a multi-state system with continuous monitoring of state

3.1 Dynamic scheduling of inspections

A very sophisticated maintenance policy would use a dynamic scheduling in the sense that it would allow the time for the next inspection to depend on the last observed state. The differing reliability characteristics of different system states could then be taken into account by choosing longer inspection intervals for more reliable states and shorter inspection intervals for less reliable ones. Such a dynamic maintenance policy will be characterized by a vector $\underline{T} = (T_0, T_1, \ldots T_N)$ with $T_0 > 0$, $T_i \geq 0$, $1 \leq i \leq N-1$, $T_N = 0$. A state i with $T_i > 0$ is to be considered as non-critical and whenever as a result of an inspection the system is found in that state the next inspection will be scheduled for T_i time units later. On the other hand $T_i = 0$ indicates to a critical state and a preventive replacement has to be performed whenever such a state is encountered. The first inspection after each system replacement is therefore scheduled for time T_0, as the process restarts. A system failure is recognized immediately and is followed by an emergency replacement (see Fig. 5).

For a given policy vector $\underline{T}$ the sequence of encountered system states defines again a markov chain with the transition probability matrix $\phi(\underline{T}) = (\varphi_{ij}(T_i))$, where

$$\varphi_{ij}(T_i) = \begin{cases} p_{ij}(T_i), & \text{if } T_i > 0 \\ \delta_{0j} & \text{, if } T_i = 0 \end{cases}, \ 0 \leq i, \ j \leq N \quad (10)$$

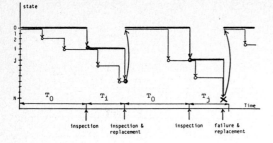

Fig. 5: Illustration of dynamic inspection scheduling for systems with no continuous monitoring of the state

If each encountered state is thought to remain unchanged until next state identification (as shown in Fig. 5) the system state defines a semi-markovian process with the mean sojourn time

$$\mu_i(T_i) = \begin{cases} [1-p_{iN}(T_i)]T_i + \int_0^{T_i} t\,dp_{iN}(t), & \text{if } T_i > 0 \\ \\ 0 & , \text{if } T_i = 0 \end{cases} \quad (11)$$

and the corresponding mean one-transition cost

$$R_i(T_i) = \begin{cases} C_I[1-p_{iN}(T_i)] & , \text{if } T_i > 0 \\ C_{Pi} & , \text{if } T_i = 0, i \neq N \\ C_E & , \text{if } i = N \end{cases} \quad (12)$$

for each state $i\epsilon\{0,1,\dots N\}$. The average maintenance costs K together with relative values V_i, $i=0,1,\dots N$ can again be obtained from the linear system of equations

$$V_i = R_i(T_i) - K\cdot\mu_i(T_i) + \sum_{j=0}^{N} \varphi_{ij}(T_i)V_j, \quad 0 \le i \le N \quad (13)$$

It can again be shown (see [13] and [17]) that the policy iteration algorithm may also here be applied to find the optimal maintenance policy T*, although we have now asymptotical convergence because of the continuous action space. The policy improvement is achieved by minimizing for each $i\epsilon\{0,1,\dots N\}$ the right hand side of (13) with respect to T_j, where K and V_j, $0 \le i \le N$ correspond to the foregoing policy. The discontinuities of the functions $\varphi_{ij}(\cdot)$, $\mu_i(\cdot)$ and $R_i(\cdot)$ at zero dont present any difficulties as long as $C_I > 0$, which is assumed here (see [13]).

3.2 Uniform scheduling of inspections

Dynamic scheduling of inspections with its irregular frequency pattern may sometimes not be applicable in real world situations because of operational considerations. In many cases one is willing to accept some higher maintenance costs and have more regular maintenance schedules. A maintenance policy to fulfill this concern is one, which is specified by only one single inspection intervall T > 0 as well as a

partitioning of states into critical and non-critical ones, represented again by a boolean vector $\underline{\rho}$ with $\rho_0 = 1$, $\rho_i\epsilon\{0,1\}$, $1 \le i \le N-1$, $\rho_N = 0$. Inspections are performed periodically at times T,2T,3T,... and whenever a critical state (with $\rho_i = 0$) is encountered the system is replaced by a new one until a system failure restarts the whole process (see Fig. 6).

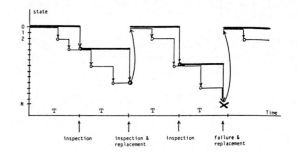

Fig. 6: Illustration of uniform inspection scheduling for systems with no continuous monitoring of the state

For a given maintenance policy $(T,\underline{\rho})$ the sequence of encountered states defines again a markov chain with the transition probability matrix $\phi(T,\underline{\rho}) = (\varphi_{ij}(T,\rho_i))$, where

$$\varphi_{ij}(T,\rho_i) = \begin{cases} p_{ij}(T), & \text{if } \rho_i = 1 \\ \delta_{0j} & , \text{if } \rho_i = 0 \end{cases} \quad 0 \le i, j \le N \quad (14)$$

Correspondingly the mean sojourn time and the mean one-transition cost of state $i\epsilon\{0,1,\dots N\}$ are given by:

$$\mu_i(T,\rho_i) = \begin{cases} [1-p_{iN}(T)]T + \int_0^T t\,dp_{iN}(t), & \text{if } \rho_i = 1 \\ \\ 0 & , \text{if } \rho_i = 0 \end{cases} \quad (15)$$

and

$$R_i(T,\rho_i) = \begin{cases} C_I[1-p_{iN}(T)] & , \text{if } \rho_i = 1 \\ C_{Pi} & , \text{if } \rho_i = 0, i \neq N \\ C_E & , \text{if } i = N \end{cases} \quad (16)$$

Again the average maintenance costs K, together with the relative values V_i, $i=0,1,\dots N$ are obtained from the linear system of equations

$$V_i = R_i(T,\rho_i) - K\cdot\mu_i(T,\rho_i) + \sum_{j=0}^{N} \varphi_{ij}(T,\rho_i)V_j,$$
$$i=0,1,\dots N \quad (17)$$

Contrary to the previous cases the optimization of the maintenance policy $(T,\underline{\rho})$ by means of policy iteration algorithm is not possible any more, as now all states have one and the same action parameter T in common. A combination of policy iteration algorithm and the method of steepest descent makes it however again possible to settle up an optimization algorithm also for this case, as it is described below.

For any given T > 0 the optimal partitioning vector $\rho^*(T)$ may still be found by policy iteration algorithm using (14) - (17). On the other hand for any given partitioning vector ρ the optimal inspection intervall $T^*(\rho)$ may be searched for by the method of steepest descent. We are namely able to calculate the derivative of the maintenance costs K with respect to T for any given policy (T,ρ) although the dependency of K on T is not explicitly known. It may easily be verified that (see [13] or [20])

$$\frac{dK}{dT} = \frac{\sum\limits_{i=0}^{N} \pi_i \left[\dfrac{dR_i(T,\rho_i)}{dT} - K\dfrac{d\mu_i(T,\rho_i)}{dT} + \sum\limits_{j=0}^{N} V_j \dfrac{d\varphi_{ij}(T,\rho_i)}{dT} \right]}{\sum\limits_{i=0}^{N} \pi_i \mu_i(T,\rho_i)} \qquad (18)$$

where $\pi = (\pi_0, \pi_1, \ldots \pi_N)$ is the stationary distribution of the underlying markov chain and V_i, i=0,...N are the relative values, both corresponding to the policy (T,ρ). For evaluation of (18) we also need the derivatives of $R_i(T,\rho_i)$, $\mu_i(T,\rho_i)$ and $\varphi_{ij}(T,\rho_i)$, i,j=0,...N with respect to T, which are readily obtained from (14), (15) and (16) using also Kolmogorov's differential equation system (2). The overall optimization, over both ρ and T, can therefore be attempted by a sequence of alternate optimizations over ρ respectively over T one at each time (see Fig 7 and [13]).

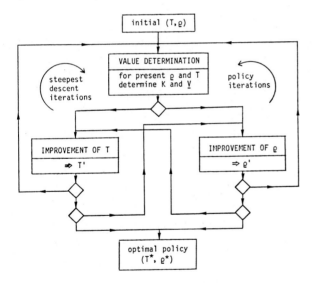

Fig. 7: Combination of policy iteration algorithm and the method of steepest descent for alternate optimization over ρ and over T.

Because of the myopic nature of the method of steepest descent it can, at least principally, happen that the described algorithm delivers only a local optimum instead of the global one. Practical computations have however shown that the dependency of the maintenance costs on the length of inspection intervall T is unimodal, so that the local and global optimum coincide. Anyway a careful choice of the starting policy can assure finding at least a good approximation to optimal policy (T^*,ρ^*) under all circumstances.

REFERENCES

[1] R.E. Barlow & F. Proschan, *Mathematical Theory of Reliability*, Wiley (1965).

[2] R.E. Barlow & F. Proschan, *Statistical Theory of Reliability and Life Testing*, Holt Rinehart and Winston (1975).

[3] F. Beichelt, *Prophylaktische Erneuerung von Systemen*, Vieweg (1976).

[4] F. Beichelt & P. Franken, *Zuverlässigkeit und Instandhaltung*, Carl Hanser (1984).

[5] A. Birolini, *Qualität und Zuverlässigkeit technischer Systeme*, Springer (1985).

[6] A. Birolini, *On the Use of Stochastic Processes in Modeling Reliability Problems*, Springer (1985).

[7] E. Cinlar, *Introduction to Stochastic Processes*, Prentice-Hall (1975).

[8] C. Derman, *Finite State Markovian Decision Processes*, Academic Press (1970).

[9] K. Fischer, *Zuverlässigkeits- und Instandhaltungstheorie*, Transpress (1984).

[10] I.B. Gertsbakh, *Models of Preventive Maintenance*, North-Holland (1977).

[11] I.B. Gertsbakh, "Sufficient Optimality Conditions for Control-Limit Policy in a Semi-Markov Process", *J. App. Prob.* (1976).

[12] H. Kawai, "An Optimal Inspection and Replacement Policy of a Markovian Deterioration System", in *Stochastic Models in Reliability Theory*, Springer (1984).

[13] K. Hazeghi, *Optimale Unterhaltspolitik für komplexe Systeme*, Haupt (1979).

[14] R.H. Howard, *Markov Provesses and Dynamic Programming*, Wiley (1960).

[15] W.S. Jewell, "Markov-Renewal Programming I and II", *J. Operations Research 11* (1963).

[16] J. Kohlas, *Stochastic Methods of Operations Research*, Cambridge University Press (1982).

[17] H. Kushner, *Introduction to Stochastic Control*, Holt Rinehart and Winston (1971).

[18] P. Pierskalla, J.A. Völker, "A Survey of Maintenance Models", *Naval Research Log.* 23 (1976).

[19] S.M. Ross, *Applied Probability Models with Optimization Applications*, Holden-Day (1970).

[20] P.J. Schweizer, "Perturbation Theory and Undiscounted Markov-Renewal Programming", *J. Operations Research 17* (1969).

Reliability Technology — Theory & Applications
J. Møltoft and F. Jensen (Editors)
© Elsevier Science Publishers B.V. (North-Holland), 1986

A QUASI EXHAUSTIVE METHOD FOR OPTIMIZED SPARES PROVISIONING

Bruno BEARZI, Giorgio TURCONI
ITALTEL - Società Italiana Telecomunicazioni
Cascina Castelletto, I-20019 Settimo Milanese, Italy

This method is presented in the aim of finding an optimal spare kit, it involves short computation times, when compared to Kettelle Method, and has high probability of getting to the true optimal solution. The procedure uses, for a first phase calculation, the Proschan-Black Method, then it disjoins from that by the execution of a second calculation phase to reach the optimal solution. Classical and proposed methods are critically analyzed and applied to examples.

1. INTRODUCTION

The problem handled regards the optimal choice of a spare kit, which is required for the logistic support in field. The spare provisioning has to satisfy part type demands, caused by failures occurring on operating units.

Characteristic parameters of a spare kit are protection level and total cost.

The protection level is represented by the probability that the demand for spares to the depot, where the kit is stored, be satisfied without delay.

The parameter describing such probability is adequacy; it will be here used as figure of merit of a spare kit.

By a steady demand rate flow, adequacy is defined on the basis of Poisson distribution, once the restocking time is fixed, that is, the time interval for which has to be assured the logistic support.

The optimization problem can be submitted in two ways:

Problem a/: find the allocation with minimal cost, once the total adequacy $\underline{A}$ is fixed.

Problem b/: find the allocation with best total adequacy, corresponding to a given cost $\underline{C}$.

In the paper two classical optimization methods are first described: Kettelle Method (KM) and Proschan-Black Method (PBM).

The methods are discussed and it is shown that:

- the exactness of KM is payed for by very long computation times;

- the quickness of PBM may be pregiudicial to the exactness of the optimal found kit arrived at, owing to the lack of completeness of the built sequence.

Then a method is proposed which makes use of PBM in a first computation phase, leading to results having high probability to be optimal and obtained through computation times comparable whith those of PBM.

2. MATHEMATICAL FORMULATION OF THE PROBLEM

Let s_j $(j = 1...k)$ represent the decision variables, that is the spares quantity for the parts $1...k$; c_j their costs and m_j the predicted average number of repair actions during a fixed period of time $m_j = d_j \times N_j \times T_j$, where d_j is the demand rate, N_j is the quantity of operating units in field and T_j the restocking time (or in case of continuous replacement, the turn around time).

A given value for the s_j's corresponds to a certain allocation of spares, whose total cost is:

$$C = \sum_{j=1}^{k} c_j \times s_j \qquad (1)$$

The adequacy of part j is:

$$A_j(s_j) = \sum_{i=1}^{s_j} \frac{(m_j)^i \times e^{-m_i}}{i!} \qquad (2)$$

and the total adequacy (for the whole kit) is:

$$A = \prod_{j=1}^{k} A_j(s_j)$$

Let us consider a sequence of allocations of the spares kit: S_1, S_2, ..., S_x,... where S_x is:

$$Sx \underline{=} (s1, s2, \ldots, sj, \ldots, sk) \quad (4)$$

with increasing costs and adequacies, that is:

$$C1 < C2 < \ldots < Cx < \ldots \quad (5)$$
$$A1 < A2 < \ldots < Ax < \ldots \quad (6)$$

When allocation Sx is such that any other allocation doesn't exist having the same cost and greater adequacy and any other allocation doesn't exist with the same adequacy and less cost, Sx is said to be an undominated allocation.

If a sequence contains all the meaningful allocations that means that no additional allocations may be interpolated, except possibly those yielding identical adequacy-cost pair, it is said to be complete.

A complete sequence of undominated allocations allows the solution of both submitted problems.

3. KETTELLE METHOD

Kettelle Method (KM) consists in an algorithm which builds complete sequences of undominated allocations (Ref.1).

The considered figures of merit, total adequacy and total cost, are strictly increasing functions of the quantities of the part types.

Then, the sequence built by KM, being a complete sequence of undominated allocations, leads to the exact solution of the optimization problem.

The generation of the sequence proceeds by steps.

Step a/ Starting from the allocation $(0,0,\ldots,0,\ldots,0)$, taking into account part types 1 and 2, consider the two complete sequences:

$(0,0,\ldots,0,\ldots,0)$,
$(1,0,\ldots,0,\ldots,0)$,
$(2,0,\ldots,0,\ldots,0)$,
$\ldots$,
$(n1,0,\ldots,0,\ldots,0)$

and:
$(0,0,\ldots,0,\ldots,0)$,
$(0,1,\ldots,0,\ldots,0)$,
$(0,2,\ldots,0,\ldots,0)$,
$\ldots$,
$(0,n2,\ldots,0,\ldots,0)$

where n1 and n2 are determinated by the cost constraints, that is n1 and n2 are the smallest integers such that:

$$n1 \geq C /c1 \text{ and } n2 \geq C /c2 \quad (7)$$

where $\underline{C}$ is the maximum total cost.

Step b/ Calculate costs and adequacies, corresponding to all possible combinations of both given sequences; costs and adequacies are arranged in two rectangular matrices of the same shape.

Step c/ Find a sequence of allocations, which is complete and undominated referring to the two considered part types. The sequence starts from the allocation $(0,0,\ldots,0,\ldots,0)$ and stops when total cost exceeds the value $\underline{C}$.

Step d/ Repeat steps b/ and c/ by setting, instead of the first part type, the complete undominated sequence found in step c/, and, instead of the second part type, the further one not yet considered in the calculation.
Computation will be stopped when the k-th part type is reached: the last found sequence is a complete sequence of undominated allocations.

In this way, both optimization problems submitted in Section 1. obtain exact solution.

4. PROSCHAN AND BLACK METHOD

An other way to find a sequence, allowing the solution of the optimization problem, is given by the Proschan and Black Method (PBM).

The basic principle of this method is to generate a sequence of undominated allocations by incrementing, starting from a given allocation, the part type j for which a function Fj(sj) of adequacy and cost obtains the maximum value.

For a given allocation $(s1, s2, \ldots, sj, \ldots sk)$, this function is defined as (being "ln" the natural logarithm):

$$Fj(sj) = \frac{\ln \dfrac{Aj(sj+1)}{Aj(sj)}}{cj} \quad (8)$$

The first term of the sequence is the allocation $(0,0,\ldots,0,\ldots,0)$: there are, then, k possible allocations with an additional spare, that is:

1st. $(1,0,\ldots,0,\ldots,0)$
2nd. $(0,1,\ldots,0,\ldots,0)$
. . .
jth. $(0,0,\ldots,1,\ldots,0)$
. . .
kth. $(0,0,\ldots,0,\ldots,1)$

The second point of the sequence is found by choosing, among the last ones, that having the greatest value of Fj(sj).

Any other following point, starting from an allocation:

$S = (s_1, s_2, \ldots, s_j, \ldots, s_k)$, is found by choosing, among the k allocations having an additional spare:

$(s_1+1, s_2, \ldots, s_j, \ldots, s_k)$
$(s_1, s_2+1, \ldots, s_j, \ldots, s_k)$
. . .
$(s_1, s_2, \ldots, s_j+1, \ldots, s_k)$
. . .
$(s_1, s_2, \ldots, s_j, \ldots, s_k+1)$

that which has the greatest value of $F_j(s_j)$.

Thus, a sequence is given which is made of undominated allocations.

5. COMPARISON BETWEEN KM AND PBM

The PBM takes very much shorter computing time than KM, as will be seen later on (Section 8.), whether on the basis of theoretical estimation or with numerical examples.

Another point to note is that PBM procedure does not depend on whether it is to solve Problem a/ or Problem b/ in Section 1.

KM, on the other hand, gives a direct solution only to Problem b/, because it requires a fixed maximum cost $\underline{C}$ to shape the sequences and the matrices in steps a/ and c/. So, in the case of Problem a/, an additional condition on $\underline{C}$ is necessary.

When the additional condition on $\underline{C}$ is not assigned, an upper bound has to be found, e.g. the cost of PBM solution.

On the other hand the PBM leads to sequences which are usually incomplete. Completeness implies that no additional undominated allocations may be interpolated between any couple of allocations, except those yielding identical adequacy-cost pairs.

This fact is not ensured by the procedure described in Section 4, as can be shown in the following example.

Let us consider the case of a system composed by 4 part types P1, P2, P3 and P4, whose costs and predicted number of demands are the following:

Part Type	Cost	N. of demands
P1	13485	3.066
P2	25440	1.533
P3	14537	0.876
P4	9441	0.569

The application of the KM, setting a maximum adequacy of 0.99, yields the following complete sequence of undominated allocations:

Allocation				Adequacy	Cost	
7	4	3	2	0.9355	258664	°
7	4	3	3	0.9521	268113	°
7	4	3	4	0.9545	277562	
8	4	3	3	0.9608	281598	
7	4	4	3	0.9619	282650	°
8	4	3	4	0.9632	291047	
7	4	4	4	0.9643	292099	
7	5	3	3	0.9669	293553	
8	4	4	3	0.9707	296135	°
8	4	4	4	0.9732	305584	
8	5	3	3	0.9757	307038	
7	5	4	3	0.9769	308090	
8	5	3	4	0.9782	316487	
7	5	4	4	0.9793	317539	
8	5	4	3	0.9858	321575	°
8	5	4	4	0.9883	331024	°
9	5	4	3	0.9889	335060	
9	5	4	4	0.9913	344509	°

The PBM, under the same conditions, gives:

Allocation				Adequacy	Cost
7	4	3	2	0.9355	258664
7	4	3	3	0.9521	268113
7	4	4	3	0.9619	282650
8	4	4	3	0.9707	296135
8	5	4	3	0.9858	321575
8	5	4	4	0.9883	331024
9	5	4	4	0.9913	344509

The complete sequence found with KM contains 18 allocations, while the PBM sequence contains only 7, all appearing also in the KM sequence, marked with a circle (°).

If the problem is to find the allocation with minimal cost, associated to an adequacy of 0.99, the solution is the same for both methods, that is allocation 9 5 4 4.

The incompleteness of PBM leads, in some cases, to a solution which is not the true optimal, for instance, the problem of finding the allocation with minimal cost and adequacy of 0.975.

Thus PBM may have solutions with relevant undesired cost increase of the spares kit or adequacy less than obtainable.

6. THE PROPOSED QUASI-EXHAUSTIVE METHOD

6.1 Preliminary considerations

KM starts from the allocation L0 $(0, 0, \ldots, 0, \ldots, 0)$. Such starting condition assures the completeness of the sequence: it is important to understand what does happen when a calculation is performed starting from a different allocation:
$L = (l_1, l_2, \ldots, l_j, \ldots l_k)$.

In this case the generated sequence could be incomplete, because it does not contain the undominated allocations having at least one $s_j < l_j$.

Such a sequence, however, contains all the undominated allocations for which $s_j \geq l_j$ for every j; thus, if $\underline{S}$ is the optimal solution obtained with KM, starting from the 'zero' allocation, and $\underline{s_j > l_j}$ for every j, $\underline{S}$ is contained in any sequence starting from L.

These considerations lead to two kinds of changes to the above mentioned methods: the choice of a suitable starting point and the use of a two-phase procedure.

6.2 Choice of a suitable starting point

In the case of Problem a/ a good starting point is the allocation $(l_1, l_2, \ldots, l_j, \ldots l_k)'$, for which the adequacy of every part type is less than, or equal to, the given fixed total adequacy: that is l_j is equal to the greatest integer for which

$$l_j \quad \text{Sum} \frac{(m_i)^i \times e^{-m_i}}{i!} \leq \underline{A} \qquad (9)$$
$$i=1$$

When Problem b/ is submitted, it can be set the same type of condition, by setting an upper bound to adequacy, or, at least, it can be set: l_j equal to the greatest integer for which $l_j < m_j$ where m_j is the predicted number of demands for the part type j (Section 2.).

6.3 The quasi exhaustive procedure (QE)

The PBM not always gives the true optimal solution, because the application of decision criterion (8) can lead to a choice which jumps over the optimal solution.

To find the true optimal allocation it will be enough to go back a suitable number R of steps in the PBM sequence and to take this latter allocation $(r_1, r_2, \ldots, r_j, \ldots r_k)$ as the starting point of a KM procedure, that will be exhaustive of all meaningful allocations having: $s_j \geq r_j$

The completeness of the sequence in its last allocations and, then, the exactness of the solution of the optimization problem, depend on R, the number of back steps.

In the case of R=1, an alternative way can be used: in the second phase of the calculation, instead of restarting with KM, a PBM can be performed, excluding from the computation the part type chosen in the last step, repeating k-1 times this operation, excluding every time the part type chosen in the previous step, and choosing the best result.

When the part type number and, then, the spare number, are very great, this restarting procedure is very helpful, requiring much less computation time than the former and leading, for R=1, to the same result.

In all considered examples, the last described procedure led to satisfactory results, obtaining cost and adequacy very near to the imposed constraints

The precision of the method is controllable, that means that it is possible, when a satisfactory result is not reached, to explore the allocation of spares in a more exhaustive way. This is done by increasing the value of R.

7. APPLICATION OF THE METHOD

To show the application of the described procedures, let us consider an optimization problem for the system of Section 6., when the imposed condition is that total adequacy has to be greater than 0.90.

The KM gives:

Allocation				Adequacy	Cost
4	2	1	1	0.4465	128806
4	2	1	2	0.4927	138255
5	2	1	1	0.5050	142291
4	2	2	1	0.5379	143343
5	2	1	2	0.5571	151740
4	2	2	2	0.5934	152792
5	2	2	1	0.6083	156828
5	2	2	2	0.6711	166277
5	2	2	3	0.6830	175726
4	3	2	2	0.6895	178232
6	2	2	2	0.7108	179762
6	2	2	3	0.7234	189211
5	3	2	2	0.7798	191717
5	3	2	3	0.7936	201166
6	3	2	2	0.8259	205202
6	3	2	3	0.8406	214651
7	3	2	2	0.8461	218687
6	3	3	2	0.8668	219739
6	3	3	3	0.8822	219188
7	3	3	2	0.8880	233224
7	3	3	3	0.9038	242673

The PBM, under the same conditions, gives:

Allocation				Adequacy	Cost
4	2	1	1	0.4465	128806
4	2	2	1	0.5379	143343
4	2	2	2	0.5934	152792
5	2	2	2	0.6711	166277
5	3	2	2	0.7798	191717
6	3	2	2	0.8259	205202
6	3	3	2	0.8668	219739
6	4	3	2	0.9131	245179

The Quasi-Exhaustive procedure with the one back step (R=1) and the second phase performed with KM (QEKM) gives:

Allocation				Adequacy	Cost
4	2	1	1	0.4465	128806
4	2	2	1	0.5379	143343
4	2	2	2	0.5934	152792
5	2	2	2	0.6711	166277
5	3	2	2	0.7798	191717
6	3	2	2	0.8259	205202
6	3	3	2	0.8668	219739
7	3	3	2	0.8880	233224
7	3	3	3	0.9038	242673

and the QE procedure, with PBM completion (QEPBM) :

Allocation				Adequacy	Cost
4	2	1	1	0.4465	128806
4	2	2	1	0.5379	143343
4	2	2	2	0.5934	152792
5	2	2	2	0.6711	166277
5	3	2	2	0.7798	191717
6	3	2	2	0.8259	205202
6	3	3	2	0.8668	219739
7	3	3	2	0.8880	233224
7	3	3	3	0.9038	242673

From these results it can be seen that the exhaustive solution has less cost and adequacy closer to the fixed value than PBM, and that both Quasi-Exhaustive procedures lead to the same result as KM.

8. NUMBER OF OPERATIONS AND COMPUTING TIME, COMPARISON AMONG THE FOUR METHODS

In the present section the number of operations and the computing times of the examined methods are compared.

The total number of operation can be estimated by considering the main features of the methods. All of them are considered starting from the same allocation, chosen accordingly to Section 6.2.

8.1 KM approach

The order of magnitude of the KM computation can be evaluated as follows:

- The shape of rows and columns of the cost and adequacy matrices is of the same order of magnitude as the total number of spares S when the procedure starts from the 'zero' allocation; if a suitable starting allocation is set, with total number of spares equal to S0, the above mentioned shape is of the order of magnitude of S-S0.

Thus, (S-S0)x(S-S0) calculations of adequacy and Cost have to be performed at every step.

- A sequence of optimizations has to be carried out, at every step, among the (S-S0) x (S-S0) elements of the matrix, and the allocations corresponding to the optimal points have to be singled out, the number of points is greater than S-S0.

- The number of steps is k-1, where k is the number of part types.

It has to be remarked that S-S0 is a lower estimate of the shapes, because the KM sequences contain, in general, much more than S-S0 allocations.

A complete computation requires, then, more than (k-1)x(S-S0)x(S-S0) calculations of adequacy and cost; (k-1) optimizations among (S-S0)x(S-S0) elements and (k-1)x(S-S0) singling out of allocations from a matrix having, in average, (S-S0) rows and columns.

8.2 PBM approach

The number of operations of PBM can be evaluated as follows:

- When the total number of spares is S, the procedure consists in performing exactly S-S0 times the following calculations: k computations of the function $F_j(s_j)$, one determination of maximum among those k numbers and one singling out of this maximum among k part types.

- The total number of operations is then (S-S0)xk calculations of adequacy and cost and (S-S0) optimizations and singling out among k elements.

The advantages of PBM on KM in computing times is essential, not only owing to the great difference between the number of operations, but also because some operations, required by KM, are much more time-consuming.

This fact will be clear from the comparison on examples, by which it can be seen that the difference between the two methods is much greater than expected from the above considerations. From the presented results this difference seems to be growing exponentially.

8.3 QEMK approach

For the Quasi-Exhaustive procedure, in which the second phase is done with KM (QEMK), the first phase is a PBM procedure and has the same operation number; the second phase adds (k-1) steps of a KM procedure having a very small matrices: the number of rows and columns will be, generally equal to 2, or, in some cases, a little integer.

Most computation time is spent in the second phase, as can be seen from the difference between PBM and QEKM computation times.

8.4 QEPBM approach

The Quasi-Exhaustive procedure with PBM completion (QEPBM) has (k-1+S-S0)xk calculations of Fj(sj) and the relative optimizations.

The comparison among computing times was performed taking into account CPU time for systems composed by 3, 4, 5, 6 and 10 part types. Kettelle Method has been performed only in the first 3 cases, because already the case of 5 part types was rather time-consuming.

The average CPU times obtained for each case are reported on the table below as a multiple of the computation time of the case PBM, k=3.

k	PBM	QEPBM	QEKM	KM
3	1.0	2.6	6.6	61
4	1.8	4.8	11.9	258
5	2.9	5.4	24.2	2033
6	3.4	8.7	29.6	
10	6.3	18.1	91.3	

9. RESULTS AND CONCLUSIONS

The proposed method bases itself on both classical methods for optimal spare provisioning.
This method is particularly suitable when part type number k is so great that KM involves too long computation time and budget considerations require a strict exactness of the results.
The developed general remarks and the simulations performed on meaningful and suitable cases show that the proposed method can be applied with satisfactory results.

The practical calculation performed on cases with a great number of part types showed, in conclusion, that:

- there is a relevant number of cases in which the QE solution is better than the PBM solution;

- adequacy and cost obtained are always very close to the required values.

REFERENCES

(1) Barlow R.E., Proschan F., Statistical Theory of Reliability and Life Testing. H.R.W. 1975
(2) Kettelle J.D., Least cost allocation of reliability investiment. Operation Research 10, 249-265, 1962
(3) Goldman A.S., Slattery T.B., Maintainability: a Major Element of System Effectiveness, John Wiley & Sons, Inc. 1964
(4) Black G., Proschan F., On Optimal Redundancy, in Operations Research, Vol. 7, No 5, September-October, 1959
(5) Proschan F., Optimal System Supply (EDL-E38), Mountain View, California: Electronic Defense Laboratories, Sylvania Electronic Systems, January 1960 (ASTIA Document No. 233591)

BIOGRAPHIES

Mr. Bruno Bearzi (born 1944) received his degree in Physics at Università degli Studi in Milan in 1969. He worked till 1972 with the Theoretical Physics Institute, teaching and making a study about the properties of anharmonic chains. With Italtel he worked on telephonic traffic and, later, on Computer Aided Design, particularly dealing with wireability problems on printed circuit boards. Afterwards he worked as reliability and LCC expert in the Reliability and Quality Department of Italtel. Now he is involved in customer training courses.

Mr. Giorgio Turconi (born 1946) received his degree in electronics engineering (telecommunications) in 1971 from Politecnico in Milan. In 1973 he was employed by Italtel and became reliability responsible for military equipment. In 1978 he undertook the responsibility for reliability methodologies for the Central Reliability Department dealing with reliability and maintainability predictions, LCC, logistic support and failure data base on complex systems. Since 1982 he has also been responsible for Quality Control methodologies and data collection procedures.

SCREENING AND LIFE TESTING

Reliability Technology — Theory & Applications
J. Møltoft and F. Jensen (Editors)
© Elsevier Science Publishers B.V. (North-Holland), 1986

STRESS SCREENING: PROGRESS, SETBACKS AND THE FUTURE

Gregg K. Hobbs, Ph.D.
Hobbs Engineering Corporation
23232 Peralta Drive, Suite 221
Laguna Hills, CA 92653, USA

The current status in state-of-the-art of stress screening of electronic and mechanical hardware is discussed. Recent advances in techniques are addressed and some examples from the author's experience and from the open literature are given. Some common misconceptions regarding effectiveness of various stimuli in stress screening are addressed, and the direction of stress screening in the next few years is discussed.

INTRODUCTION

Until the mid 1970's, little public information on stress screening was available in the U.S.A. and the few companies which were active in the field were relatively unwilling to share information with each other or the public in general. The Institute of Environmental Sciences (IES) held conferences in 1979, 1981, 1984, and 1985 on the subject of stress screening, and published guidelines on screening of assemblies in 1981 and 1984, and on screening of parts in 1985. These guidelines, if taken as guidelines only, and read very carefully, would allow one to begin a stress screening program with some degree of confidence of a rational approach. Unfortunately, the fine print is seldom studied carefully and many cursory readers are left with incorrect notions.

Many screening programs have emphasized defect detection and repair instead of elimination of future defects by corrective action.

By late 1984 a dichotomy had formed between military and commercial screening in the U.S.A. [1] As of this writing, that dichotomy still persists, but there is some hope that the military screening will begin to focus on creative corrective action instead of irrational perserverence to following some "MIL SPEC" screen which is tantamount to an acceptance test. This approach misses the greatest cost and quality benefits of a properly run stress screening program.

Some of the main points missed by the "MIL SPEC" approach are addressed in the following paragraphs.

FLAW-STIMULUS RELATIONSHIPS

First, two definitions: A latent defect or a flaw is some irregularity due to manufacturing processes or materials which will advance to a patent defect when exposed to environmental or other stimuli. A subset of this is a design defect which should be corrected during design maturation. A patent defect is a flaw which has advanced to the point where an anomaly actually exists. For example, a cold solder joint represents a flaw. After vibration and/or thermal cycling, the joint will (we assume) crack. The joint would now have a detectable defect. Note that not all defects are actually detected during or after the screen because we may not know how to stimulate the product so that a defect actually shows up or we may not properly monitor the product in order to detect the defect when it does occur, perhaps intermittently.

Some flaws or latent defects can be stimulated into patent defects by thermal cycling, some by vibration and some by voltage cycling (or some other stimulus). Not all flaws respond to all stimuli. This effect can be shown in table form as has been done in many publications. One such table is shown as Figure 1 below, which is taken from MIL-STD-883 for integrated circuits.

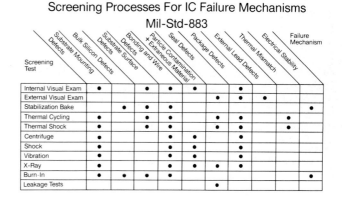

Screening Processes For IC Failure Mechanisms
Mil-Std-883

Screening Test	Substrate Mounting Defects	Bulk Silicon Defects	Substrate Surface Defects	Bonding and Wire Defects	Particle Contamination + Extraneous Material	Seal Defects	Package Defects	External Lead Defects	Thermal Mismatch	Electrical Stability
Internal Visual Exam	●		●	●	●			●		
External Visual Exam						●	●	●		
Stabilization Bake		●	●	●						●
Thermal Cycling	●		●	●		●	●		●	
Thermal Shock	●		●	●		●	●		●	
Centrifuge	●			●	●		●			
Shock	●			●	●		●			
Vibration	●			●	●		●			
X-Ray	●			●	●	●	●			
Burn-In	●	●	●	●						●
Leakage Tests						●				

Table 1

The black dot indicates that the stimulus at the top of the column is effective in precipitating the flaw type listed on the left.

Another such table is taken from the 1985 IES Guidelines [2] and addresses the burn in types which are effective in screening of IC's.

Burn-In Screening Effectivity as a Function of Failure Mechanisms Detected

Failure Mechanism	Dynamic Burn-In	Steady State Burn-In	High Voltage Cell Stress Test
Surface Defects	X	X	
Oxide Defects		X	X
Metallization Defects	X	X	
Contamination/Corrosion-Intermetallic		X	
Junction Anomalies (Breakdown)	X		X
Wire Bond	X	X	X

Table 2

The concept of flaw-stimulus relationships can also be shown in Venn diagram form as in Figure 1 below for a hypothetical but <u>specific</u> product. It would be different for a different product.

Note that, for the hypothetical example given, there are many latent defects that will not be transformed into a patent defect by any one stimulus. For example, a solder splash which is just barely clinging to a circuit board would probably not be broken loose by high temperature burn in or voltage cycling, but vibration or thermal cycling would probably break the particle loose. Also, note that in order to find that the defect exists, the particle must be seen by eye, heard by sound (as in PIND) or by the electrical or mechanical function of the device being screened showing some sort of a change.

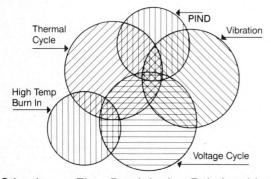

Stimulus — Flaw Precipitation Relationships

Figure 1

Another example might be that of a latent defect of a chemical nature where a high temperature bake would cause a reaction to proceed and a defect to show up. Applying vibration, mechanical shock or centrifuging would be completely ineffective in precipitating such a defect. The stimulus <u>must</u> be chosen to precipitate flaws into defects which can be found during or after the screen.

In 1981 the IES guidelines for stress screening [3] presented Figure 2 as a result of a poll done to determine what was being done in stress screening. The statement requiring an answer in that poll was (question 7, page A-3 of [3]) "RANK THE OVERALL EFFECTIVENESS OF SCREENING ENVIRONMENTS, WITH 1 BEING THE MOST EFFECTIVE, 2 NEXT EFFECTIVE, ETC. INCLUDE ONLY THOSE SCREENS YOU HAVE USED." The figure is reproduced below with the permission of the IES.

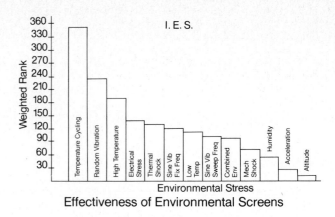

Effectiveness of Environmental Screens

Figure 2

The figure gives valuable information <u>in the context of the survey</u>, but taken out of context, it is not correct at all. Unfortunately, many novices have simply found the figure and proceeded to draw conclusions from it with no other facts as to assembly levels or defect types precipitated by the screens being considered. To further compound the potential confusion, the French published in [4] a similar result, Figure 3, which is reproduced with the permission of the IES.

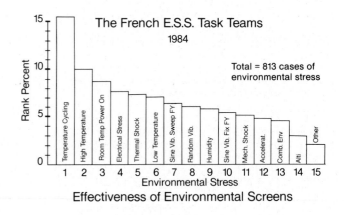

Effectiveness of Environmental Screens

Figure 3

Again, the figure is valuable when taken in context, but very misleading if taken alone, at face value. Note that the ordering of "effectiveness" is different in the USA and Europe. This, of course, is due to the question asked which required the respondent only to include those stimuli that he had used. In the USA, it is widely accepted that room temperature power on is very ineffective in screening, yet it is "rated" number three in "effectiveness" in Europe. Also, compare the relative position of random vibration as number two in the USA and as number eight in Europe. This is probably because of the military requirement for random vibration in the qualification, acceptance and screen testing on military programs in the USA. Hence, random vibration equipment is found in many companies and is readily available and familiar.

The author has named figures 2 and 3 "Popularity Contests" since this title really seems to be more fitting than "Effectiveness."

The author has seen many very serious mistakes made by simple interpretation of the two "Effectiveness Rankings". For example, one company tried thermal cycling of 100% of their production for nearly a year and improved their field failure rate hardly at all. The troublesome defects were not susceptible to thermal cycling, but probably would have been found by vibration. The company concluded that it was much cheaper to just accept field failures and fix them when they occured. Another company replaced 100% burn in with a 100% vibration screen and, of course, found different defect types and could then not correlate cost and effectiveness in decreasing field failures.

These types of failures of stress screening to be effective, and the IES emphasis on 100% screening instead of sampling, have led many to conclude that stress screening was simply not worth doing.

If the stimulus-flaw precipitation relationships had been taken into account, and corrective action emphasized so that the number of products screened could be a small sample, then the cost effectiveness probably would have been proven.

VIBRATION SPECTRA

It has been well known by a few investigators for many years, that there are usually many different vibration spectra that will perform a good screen on a given specimen [5]. Most new people in the stress screening field, particularly those with a MIL-SPEC test background, will look for a MIL-SPEC and find NAVMAT P-9492 [6], which is not a spec at all, but a guideline. This document is sometimes then incorrectly taken to be "the test" for vibration screening and used blindly with often catastrophic results. Sometimes the results are much better and are just completely ineffective screens without damage to the specimen. The author has consulted (after the fact) on four programs where every item vibration screened had been damaged or destroyed by the customer-specified NAVMAT level and profile. (The temperature profile shown in Figure 3 of [6] is also often frequently used even though it says on pg. 7 "The AGREE cycle in Figure 3 is NOT to be used.")

A truly good screen is usually developed by developing the profile and level for the product along with development of the product. Some products are not screenable without design changes to accomodate screening. This applies to all of the screen stimuli, not just vibration. This is a very important fact often missed by the novice. Screenability must be built in.

Hewlett Packard, in Santa Rosa, California, developed a vibration screen for a given product and compared the screen to the (in)famous NAVMAT profile and got much better results with the developed screen than with the "baseline" NAVMAT. The profile developed was a flat PSD from 5 Hz to 380 Hz and then a -3db/octave roll off to 500 Hz, where the input was truncated. The overall level was 3.5 grms. The developed screen at 3.5 grms was found to be more effective than the NAVMAT profile with an overall level of 8 grms! Incidentally, many more flaws were found using this vibration screen than were found in thermal cycling. Both the vibration and thermal cycling are product unique. That is, there may be many good screens for a given product built with a given selection of parts and processes, but these screens are not necessarily good for other products.

Notching of vibration spectra, that is, reduction of the acceleration spectral density over small frequency ranges, is often performed to prevent damage by reducing certain resonant responses. In most cases, this approach reduces screen effectiveness and, in some cases, ignores real design problems. In addition, a substantial portion of modal response is due to inputs outside the area of peak response. Notching during vibration screening should only be done after a thorough step-stress-to-failure program including an under-standing of the failure modes coupled with design changes for enhanced flaw tolerance has been performed. If, after performing these steps, it is desirable to use higher vibration levels for enhanced screening and the design is known to be adequate for survival in the transportation and field environments, notching can be performed in order to allow higher vibration levels at frequencies other than those known to be limiting by the proven design.

TAPED (PSEUDO) RANDOM VIBRATION

With the advent of NAVMAT P-9492, many companies were faced with purchase of expensive control systems and some elected to use taped random (which was discussed in P-9492) in order to reduce cost. Taped random is run open loop and depends on the temporal stability of vibration system and specimen properties for reproduction of the "equalized spectrum", i.e., the desired input from the shaker to the specimen.

Several little-publicized disasters occured using this technique when changes in the shaker, power amp, tape player, preamp, fixture or specimen resulted in an overtest and led to damage or destruction of the product.

An example of a type of change which can occur [7] is reproduced below with

permission of the IES. The figure shows a
spectrum resulting when a system equalized
at ambient is then run open loop at -40°C.
Note the significant change in spectrum.

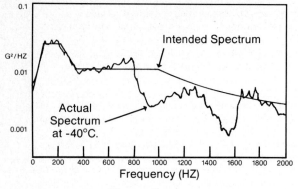

Figure 4

The above change would probably not be
damaging, but what might happen if the
temperature were to be raised to 80°C?
Szymkowiak (Reference [7]) points out that
the taped random can produce excellent
results if the thermal prehistory is taken
into account, i.e., the same thermal cycling
profile should be done before equalization
as is to be done during the actual test.
Simplified, this means that all components
involved in the test should always be at the
same temperature as when the tape was
equalized.

Other considerations are currently used
with taped inputs including protective
features on the shaker and amplifier and
overall grms limit protection. Digital
tapes allow vibration reproduction
essentially down to zero frequency.

One equipment manufacturer reports a
tape in use for 3 years and accumulating
2900 operational hours with performance
being "completely satisfactory".

Taped random is now recognized as cost
effective and safe if the appropriate
limiters are included in the control system
and equalization is done under the same
temperature and dynamic loading as is the
actual test.

ACOUSTIC EXCITATION

Excitation of large structures using
acoustic excitation has been used for many
years for MIL-SPEC testing. Until the last
year or so, only broad band (1/3 octave)
control was possible, and then only by
manual control systems with a closed loop
control being supplied by a very skilled
technician. A new digital acoustic
excitation control system is now available
which will supply automatic narrow band
control in a fashion similar to that of
conventional vibration control systems.

The acoustic control system is based on
an existing random vibration control system
with new software. Several microphones can
be used for averaging of weighted inputs

(just as in vibration control) or response
accelerometers mounted on the speciman can
be used for control. The system is capable
of 72 db of dynamic range and can accomodate
multiple microphone failures during a test
period without test abort as long as a
specified number of microphones remain
operative. The latter capability has been
included since the sound pressure levels
required in some tests exceed the capability
of the microphones! Spectral resolution and
control of up to 800 lines per exciter
(maximum of four) is available as is a base
band of up to 20 kHz. The four exciters
cover progressive ranges in frequency.
Other capabilities match those of current
state-of-the-art random vibration control
systems.

One of the problems encountered in
vibration stress screening is distribution
of the energy into the specimen, or even
transmission at all, particularly at high
frequencies where electronic boxes just do
not transmit vibration uniformly.
Acoustic screening may just be a way around
this problem. Also, multi-axis response
always results from acoustic excitation, but
independent axis control is not possible
since a reverberant chamber is used.

OMNI AXIAL VIBRATION

The use of pneumatic vibration for
stress screening had been shown to be
effective in the mid to late 1970's by
General Dynamics, Hughes Aircraft and
Westinghouse.

An investigation of a particular quasi
random vibration system [8] showed that the
linear accelerations in orthogonal
directions were non-coherent and that
angular accelerations about those axes were
present and were broad band. This shaker
was therefore described as "omni axial" by
the author. It was mentioned that some
defects had been uncovered by the omni-axial
vibration which could not be found in a
rather exhaustive attempt on a single axis
shaker. Similar results using a different
type of pneumatic shaker, which may also be
omni-axial, were reported to the author by
another investigator, Bill Silver of
Westinghouse Corporation, Baltimore, MD. It
is not, to the author's knowledge, known
quantitatively just how much better
omni-axial vibration is than single or dual
axis excitation, but it is quite clear that
six degree of freedom excitation will excite
more modes of vibration to suitable
screening levels than will excitation of
fewer degrees of freedom. The author has
not found the lack of spectrum or axis
balance control on the quasi random or other
multi axis systems to be limiting in
practical cases, but one can hypothesize
cases wherein lack of spectrum or axis
balance control could severely limit the
screening effectiveness by requiring low
vibration levels in order to prevent damage
by the response of a particular mode.

DISTRUBITED VIBRATION, THERMAL, AND ELECTRICAL EXCITATION

Distributed excitation refers to excitations at more than one point and, in general, in more than one direction. The excitations can include force, vibration, voltage, current, temperature or other variable.

These disparate stimulations are lumped together for discussion because the philosophy behind them is the same: Control system response to the desired level by distributing the inputs which allows better control of a specific response, and more generally, of all responses.

Paraphrasing the theorem in controllability from control theory as "a system is completely controllable if one can independently control each and every mode of the system", it is obvious that single point (or input) control can not in general provide the responses required for a good screen. This applies to either vibration, thermal or electrical inputs and to other input stimuli as well.

These techniques of distributed inputs are not new but have not yet received much attention except for limited technical areas, such as experimental modal analysis where multiple shaker inputs are common and by a limited number of investigators.

In [9], the power of the distributed input techniques is illustrated for vibration. Classical single axis excitation could not excite the fundamental resonant frequency of the circuit boards, but multiple inputs into the corners of the box provided excellent response and therefore supplied a good vibration screen.

The author is currently involved in developing a vibration screen on a main frame computer wherein forty (40) independently controlled multi axis shakers are used for stimulation of flaws and then for detection of the resultant defects. This approach shows great promise and results will be reported in the open literature when approved by the computer manufacturer.

In another case, the author is also utilizing distributed thermal excitation to provide very different thermal environments in various areas in a "black box". The different thermal environments have been developed to provide a rapid and effective screen accounting for the different fragility levels and screen strengths required of the various assemblies.

FLAW TOLERANT DESIGN & ENHANCED SCREENING

A flaw tolerant design is one which will allow less than perfect processes and assemblies to survive in the normal in-use environment. Such a design is the usual result of a screen development program wherein step stress testing and design ruggedization are done in order to allow enhanced screening.

Enhanced screening is screening which is done at stimuli levels above those experienced in the normal use environment. This, of course, implies that the device being screened is not excessively damaged (in the fatigue sense) by the screen.

Note that a screen, other than an inspection, usually generates fatigue damage and our goal is to actually break, during the screen, those defective parts or assemblies which would eventually break in normal use. Our goal is to also pass, without excessive fatigue damage, those parts or assemblies which are sufficiently defect free as to survive a normal lifetime of service after the screen.

The techniques of enhanced screening, design ruggedization to allow enhanced screening and the resulting flaw tolerant designs have been used on many military and commercial programs by the author and have been found to be very cost effective if the screens are developed during a time frame when the design is not frozen. These techniques are not applicable to programs after a design freeze.

An example of the success of the application of such techniques is given in [10].

By the use of the enhanced screening techniques, one commercial manufacturer saved an estimated 3.4 million dollars after the expenditure of only 200 thousand dollars on screen and product development for screen enhancement. It was further estimated that over the five year life of the product, the total expenditure on screening would be 6 million dollars and the savings would be 34 million dollars.

AUTOMATED SCREENING

In the high production environment of commercial manufacturing, rapid screening using enhanced screening techniques and flaw tolerant design methods are inevitable. It is also inevitable that completely automated screening would occur.

Such screening (using vibration) is now a reality including product handling, fixturing, electrical powering and monitoring under either local or remote computer control. This technique allows an omni axial vibration screen on multiple products to be accomplished in only 140 seconds including loading into the shaker, powering up and vibration screening while monitoring. This technique could easily also include distributed thermal excitation in susceptible areas.

Fully automated screening using distributed thermal and vibration excitation is now a reality, but to the author's knowledge, no reports have yet been published in the open literature.

CONCLUSION

Many ill-conceived attempts at stress screening have resulted in destruction of whole lots of products, reduced field reliability and increased costs. Well conceived attempts have resulted in lower production costs and higher field reliability. The results obtained depend on taking an intelligent approach, considering the flaw-stimulus relationships, and not seeking a "cook book" or "MIL-SPEC" approach to screening. There is not now and never will be one good screen for all equipment. A screen must be developed and proved for each assembly to be screened. The screen development should be done during product development so that the power of enhanced screening and flaw tolerant design techniques can be brought to fruition.

The secret to success in the stress screening process lies in corrective action when problems are found and in good common sense engineering.

The military services, particularly the U.S. Navy, promoted and forced attention on stress screening in the late 1970's. By the mid 1980's, commercial industry had pulled ahead in techniques and efficiency, particularly in terms of corrective action and in enhanced screening and flaw tolerant design techniques.

In the author's view, the leaders in stress screening will, in a few years, be screening during full scale production on a sample basis, using distributed input excitation (electrical, thermal, vibration) at enhanced levels on flaw tolerant products and will be finding very few defects. We will then have arrived at our goal - built in quality.

REFERENCES

[1] Hobbs, G. K., Comparing Military to Commercial Screening - A Dichotomy Forming, editorial in Sound and Vibration Magazine, April 1985, pg. 5.

[2] Institute of Environmental Sciences, Environmental Stress Screening Guidelines for Parts, Sept. 1985, pgs. 2-11.

[3] Institute of Environmental Sciences, Environmental Stress Screening Guildlines, 1981, p. A-3

[4] Alis, Bernard, The French Environmental Stress Screening Program, IES 1985 Proceedings, 31st Annual Technical Meeting, pgs. 439-442.

[5] Curtis, A. J., On the Development of Effective Vibration Screens, Journal of Environmental Sciences, Jan/Feb 1982, pgs. 16-18.

[6] Department of the Navy, Naval Material Command, NAVMAT P-9492, May 1979.

[7] Szymkowiak, E. A., An Implementation of a Taped Random Vibration System for CERT, Proceedings of the Institute of Environmental Sciences, 31st Annual Technical Meeting, April 30-May 2, 1985, pgs. 524-527.

[8] Hobbs, G. K. and Mercado, R., Six Degree of Freedom Vibration Stress Screening, Journal of Environmental Sciences, Nov/Dec 1984, pgs. 46-53.

[9] Silver, W. and Caruso, H., A Rational Approach to Stress Screening Vibration, 1981 Proceedings Annual Reliability and Maintainability Symposium, pgs. 384-388.

[10] Chesney, Ken, Step Stress Analysis of a Printer, Proceedings of the Reliability and Maintainability Seminar, Las Vegas, Nevada, 28-30 January 1986.

BIOGRAPHICAL SKETCH

Dr. Gregg K. Hobbs has been a consulting engineer since 1978, specializing in the fields of stress screening, ruggedized design, and structural dynamic analysis and testing. He has been employed as a specialist in these areas by many companies in the aerospace, commercial, military, and industrial fields, and has taught courses on these subjects in private companies and at open seminars in Europe and the U.S.A. He has also taught Mechanical Engineering courses at the Universities of California and Minnesota.

Dr. Hobbs has been employed by TRW Systems, Corp., Hughes Aircraft Co., Astro Research Corp., and MTS Systems Corp.

Dr. Hobbs earned his BS and MS degrees in Structural Dynamics at the University of California, Los Angeles, and his Ph.D. in that subject at the University of California, Santa Barbara. His professional registrations include Control Systems Engineering, Civil Engineering and Mechanical Engineering.

Reliability Technology — Theory & Applications
J. Møltoft and F. Jensen (Editors)
© Elsevier Science Publishers B.V. (North-Holland), 1986

OPTIMIZATION OF RELIABILITY TEST CONDITIONS

Povl K. Birch
ELEKTRONIKCENTRALEN, Hoersholm, Denmark

It is desirable to perform reliability testing in early phases of product development on few items in reliability short time. This paper describes a procedure for optimization of reliability test conditions minimizing the total uncertainty of the test result. It is concluded that it is often beneficial to sacrifice degree of simulation for an increased time compression.

1. INTRODUCTION

1.1 The need for reliability measurements

Electronic equipment plays a more and more vital role in everyday life. The reliability of these products, therefore, is a more and more important parameter and is often incorporated as a target specification on the same line as other important characteristics from the very beginning of the development phase.

This fact creates a need for reliability measurements during the early phases of development in order to be able to decide on proper measures in due time.

1.2 Making reliability measurements

Measuring reliability is a well established technique so far. In principle it consists of the observation of the performance of a number of representative items under representative conditions of use.

Problems arise, however, because it requires a reasonable number of items and a reasonable amount of time. First of all because of the statistical uncertainty related to a small number of events occuring at random points of time, but also due to the fact that the failure intensity cannot be expected to be constant. The time involved should therefore be comparable to the time for which the reliability information is desired.

These demands of items and time are in the highest degree in contradiction with any development programme, where the number of prototypes and the time for testing are about the most precious matters to get.

Adding furhter to the complexity comes the fact that field trials under real use conditions are for many reasons often not practical during the early phases of development. Such reasons may be:

1) Not all units for making up a functional system may be available at the same time.

2) The coming users may not be available for this purpose, either due to lack of a suitable organization, lack of training, lack of interest or other reasons.

3) The feedback of information from the users to the development laboratory concerning performance in the field is in general difficult and often not sufficiently precise.

4) Secrecy in relation to possible competitors may be vital.

This means that some kind of artificial test environment, simulating real use conditions will have to be used. The proper design of these test conditions is most important in order to ensure that those precious samples and that sparse time, which could be obtained, are used in the best possible way. As it shall be demonstrated in this paper, it is possible, for a fixed number of items and a fixed testing time, to optimize test conditions for minimum uncertainty of the reliability determination.

2. RELIABILITY MEASUREMENT UNCERTAINTIES

The uncertainties in determination of reliability by testing are due to the statistical uncertainties and to the possible dissimilarities between test conditions and use conditions. These two are further found to be mutually dependent.

2.1 Statistical uncertainty

The statistical uncertainty is usually expressed as a confidence interval realted to a confidence level, C. In this paper only two-sided confidence intervals for the failure intensity will be used. This means that with probability C, the true failure intensity lies between the lower limit and the upper limit of the confidence interval.

The values can be found using the formulae:

$$\text{Lower,} \quad \lambda_1 = \frac{\chi^2_{(0.5(1-C)\;;\;2r)}}{2T} \quad (1)$$

$$\text{Upper,} \; \lambda_u = \frac{\chi^2 (1-0.5(1-C) \; ; \; 2r+2)}{2T} \quad (2)$$

This is well known form any textbook in reliability testing and shall not be discussed any further in this paper. A convenient measure of this statistical uncertainty is the ratio between upper and lower limit of the confidence interval. This ratio will be referred to as the "statistical determination ratio".

2.2 Test environment uncertainty

In order to test the reliability, the equipment has to be exposed to conditions simulating the anticipated use conditions. This could be a field trial under real use conditions. In this case, there is no uncertainty in that part of the reliability measurement.

As mentioned in the introduction, however, it is most likely that some kind of artificial test environment is used. It is therefore appropriate to define the degree of simulation achieved by a test environment.

The degree of simulation describes the relationship between failure intensity during real use and during the test. Further, test environments could advantageously be designed to exhibit a more rapid time history than the real use. This feature is referred to as time compression. The ratio between the failure intensity during the test and the failure intensity in real use is referred to as the "time compression factor".

It is assumed that the test environment exhibits some known distribution of the probability of the time compression factor to be equal with some intended value, being the basis for its design. The degree of simulation will therefore be described by the confidence interval of the time compression factor and the related confidence level.

The test environment is often arranged as a repeated cycle of simulated environmental conditions in combination with simulated use. The measure for the test environment uncertainly will be the ratio between upper and lower limit of the time compression factor confidence interval. This ratio will be referred to as the "test cycle determination ratio".

2.3 Application uncertainty

The use of the equipment implies some stresses acting upon it. These stresses are made up of periods of use in different modes and with various levels of functional load in combination with different environmental stresses. The use of the product may be more or less well defined, but some model application will have to thought upon in order to define the reliability test conditions. This fact introduces a third source of uncertainty, because the real use implies a range of applications and the model application is just one single point, selected within this range.

This application uncertainty is of course important if the measured reliability has to be related to some observed reliability for a specific application. However, it is considered less important for the optimization of the test conditions and therefore not incorporated in the following considerations.

3. OPTIMIZATION

The principle of optimization is to find the test cycle minimizing the total uncertainty in the reliability determination test. The application uncertainty is not considered (alternative test cycles are designed for the same model application) and the statistical uncertainty and the environmental uncertainty are dependent of the selected test cycle.

The best reliability determination is achieved when the product of "statistical determination ratio" and "test cycle determination ratio" is minimum. The confidence level of the overall reliability determination will be the product of the two confidence levels. In order to end up with an overall confidence level of, e.g. 90%, the two confidence levels could be approx. 95% (94,87%). The share between the two could possibly be different, but this has been considered too theoretical at present, and it is believed that allocating 95% to both the statistical and the test cycle determination ratio is sufficiently accurate.

3.1 Statistical determination ratio

The statistical determination ratio depends on the expected number of failures, which in turn depends on the target failure intensity, the number of items to test, the test time and the time compression factor of the test cycle. Under the assumption of a constant failure intensity applies:

$$r_{exp.} = \lambda_{target} \cdot N_{test} \cdot T \cdot F_{time comp.} \quad (3)$$

Table 3.1.1 below has been calculated according to formula (1) and (2) for a 95% confidence level and test time equalling unity.

3.2 Test cycle determination ratio

The test cycle is preferably to be designed according to the systematic procedure described in the coming IEC standard: DRAFT PUBLICATION 605: EQUIPMENT RELIABILITY TESTING - PART 2: GUIDANCE FOR THE DESIGN OF TEST CYCLES. It is at present available as 56(Central Office)106, dated June 1984.

The principle of this procedure is an analysis of the use conditions concerning operation and environmental conditions. They are subdivided in fractions of the life history, where the influencing parameters are characterized one by one or in combination as they occur. As each parameter has a certain interval of variation, the number of fractions depends on how fine the subdivision is made for each parameter. To each fraction is then allocated a representative se-

Table 3.1.1
Statistical determination ratio

r	Chisq (.025)	Chisq (.975)	Failure intensity Lower limit	Upper limit	Determination ratio
1	.0506	9.35	.0253	4.675	184.78
2	.484	12.8	.1210	3.200	26.45
3	1.24	16.0	.2067	2.667	12.90
4	2.18	19.0	.2725	2.375	8.72
5	3.25	21.9	.3250	2.190	6.74
6	4.40	24.7	.3667	2.058	5.61
7	5.63	27.1	.4021	1.936	4.81
8	6.91	30.2	.4319	1.887	4.37
9	8.23	32.9	.4572	1.828	4.00
10	9.59	35.5	.4795	1.775	3.70
11	11.0	38.1	.5000	1.732	3.46
12	12.4	40.6	.5167	1.692	3.27
13	13.8	43.2	.5308	1.662	3.13
14	15.3	45.7	.5464	1.632	2.99
15	16.8	48.2	.5600	1.607	2.87
16	18.3	50.7	.5719	1.584	2.77
17	19.8	53.2	.5824	1.565	2.69
18	21.3	55.7	.5917	1.547	2.62
19	22.9	58.1	.6026	1.529	2.54
20	24.4	60.6	.6100	1.515	2.48
21	26.0	63.0	.6190	1.500	2.42
22	27.6	65.4	.6273	1.486	2.37
23	29.2	67.8	.6348	1.474	2.32
24	30.8	70.2	.6417	1.463	2.28

verity of the actual parameters and a test cycle is designed based on these representative severities.

Time compression of the test cycle can then be obtained by excluding those fractions, judged to contribute least to the occurence of failures, i.e. the low-stress periods.

The choice of representative severities and especially the exclusion of fractions of the life history will influence the test cycle's degree of simulation. This means that the test cycle determination ratio is a function of the time compression factor. This function can be found by analysing the evaluations made in connection with the design of the test cycle.

This analysis is very complicated and has no influence on the principle of test condition optimization. In the following, it will therefore be assumed, that the confidence interval and thus the test cycle determination ratio is a known function of the time compression factor. For use in the invented example below, the principle is demonstrated using only a very simple test environment: Elevated temperature.

4. EXAMPLE

Three months, approx. 2000 hours, are available for a reliability test and the target MTBF is 10000 hours.

You wish to optimize test conditions for 5, 10

or 20 items selected for the test.

The product is intended for constant operation at a constant temperature of 25°C and the failure intensity follows Arrhenius law with an estimated activation energy of 1 eV. The activation energy is known with 95% confidence to be between 1.4 eV and 0.6 eV.

This is of course an extremely special case, but useful for a simple demonstration of the principle. Table 4.0.1 below shows the test cycle determination ratio as a function of the time compression ratio for a "test cycle", which consists of an increased temperature only.

4.1 Five items to test

The expected number of failures is 1, calculated according to formula (3), for a time compression factor of unity. For increasing time compression factors, the test cycle determination ratio (ref. table 4.0.1), the expected number of failures (ref. formula (3)), the statistical determination ratio (ref. table 3.1.1) and the total determination ratio appear from table 4.1.1 below.

If the failure intensity equals the target value, it is seen that the least total uncertainty is achieved with a test cycle with a time compression factor of 7 with no substantial increase in uncertainty for test cycles with time compression factors in the range 5 to 14.

Table 4.0.1
Test cycle determination ratio

Test temperature (^{o}C)	Time compression factor values			Test cyc. determination ratio
	Estimated (1.0 eV)	Upper limit (1.4 eV)	Lower limit (.60 eV)	
25.00	1.00	1.00	1.00	1.00
28.15	1.50	1.77	1.28	1.39
30.39	2.00	2.63	1.51	1.74
33.64	3.00	4.65	1.93	2.41
36.00	4.00	6.96	2.30	3.03
37.84	5.00	9.51	2.63	3.62
39.37	6.00	12.28	2.93	4.19
40.67	7.00	15.24	3.21	4.74
41.81	8.00	18.38	3.48	5.28
42.82	9.00	21.68	3.74	5.80
43.73	10.00	25.13	3.98	6.31
45.31	12.00	32.42	4.44	7.30
46.66	14.00	40.22	4.87	8.26
47.84	16.00	48.48	5.28	9.19
48.89	18.00	57.19	5.66	10.10
49.84	20.00	66.30	6.03	10.99

Table 4.1.1
Total determination ratio, five items

Time compress. factor	Test cyc. determin. ratio	Expected no. of failures	Statist. determin. ratio	Total determin. ratio
1.00	1.00	1	184.78	184.78
2.00	1.74	2	26.45	46.00
3.00	2.41	3	12.90	31.04
4.00	3.03	4	8.72	26.42
5.00	3.62	5	6.74	24.40
6.00	4.19	6	5.61	23.53
7.00	4.74	7	4.81	22.82
8.00	5.28	8	4.37	23.07
9.00	5.80	9	4.00	23.19
10.00	6.31	10	3.70	23.36
12.00	7.30	12	3.27	23.90
14.00	8.26	14	2.99	24.66
16.00	9.19	16	2.77	25.45
18.00	10.10	18	2.62	26.40
20.00	10.99	20	2.48	27.29

4.2 Ten items to test

The expected number of failures is 2, calculated according to formula (3), for a time compression factor of unity. For increasing time compression factors, the test cycle determination ratio (ref. table 4.0.1), the expected number of failures (ref. formula (3)), the statistical determination ratio (ref. table 3.1.1) and the total determination ratio appear from table 4.2.1. below.

If the failure intensity equals the target value, it is seen that the least total uncertainty is achieved with a test cycle with a time compression factor of 4 with no substantial increase in uncertainty for test cycles with time compression factors in the range 3 to 7.

4.3 Twenty items to test

The expected number of failures is 4, calculated according to formula (3), for a time compression factor of unity. For increasing time compression factors, the test cycle determination ratio (ref. table 4.0.1), the expected number of failures (ref. formula (3)), the statistical determination ratio (ref. table 3.1.1) and the total determination ratio appear from table 4.3.1 below.

If the failure intensity equals the target value, it is seen that the least total uncertainty is achieved with a test cycle with a time compression factor of 2 with no substantial increase in uncertainty for test cycles with time compression factors in the range 1.5 to 4.

Table 4.2.1
Total determination ratio, ten items

Time compress. factor	Test cyc. determin. ratio	Expected no. of failures	Statist. determin. ratio	Total determin. ratio
1.00	1.00	2	26.45	26.45
2.00	1.74	4	8.72	15.16
3.00	2.41	6	5.61	13.50
4.00	3.03	8	4.37	13.25
5.00	3.62	10	3.70	13.41
6.00	4.19	12	3.27	13.73
7.00	4.74	14	2.99	14.16
8.00	5.28	16	2.77	14.62
9.00	5.80	18	2.62	15.17
10.00	6.31	20	2.48	15.67

Table 4.3.1
Total determination ratio, twenty items

Time compress. factor	Test cyc. determin. ratio	Expected no. of failures	Statist. determin. ratio	Total determin. ratio
1.00	1.00	4	8.72	8.72
1.50	1.39	6	5.61	7.78
2.00	1.74	8	4.37	7.60
3.00	2.41	12	3.27	7.88
4.00	3.03	16	2.77	8.40
5.00	3.62	20	2.48	8.99
6.00	4.19	24	2.28	9.55

5. DISCUSSION

These results demonstrate that even when the use of time compressed test cycles add a considerable uncertainty, because such test cycles generally have a low degree of simulation, the total uncertainty in reliability determinations with a small number of items tested in short time, will be lower than if test cycles with high degree of simulation and no time compression are used.

These results were seen under the assumption of a constant failure intensity, which will not normally be the case. In the general case of a time-varying failure intensity it is important that the effective test period, taking into account the time compression, is comparable to the intended use period.

In the three cases of the example, this effective test period corresponds to:

5 items: 15 to 42 months

10 items: 9 to 21 months

20 items: 4 to 12 months.

This might dictate a higher time compression factor for the larger number of items than immediately arrived at by the method of optimization described in this paper.

BIOGRAPHY

Povl K. Birch was born in Copenhagen 1940.

He has studied electrical and electronics engineering at the Technical University of Denmark (DTH), from which he graduated 1965.

Until 1976 he has worked with development of electronic equipment in different Danish companies.

From 1976 he has been with ELEKTRONIKCENTRALEN, where he is now responsible for the area reliability- and environmental testing.

Reliability Technology — Theory & Applications
J. Møltoft and F. Jensen (Editors)
© Elsevier Science Publishers B.V. (North-Holland), 1986

METHODS FOR A SYSTEM'S LIFE ESTIMATION FROM ACCELERATED TESTS

Włodzimierz LEWIN
Systems Research Institute of Polish Academy of Sciences
Newelska 6, 01-447 Warsaw, Poland

Accelerated life testing of a product is often the only way to get operative information on its reliability at the use-stress. Most methods for reliability assessment from accelerated tests assume that the product of interest is a single component system or, equivalently, only a single failure mechanism is considered. In the paper a controversial problem of accelerated life testing of a N-component system is discussed. Several approaches to this problem are compared. This includes, in principle, the methods of "equal percentiles" and "competing risks". Some properties of the N-component series system are shown. They deal with asymptotic (when $N \to \infty$) correlation between system's failure times at different stress levels, suggested by Kartashov theorem, and with the possibility of composition of similar type acceleration models. A simple procedure for two-sample accelerated test based on these properties is presented and illustrated with numerical example.

1. INTRODUCTION

In reliability testing it is a common practice to subject "long life" items to larger then usual stresses so that failures can be observed in a short amount of test time. Such tests are called accelerated tests. The problem is how to infer reliability characteristics at the use-stress using life data from accelerating-stress.

Usually, only elements (components) are considered in the accelerated test theory. Numerous statistical methods are available for estimating diverse reliability characteristics in this case. In practice, to get operative information on N-component system reliability, similar methods are also needed, but there exist controversial opinions on the possibility of development of such methods for systems because of:

- diverse performance of components under the same stress,

- the lack of acceleration models for systems,

- smaller and more costly samples of systems to be tested.

In spite of that, some approaches to the problem were proposed. A nonparametric approach was presented by Basu and Ebrahimi {1}. Nelson assumed that for a given stress s each system's component follows an independent log-normal distribution with parameters $\mu_i(s) = \alpha_i + \beta_i s$ and σ_i which is constant with respect to s (i=1,...,N). He obtained graphical {6} and maximum likelihood {7} estimates of α_i, β_i and σ_i. Klein and Basu {5} assumed that components

(each with its own acceleration model) compete to cause the system failure. They used the maximum likelihood method to estimate parameters of acceleration models for system's components. They also elaborated accelerated life tests in the competing risks framework for the Weibule distribution and diverse censoring schemes. In their model the number of unknown parameters is proportional to the number of system's components. For actual many-component systems the use of such methods requires a large number of systems to be tested at each accelerating stress level (usually there are 3 or 4 levels). In many practical situations systems have more than 100 components and therefore it is almost impossible to apply the above methods.

For many-component systems another kind of accelerated tests {2,4,8} seems to be more useful. These tests require only one accelerating-stress level. Life times obtained at such a level are then transformated to the use-stress according to a transformation function which can be determined from preliminary comparative tests or from a two-sample test which will be described in section 4.

To perform the last kind of test one should assume that systems possess certain properties. It is reasonable to expect these properties in many cases - as it will be shown in sections 2 and 3.

2. COMPOSITION OF ACCELERATION MODELS

Consider an N-component series system. All components are assumed to have independent failure times which follow a Weibull survival function of the form

$$R_i(t, \alpha_i, \beta_i) = \exp(-\alpha_i t^{\beta_i}) \qquad (1)$$

where α_i and β_i are the scale and shape parameters.

Assume the proportional hazard rate model {3} as an acceleration model for each component. It means that

$$\alpha_i^{(a)}(\underline{s}_i, \underline{\gamma}_i) = \alpha_i^{(u)} \exp(\underline{\gamma}_i \underline{s}_i), \qquad (2)$$
$$i = 1, \ldots, N$$

where

$\underline{s}_i$ is a vector of relative values of different stress types $(s_{i1}, s_{i2}, \ldots, s_{ik_i})$. These values are of such nature that vector $\underline{0}$ corresponds to the use-stress,

$\underline{\gamma}_i$ is a vector of unknown model's parameters $(\gamma_{i1}, \gamma_{i2}, \ldots, \gamma_{ik_i})$,

(a)   is the index of accelerating-stress,
(u)   is the index of use-stress.

For $\underline{s}_i = \underline{0}$ $(s_{i1} = 0, s_{i2}, \ldots, s_{ik} = 0)$ we have from (2) $\alpha_i(\underline{0}_i, \underline{\gamma}_i) = \alpha_i^{(u)}$, i.e. the value of scale parameter at the use-stress.

This model of component performance seems to be general enough to describe multi-stress situations and to include well known single stress standard models such as "Arrhenius" or "inverse power". In this section we shall discuss the performance of a system composed of such components.

According to the assumed proportional hazard rate model, the Weibull cumulative hazard rate for i-th component, $i = 1, \ldots, N$, can be written as

$$\Lambda_i^{(a)}(t, \underline{s}_i) = \Lambda_i^{(u)}(t) \exp(\underline{\gamma}_i \underline{s}_i) \qquad (3)$$

where

$$\Lambda_i^{(u)}(t) = t^{\beta_i} \alpha_i^{(u)} \qquad (4)$$

The survival function for i-th component is

$$R_i^{(a)}(t) = \exp\left[-\Lambda_i^{(u)}(t) \exp(\underline{\gamma}_i \underline{s}_i)\right] \qquad (5)$$

and for the whole system

$$R^{(a)}(t) = \exp\left[-\sum_{i=1}^{N} \Lambda_i^{(u)}(t) \exp(\underline{\gamma}_i \underline{s}_i)\right] \qquad (6)$$

Our prime interest is in obtaining the relationship between system failure times corresponding to equal percentiles of life distribution at the use - and accelerating-stress.

Consider first a special case: a single component system (N=1). The indices "i"

can be ommitted at this point. The survival functions at the accelerating-stress $(\underline{s} > \underline{0})$ and at the use-stress $(\underline{s} = \underline{0})$ are

$$R^{(a)}(t) = t^{\beta} \alpha^{(u)} \exp(\underline{\gamma}\,\underline{s}) \qquad (7)$$

and

$$R^{(u)}(t) = t^{\beta} \alpha^{(u)} \qquad (8)$$

The lives $t_{R^{(a)}}$ and $t_{R^{(u)}}$, corresponding to given values $R^{(a)}$ and $R^{(u)}$ of the survival function, are

$$t_{R^{(a)}} = \left[-\ln R^{(a)} / \alpha^{(u)} \exp(\underline{\gamma}\,\underline{s})\right]^{1/\beta} \qquad (9)$$

and

$$t_{R^{(u)}} = \left[-\ln R^{(u)} / \alpha^{(u)}\right]^{1/\beta} \qquad (10)$$

For $R^{(a)} = R^{(u)} = r$ the proportion $t_r^{(u)} / t_r^{(a)}$ is

$$C_r(\underline{\gamma}, \underline{s}) = \frac{t_r^{(u)}}{t_r^{(a)}} = \left[\exp(\underline{\gamma}\,\underline{s})\right]^{1/\beta} = c \qquad (11)$$

It follows from (11) that between lives $t_r^{(u)}$ and $t_r^{(a)}$ (corresponding to equal percentiles) there exists the linear relationship $t_r^{(u)} = t_r^{(a)} \cdot c$, and the coefficient $C_r(.)$, which may be called "acceleration coefficient", is independent from percentiles r of life distributions at the use- and accelerating-stress. This property is very useful in performing and analyzing accelerated life test. We shall see that the relationship of the form similar to (11) can be preserved for a series system.

The survival function for the N-component series system, when its components run under stresses $\underline{s}_1, \underline{s}_2, \ldots, \underline{s}_N$, is

$$R^{(a)}(t) = \exp\left[-\sum_{i=1}^{N} \Lambda_i^{(u)}(t) \exp(\underline{\gamma}_i \underline{s}_i)\right] \qquad (12)$$

At the use-stress (all $\underline{s}_i$ are equal to $\underline{0}$) the survival functions is

$$R^{(u)}(t) = \exp\left[-\Lambda^{(u)}(t)\right] \qquad (13)$$

where $\Lambda^{(u)} = \sum_{i=1}^{N} \Lambda_i^{(u)}(t)$.

Expanding $\exp(\underline{\gamma}_i \underline{s}_i)$ in the Taylor series we can obtain the following form of $R^{(a)}(t)$ -

$$R^{(a)}(t) = \exp\left\{-\sum_{i=1}^{N} \Lambda_i^{(u)}(t)\left[1 + \right.\right.$$

$$+ \frac{\sum_{j=1}^{N} \Lambda_j^{(u)}(t) \cdot \underline{\gamma}_j \underline{s}_j}{\Lambda^{(u)}(t)} + \frac{\sum_{j=1}^{N} \Lambda_j^{(u)}(t) \cdot (\underline{\gamma}_j \underline{s}_j)^2}{2\Lambda^{(u)}(t)} +$$

$$+ \cdots + \frac{\sum_{j=1}^{N} \Lambda_j^{(u)}(t) \cdot (\underline{\gamma}_j \underline{s}_j)^k}{k! \; \Lambda^{(u)}(t)} + \cdots] \} \qquad (14)$$

For well balanced systems, i.e. when the differences between components' reliability are not too vast, we can write for large N -

$$\frac{\Lambda_j^{(u)}(t)}{\Lambda^{(u)}(t)} = \frac{\Lambda_j^{(u)}(t)}{\sum_{j=1}^{N} \Lambda_j^{(u)}(t)} \approx \frac{1}{N} \qquad (15)$$

Now, we define the "mean stress" S -

$$S = \frac{1}{N} \sum_{i=1}^{N} \underline{\gamma}_i \underline{s}_i \qquad (16)$$

and we can calculate the difference D between the expression in square brackets in (14) -

$$1 + \frac{1}{N} \underline{\gamma}_1 \underline{s}_1 + \frac{1}{2N} (\underline{\gamma}_2 \underline{s}_2)^2 + \cdots +$$

$$+ \cdots + \frac{1}{k!N} (\underline{\gamma}_k \underline{s}_k)^k + \cdots \qquad (17)$$

and exp(S) expanded in the Taylor series -

$$1 + S + \frac{S^2}{2} + \cdots + \frac{S^k}{k!} + \cdots \qquad (18)$$

This difference can be written as follows

$$D = \sum_{k=2}^{\infty} \frac{S^k}{k!} \left[\frac{1}{N} \sum_{i=1}^{N} b_i^k - 1 \right], \qquad (19)$$

where $\quad b_i = \dfrac{\underline{\gamma}_i \underline{s}_i}{S}$.

Finally, $R^{(a)}(t)$ can be expressed as

$$R^{(a)}(t) = \exp\{-\Lambda^{(u)}(t) [\exp(S) + D]\} \qquad (20)$$

where

$$\exp(S) \leqslant \exp(S) + D \leqslant$$

$$\leqslant \exp(S \cdot b_{max}) - S(b_{max} - 1), \qquad (21)$$

$$b_{max} = \max_{1 \leqslant i \leqslant N} b_i$$

Since $S(b_{max} - 1) << \exp(Sb_{max})$ for actual systems, the inequalities (21) can be simplified -

$$\exp(S) \leqslant \exp(S) + D \leqslant \exp(Sb_{max}) \qquad (22)$$

When all components have the same stress dependence, i.e., when all b_i, $i = 1, \ldots, N$, equal each other, then the equalities arise in (22) and the system's survival function $R^{(a)}(t)$ has exactly the same form as for a single component. In general, for actual many-components systems with components' lives following the Weibull proportional hazard rate model, one can expect an approximately linear relationship between system lives corresponding to equal percentiles of life distribution at the use- and accelerating-stress.

The acceleration coefficient c can be directly determined from (20) for given components' models only when all shape parameters β_i, $(i = 1, \ldots, N)$, equal each other.

3. MONTE CARLO EXPERIMENT

The method

In order to verify the truthfulness of the expected liearity for actual systems, a few Monte Carlo experiments were performed. The objective was to both check the liearity arising from the consideration of the previous section, and to examine the relationship between lives of equivalent systems at the use - and accelerating stress ("equivalent systems" means: theoretically exactly the same systems). In the last case the linearity is sugested by Kartashov's theorem {4} dealing with asymptotic correlation between life times at the use - and accelerating-stress.

The system's life times were simulated according to the following principles. For each of N components two kinds of stresses were assumed. The relative values for stress levels and parameters of stress-life dependence were randomly settled. They were selected to obtain mean acceleration coefficient (for a system) of about 5-15. Such values were observed for actual systems {2}. Systems' lives were simulated first at the use - stress. Two ways were followed to obtain system's lives at the accelerating-stress level:

a) assuming for each component a linear relationship between its life at the use-stress and accelerating-stress -

$$t_i^{(a)} = c_i^{-1} \cdot t_i^{(u)}, \quad i = 1, \ldots, N \qquad (23)$$

where

$$c_i = [\exp(\gamma_{i1} s_{i1} + \gamma_{i2} \cdot s_{i2})]^{1/\beta_i}, \text{ and}$$

$t_i^{(u)}$ is simulated ("observed") failure time at the use-stress level,

b) assuming for each component a linear
   relationship between its acale parameter at the use - and accelerating
   stress -

$$\alpha_i^{(a)} = c_i^{-1} \cdot \alpha_i^{(u)} \qquad (24)$$

In this case lives of i-th component
were simulated from Weibull distribution with the scale parameter $\alpha_i^{(a)}$.

Results

A considerably modest number of Monte
Carlo experiments were performed up to
now because of a long duration of simulations executed on a minicomputer MERA
400. All simulated samples had 20 systems at each stress level, and the numbers of system components were taken
from 1 to 300 (1,2,10,20,50,100,200,
300). After each experiment the linearity in question was examined and correlation coefficient was calculated.

In spite of a small volume of experiments the qualitative results were clear:

- There is no linear relationship between lives of equivalent systems (in
  the experiment discussed - of the same systems) even if there exists such
  a relationship for each component of
  this system. No tendency towards correlation growth with growing sample
  size was observed.

There are at least two reasons for such
a performance of the system. The first
is due to the assumptions of Kartashov
theorem. In reality, it is hard to assume that if a system fails in the first
accelerating-stress cycle it has to fail
also in the first cycle at the use -
stress (cycle periods being the same).
The second reason is due to an asymptotic property which requires that random variables-life times of a system -
be close to zero. That is unacceptable in
reliability practice and therefore, in
simulations, life times were much greater than 0.

- In both cases (a and b) it was observed in all experiments for $N > 10$ that
  there is an approximate linear relationship between lives corresponding
  to equal percentiles of distributions
  at the use- and accelerating-stress.

The following reasoning may be profitable in accelerated life testing. Suppose that we have two samples of size
n (n system to be tested at each stress
level). Practically, it is imposible to
identify equivalent systems at each
stress level. But it is possible to set
in order the observed life times from
the smallest to the biggest separately
at the use - and at the accelerating-
stress. We can treat the established

pair of lives -

$$(t_1^{(u)}, t_1^{(a)}), (t_2^{(u)}, t_2^{(a)}), \ldots, (t_n^{(u)}, t_n^{(a)})$$

as n observations of two life times of
equivalent systems (first at the use-
stress and then at the accelerating-
stress). That would imply the existence
of a good correlation between lives
$t^{(u)}$ and $t^{(a)}$ of the system. In fact,
that was observed for $N > 10$ and it was
practically independent from the kind of
relationship (a or b) assumed for components. The observed values of correlation coefficient were 0,95-0,99 which
corresponds to the expected linear relationship between lives for equal percentiles. The above autorizes us to use
a simple two-sample method of accelerated life test that will be described in
the following section.

4. TWO-SAMPLE ACCELERATED TEST

The two-sample accelerated life test
can be performed as follows.

Each systems' sample should have the
same size n. Systems from the first
sample run under accelerating-stress.
At the same time systems from the second sample rum under use-stress. The
accelerating stress level should be chosen in order to assure observations of
all failures from the first sample and
about 20% of failures from the second
one. This may be difficult to achieve
when single components are tested (in
particular - semiconductors) but it is
fully obtainable for many-component systems.

The acceleration coefficient can be calculated from the following expression

$$c = \frac{1}{k-1} \sum_{j=1}^{k} c_j , \quad k < n \qquad (25)$$

where r_j denotes given percentile of
life distribution in

$$c_j = \frac{t_{r_j}^{(u)}}{t_{r_j}^{(a)}}$$

It is safer to start the summation in
(25) from k=2 because it was observed
in Monte Carlo experiments that lives
corresponding to extreme percentiles
sometimes do not follow the same relation as lives corresponding to the remaining percentiles.

Thus, the whole life distribution at
the use-stress can be reconstructed
using acceleration coefficient c, and
the point estimation of the required

life characteristic at this stress can be determined. The problem of interval estimation in this case is as yet unsolved and will be considered in the nearest future.

Numerical example

For this example life data were simulated for systems each composed of 20 components, each component having different parameters β_i, $\alpha_i^{(u)}$, γ_{i1}, γ_{i2}, s_{i1}, s_{i2} (two kinds of stresses were assumed).

The relationship of the type (b) between scale parameter of the Weibull life distribution at the use- and accelerating-stress was assumed for each component.

The resulting life data are presented in table 1. Sample size is 20 (systems).

Table 1

Simulated Life Test Data

No.	Percentile of life distribution $\|\%\|$	Failure time under	
		use-stress	accelerating-stress
1	5	170.40	29.00
2	10	202.57	46.71
3	15	298.73	50.06
4	20	300.79	56.44
5	25	361.89	60.89
6	30	470.37	82.41
7	35	486.25	87.44
8	40	493.40	93.49
9	45	499.48	100.94
10	50	508.72	113.25
11	55	521.85	115.63
12	60	637.37	122.75
13	65	695.35	143.30
14	70	744.38	147.91
15	75	766.74	169.14
16	80	834.02	169.95
17	85	858.47	184.26
18	90	991.95	184.52
19	95	1065.64	199.56
20	100	1324.12	298.62

The hypothetic test was terminated when all systems failed the accelerating stress, that is, after ca 300 hours.

At the use-stress, about 20% of failures from 20 system should be observed. It gives 4 failure times to be observed. The 4-th failure time was 300.79 hours.

According to (25), the acceleration coefficient is calculated as follows:

$$c = \frac{1}{3}\left(\frac{202.57}{46.71} + \frac{298.73}{50.06} + \frac{300.79}{56.44}\right) = 5.21$$

Suppose now that we are interested in the mean life $\overline{T}^{(u)}$ at the use-stress.

The observed mean time at the accelerating-stress is 122.81 hours, so at the use-stress -

$$\overline{T}^{(u)} = 122.81 \times 5.21 = 639.84 \text{ hours.}$$

From all simulated life times at the use-stress we can calculate the actual value of the mean time $\overline{T}^{(u)}$ which equals 611.62 hours.

The observed relative error of estimation is in this case 4,6%. Test time is about 4,4 times shorter compared to the test time required for observation of all failures from the sample tested at the use-stress.

REFERENCES

{1} A.P. Basu, N. Ebrahimi, Nonparametric accelerated life testing, IEEE Trans. on Reliability, vol. R-31, no.5, 1982, 432-434.

{2} E.T. Davydov, Ispolzovanye metoda ravnykh veroyatnostey pri forsirovannykh ispytaniyakh apparatury na nadejhnost, Nadejhnost i Kontrol Kachestva, no.1, 1985, 39-42.

{3} J.D. Kalbfleisch, R.L. Prentice, The Statistical Analysis of Failure Time Data, J. Wiley, New York 1980.

{4} G.D. Kartashov, Forsirovannye ispytaniya apparatury, Nadejhnost i Kontrol Kachestva, no.1, 1985, 18-24

{5} J.P. Klein, A.P. Basu, Accelerated Life Tests under Competing Weibull Causes of Failure, Communications Statist. - Theory Meth., vol. 11, no.20, 1982, 2271-2286.

{6} W.B. Nelson, Graphical Analysis of Accelerated Life Test Data with Different Failure Modes, General Electric TIS Report, no. 73CRD001, Oct. 1973.

{7} W.B. Nelson, Analysis of Accelerated Life Test Data with a Mix of Failure Modes by Maximum Likelihood, General Electric TIS Report, no. 74CRD160, Dec. 1974.

{8} L.Ya. Peshes, M.D. Stepanova, Osnovy teorii uskorennykh ispytaniy na nadejhnost, Izd. Nauka i Tekhnika, Minsk 1972.

BIOGRAPHY

Dr. Włodzimierz LEWIN was born in Warszawa, on April 5th 1947. He received the M.Sc. degree in electrical engineering from the Technical University of Warsaw in 1972 and the D.Sc. degree in reliability from the Systems Research Institute of Polish Academy of Sciences in 1983. At present he works as an Assistant Professor at the Systems Research Institute and at the Electrical Engineering Institute. His research interests include areas of reliability, quality control and engineering statistics.

MICROELECTRONIC COMPONENTS I

Reliability Technology — Theory & Applications
J. Møltoft and F. Jensen (Editors)
© Elsevier Science Publishers B.V. (North-Holland), 1986

Evaluation of Deformations in Chip Surfaces with a
Temperature-Compensating Piezoresistive Sensor

Otmar Selig; Günther Haug; Hermann Weidner; Kurt Wohak
Siemens AG, Corporate Production Engineering, Munich, GFR

A temperature-compensating piezoresistive sensor is described which has been used to characterize mechanical strain in chips as induced by assembly procedures. The stress distribution across the surface due to chip mounting and chip encapsulation has been determined. In the experiments discussed, mounted chips are bent convex (upper surface under tensile strain), encapsulated ones slightly concave. The relaxation behaviour of thermally induced stresses during temperature shocks has been recorded.

1. INTRODUCTION

Up to this decade corrosion has been considered to constitute the most severe reliability hazard in plastic encapsulated devices. This is reflected in the abundance of humidity tests in current standards. Few years ago, however, epoxy resins have been introduced that provide a reasonable shielding of the IC against moisture penetration. Their great impact on reliability testing can be judged from the efforts to create new highly accelerated temperature-humidity stress tests [1, 2].

Apart from lowering the amount of corrosive ions like chlorine in the resins, much of the success of the new molding compounds is due to its tight adherence to the surface of the chip. As a consequence, mechanical stress is transferred to the IC because of the thermal mismatch of the materials. Whereas there should be no problems with small ICs, the strain of the chip may impair the quality of VLSI ICs. In order to cope with this hazard several authors have introduced piezoresistive stress sensors as a probe of chip strain [3-6].

Restricting themselves to plane stresses, Groothuis et al. [7] recently calculated the stress distribution in plastic IC-packages by finite element analysis. Their results - compressive stress in the chip, tensile stress in the epoxy compound, and shear stress along the interface - agree well with early data of this group [3] and explain recent observations of shifts in metallic structures [6, 8].

During our work we observed stress distribution profiles somewhat different to the one reported by [3, 4, 7]. We interpret our data by some bending of the chip which is superimposed to the plane stress. This view is corroborated by X-ray projection imaging.

2. STRAIN GAUGE

2.1 Layout

Like other authors we employed the highly strain-sensitive piezoeffect of ion-implanted semiconductor resistances. The layout has been designed to compensate their large inherent temperature dependence.

The basic sensor structure is a parallelogram of 4 boron-implanted (p-type) 12 kΩ-resistances (180 microns x 30 microns) in a phosphorous-doped n-Si(100)-substrate as shown in fig. 1. Within this crystal plane, piezoresistance coefficients are greatest in the [110]-directions (horizontal and vertical lines in fig. 1) and vanish along the [010]-crystal axis (diagonal lines) [9]. Therefore the diagonally positioned resistance lines do not contribute to the piezosensitivity of the sensor while their dependence on temperature can be employed to compensate the temperature drift of the stress-sensitive resistances.

Fig. 2 presents the layout of our testchip which is densely packed with sensor parallelograms. 24 each of them are anchored to a metallization frame of 4.5 mm x 3.0 mm so that several of them may be connected to the lead frame with one single bond contact. For tests with dies larger than 4.5 mm x 3.0 mm, still greater areas are cut out of the 4"-wafer and individual frames are internally connected by wire bonds.

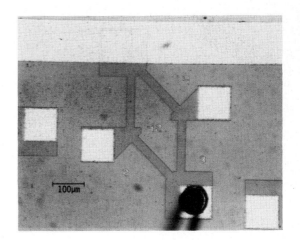

Fig. 1: Single stress sensor consisting of 2 piezoresistive active and 2 inactive branches of 12 kΩ each.

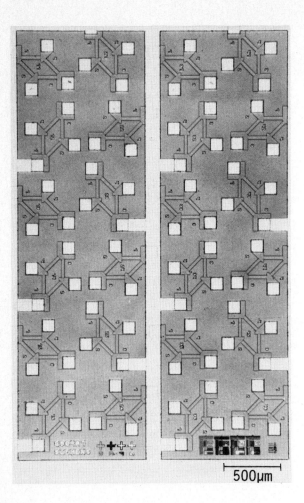

500μm

Fig. 2: Group of 24 stress sensors within one metallization frame.

2.2 Physical properties and calibration procedure

In the (100)-plane of p-silicon the piezoresistances for longitudinal and transverse tensions (with respect to the direction of the resistance path) are comparable in magnitude yet of opposite sign [9]. Therefore the signals of equal compressions in longitudinal and transverse direction (e. g. diagonal strain) almost cancel, while the piezoeffect of longitudinal tension would be enhanced by transverse compression. In other words, the resistance variation of any single [110]-piezosensitive line is the difference of its longitudinal and its transverse strain. Therefore pairs of orthogonal strain gauges do not yield additional information. This should hold for the stress sensor of Spencer et al. [3] as well.

The wiring of the piezoresistances to a Wheatstone bridge provides an effective means of compensating the temperature dependence of the semiconductor resistances. While a temperature rise of 100 K increases the individual resistances by about 50 % it does not influence the voltage across the bridge. The total resistance of the bridge itself (as indicated by the supply current) may even be used as a measure for the chip temperature as it is only slightly influenced by the strain.

There is, however, an additional influence of temperature on the piezoresistance effect itself which reduces the sensitivity of the bridge by about 20 % when the temperature is increased by 100 K. This effect has been calibrated and is readily dealt with compared to the varying output of an individual strain gauge resistance even in the absence of external stress.

Rather than calculating the piezoeffect we calibrated the sensor with stripes of 33 mm x 3 mm. They were mounted on a sharp-edged support, protruding like a lever. When the free end of the beam is moved up or down by a micrometer screw the upper surface of the beam is strained by

$$\Delta l/l = d/2 \rho \qquad (1)$$

where d is the thickness of the beam (e. g. 380 microns) and ρ is its radius of curvature at the position of the sensor parallelogram to be calibrated [10]:

$$\rho = \frac{[1 + x^2 s^2 \cdot (1-u/2)^2]^{3/2}}{s \cdot (1-u)} \qquad (2)$$

with

$$s = 3 \, f/L^2 \qquad (3)$$

and

$$u = x/L \qquad (4)$$

where x is the distance of the sensor element from the edge of the support, L is the free length of the beam between support and micrometer screw, the displacement of which is denoted by f.

Under the conditions used, the numerator in (2) is very close to 1, so that

$$\Delta l/l = d \cdot s \cdot (1-u)/2 \qquad (5)$$

The sensitivity of the sensor is given by the factor K:

$$\Delta R/R = K \cdot \Delta l/l \qquad (6)$$

where $\Delta R/R$ denotes the relative increase of the piezoresistance in response to the strain. The K-value of our sensor is about 50 (depending on temperature) and constant up to strain levels of $\Delta l/l = 10^{-3}$.

2.3 Circuit wiring

In principle, use of the common frame should allow to bond 5 sensor parallelograms to a 16-pin DIL-package. In practice, however, no more than 4 parallelograms can be contacted because of mutual sterical hindrances of the bond wires. Yet it is possible to measure the distribution of strain across the chip surface in even greater detail if one regards each parallelogram as a single piezoresistance composed of 2 parallel tracks. Then 2 such structures may be combined with 2 external 12 kΩ resistances to yield another bridge which responds to strain differences. In fig. 3 the

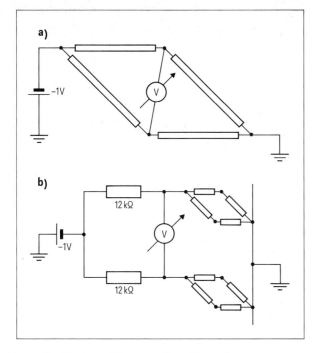

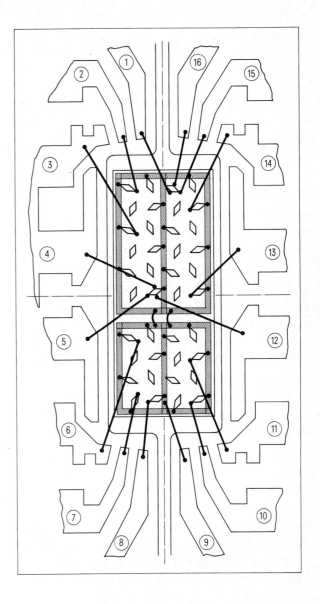

Fig. 3: Wiring diagram for the stress sensor:
Wheatstone bridges for measuring
a) the absolute level of strain
b) the strain differences between
2 stress sensor positions.

Fig. 4: Example of a bond map. Absolute strain
levels are recorded for pin positions 5
and 15, the other sensors may be
related to them.

detailed wiring diagrams for a single sensor
and for a pair of sensors are given. The
sensitivity of the pair is reduced by a factor
of 4, but at the most commonly used supply
voltage of 1 V the signals are still in the
mV-range. Note that again the temperature
dependence of the individual resistances is
compensated by the bridge. Usually we bond
1 or 2 sensor parallelograms in full to obtain
absolute strain values at reference positions.
The remaining contact pins are reserved for
difference measurements. Fig. 4 presents such a
bond map.

3. DEFORMATION OF MOUNTED CHIPS

In a first series of experiments we
investigated the effects of die bonding.
Testchips were fixed with conductive adhesives
to DIL-16 lead frames. After curing and bonding
of the sensor elements the pins were soldered
to ceramic substrates. With a drop of silicone
glue, the frame centres carrying the dies were
flexibly fixed along their short edges to the
adjacent pins. Then all the pins of the lead
frame could be separated from each other by
laser cuts. Such a testchip is shown in fig. 5.
It is both electrically accessible and free of
external stress.

3.1 Distribution of strain

In general the chip surface is strained by
longitudinal tension (with respect to the chip
length). We liken this strain to the bending of
a bimetallic bar where both metals are glued
together at elevated temperature. Upon cooling
the silicon chip with its smaller coefficient
of thermal expansion is bent convexly.

Details of the die bonding process influence
the stress profile greatly. In fig. 6 we
present a map of strain gauge signals of some
experimental samples. There the long edge seems
to be strained more vigorously than the central

line. This is an artefact, however, which can
be explained by the properties of the piezoresistances (see section 2.2): any transverse
tension at the chip centre will reduce the
sensor's output there.

Interpreting these data according to the bimetallic bar model we may try to construct a
grid of bidirectional strains (e. g. curva-
tures). In each point the difference of
longitudinal and transverse tension is given by
the sensor signal. The second condition is
derived from the requirement that the
calculated chip deformations have to be
consistent throughout the grid. The evaluation
yields the deformation profile of fig. 7a where
the longitudinal strains are almost constant
along any transverse line across the chip. This
is consistent to the distortion of a beam under
load.

Mounted dies that are not packaged are readily
accessible for other deformation tests. By
scanning surface topographies of samples from
the same die bonding lot (which had been left
without bond wires) with a surface profiler
(Sloan Dektak IIa) we got similar results.
While fig. 7a represents an idealized view of a
mounted chip based on averaged data, fig. 7b
demonstrates that real distortions need not be
symmetrical with respect to the chip centre.

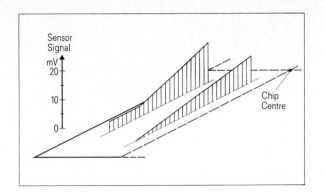

Fig. 6: Strain field within the surface of an
experimentally mounted die (sensor
readings), averaged to one chip
quadrant from the data of 3 samples.

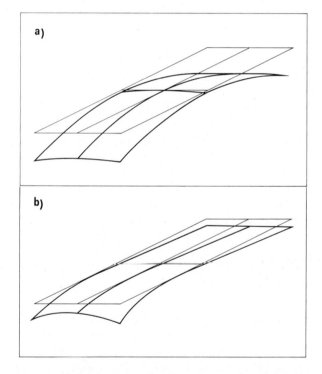

Fig. 7: Chip distortions of experimentally
mounted dies:
a) reconstruction from fig. 6
b) measured by direct surface
scanning of an individual sample.

3.2 Temperature dependence and relaxation behaviour

The variation of the chip strain with temperature is an indicator of the mechanical properties of the die bond. When our samples were
cooled from room temperature down to -55 °C,
chip strains increased by about 25 - 30 %
without gross changes of the overall stress

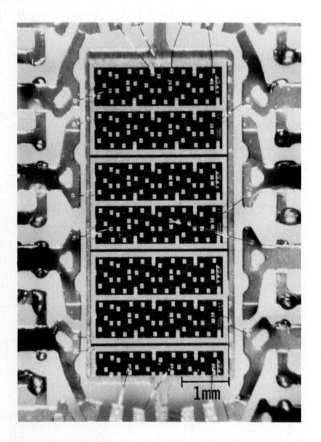

Fig. 5: Testchip as prepared for stress
measurements on mounted chips.

distribution. Upon heating to +150 °C the strains almost vanished.

With time any die bond will exhibit some relaxation. For each die mounting procedure under investigation, we record the strain variations of 5 to 10 piezoresistive sensor chips in the course of several hundred temperature shock cycles (-55 °C/150 °C; dwell time 10 min). Two such relaxation curves are plotted in fig. 8. In these examples, we find a gradual decrease of strain, reaching a plateau of 40 - 60 % of the initial strain values after about 50 to 100 cycles (curve a). Curve b is interpreted by a strong relaxation of the longitudinal tensile strain without an adequate decrease of the transverse strain (the sensor output changes sign). In view of the additional stress which is exerted on the chip by the plastic package, those probably weaker die bonds may ultimately prove to be the more suitable ones.

stress, however, we cannot give a general rule: even within a single lot we observed both local maxima and minima for the strain readings in the chip centres. This would mean that transverse tension or compression, respectively, may be superimposed to the longitudinal compression.

The longitudinal strain distribution is shown in fig. 9a. It agrees well with the results of Spencer et al [3]. Its interpretation by chip bending analogous to section 3.1 would yield distortions as indicated in fig. 9b. Compared to the deformations of mounted, unpackaged ICs they would be larger and in opposite direction.

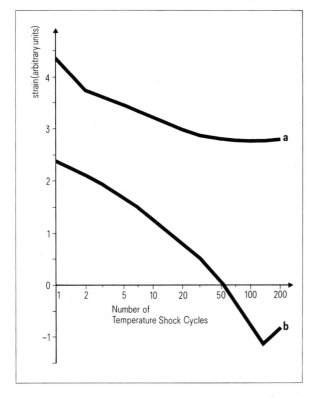

Fig. 8: Long term stress relaxation of mounted dies. Curves:
a) tight adhesion
b) preferential relaxation of longitudinal strain.

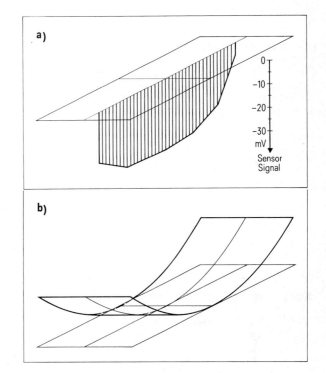

Fig. 9: Longitudinal strain within an encapsulated IC:
a) $\Delta l/l$ sensor-readings along the long axis of the chip
b) resulting chip distortion if all the stress were attributed to chip bending (neglection of isotropic compression)

4. DEFORMATIONS OF PLASTIC-ENCAPSULATED ICs

4.1 Stress distribution

Our most extensive data were obtained when we investigated the influence of several parameters of the encapsulation process. All our samples shared one common feature: the plastic package overrides the tensile stress of the chip surface which has been induced by the die bonding process, thus conversing the stress to a longitudinal compression. For the transverse

While mounted dies may readily be likened to bimetallic bars, the epoxy encapsulant is expected to exert some isotropic compression to the chip. We therefore cannot assume the interpretation of fig. 9b to be realistic. On the other hand, we do not see the reason why some samples should be without transverse compressive strain. It is conceivable then, that at least part of the observed strain of the chip surface has to be attributed to concave chip bending. In fact we could visualize the chip deformation by X-ray microfocus projection imaging. Upward bending of the chip corners seems to occur to about half the distortion that would be suggested by the sensor measurements.

4.2 Temperature dependence and relaxation behaviour

As in mounted chips the strain in encapsulated ICs increases with decreasing temperature, reaching levels of 0.1 % in extreme cases for Δ l/l at -55 °C. Again the samples were almost free of stress at 150 °C.

Stress relaxation was followed by continuous monitoring during more than 1000 temperature shock cycles (-55 °C/+150 °C; dwell time 10 min). As plotted in fig. 10, we observed some increase of strain by 3 - 20 % during the first 8 cycles. After cycle No. 20 the strain started to decrease again, crossing the initial level after about 60 cycles. Finally, between cycles No. 200 and 1000 the strains leveled off at plateaus another 15 - 30 % lower. One low-stress epoxy compound should be mentioned as an exception (curve d in fig. 10): starting strain levels were lower by 40 % compared to the other materials, yet during the first few cycles the strain increased more rapidly without relaxing as much subsequently. Therefore, at the end of the test the initial advantage was reduced by about 50 %.

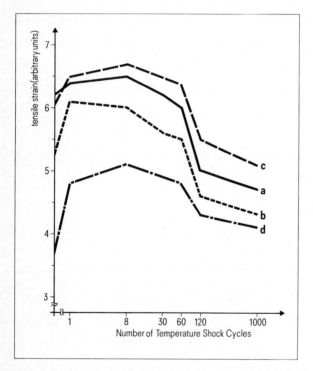

Fig. 10: Stress relaxation at the centre of encapsulated ICs. Curves: a) epoxy A, b) epoxy A, slow curing at 100 °C, c) epoxy B, d) low-stress epoxy compound.

5. DISCUSSION

ICs in plastic packages are submitted to two different mechanical stresses.

First, due to its bonding to the lead frame, the chip is bent to a convex shape. ICs that are sealed in hermetic packages are left with their active surface under tension. This surface strain is related to some shear stress in the adhesion layers where splintering of the chip might then originate.

Matters change completely when the IC is encapsulated within an epoxy package. Its higher coefficient of thermal expansion transfers compressive stress from all directions to the chip. In contrast to Spencer et al. [3], however, we did not find it to generally culminate towards the chip centre. While there is always a maximum of longitudinal compressive strain, we did find several samples where the transverse compressive strain seems to decrease towards the chip centre (the argumentation is somehow indirect - as it should be with Spencer's data - because of the physics of the piezoresistive effect). This observation led us to check for some chip bending in addition to the isotropic compressive stress. We could actually visualize it by an elaborate X-ray projection technique. It may account for about half of the measured strain levels.

At the first glance it may seem odd why the chip should be bent by the epoxy which surrounds it almost symmetrically. There are, however, differences in the bonding strength of the epoxy resin to the passivation layer of the upper chip surface and to the metal of the lead frame underneath: it is generally much easier to fracture the IC-package along the rear side of the lead frame than to mechanically separate the epoxy from the chip. Accordingly the longitudinal compression due to the epoxy's thermal contraction is primarily transferred to the upper chip surface. Transverse stress patterns are expected to be more complex because of the large ratio of package to chip width and because of the shapes and positions of the pins within the package.

ACKNOWLEGDEMENT

We thank Mr. v. Borcke, Mrs. Sprenger and Mr. Römer for the mounting and encapsulation of the testchips and for many stimulating discussions.

REFERENCES

[1] J.E. Gunn, S.K. Malik and P.M. Mazumdar, Highly Accelerated Temperature and Humidity Stress Test Technique (HAST), IEEE Ann. Proc. Reliability Physics, vol. 19, 1981, 48-51.

[2] T. Yoshida, T. Takahashi and S. Koyama, A New Accelerated Test Method for Moisture Resistance of Plastic Encapsulated LSIs, IEEE Ann. Proc. Reliability Physics, vol. 20, 1982, 268-271.

[3] J.L. Spencer, W.H. Schroen, G.A. Bednarz, J.A. Bryan, T.D. Metzgar, R.D. Cleveland and D.R. Edwards, New Quantitative Measurements of IC Stress Introduced by Plastic Packages, IEEE Ann. Proc. Reliability Physics, vol. 19, 1981, 74-80.

[4] W.H. Schroen, J.L. Spencer, J.A. Bryan, R.D. Cleveland, T.D. Metzgar and D.R. Edwards, Reliability Tests and Stress in Plastic Integrated Circuits, IEEE Ann. Proc. Reliability Physics, vol. 19, 1981, 81-87.

[5] R.J. Usell and S.A. Smiley, Experimental and Mathematical Determination of Mechanical Strains within Plastic IC Packages and Their Effect on Devices During Environmental Tests, IEEE Ann. Proc. Reliability Physics, vol. 19, 1981, 65-73.

[6] K. Kuwata, K. Iko and H. Tabata, Low Stress Resin Encapsulants for Semiconductor Devices, Proc. Electronic Components Conference, vol. 35, 1985, 18-22.

[7] St. Groothuis, W. Schroen and M. Murtuza, Computer Aided Stress Modeling for Optimizing Plastic Package Reliability, IEEE Ann. Proc. Reliability Physics, vol. 23, 1985, 184-191.

[8] R.E. Thomas, Stress-Induced Deformation of Aluminum Metallization in Plastic Molded Semiconductor Devices, Proc. Electronic Components Conference, vol. 35, 1985, 37-45.

[9] Y. Kanda, A Graphical Representation of the Piezoresistance Coefficients in Silicon, IEEE Trans. Electronic Devices, vol. ED-29, 1982, 64-70.

[10] G. Berger, Festigkeitslehre im Maschinenbau, Vogel-Verlag, Würzburg, 1976, 82-83.

Otmar Selig was born in 1933 in Neuhof (CSSR) and studied physics at the Technical University of Munich, where he received the PhD in 1963. Afterwards he had been engaged at the Institute for Radiochemistry there with the application of radioactive nuclids for technical research. 1965 he went to Farbwerke Höchst AG, where he was concerned with problems of the fabrication of PVC-foils. Since 1971 he is working on materials for the electronic industry of Siemens AG, Munich.

Günther Haug graduated from the University of Würzburg in 1960 with Dipl. Phys. and in 1964 with Dr. rer. nat. and has been employed by Siemens AG since 1964. In different departments for technology development he worked in the field of electronic engineering.

Hermann Weidner studied physics at the University of Munich where he received a PhD in 1973. He did research in physical biochemistry at the Universities of Basel, Yale and Munich before he joined the Siemens Company in 1978. Since 1982 he heads a laboratory there which is concerned with reliability aspects of plastic encapsulated devices.

Kurt Wohak graduated from the University of Vienna in 1958 with Dr. phil. as a physicist and has been employed by Siemens AG since 1958. In different departments for technology development he worked in the field of electronic engineering.

Reliability Technology — Theory & Applications
J. Møltoft and F. Jensen (Editors)
© Elsevier Science Publishers B.V. (North-Holland), 1986

RELIABILITY METHODOLOGY ON TEST VEHICULE
TO QUALIFY PROCESS AND CELL LIBRARIES

Michel POTIN
MATRA HARRIS SEMICONDUCTEURS (M.H.S.)
BP 942 ROUTE DE GACHET
44075 NANTES CEDEX

A novel approach is used to qualify VLSI gate arrays and standard cells. It becomes very long and expensive to qualify each new design done in a gate array or standard cell family. A methodology is proposed. Main aspects are process qualification through test vehicule and validation of the cell library.
Test structures are used to evaluate failure mechanisms. This study enables to establish acceleration laws versus several parameters. This paper presents the different steps of the methodology - test vehicule design - process validation - life estimation and parameters variations introduced in electrical rules.
This proposed methodology shortens the study cycle of a new product. Based on mutual confidence and data, the circuit is considered as being qualified after electrical validation.

1. INTRODUCTION

The development of a new generation of custom or semi-custom possibilities (high level gate arrays, standard cell or silicon compiler) increases drastically the number of new designs and monolothic circuits.
As a consequence, it seems imperative to rethink the procedures for assessment of reliability. The previous approach was based on standard test like life test at 125°C or other high temperature, under bias conditions. To perform such reliability testing on each circuit becomes each day more and more difficult and costly to implement. Several factors as the device complexity request large dynamic pattern to cover the maximum number of the transistors. The number of pins increases, so the number of lines on printed board follows and the density on board is small.
Such a development in this hypothesis must be done by circuit, inducing a set of printed board for each design.

We see that this approach is not realistic on custom or semi custom design. A confident failure rate, established from life test at 125°C during 2000 hours implies a minimum of 200 devices (i.e 400 000 device-hours) to get a significant number.

In another way, this testing is a constat but does not cover the weakness of the products (in term of technology, design) so confidence is reduced. It is now well known that high temp is not the only critical parameter. Device scaling induces new effects as hot electrons, accelerated at low temperature. So a change in reliability approach is necessary.

2. APPROACH
2.1 General

We have seen that the classical approach will no longer be valid on such VLSI products.

Our proposition is developped in the following sheets.

The suggested approach is a supplier view. The objective is to anticipate the problems by introducing the reliability as a starting point of the project.

Our purpose is reliability assurance by evaluation of potential failure mechanism on a process test vehicule and by validation of the cell libraries on another vehicule representative of the design.

This method is necessary. The customer wants the shortest study and qualification cycle. Our purpose is :
- First, to try to prevent any potential reliability concern
- Second, to supply, in the discussion with the customer, a complete set of reliability data, with sufficient strength to get the mutual confidence.
The customer circuit qualification could be completed by electrical characterization.

2.2 Application

This approach is developped around a new gate array family (2700, 4000, 5000 gates) processed on a 2 microns CMOS technology with two metal layers.

The methodology described addresses process and design aspects.

The reliability strategy covering package, die and package associativity, testability and software is not included in this paper.

3. METHODOLOGY

Diagram of the methodology is layed out figure n°1.

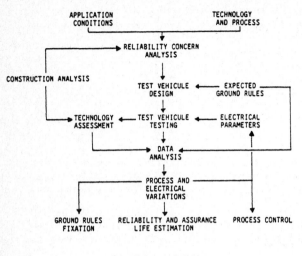

APPLICATION   TECHNOLOGY
CONDITIONS   AND PROCESS

RELIABILITY CONCERN
ANALYSIS

CONSTRUCTION ANALYSIS

TEST VEHICLE   EXPECTED
DESIGN   GROUND RULES

TECHNOLOGY   TEST VEHICLE   ELECTRICAL
ASSESSMENT   TESTING   PARAMETERS

DATA
ANALYSIS

PROCESS AND
ELECTRICAL
VARIATIONS

GROUND RULES   RELIABILITY AND ASSURANCE   PROCESS CONTROL
FIXATION   LIFE ESTIMATION

FIGURE N° 1 : METHODOLOGY

3.1 Reliability concern analysis

This work is very important. From the basic technology choice and application condition, helped by first physical and construction analysis, the reliability man must fix the potential concern of the product family. The background on similar technologies and litterature is helpfull to determine concerns and failure mechanisms attached.

Basic failure mechanisms like contamination, oxide integrity and electromigration are taken into account . Specific phenomena should be reviewed :
- Hot electrons degradation
  Hot electron injection is relative to small dimensions. Origin could be the channel (channel hot electrons) or the substrate (generated by the substrat current).
- Contact degradation - smaller contact size increased contact resistance and new materials are used.
- This is a double metal layer structure. So via reliability is also a concern.
- Metal : we perform analysis of electromigration and also of internal stresses due to several factors like small width (same order of magnitude than grain size) and topology.
- Oxide reliability
  . surface leakage or inversion under thin or thick oxide
  . inter level leakage by material degradation, charge movement

This analysis gives us informations for test vehicle design and determines the choice of structures.

Two ways are chosen : individual structures for failure mechanism understanding and simulation die with cells from libraries to confirm trends on physical designs.

3.2 Test vehicle designs

Two test circuits are drawn, the first one, process orientated, consists in :
- transistors (drawn with several lengths and widths)
- capacitors (thin oxide and thick oxide)
        maximum size for contamination
        minimum size in parallel
- metal lines of different widths and on various topographies
- several contacts or vias configurations
- parasitic structures

Process test vehicule - figure n°2

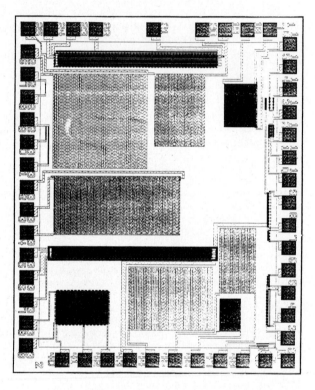

Second part of the study is implemented on a test vehicle, representative of the final family. It is the basic matrix of the GA 5000 with all I/O cells implemented.

The major cells of the libraries are lay-outed on this silicium.

3.3 The test

3.3.1 The test of process vehicule
Now, the objectives are :
- to determine sensitivity of device to failure mechanism

- if any exists, to extract the acceleration factors

We assume test vehicle design representative of process/design variations.

Tests are performed at different levels of the stress parameters following a matrix.

For instance, if we take the electromigration, we will build a two factors matrix (temperature, current density).
The model used is the Black Law (1)

$$\tau = AJ^n .e^{\frac{-EA}{RT}} \quad (1)$$

J = current density in A/cm2
n = current acceleration parameter
EA= activation energy
A = cste
The matrix is :

J \ T	J1	J2
T1		X
T2	X	X

Three conditions are enough to verify the n and EA parameters of the law (1). If we wish to verify the model, the matrix must be extended to two extra conditions (T3, J2) and (T2, J3).

The stress levels are chosen above use conditions to accelerate the failure mechanism but must be representative of correct behavior of the device.

- Temperature is generally taken in the 125°-175°C range, compatible with device junction temperature.
- Voltage and/or current are determined from characterization. Value of stress must be representative of the normal behavior of the device (i.e, transistor drain voltage kept under the avalanche value).

In each matrix condition, the sample size is 30 devices from 3 different diffusion lot numbers to allow statistical analysis.

So far, 2000 hours stress time is arbitrary chosen. But this duration could be extended if no sufficient number of failures occur during this fixed time.

Tests of hypothesis are performed to compare stress conditions and assure that failures are created by the same failure mechanism

Failure analysis could confirm failure mechanism.

These assumptions verified, we can compute the specific parameters of the mechanism like the activation energy and calculate the acceleration factors (temperature, voltage, current).

The following figure (n°3) summarizes the different phases of the study.

FIGURE N°3

FAILURE MECHANISM EVALUATION FLOW CHART

FIX POTENTIAL CRITICAL PARAMETERS (temperature, tension)

AJUST PARAMETER LEVELS

BUILT A TEST MATRIX

CONDUCT TEST

PLOT FAILURE TIMES ON DISTRIBUTION GRAPHIC PAPER

TEST THE LAW - EXTRACT LAW PARAMETERS

VERIFY ACCELERATIONS OF MECHANISM EQUAL FOR EACH STRESS LEVEL

COMPUTE ACCELERATION OF FAILURE RATE IN USE CONDITION

3.3.2 Test of design vehicule

This testing is complementary of the analysis of structure. It allows to verify the functionnality of cells over time without drift of characteristics or failures.
To verify the different assumptions, this test vehicule is stressed (see figure n°4).

FIGURE N° 4

DESIGN TEST VEHICULE AND LIFE TEST CONDITIONS

STRUCTURE	BIAS
BUF IN CMOS TTL CMOS SCHMIT TRIGGER TTL	- Stuck to Vss
BUF OUT STD 3 STATES POWER	- Square wave on input
N CH TX P CH TX	- Output connected for Iout = 4mA and 8mA for power
Ring oscillator	- High frequency
RAM - ROM	- Dynamic signals but output non connected
ALU	- Idem
Operator chains	- Drive by internal oscillation

Objectives of design vehicle testing are :
- validation of cells (simple or macro)
- qualification of design and lay-out
- qualification of process and ground-rules at 125°C, 150°C and low temperature.

A photo of the design test vehicle is supplied figure n°5.

4. DATA ANALYSIS

Two separate ways of analysis are investigated to answer the following needs :
- to calculate at use conditions a failure rate of each individual mechanism
- to fix for circuit generation and further production, the absolute limits of critical parameters and take into account drifts in electrical design package to insure device life compatible with objectives (10/20 years).

4.1 Failure rate calculation

From raw data, we match the failure/time point with a repartition function F(t). Each failure mechanism follows a statistical distribution. A log normal distribution (parameters μ, σ) assumes a wear-out mechanism and a exponential distribution (parameter m) a random phenomenon.

This could be done by calculations. The use of gausso-log normal plotting paper (for log normal low) is easier. On it, the percentage of failure is linear with the log of the time and distributions of parameters are achieved graphically (t50% failures) indicate the u parameter and the sigma of the distribution is equal to:

$$\sigma = \log \frac{(t50)}{(t16)}$$

After hypothesis tests and if the hypothesis of same mechanism is verified at each stress level, we can determine the different acceleration factors.

$$Facc : \frac{t1\ 50\%}{t2\ 50\%}$$

All these parameters allow to calculate the failure rate in the use conditions by using the calculation formula.

For tu $\ll$ tu50,

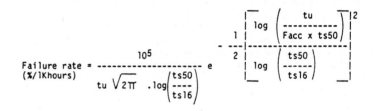

$$\text{Failure rate} \atop (\%/1Khours) = \frac{10^5}{tu\ \sqrt{2\pi}\ .\log\left(\frac{ts50}{ts16}\right)}\ e^{-\frac{1}{2}\frac{\left[\log\left(\frac{tu}{Facc\ \times\ ts50}\right)\right]^2}{\left[\log\left(\frac{ts50}{ts16}\right)\right]}}$$

tu   = lifetime in use
tu50 = time at 50 % failure in use
ts50 = time at 50 % failure in test
ts16 = time at 16 % failure in test

Next step is the summarization of failure rates of all identified failure mechanisms to compare to objectives.

4.2 Research of absolute limits

This approach described above informs on the reliability level of the technology. However, the purpose of this study, in case of non success of test is to understand the failure. The mechanism could be general. In this case, the technology has to be reviewed.

But often, failures could be correlated to a part of the characteristics of the distributions. In this case, the work is to find and assure limits of design rules or electrical/process rules to reach the reliability objective.

For this purpose, a device identification is done during the testing. This allows to follow individual parametric evolution. The knowledge of parameter value (physical or electrical dimension) inducing inacceptable drifts or failures allows to bring corrective action on the process or design electrical package and fix dimensions or parameters limits to enhance the reliability of the product.

As an illustration of this methodology, we can supply some preliminary informations regarding hot electrons. Hot electrons are considered as a reliability risk in small length transistors. A test matrix, built after a theoritical analysis and review of litterature on the subject shows the devices are not affected by any drift for an effective length minimum estimated for the process under review.

From this result, we have two major conclusions :
1. Minimum polysilicon dimension and effective length must be measured by process control and central lines and absolute lower limits (statistics) of these parameters must be kept over the risk values.
2. In these conditions, useful life is superior to 20 years.

4.3 Design vehicule analysis

Failures are analysed and data from process test vehicule (specific acceleration factor by mechanism) are used to estimate cell reliability. This part of the program is now in progress. Evolution will be given later.

CONCLUSION

We have described our methodology for reliability assessment of VLSI built on a gate array family or more generally on a circuit library (standard cell, silicon compiler). Through this approach, linked with parallel qualification of packages, for these large dies, we think we could assure product reliability without long and costly unitary qualification of the customer circuit.

The development of a test vehicule representative of the technology/design and his reliability qualification must be the support of the supplier/customer mutual confidence. Access to the results and conclusions of this qualification conforts the customer in his choice by a knowledge of physical testing of same structures built in the technology. In the next future, our intention is to provide from design vehicule analysis a confident way for reliability prediction of customer circuit.

BIBLIOGRAPHY

- Les techniques d'homologation de circuits intégrés VLSI (MOS) - L'approche IBM - Colloque International Qualité Composants Bordeaux 1983 - Michel Bianchini

- Méthodes d'évaluation de la qualité des circuits VLSI et en particulier des réseaux prédiffusés -
A. Baiget, A. Labarthe CNES Toulouse -
4è International Conference Reliability and Maintenability - Perros-Guirec - 1984

MICROELECTRONIC COMPONENTS II

Reliability Technology — Theory & Applications
J. Møltoft and F. Jensen (Editors)
© Elsevier Science Publishers B.V. (North-Holland), 1986

DIELECTRIC RELIABILITY AND BREAKDOWN OF THIN DIELECTRICS

D.R. Wolters and A.T.A. Zegers-Van Duynhoven
Philips Research Laboratories 5600 JA Eindhoven - The Netherlands

ABSTRACT

The use of extreme value statistics, i.e. Gumbel I and III distributions, is advocated for the characterization of dielectric breakdown of thin dielectrics. It is shown that for defect-related breakdown an excellent fit can be observed when the mean number of defects present in each test sample is sufficiently large.

A strong correlation is found between distributions of defective capacitors measured with different testing techniques. The results of the testing techniques can be compared when they are plotted with the injected charge as variate. The dependence of the lifetime on injection currents shows furthermore that charge transport through the capacitors is the driving force for breakdown.

Using this concept, the hazard rates for field-dependent and time-dependent breakdown can easily be derived and they are shown to fit the requirements of the Gumbel I and Gumbel III extreme value distributions. Their applicability is further corroborated by the observations that the dielectric strength increases semi-logarithmically with the ramp rate of the field and the charge for breakdown depends on a power law of time.

Furthermore, it is found that the hazard rate for breakdown by constant current does not depend on time but only on the magnitude of the current. It is proposed to perform screening tests with constant currents.

I. Introduction

The increasing demand for automation in the post-industrial society lays a heavy load on the micro electronic industries to produce low-cost and reliable integrated circuits.

The drive toward lower-cost production of multicomponent devices with increasing complexity forces the manufacturer to continually reduce the device dimensions [1]. This holds for the lateral dimensions of the circuits on the semiconductor substrate but also for the layer thickness of the dielectric insulators. These have been scaled down by a factor of five in the last ten years.

Nowadays devices with a dielectric layer thickness of 20 nm or thinner are commercially available. With a standard operation voltage of 5 Volts a field of 2.5 MV/cm is applied across these devices. The extremely high field is about 50 times as high as in high-voltage cable insulators. This results in yield loss and reliability problems and is therefore of great concern to any manufacturer.

It is of utmost importance to thoroughly unders-

tand the failure mechanism at these high fields. A sound model leads to the proper statistical tools necessary to predict the yield and the end of life of integrated circuits.

In many investigations the log-normal failure distribution is used and although often a satisfactory fit of the data is observed, these statistics are less appropriate [1-16]. The main points of criticism are [13]:

1) The hazard rate h of the time-dependent failures obeys the power law

$$h \propto t^{\alpha}, \qquad (1)$$

(where α is a constant) rather than the hazard rate for the log-normal distribution which is in good approximation given by [17]:

$$h \propto t^{-1} \cdot \log t \qquad (\log \frac{t - \bar{t}}{\sigma} > 2). \qquad (2)$$

Of course if $\alpha \cong -1$, Eq. (1) and Eq. (2) will be both in close agreement at large t.

Crook [2], who advocates the use of the log-normal statistics, however presents graphs of the hazard rate versus time that demonstrate the validity of Eq. (1) over many orders of magnitude (e.g. $\alpha = -0.9$ and $\alpha = -0.8$ over 12 orders of magnitude).

2) More important is the fact that the log-normal statistics fail to offer a physical model for breakdown. Normal distributions are expected in the case where failures are caused by a large number of stochastic variables. In log-normal distributions we must consider the logarithms of the variables. Therefore the failure is expected to be caused by stochastic variables which multiply [17]. A realistic model predicting such a multiplication of variables has not been presented. A model presented by Metzler [18] does not yield a proper fit without assuming a tailored distribution of injection sites.

It will be shown here that the use of extreme value statistics is justified by a realistic model as well as by experimental results. It will be shown that the hazard rates for time- and field-dependent breakdown exhibit the proper time and field dependence.

The condition of unlimited variates necessary for extreme value distribution will be verified. The stability postulate as predicted by the extreme value statistics is demonstrated and it has proved to be a convenient tool in the prediction of yield and reliability [12-15, 19].

II. Testing techniques

The samples used in the following are all made by thermally oxidising mono-crystalline silicon at 950 °C in an ambient of oxygen. By varying the duration of the oxidation the thickness of the grown silica can be varied in a well-defined way. The homogeneous layer thickness showed variations of less than 1%. The electrodes deposited on the dielectric consist of magnetron sputtered Al or of vapour-deposited poly-crystalline silicon. Testing occurred in an automatic testing station (Fairchild or Keithley 350 system) directly on the silicon wafer. Electrodes of 0.5 mm^2 and of 2.10^{-2} mm^2 were defined by lithographic means. Three essentially different test methods are commonly used to induce a dielectric breakdown of the sample:

Field-dependent breakdown
The voltage of the source is ramped in steps and at each step the current is measured. At an abrupt increase above a preset level of the current the sample is regarded as broken down. Often a second low voltage test is performed to check whether the sample is short-circuited or not.

Time-dependent breakdown
A constant voltage (applied field) is sustained and the current is registrated until it increases above a certain preset level.

Current-induced breakdown
A constant current is sustained through the device and the voltage (applied field) is registrated until it drops abruptly and the sample is regarded as being broken down.

III. The use of extreme value statistics

Extreme value statistics is expected to be useful in the case were the distributed property depends on the largest or smallest value of a whole distribution of values. Mechanical fracture of materials but also dielectric breakdown are typical weakest-link processes and are therefore appropriate candidates. As with mechanical fracture, dielectric breakdown depends on intrinsic material properties but also on the occurrence of defects or weak spots.

It will be shown that in most cases sufficiently large-sized distributions are composed of two distinct components which we shall indicate by "defect-related" and "intrinsic" distributions. In the following it will be indicated that for both components a justification of the extreme value statistics can be furnished.

III. 1. *Defect related breakdown*
Let us assume that:
A) A device will fail at some defect which acts as weakest spot.
B) The average number of defects is so large that each device may contain a number of defects.
C) The properties of the defects are distributed along a variate x which is not limited, hence $-\infty < x < \infty$.

The probability that a sample stressed under the condition given by the variate x will fail in the next interval dx is given by the hazard rate,

$$h = \frac{1}{1 - F(x)} \frac{dF(x)}{dx}, \qquad (3)$$

where $F(x)$ is the cumulative probability of failure at x. For a sample containing a whole initial distribution of defects $Di(x)$ where each defect fails at a specific value of x, h will be equal to the probability that the defect with the smallest value will give a failure between x and $x + dx$. The crucial point is now that in most cases the tail of $Di(x)$ at low values of x is given in good approximation by exp(x), independent of the shape of Di(x). The proof of this statement is given by Gumbel [20] for a number of distributions like the normal, log normal, exponential etc. Assuming that this holds we can write

$$\frac{1}{1 - F(x)} \frac{dF(x)}{dx} \cong \exp(x) \, , \qquad (4)$$

and upon integration and conversion we find

$$\ln \ln (1 - F(x))^{-1} \cong x \, . \qquad (5)$$

When Di(x) is small, Eq (4) will be only a crude approximation. An essential condition is that x is unlimited hence the hazard rate is zero at $x = -\infty$ [20].

A convenient property of extreme value statistics is that they obey the so-called "stability postulate". The shape of F(x) is independent of the size of Di(x) except for a linear transformation in x. The size of Di(x), for instance, can be changed by increasing the area of a test capacitor or by taking a number of test capacitors in parallel.

This property is indicated in literature on breakdown as the area effect [19]. It causes the decrease of dielectric strength with increasing electrode area. The property is unique for extreme value distributions.

III. 1.1 *Field-dependent breakdown*

An example of the use of Eq. (5) and of the stability postulate is given in fig. 1. A batch of

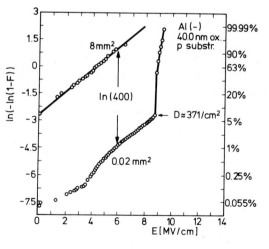

Figure 1.

Field-dependent breakdown distribution of 12000 capacitors (lower curve) and the smallest value of each group of 400 (upper curve). The full line of the upper curve is calculated from the values of the mean and the standard deviation of the 30 minima.

12000 capacitors has been tested with a field-dependent breakdown procedure. The aquired data are plotted according to Eq. (5). Where the cumulative probability F(E) is taken equal to the cumulative failure fraction F'(E) given by [20]:

$$F'(E) = \frac{n}{N + 1} \, , \qquad (6)$$

where N is 12000 and n the number of failed capacitors at or below field E. The distribution of 12000 capacitors shows up as a kinked curve in fig. 1. The steep part of F' has practically no dispersion and corresponds to the fraction of samples where an intrinsic breakdown takes place. The lower part is widely distributed and is believed to correspond with the fraction of samples where a defect-related breakdown takes place.

Obviously the lower part is not quite a straight line although deviations are only small. The data of the initial distribution have been taken together in 30 groups of 400 and the lowest value of each of the 30 groups is taken as representative for the group. A device containing 400 capacitors will fail when this lowest value is reached.

Plotted in the same fig. (1), the points are ordered along the theoretical line given by

$$\ln \ln (1 - F)^{-1} = \frac{\sigma_N}{S_v} (E - \bar{E}) - y_N \, , \qquad (7)$$

where E is the applied field at breakdown, $\bar{E}$ the mean value of E, S_v is the standard deviation of the 30 values, σ_N and Y_N are theoretical quantities, only dependent on N and given in table 6.2.3. by Gumbel [20]. For $N \to \infty$, $Y_N \to 0.577$ (Eulers Number) and $\sigma_N = \pi/\sqrt{6} = 1.28....$

The fit of the 30 points to the theoretical line is so good that no control curves are necessary. The shift between the lower and the upper curve is about ln(400). This corresponds with the increase of the probability of finding a defect among the group of 400. It holds as long as the defects are distributed homogeneously. When defects are clustered the shift is less [15].

This shift of the distribution is a convenient tool since it can be used when the yield or reliability of a multi-component device or large-area device has to be predicted. The shift of the straight line in horizontal direction is predicted by the stability postulate.

The kink in the lower curve can be used to calculate the defect density. Let us first assume that defects are distributed randomly over the area. The probability that defects, failing at or below x (E or ln(t)), are present in a certain area is given by the Poisson distribution

$$F = \sum_{m=1}^{\infty} \frac{\lambda^m \exp(-\lambda)}{m!} \, , \qquad (8)$$

where λ is the mean number of defects in each sample. The probability that no failure occurs at x is then

$$1 - F_{(m \neq 0)} = F_{(m=0)}$$

or

$$1 - F = \exp(-\lambda) \, . \qquad (9)$$

Taking $\lambda = A.D$ where A is the area and D is the defect density, Eq (8) can be conversed into

$$\ln \ln(1 - F)^{-1} = \ln(A.D) \, . \qquad (10)$$

Of course the number of defects depends on the value of x at which they are detected. The highest value of D which is still measurable is at the point where the intrinsic distribution starts. Although D might be defined at an arbitrary value of x it is convenient to take this maximum. For a field ramp as used in fig. 1 we take

$$D = - \frac{1}{A} \ln(1 - F_i) \, , \qquad (11)$$

where F_i is the probability of failure at the kink. For the distribution of 12000 capacitors with $A = 2.10^{-4}$ cm^2 we find $D = 361$/cm^2. The mean number of defects in the single capacitors $\lambda_1 = 0.07$ is too small to expect a straight line. However, for the groups of 400 capacitors we find $\lambda_2 = \lambda_1 \cdot 400 = 28$ defects per capacitor.

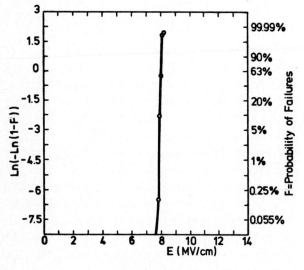

Figure 2.
Field-dependent breakdown distribution of 16000 capacitors without a low field breakdown tail. The fact that not always a tail is present illustrates that these are probably due to defect-related breakdown.

As is illustrated by the excellent fit of Eq (7), λ_2 in this case is sufficiently large.

The occurrence of defects is presumed but not really proved. The best way to make this at least plausible is to show that under some conditions these tails of the distributions vanish. This is demonstrated in fig. 2 where the results of field-dependent breakdown tests of N = 16000 capacitors tested on one wafer are plotted.

The number of defective capacitors is zero and therefore the value of D < 0.3 /cm^2 (since AD.N < 1).

III. 1.2 *Time-dependent breakdown*

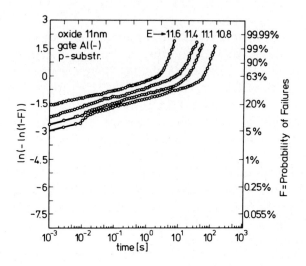

Figure 3.
Time-dependent distributions for four different applied fields. Each distribution consists of 900 test samples.

Fig. 3 shows data resulting from time-dependent breakdown tests. Four different constant fields are applied across four sets of 900 capacitors. The results in fig. 3 are ordered along four kinked distributions which all appear to have approximately the same defect density D as defined by Eq. (11). The value of D ranges between $3.5 \cdot 10^3$ and 5.10^3/cm^2. As a check 16000 of the capacitors on the same wafer were submitted to a field-dependent breakdown test as well. The results are shown in fig. 4. As can be seen the value of $D = 2.7 \cdot 10^3$/cm^2 from the last-mentioned test is close to those for the time-dependent breakdown. This correspondence shows that defects in field-dependent tests cause early breakdown in time-dependent tests and vice versa.

Furthermore, it is remarkable that the intrinsic part in fig. 3 shifts over almost one and a half order of magnitude when the field is changed over

0.8 MV/cm. This indicates that for lower fields down to 5 or 6 MV/cm the intrinsic componentsof the distribution will require impractically long stress times for each breakdown. This is the reason that the intrinsic time-dependent distributions are seldomly observed at low constant fields.

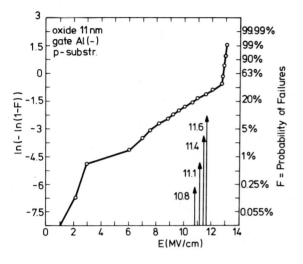

Figure 4.
Field-dependent distribution of 16000 capacitors from the same batch (wafer) as tested in fig. 3. Note that approximately the same defect density is observed.

III. 1.3 *Current-induced breakdown*
Sustaining a constant current through a capacitor is the most convenient way for a breakdown test.

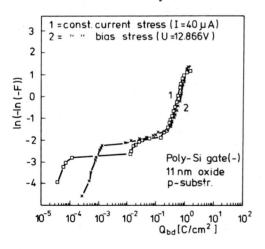

Figure 5.
Distributions of breakdown results tested on the same wafer. The value of the total charge transported through the capacitor is chosen as the variate. 1) Breakdown by constant current stress. 2) Breakdown by constant bias stress.

Fig. 5 shows two distributions of capacitors. Those of curve 1 were submitted to a constant-current stress and those of curve 2 were submitted to a constant-field stress. In order to make it possible to plot them in the same figure, the time integral of the current is plotted. The variate is therefore the charge needed for breakdown and it is seen that both distribution functions closely match. This indicates that here also the mechanism for breakdown does not depend strongly on the test technique. Furthermore it shows the importance of the integrated charge injection for breakdown. This charge injection is very easily acquired when the constant current technique is used. The value of the injected charge Q is directly proportional to the total stress time t_{bd}. For a large number of Si wafers this constant-current technique has been used to measure distributions as in fig. 5 curve 1. The values of the intrinsic t_{bd} distributions have been plotted against the constant current density in fig. 6. The general trend of the t_{bd} value is to increase inversely with J. From

$$J \propto t_{bd}^{-1},$$

we may conclude that probably

$$J \cdot t_{bd} = Q_{bd} = \text{constant} .$$

As has been checked for many wafers the value of Q_{bd} is indeed practically constant at low values of J or E, although above a critical value of J(E), Q_{bd} can be substantially lowered [15]. This can be inferred from the kinks in the curves in fig. 6.

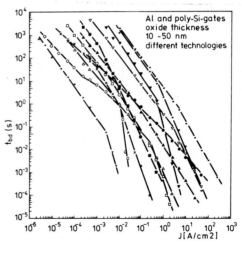

Figure 6.
Compilation of t_{bd} measurements at constant injection current J plotted on a double logarithmic scale. The parameters are the various electrode materials, oxide thickness and the technological treatments. Note the slope $\alpha \cong -1$, at low values of J.

III. 1.4 Driving force for breakdown

Without going into details we can state here that breakdown is a typical result of charge transport through the dielectric [15]. The energy dissipated by the charge is damaging the dielectric in a continuous way resulting in a low-ohmic path way after sufficient transport has taken place. By this low-ohmic path way the capacitor can discharge its stored energy and at the moment this happens the capacitor is burnt partly away [15].

III. 2 Intrinsic breakdown

In section III.1 we have given a plausible derivation for the use of extreme value statistics in the case of defect-related breakdown distributions. Here we shall indicate that also for the intrinsic distributions extreme value statistics is "appropriate". The explanation, however, will be based on the aforementioned charge needed for breakdown Q_{bd} instead of on defects with varying properties. Connecting the probability of breakdown to the injection of charge we redefine the hazard rate by

$$h = P(Q \geq Q_{bd} | Q) , \qquad (12)$$

i.e. the probability that a device fails (and $Q \geq Q_{bd}$), knowing that already Q has been transported. In terms of distributions we can write similarly to Eq. (3)

$$h = \frac{1}{1 - F(Q)} \cdot \frac{dF(Q)}{dQ} = \frac{d \ln(1 - F(Q))^{-1}}{dQ} . \qquad (13)$$

At constant current injection $Q = Q(t)$ we can write:

$$h = \frac{d \ln(1 - F)^{-1}}{dt} \cdot \left(\frac{dQ}{dt}\right)^{-1} = h_J \cdot J^{-1}. \qquad (14)$$

When $Q = Q(E,t)$ at ramping or constant fields we must use partial differentiation and

$$h = \frac{\delta \ln(1 - F)^{-1}}{\delta E} \cdot \frac{dE}{dt} \cdot \left(\frac{dQ}{dt}\right)^{-1} +$$

$$+ \frac{\delta \ln(1 - F)^{-1}}{\delta t} \cdot \left(\frac{dQ}{dt}\right)^{-1} \qquad (15)$$

or $\qquad h = h_t \cdot R \cdot J^{-1} + h_E \cdot J^{-1}. \qquad (15a)$

where $R = dE/dt$ is the ramprate of the field and h_t and h_E are the hazard rates for field- and time-dependent breakdown, respectively. For field-dependent breakdown distributions we can thus express the hazard rate by

$$h_t \propto \frac{hJ}{R} \qquad \qquad \text{for ramping fields (16)}$$

while for time-dependent breakdown distributions we find

$$h_E \propto hJ \qquad \qquad \text{for constant fields. (17)}$$

Furthermore for constant current breakdown

$$h_J = h \cdot J . \qquad (18)$$

To find the field-, time- and current-dependence of the hazard rates we have to insert the empirical relation between these three variables.

In dielectrics current- voltage-time characteristics can be approximated by the Curie-von Schweidler law [21,22]

$$J = J_0 \exp\left(\frac{E}{E_0}\right) \cdot \left(\frac{t}{t_0}\right)^{\alpha} , \qquad (19)$$

where J_0, E_0, t_0, and α are constants.
Inserting Eq. 19 into Eq. 16 and Eq. 17 yields

$$h_t \propto \frac{h}{R} \cdot \exp\left(\frac{E}{E_0}\right) \qquad (20)$$

$$h_E \propto h \cdot \left(\frac{t}{t_0}\right)^{\alpha} \qquad (21)$$

$$h_J \propto h , \qquad (22)$$

where the terms of Eq. 19 that are constant have been left out for clarity. Eq. 20 and 21 are the essential equations for the hazard rates of the first and third extreme value distributions Gumbel I and Gumbel III.

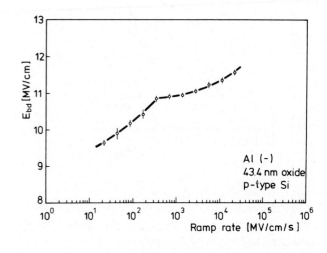

Figure 7

The variation of the intrinsic breakdown field with the ramp rate of the electric field. The error bars give the width of the intrinsic components only.

A further corroboration of the model and Eq. 20 can be found by plotting E versus R. Since Eq. 20 predicts that if h is constant (this is not always the case), then $E_{bd} \propto \ln R$. This is shown in fig. 7 for the intrinsic parts of the distributions measured at different ramp rates. These results make clear that the dielectric strength as found by ramping fields is not a real material property.

The validity of Eq. 21 can be checked as follows. By integration it follows that when h is taken constant

$$\int_0^{tbd} h_E dt = \int_0^{tbd} h \cdot J \cdot dt \propto \int_0^{tbd} h \left(\frac{t}{t_0}\right)^\alpha dt$$

$$Q_{bd} \propto \left(\frac{t_{bd}}{t_0}\right)^{1+\alpha} \qquad (23)$$

For three intrinsic distributions Q_{bd} has been plotted versus t_{bd} in fig. 8. The predicted correlation is in excellent agreement with Eq. 23. We have shown elsewhere [21] that α in Eq. 19 is dependent on the trap concentration. When poly-Si electrodes are used α is close to zero. When Al electrodes are used α is close to $-0,25$. Although these values are different for both types of electrodes, in both cases, the value of α for injection (Eq. 19) is equal to that for breakdown (Eq. 23) [15]. Trivially, for constant current injection there is a linear correlation between Q_{bd} and t_{bd}.

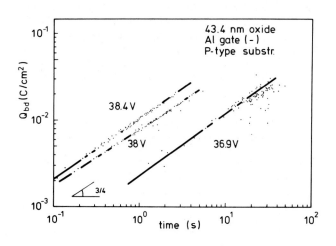

Figure 8.
A plot of Q_{bd} and t_{bd} values at constant bias. The linear relationship is predicted by Eq. 23.

In the treatment above h has been regarded constant which is generally true for the intrinsic part and for relatively low values of J or E. However, above a certain level of J and E, h often increases very fast with injection conditions and the value of Q_{bd} decreases very drastically. It has been shown elsewhere that in that case inhomogeneous injection takes place which causes a local breakdown. This local breakdown clearly must happen at a Q_{bd} value which is smaller over orders of magnitude. This effect causes the kinks in fig. 6 and 7.

IV. Discussion

As is illustrated above, the extreme value distributions are useful to fit results of dielectric breakdown. It has been made plausible that the defect-related breakdown distributions as well as the intrinsic breakdown distributions can be described using extreme value statistics.

This is not new since Weber and Endicott [19] advocated the statistics as early as (1956) to describe the electrical breakdown of transformer oil. The Gumbel I statistics fitted better than those of Gumbel III but were seriously commented since the hazard rate does not vanish at zero field. The required probability of breakdown at $E = 0$ was regarded to be a strong argument against the physical justification of the extreme value statistics. Weber and Endicott hesitated whether they should not instead use the Gumbel III distribution which uses $\ln(E)$ as the variate and therefore avoids negative values.

We can solve this dilemma now. Since we know that the mechanism for breakdown is driven by charge transport through the dielectric and not by the field, we can state that the hazard rate is not zero at zero electric field.

The fact is that although at $E = 0$ the charge injection is very small it is not quite zero.

Values of $J = 10^{-18} - 10^{-20}$ A/cm² can be extrapolated from experimental J/E curves [22]. This injection current is governed by thermionic emission of charge carriers. With an average Q_{bd} value of 1C/cm² [15], it takes therefore at least $10^{10} - 10^{12}$ yr before the injection is sufficient for a breakdown. While this is impossible to check, experiments with energetic or "hot" charge carriers have been performed that prove that indeed at zero field a breakdown-like phenomenon takes place [23].

To reduce the probability of breakdown we must reduce the injection current by applying a negative field. In the asymptotic case we get

$$J(+) \rightarrow 0 \text{ for } E \rightarrow -\infty$$

and therefore

$$h(+) \rightarrow 0.$$

Where $J(+)$ is the current density in the same direction as at positive field and $h(+)$ is the hazard rate for a breakdown in that direction. It must be clear that this includes that a breakdown in one direction is different from that in the opposite direction. This has been established experimentally. In general $Q_{bd}(+) \neq Q_{bd}(-)$ where $+$ and $-$ stand for the direction of breakdown [15].

For comparison the techniques, variates, hazard rates and distribution laws are summarized in table I.

testing technique	variates	hazard rate h	distribution $\ln\ln(1-F)^{-1}$
Field-dependent test	E	$\propto \exp(E)$	Gumbel I
Time-dependent test	lnt	$\propto t^{\alpha}$	Gumbel III
Constant current test	lnt lnQ	constant	Gumbel III Gumbel III

Table I.

Compilation of statistics in use for the analysis of the results of different testing techniques.

Screening tests or burn-in tests

These tests are useful when the hazard rate is large at low values of the variate and then decreases. This occurs when, for instance, a number of low-voltage defects or early failure defects are present. These failures can be traced by using accelerated tests. By removing devices having these failures often the average properties and reliability can be improved considerably. Screening conditions are nowadays mostly based on empirical investigations combining a certain bias and temperature stress during a fixed period. However, by applying a constant bias, the injection current often decreases with time (Eq. 16 with $\alpha < 0$). The total amount of injected charge given by

$$Q = \int_0^\infty J dt \propto \int_0^\infty t^\alpha dt$$

convergents to a limit when $\alpha < -1$. This often happens when severe trapping takes place [22]. Therefore the total amount of charge needed for the early failure may not be reached. An expensive long term stress then has not the wanted result. Screening tests can be improved, however, and made cheaper by adapting the stress conditions. For breakdown a certain quantity of charge is needed. Instead of using a constant field stress, it is more convenient to use a constant-current stress. A fixed time of stress gives the required injection of charge. It is advisable to restrict this quantity to values below $10^{-6} - 10^{-5}$ C/cm^2

since beyond this range the properties of the capacitors tend to change.

V. Conclusions

The use of extreme value statistics has been demonstrated. The area effect is a convenient tool to predict yield and reliability of large devices. The hazard rate of field-dependent and time-dependent distributions obey the required dependence in the case of intrinsic breakdown. Screening tests can be more profitable when using constant- current techniques.

References

[1] D.A. Baglee, IEEE/IRPS Conf. on Reliability Physics (1984) 152.

[2] D.L. Crook IEEE/IRPS Conf. on Reliability Physics (1979) 1;
D.L. Crook and W.K. Meyer IEEE/IRPS conf. on Reliabilty Phys. (1981) 1.

[3] E.S. Anolick and G.R. Nelson IEEE/IRPS conf. on Reliability Phys. (1979) 8;
E.S. Anolick and L.Y. Chen IEEE/IRPS conf. on Reliability Phys. (1981) 23;
E.S. Anolick and L.Y. Chen IEEE/IRPS conf. on Reliability Physics (1982) 238.

[4] K. Yamabe Conf. on Solid State Devices and Materials Tokyo (1985) 261-266.

[5] C. Hu IEEE/IEDM **85** (1985) 368.

[6] Y. Hokari, T. Baba and N. Kawamura, IEEE Trans. on Electr. Devices **ED 32** (1985) 2485.

[7] M. Hirayama, T. Matsukawa, N. Tsubouchi and H. Nakata, IEEE/IRPS Proceed. on Reliability Physics (1984) 146.

[8] M. Schatzkes, M. Av-Ron and K. V. Srikrishnan IEEE/IRPS Proceed. on Reliability Physics (1984) 138.

[9] J.R. Monkowski and R.T. Zahour IEEE/IRPS Proceed. on Reliability Physics (1982) 244.

[10] H. Ishiuchi, Y. Matsumoto, T. Mochizuki, S. Sawada and O. Ozawa, IEEE/IRPS Proceed. on Reliability Physics (1982) 228.

[11] E. Domangue, R. Rivera, C. Shepard, IEEE/IRPS Proceed. on Reliability Physics (1984) 140.

[12] R.M. Hill and L.A. Dissado, J. Phys. C. **16** (1983) 2145.

[13] D.R. Wolters and J.F. Verwey Instabilities in Integrated Devices eds. G. Barbottin and A. Vapaille, (North Holland Publishing Co. Amsterdam 1986).

[14] D.R. Wolters, T. Hoogestijn and H. Kraay, Physics of MOS insulators eds. G. Lucovsky, S.T. Pantelides and F.L. Galeener. Proc. Int. Top. Conf. Raleigh N.C. (1980) 349.

[15] D.R. Wolters and J.J. Van der Schoot, Philips J. of Res. **40** 115, 137, 164 (1985).

[16] D.R. Wolters, Insulating Films on Semicond., eds. M.Schulz and G. Pensl, Springer Series in Electrophysics **7**, 180 (Springer N.Y., 1981).

[17] W. Rey, Personal communications.

[18] R.A.Metzler IEEE/IRPS Proc. on Rel.Physics. (1979) 233.

[19] K.H. Weber and H.S. Endicott A.I.E.E. Trans. **75** Power App. Syst. (1956) 371.

[20] E.Gumbel. Statistics of extremes. (Colombia Univ.Press. N.Y., 1958).

[21] A.K. Jonscher, Dielectric relaxation in solids, (Chelsea Dielectrics Press Ltd., London 1983).

[22] D.R. Wolters and J.J. Van der Schoot, J. Appl. Phys. **58** 831 (1985).

[23] T.Poorter and D.R.Wolters, Insulating films on Semiconductors, eds. J.F.Verwey and D.R.Wolters, (North Holland Publishing Co. Amsterdam 1983) p. 270.

Reliability Technology — Theory & Applications
J. Møltoft and F. Jensen (Editors)
© Elsevier Science Publishers B.V. (North-Holland), 1986

OXIDE BREAKDOWN IN MOS STRUCTURES UNDER ESD

AND CONTINUOUS VOLTAGE STRESS CONDITIONS

E.A. AMERASEKERA and D.S. CAMPBELL
Electronic and Electrical Engineering Department
Loughborough University of Technology, Loughborough, Leicestershire, UK

Investigations have been conducted into the oxide breakdown mechanisms caused by the application of large voltage stresses in the form of both Electrostatic Discharge (ESD) pulses and continuous d.c. voltages, on metal-oxide-semiconductor (MOS) structures.

The MOS structures had been fabricated on p-type silicon wafers and had an oxide thickness of $\cong 4000\text{Å}$. Both enhancement-type and depletion-type n-channel MOSFETs and MOS capacitors were constructed on the same wafer.

Earlier experiments on the ESD susceptibility of n-channel enhancement mode MOSFETs on the same wafer have shown there to be no temperature dependence of the breakdown although a significant voltage dependence was observed.

The experiments presented in this paper have investigated the correlation between the ESD sensitivity of MOS oxides and the continuous d.c. voltage breakdown strength, with the intention of aiding the identification of the principle mechanisms associated with each type of applied stress.

MOS capacitors were subjected to both ESD pulses, in accordance with the "human-body" model as presented in MIL-STD-883C Method 3015.2, and continuous d.c. voltages. The breakdown of MOS capacitors upon application of the ESD pulse occurred at 200 volts. (Breakdown being taken as the condition when the parallel resistance of the capacitor becomes finite, i.e. reduced by a factor of approximately 100.)

Identical capacitors were then subjected to a continuous voltage for approximately 300 sec. Breakdown then occurred at 36 volts. This is a factor of approximately 5 less than the sensitivity to ESD pulsing, indicating that two different oxide breakdown mechanisms are involved.

Such observations are in keeping with the time-independent nature of the ESD stress voltage, and the time-dependent nature of the continuous stress voltage. The time-dependent breakdown implies that the mechanism involved is that of impact-ionization of charge carriers within the oxide under high-field conditions which leads to thermal runaway and then burnout. This mechanism is temperature-independent. Time-dependent breakdown suggests that the generation of electron-traps leads to ultimate failure. As this mechanism is temperature-dependent, tests were carried out to investigate the temperature-dependence of continuous voltage breakdown. The results showed that continuous-voltage breakdown is significantly temperature dependent between 25°C and 200C, the breakdown voltage for a 400Å oxide decreasing from 36V to 28V over this range.

1.   INTRODUCTION

The sensitivity of MOS devices to Electrical Overstress (EOS)/Electrostatic Discharge (ESD) damage has long been established. However, the extent of device susceptibility is not a well defined area because of the difficulties in specifically locating and identifying a particular breakdown mechanism.

It is intended in this paper to aid this identification of the principle physical processes by which EOS and ESD manifest themselves in metal-oxide-semiconductor (MOS) devices. Once the particular mechanism by which breakdown takes place is understood, it becomes easier to predict the effect of the process on the different characteristics of the devices.

In an earlier paper[1], experiments on the sensitivity of NMOS devices to ESD were described. As a result of these experiments it was shown that the ESD sensitivity of enhancement mode NMOS devices was not temperature dependent. This is a result which shall be referred to later in this paper in context with the results presented here.

The experiments described here were conducted on both Enhancement-mode and Depletion-mode MOS structures. Devices have been subjected to both ESD and continuous voltage stress and the temperature dependence of the oxide breakdown strength has been determined. It is therefore possible to compare the principle physical processes of the respective breakdown mechanisms.

2. THE EXPERIMENTS

2.1 The MOS Devices

MOS Capacitors were used in these experiments.
The capacitors had been fabricated in chips on
complete p-type silicon wafers and have an
oxide thickness of $\cong 400\text{Å}$. Both enhancement
(E) type and depletion (D) type capacitors
were available on the same chip on the wafer.
The wafers themselves were provided by Plessey
Research (Caswell).

The capacitors consist of:

(a) polysilicon on thermal SiO_2 on enhance-
    ment (p-type) silicon (C1).

(b) polysilicon on thermal SiO_2 on depletion
    (n-type) silicon (C2).

Figures 1(a) and 1(b) show cross sections of
the two capacitors. In particular, the
different charge concentrations in the E-type
and D-type capacitors are emphasised.

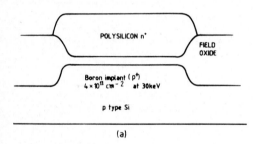

(a)

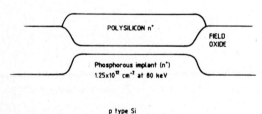

(b)

Fig. 1(a) E-Type Capacitor. The Boron
           implant ensures a positive thresh-
           hold voltage. Charge density at
           the surface is $4 \times 10^{11} cm^{-2}$ holes
           (p).

Fig. 1(b) D-Type Capacitor. The Phosphorous
           is implanted over the Boron.
           Hence the residual charge density
           at the surface is $8.5 \times 10^{11} cm^{-2}$
           electrons (n).

The dimensions of the devices were as follows:

The area of the capacitors = 49×10^3 $(\mu m)^2$
Peripheral length of the capacitors = $910 \mu m$
Oxide thickness = 400Å

2.2   The Procedure

2.2.1   ESD

A set of capacitors was subjected to a single
ESD pulse generated using the "human-body"
model as described in MIL-STD 883 C, Method
3015.2. Voltages of magnitudes ranging from
75V to 250V were applied to the devices. The
parallel resistance across the capacitor was
measured before applying the pulse and after
application of the pulse.

A typical resistance across the gate oxide
of an undamaged MOS capacitor is of the order of
$10^{14}\Omega$. The resistance bridge used in this
experiment however had a maximum measuring
capacitance of $10^9\Omega$. Breakdown was deemed to
have occurred when a finite parallel resistance
was measured across the capacitor. A typical
value of such a resistance is of the order of
$10^5\Omega$ which is very much less than that of an
undamaged capacitor.

In order to evaluate the effect of ionic
impurities in the oxide on the capacitor breakdown strength, another set of capacitors was
subjected to a High Temperature Reverse Bias
(HTRB) screen. The HTRB screen consisted of
applying a voltage of -12V at a temperature of
150°C across the capacitors for approximately
5 minutes. This has the effect of sweeping
impurities out of the oxide. These capacitors
were then subjected to the ESD pulses described
above.

2.2.2 Continuous Voltage Stress

MOS capacitors of both E-type and D-type
structures were subjected to a voltage of
increasing magnitude of the form shown in
Fig. 2. Each voltage step lasted approximately
5 minutes. The leakage current was measured at
each voltage step, and a plot made of leakage
current as a function of the applied voltage.

The temperature of the wafer was increased in
steps from room temperature (25°C) to 75°C,
125°C, 150°C and 200°C, the experiment being
repeated at each temperature. It was therefore
possible to plot the continuous voltage breakdown strength as a function of the ambient
temperature.

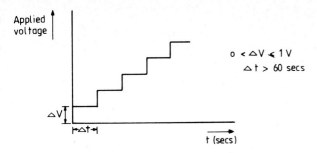

$$0 < \triangle V < 1 V$$
$$\triangle t > 60 \text{ secs}$$

Fig. 2   Waveform of the voltage used in the continuous voltage stress experiments.

3.   RESULTS

3.1   Electrostatic Discharge

Table 1 gives the results of the ESD experiments on E-type and D-type capacitors that have not been subjected to HTRB screens. Typical values of capacitance and resistance are given for a single MOS capacitor. However, some capacitors did show resistance values as low as 1kΩ with a capacitance of ≅ 0pF when damaged. The upper limit was about 300kΩ. Breakdown was always seen to occur at 200 volts.

Table 2 gives the results of the ESD experiments on E-type capacitors that have been subjected to an HTRB screen. The lower limit of resistance is still around 1kΩ with C = 0pF. The upper limit now is about 3MΩ.

These results can be compared with those obtained in ref. 1. The slight degradation characteristics of the NMOS transistors in that paper correspond to the upper limit value of resistance in the temperature, associated with a still functioning capacitor. The catastrophic breakdown characteristic corresponds to the lower limit of resistance which is ≅ 1kΩ associated with zero capacitance:-

3.2   Continuous Voltage Stress

These results are summarised in Figures 3, 4 and 5. 5 capacitors of each type were subjected to the stress. Figures 3 and 4 show the oxide leakage current as a function of applied voltage for a typical E-type capacitor and a typical D-type capacitor respectively. The oxide leakage current is seen to increase at around 32V in both graphs and rises to approximately 5μA at 36V. Resistance and capacitance measurements made at this stage for the E-type capacitor in Fig. 3 gave:-
> parallel resistance = 197 kΩ
> parallel capacitance = 18pF

and for the D-type capacitor in Fig. 4,
> parallel resistance = 4 MΩ
> parallel capacitance = 10 pF

These values are comparable with the breakdown figures given in Table 1.

Increase in voltage just beyond 36V resulted in a rapid increase in leakage current and burn-out occurring. 36V can therefore be considered to be the oxide breakdown voltage due to the applied continuous voltage stress.

Fig. 5 shows the oxide breakdown voltage of an E-type capacitor as a function of temperature. As can be seen, the breakdown strength drops sharply from +36V at ≅ 25°C to 28V at 200°C.

TABLE 2

RESULTS OF APPLYING ESD PULSES TO MOS CAPACITORS
WHICH HAVE BEEN HTRB SCREENED

CAPACITOR TYPE	NO: OF DEVICES	AREA x10³ (μm)²	APPLIED VOLTAGE (v)	TYPICAL CAPACITANCE (pF)		TYPICAL RESISTANCE (kΩ)	
				before	after	before	after
Enhancement	10	49.0	150	20.0	20.0	o/c	o/c
Enhancement	10	49.0	200	20.0	20.0	o/c	o/c
Enhancement	10	49.0	250	20.0	17.4	o/c	900

TABLE 1

RESULTS OF APPLYING ESD PULSES TO MOS CAPACITORS

CAPACITOR TYPE	NO: OF DEVICES	AREA x10³ (μm)²	APPLIED VOLTAGE (v)	TYPICAL CAPACITANCE (pF)		TYPICAL RESISTANCE (kΩ)	
				before	after	before	after
Enhancement	10	49.0	150	20.0	20.0	o/c	o/c
Enhancement	10	49.0	200	20.0	16.5	o/c	150
Depletion	10	49.0	150	13.0	13.0	o/c	o/c
Depletion	10	49.0	175	13.0	13.0	o/c	o/c
Depletion	10	49.0	200	13.0	9.0	o/c	50

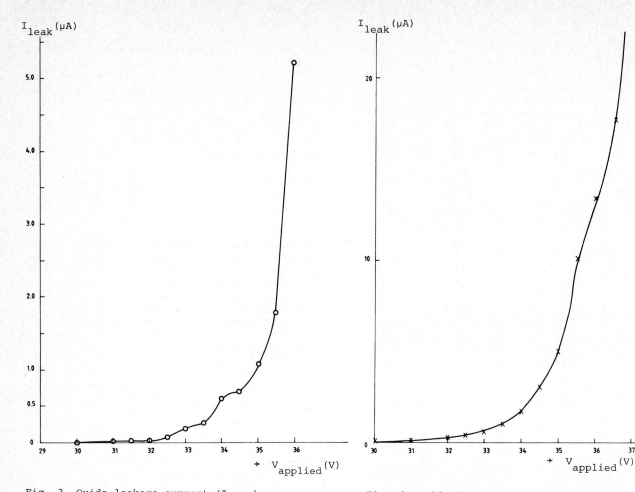

Fig. 3 Oxide leakage current (I_{leak}) vs
Applied voltage (V_{appl}) for a typical E-type
MOS capacitor. The parallel resistance across
the gate of the capacitor measured after 36V
had been applied was 197kΩ, indicating that
breakdown had occurred.

Fig. 4 Oxide leakage current (I_{leak}) vs
Applied voltage (V_{appl}) for a typical D-type
MOS capacitor. The parallel resistance across
the gate of the capacitor measured after 36V
had been applied was 4MΩ, indicating that
breakdown had occurred.

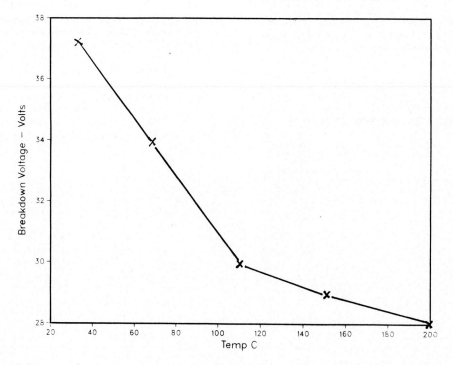

Fig. 5 Oxide breakdown voltage as a function of temperature under continuous voltage stress
        conditions.

4.   DISCUSSION

4.1   Oxide Breakdown

The breakdown voltage of 200V for an ESD pulse is $\cong 5$ times greater than the 36V observed with a continuous voltage stress. The strong temperature dependence of the continuous voltage stress breakdown is in direct contrast to earlier results establishing the temperature independence of ESD breakdown[1]. Therefore it is possible to conclude that two different oxide breakdown mechanisms are involved. As discussed in ref. 1, these two mechanisms are:

   a) impact ionization
   b) electron-trap generation.

4.1.1   Impact Ionization[2,6]

Impact ionization occurs when free electrons in the conduction band have gained sufficient energy to enable them to transfer enough energy to a valence electron upon collision such that that electron is elevated to the conduction band. The whole process occurs in about 10^{-8} secs, which for the purposes of this paper can be considered time independent. High electric fields ($\cong 10^6$V/cm) are necessary to enable the electrons to gain the energies required.

4.1.2   Electron-trap Generation[3,4]

The fabrication process can result in dangling (unattached) Si or 0 bonds being present in the oxide which then become traps for charge carriers. However, in the presence of an electric field, it is possible that the lattice structure can be deformed suffiently that thermal vibrations would create new electron traps. The generation of new electron traps has been experimentally shown by Harari as a function of both electric field and temperature[3]. Oxide breakdown in these experiments are time-dependent ($\cong 30$ secs).

Once traps are established within an oxide, the application of an electric field sufficiently distorts the SiO_2 conduction band enabling charge to hop from one trap to another. This process is known as hopping conduction[4].

Traps can also result in high electric fields occurring in localized areas of the SiO_2 thereby enhancing the prospect of impact ionization occurring.

4.2   Electrostatic Discharge

The ESD pulse is of very short duration (10^{-6}secs) and any change produced by such a pulse can therefore be considered time-independent. The ESD breakdown process is also temperature independent. It can, therefore, be concluded that impact ionization is responsible for oxide breakdown.

In Table 2 it was shown that devices which had less ionic contamination in the oxide had a higher ESD breakdown voltage at 250V. It is therefore possible for electron traps already available in the oxide before the application of the pulse to enhance the probability of impact ionization and thereby reduce the breakdown voltage of the devices.

4.3   Continuous Voltage Stress

The breakdown observed with continuous voltage stress is in keeping with Harari's results.[3] The temperature dependence and the low electric fields (cf: ESD) signify that electron-trap generation and the associated conduction processes is the mechanism concerned.

4.4   Implications

ESD is shown to be due to a high-energy impact ionization breakdown mechanism. It is significant that electron traps already available in the oxide can contribute to ESD breakdown. Hence the device sensitivity to ESD would be a function of the number of electron-traps in the oxide. This has far reaching implications when considering device reliability at the fabrication stage. It is important to reduce the number of traps in an oxide at this stage and processing techniques such as those used in the manufacture of radiation hardened devices along with thermal annealing techniques, may provide the solution.[7,8]

However, processing techniques such as ion implantation may result in a high number of electron traps being created in the oxide. Energies in the range 20keV to 100keV are presently quite common. With the requirements for higher ion implant doses at higher energies being used in the manufacture of sub-micron devices, it is possible that the number of traps thus created could have a significant effect on device reliability.

5.   CONCLUSIONS

1. A large disparity was observed in the breakdown voltages of MOS capacitors due to
   (a) Electrostatic Discharge pulsing (200V)
   (b) Continuous d.c. voltage stress (36V).

2. A strong temperature dependence was observed for oxide breakdown under continuous voltage stress compared with earlier results showing that oxide breakdown under ESD pulsing was not temperature dependent.

3. It is therefore concluded that two different mechanisms are responsible for oxide breakdown due to ESD pulsing and continuous voltage stress.

4. The time-independent property of the ESD pulse suggests that impact ionization is the mechanism through which breakdown takes place. High localized electric-fields caused by electron traps already in existence within the oxide enhance the prospect of breakdown taking place. This has direct implications on device fabrication techniques involving high energy ion implantation.

5. It is postulated that oxide breakdown
    under continuous voltage stress is
    related to the generation of new electron
    traps within the oxide. The electric
    fields required for breakdown are, there-
    fore, not as great as those required in
    the case of ESD.

6. ACKNOWLEDGEMENTS

Thanks are due to Plessey Research (Caswell)
for the wafers used in the study. This work
has been carried out with the support of the
Procurement Executive, Ministry of Defence.

REFERENCES

1. E.A. AMERASEKERA, D.S. CAMPBELL; Electro-
static Pulse Breakdown in NMOS Devices.
(To be published) Quality and Reliability
Engineering Journal, 1986.

1. J.J. O'DWYER; The Theory of Electrical
Conduction and Breakdown in Solid Dielectrics.
Clarendon Press (Oxford), 1973.

3. E. HARARI; Dielectric Breakdown in Electrically Stressed Thin Films of Thermal SiO_2
J.Appl.Phys., $\underline{49}$(4), pp 2478-2489, April 1978.

4. A.J. JONSCHER, R.M. HILL; Electrical Conduction in Disordered Non-Metallic Films.
Physics of Thin Films. (Ed. G. Hass, M.H.
Francombe and R.W. Hoffmann.) $\underline{8}$, pp 169-249,
1975.

5. P. OLIVO, B. RICCO, E. SANGIORGI; Electron
Trapping/Detrapping within Thin SiO_2 Films
in the High Field Tunnelling Regime. J.Appl.
Phys., $\underline{54}$(9), pp 5267-5276, September 1983.

6. H. FROHLICH; Theory of Electrical Breakdown of Ionic Crystals. Proc. R.Soc., A $\underline{160}$,
pp 230-241, 1937.

7. S.S. COHEN; Electrical Properties of Post-
Annealed Thin SiO_2 Films. J.Electrochem. Soc.
$\underline{130}$, pp 929-932, 1983.

8. P. BALK, M. ASLAM, D.R. YOUNG; High Temperature Annealing Behaviour of Electron Traps
in Thermal SiO_2. Sol. St. Elec., $\underline{27}$,
pp 709-719, 1984.

BIBLIOGRAPHY

AJITH AMERASEKERA
Received a B.Sc.(Hons) degree in Electronic
Engineering and Physics from Loughborough
University in 1983. Is at present researching
into the physics of failures of small dimension
semiconductor devices at Loughborough
University, UK, leading to a Ph.D. in 1986.

DAVID S. CAMPBELL
After a career in industry over 17 years,
culminating in being the Technical Manager of
a large Capacitor factory within the Plessey
Group of companies, he has been Professor of
Electronic Component Technology in the Department of Electronic and Electrical Engineering
at Loughborough University of Technology since
1971. He is now in charge of an active group
in this area which has special interests in
Interconnection Technology, particularly thick
film systems and also Reliability Analysis.
The reliability work involves a major study on
Field Failure of Electronic Components in
Equipment, a study which involves the
University with major electronic companies in
the U.K. and also with companies in Denmark
through the involvement of the Danish
Engineering Academy at Lyngby. Also studies
are in hand on the reliability of components
which are only now coming into use and these
include Surface Mounting, Electrostatic Breakdown in MOS Silicon, the effect on Failure
Mechanisms of using Submicron Silicon Devices
and finally GaAs structures.

Reliability Technology — Theory & Applications
J. Møltoft and F. Jensen (Editors)
© Elsevier Science Publishers B.V. (North-Holland), 1986

LIMITATION OF ELECTROSTATIC SENSITIVITY OF IC'S BY
RELIABILITY EVALUATION OF ON-CHIP COMPONENTS

Keith Beasley

Plessey Research (Caswell) Ltd., Caswell, Towcester, Northamptonshire, England.
Tel. 0327 50581

In order that reliability can be considered during IC design, it is essential that the appropriate failure mechanisms are understood. These mechanisms can be determined by performing well defined test programmes involving 'on-chip' components. Subsequent failure analysis and physical modelling provide accurate data for the definition of design rules for a wide range of new ICs. The combination of these activities constitutes a valuable Design Assurance activity. An example of such an activiity is the investigation of the component parts of a static protection circuit. Such a study allows the design of ICs with known immunity to damage by Electrostatic Discharge (ESD).

1. INTRODUCTION

To obtain a reliable IC product, there must be a firm commitment to this aim during the products design. Consideration must be given not only to long term life requirements, but also to other measures of reliability such as sensitivity to electrostatic discharge (ESD). No where is this more true than for VLSI IC's; as device geometries decrease, so limitation of sensitivity to ESD damage becomes increasingly more difficult.

In order to 'design in' this necessary static immunity and hence product reliability, the mechanisms which cause in-life failure must be understood. This understanding should be extensive enough to model new processes and circuit configurations, sufficient to gauge their performance in use.

To understand a failure mechanism to this extent, it is necessary to consider the inherent physical and thermal properties of the materials and structures used. This is most effectively completed by designing 'test chips' of the structure(s) under investigation. These test chips can then be subjected to rigorous testing e.g. 'Zap' testing or electromigration trials. Test conditions, physical dimensions and material compositions are thus all known. To obtain an accurate and inherent model, it is then necessary to perform simple failure analysis on the failures and statistical analysis on the results.

This paper examines the wider aspects of Reliability Engineering, highlighting the importance of Failure Analysis and its

co-ordination with other Quality functions. If significant improvements in reliability are to result, attention must be given to this 'Design Assurance' activity.

2. DESIGN ASSURANCE

One of the main aims of a Design Assurance activity is to ensure that ALL customer requirements are considered during a product design. Whilst the design team department will be concentrating on performance, the customer is likely to be asking for high reliability and static immunity. Having been involved in the testing and failure analysis of test structures, the design assurance engineering will be in an excellent position to advise the designer on appropriate design rules. Experience of similar designs and of the modelling and testing of (for example) ESD susceptibility, will also enable a design assurance engineer to help the customer decide on appropriate standards and specifications.

Compared to 'Quality Control' and 'Reliability', 'Design Assurance' is a new topic in the field of Quality Assurance. However, with Test Structures and Failure Analysis as its tools, the aim to design all products 'right first time' can be addressed, even for VLSI ICs. Despite the increasing move to Semi-custom design methods (ref. 1), the time and cost of IC designs and redesigns is a strong incentive to look closely at the 'right first time' approach. Failure analysis and appropriate feedback can do much to 'build in' quality and reliability of new designs. A few examples are summarised in Table 1.

TABLE 1

FAILURE ANALYSIS RESULT	DESIGN ASSURANCE RECOMMENDATION
Shorting Bond Wires	Improved Bonding Layout
Electromigration Down Contacts	Tighter Current - Density Design Rules
ESD Damage	Revised Protection Circuit

Much of this Design Assurance work revolves around the continuing update of models for the performance and reliability of different circuit configurations. As processes develop, so new failure mechanisms become prevalent, and models have to be refined. For this understanding, failure analysis is essential.

3.   FAILURE ANALYSIS AND FEEDBACK

Failure Analysis is not just the physical analysis of failures to determine the failure mechanism. All good Failure Analysis practitioners agree that a Failure Analysis investigation is not complete until the results of the analysis have been reported to the 'requestor' of the analysis. Not only should the analysis itself be fully understood, but the initiator should be made fully aware of the implications. Feedback can take any or all of the following forms:-

a)   To Sales/Marketing - so that liability for failure can be established.

b)   To the 'end user' - to advise on limitations of use that will eliminate further similar failures.

c)   To system and/or component designers and manufacturers - to suggest improvements to design and/or manufacturing techniques.

d)   To Reliability Engineers - to update the reliability model of the components analysed.

This latest point is particularly important, since reliability calculations based on life test and/or field failure information depend totally on the accuracy of the information about the failures. Failure rate or MTTF (Mean Time to Failure) figures are often derived by very simplified calculations taking a 'total number of failures', irrespective of the failure mechanism. The difference between an infant mortality and a wearout mechanism should be clearly indicated. and reliability predictions made accordingly. The dangers of over simplifying reliability models are now being brought to our attention (ref. 2). Clearly the identification of failure mechanisms together with a sound physical understanding is necessary for reliable reliability models. This is only possible by linking modelling exercises with appropriate testing and failure analysis. The degree of coordination required is illustrated in Fig. 1. A break in any of the feedback loops could considerably limit the value of much of the work performed.

Whether they be under the control of the Quality Department, Reliability function or some other department of an organisation, it is the bringing together of ALL work on testing, analysis and modelling that leads to improved product reliability.

4.   SOURCES OF RELIABILITY INFORMATION

Failures for analysis, and other information on reliability performance come from one of two sources 'in-house' or 'the field'. Although 'Field Failure' information offers the only data on true usage, even if obtainable, its dependability is usually questionable. As shown in Table 2, 'In-house', data has considerable advantages - particularly to the Failure Analyst.

Many failures from sources 'external' to the failure analyst suffer inevitably from lack of access to information relating to the device usage and circumstances of failure. This situation is much improved when the source of failures is 'in-house' i.e. the failure analyst has direct access to all of the background - design, processing,

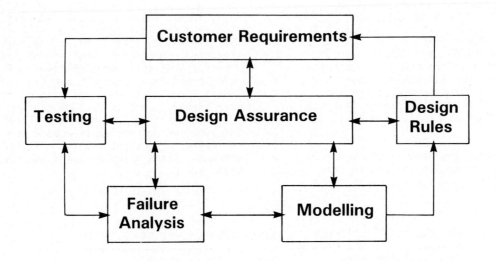

FIG 1   FEEDBACK FOR IMPROVED RELIABILITY

TABLE 2

COMPARISON BETWEEN FIELD FAILURES AND TEST CHIP FAILURES

FACTOR	FIELD FAILURE	'TEST CHIP' FAILURE
Cause of failure	Environmental? Misuse? Many possibilities	All conditions at time of failure known
Ease of Retest	Often difficult or impossible	Equipment and results of 'pre-failure test' available
Ease of Analysis	Likely problems in decap, fault analysis etc.	Structure designed to aid later analysis
History of Component	Often totally unknown	Readily available

assembly, test and 'time of failure' information. Even with some 'in-house' failures,
however, the investigative task of determining the failure mechanism can be severely
hampered by chip complexity and conflicting
evidence. It is here that test structures or
for the semiconductor industry 'test chips'
come into their own.

In the area of microelectronics, the use of
'simple' test components is well established.
Drop-in's containing single transistors and
other on-chip components are used for process
control and as yield monitors. By including
new circuit configurations and using new
process techniques, test structures are also
used to evaluate the performance and
reliability of new features. As indicated
above, it is failures from the testing of
such 'test chips' that give failure analysts
the chance to improve their feedback. With
the information very much more under the
control of the failure analyst, far more
reliable models can be developed - and used
in the later analysis of complex devices.

5.   DESIGN ASSURANCE EXAMPLE - STUDY OF ESD
      PROTECTION CIRCUITS

Failures due to Electrostatic Discharge
(ESD) are giving increasing concern as chip
complexity increases. With high speed ICs,
there is often a trade-off between static
protection and performance - i.e. the
protection circuitry literally 'slows down'
the device. The ability to model a
protection circuit thus becomes crucial in
deciding the degree of protection to be
afforded. To support theoretical models,
tests need to be carried out to compare the
performance of protection structures with
these models. For the reasons described
above (Section 4), 'Test Chips' are sued for
these comparison exercises. An obvious test
chip design would be one containing a range
of complete protection structures and indeed
such test chips are in use by Plessey.
However, even these chips do not give
sufficiently detailed results to enable
modelling of the physical mechanisms
associated with ESD failures. Plessey
Research (Caswell) is thus also using test
chips containing only the constituent parts
of protection structures.

Figure 2 shows a typical input protection
circuit for a CMOS process. It depends upon
a polysilicon resistor linking the bond-pad
to a protection diode (formed by a lateral
PNP).

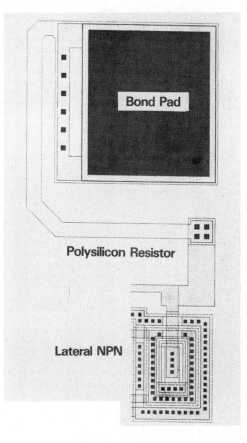

FIG 2   INPUT PROTECTION CIRCUIT

Static sensivity studies have shown that the
polysilicon resistor if often the 'weak link'
in the protection circuit i.e. the component
that fails first. By testing these resitors
in isolation from the rest of the circuit,
their true behaviour can be determined. In
particular the relationship between the size

of polysilicon resistors and performance on Zap test can be investigated. By complementing the practical work with circuit and thermal modelling, a basic understanding of polysilicon under static discharge conditions is established.

The test chip used for the polysilicon resistor testing at Plessey, is shown in Fig. 3, and the detail of two 'blown' resistors in Fig. 4. Being simple structures, failure analysis of these resistors could be

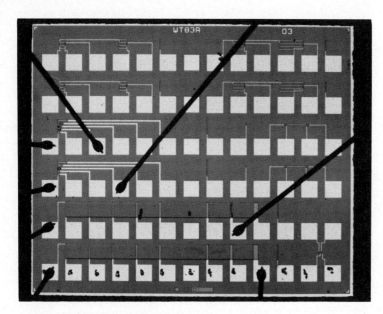

FIG 3

POLYSILICON

RESISTOR

TEST CHIP

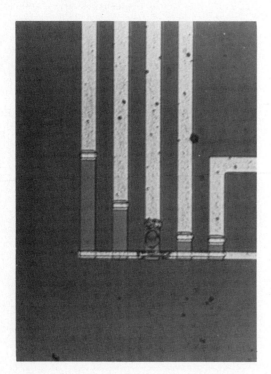

FIG 4a   FUSED RESISTOR   FIG 4b   DEGRADED RESISTOR

performed with only a microscope. Fig. 4a shows a totally evaporated resistor - this could be caused not only by a large Zap, but by simple electrical overstress (ESO), clearly indicating that in some circumstances EOS and ESD can NOT be distinguished.

Of even more concern that catastrophic failures due to Zap's are latent failures. As many researchers are reporting (e.g. ref. 3), degradation of a device due to static discharge may result in reduced life. Take for example the Polysilicon resistor shown in Fig. 4b. Here, a static discharge has caused an increase in resistance of 10%; sufficient in some usages to cause a parametric degradation which could result in premature device failure. Although the exact effect will be design dependent, its observation on a test chip enables simple failure analysis. Visual inspection clearly shows a fused track down the centre of the polysilicon. By correlating the Zap voltages with effect, it is possible to determine the relationship between power dissipated in the resistor and size (resistance value and physical size). As expected, simple heat flow theory can be utilised, but only by accurately modelling the complete physical situation. Although a simple structure, many factors have to be considered, for example the temperature coefficient of resistivity of polysilicon and the thickness and composition of the layer beneath the resistor.

At the time of the above testing there was considerable discussion about the electrical model for static sensitivity testing. It was becoming clear that the Human Body Model (HBM) of MIL STD 883C (ref. 4) represented only one of a number of possible discharge situations. Recent work (ref. 5) suggests that the Charged Device Model (CDM) and Charged Board Model (CBM) can be more stringent than the HBM. Comparing these models (Table 3) whilst considering a traditional protection circuit (as in Fig. 2), it can be seen that the discharge circuit significantly effects the performance during Zap testing. For the CDM Zap, the Polysilicon resistor is now having to dissipate all of the energy in the Zap - rather that a share of the energy with the

1.5K discharge resistor of the HBM. Returning to the physical model of the Polysiicon resistors derived from earlier testing, calculations were performed which soon identified that no reasonably sized Polysilicon resistors could survive even a 500V Zap.

Only by linking Zap testing, physical and thermal modelling and failure analysis, can a true picture of even a simple resistor be accurately determined. As Table 4 shows, the benefits are significant. Not only is it possible to predict whether a circuit will survive a given zap without testing it, but the designer can confidently discuss the trade-off between static sensitivity and circuit performance. The extension of a failure analysis into a basic physical model, has the added advantage of being applicable not only to current processes but also to the next generation.

6.   CONCLUSIONS

Improvements to product reliability can only be obtained by the coordination of many activities. These include not only reliability trials and mathematical modelling but also failure analysis and design assurance activities on appropriate test structures.

In the case of ESD protection circuits for IC's the modelling of protection circuit components, together with a reasessment of requirements and failure analysis, has led to a significant change in Design Rules for Polysilicon Resistors i.e. their elimination!

7.   ACKNOWLEDGEMENTS

Part of this work has been carried out with the support of the Alvey Directorate.

TABLE 3

COMPARISON OF ZAP MODELS

	MODEL	SITUATION REPRESENTED	DISCHARGE TEST CIRCUIT
HBM	Human Body Model	Device touched by charged body	100 pF via 1500Ω into DUT
CDM	Charged Device Model	Charged device suddenly grounded	Charged DUT via 1 ohm to ground
CBM	Charged Board Model	Charged Board suddenly grounded	Charged board via 1 ohm to ground

TABLE 4

RESULTS OF 'ZAP' TESTING OF POLYSILICON RESISTORS

1. Understanding of behaviour, based on basic physical principles

2. Design Rules to ensure 'acceptable' circuits, both on current and future generations of technology

3. Physical example of failures of known causes - for reference in analysis of field failures

8. REFERENCES

(1) Beasley K., Semi-custom IC's - Reliable and Productive, Proceedings RAMs 1983 pp469-474.

(2) Jensen F., Activation Energies and the Arrhenius Equation, 'Quality and Reliability Engineering International' Vol. 1 (1985) pp13-17.

(3) Whitehead A.P. and Lynch J.T., The Effect of ESD on CCD Reliability, Proceedings 5th EOS/ESD Symposium (1983).

(4) MIL STD 883C Method 3015 'Electro-static Discharge Sensitivity Class-ification'.

(5) Shaw R.N. and Enoch R.D., An Experimental Investigation of ESD-Induced Damage to IC's on PCB's, Proceedings 7th EOS/ESD Symposium (1985).

9. BIOGRAPHY

After graduating in 1979 from the University College of North Wales (UCNW) in Bangor, Keith joined the Quality Assurance Department at Plessey Research (Caswell). He is now responsible for Design and Product Assurance of silicon IC developments and has written a number of papers on the topics of reliability and failure analysis.

Reliability Technology — Theory & Applications
J. Møltoft and F. Jensen (Editors)
© Elsevier Science Publishers B.V. (North-Holland), 1986

CMOS LATCH-UP FAILURE MODE ANALYSIS

F. Fantini[+] , M. Giannini[++] , G. Simeone[°] , E. Zanoni[°°] and G. Enrico[++]

[+]Telettra S.p.A., Quality and Reliability Department, Via Capo di Lucca 31, 40123 Bologna, Italy.
[++]Ing. C. Olivetti S.p.A., Via Montalenghe 8, 10010 Scarmagno (Torino), Italy.
[°]CSATA, Centro Studi e Applicazioni Tecnologie Avanzate, Tecnopolis, 70010 Valenzano (Bari), Italy.
[°°]Dipartimento di Elettrotecnica ed Elettronica, Universita' di Bari, Via Re David 200, 70125 Bari, Italy.

ABSTRACT

CMOS ICs are prone to the "latch-up" failure mode, i.e. to the firing of a parasitic SCR structure, intrinsic in the device technology. The areas of the device most exposed to the latch-up phenomenon are in the input/output circuitry, where external noise pulses may be applied. Identification of the most critical layout points is fundamental for improving device resistance; this can be obtained only by analyzing the real device behaviour when subjected to external stimuli. In this paper the SEM has been used to fully characterize the latch-up failure mode: two techniques have been implemented: the first one is called Digital Differential Voltage Contrast and enables identification of the latching structure in steady-state; the second one uses the stroboscopic voltage contrast method to follow the dynamic firing event. The techniques are applied to two examples referring to a custom and a gate-array device.

1. INTRODUCTION

When compared to nMOS, CMOS ICs show reduced power dissipation, lower susceptibility to single-event soft errors, better noise margins (1). These factors make CMOS technology very attractive for VLSI circuits. As lateral dimensions are scaled, however, the sensitivity of CMOS ICs to latch-up phenomena (2) is greatly enhanced, thus leading to possible reliability hazards.

Latch-up is caused by the parasitic SCR structure inherently present in CMOS integrated circuits. When this structure is switched on by an external cause (for example an input or output voltage greater than Vdd or less than ground), Vdd and Vss contacts are connected by a low resistance path, the circuit no longer meets its functional specifications and a large current flows through the parasitic structure, permanently damaging the device.

Latch-up effectively limits CMOS reliability in practical applications. In fact, according to statistics based on life test and field results, 17% of CMOS total failures are caused by latch-up (3).

Both technological improvements (4 - 6) and layout upgrades (2) are needed to solve the latch-up problem. Particular precautions are necessary for custom ICs, because it is very difficult to quantify layout rules for latch-up prevention and high susceptibilities may arise from incorrect designs. For the purpose of layout correction, an observation technique capable of identifying latch-up sites is sorely needed.

The technique should allow (a) identification of various latch-up current paths in steady-state conditions and (b) observation of the time evolution of latch-up from the firing event to the steady-state condition. Information obtained by identifying latch-up site in steady-state may actually be insufficient for the purpose of IC layout correction because the latch-up firing point is generally different from the latch-up site in steady-state, and because latch-up tends to spread from one site to others in complex VLSI CMOS.

This paper discusses the application of SEM voltage contrast techniques to the study of latch-up failure modes in CMOS ICs. Observation of latch-up current paths in steady-state is achieved by digital image acquisition and topography subtraction; the temporal and spatial evolution of latch-up is followed by means of stroboscopic voltage contrast techniques.

Chapter Two briefly discusses the latch-up phenomenon. Description of experimental apparatus and techniques is given in Chapter Three, while two examples of applications are presented in Chapter Four, followed by the conclusions.

2. THE LATCH-UP FAILURE MODE IN CMOS IC

A parasitic SCR structure is inherently present in CMOS integrated circuits, as shown in Fig. 1a, where the cross section of a bulk, p-well, junction insulated CMOS inverter is shown. Fig. 1b reports the first order equivalent circuit of parasitic SCR structure.

The two parasitic transistors Q1 and Q2 are normally in the OFF state; if one of the two transistors is brought into the ON state (Fig. 1c), and if the current gain product $\beta_1 \cdot \beta_2$ is sufficiently high, latch-up is fired and both transistors are kept in the ON state until the device burns out or power supply is turned off.

In general, β_1 and β_2 are functions of collector current; a minimum current Idd should therefore flow through the device to achieve regeneration; the latch-up path is usually characterized by the value of the holding current, i.e. the minimum supply current which maintains the latched state.
In a CMOS device two parasitic resistances, Rs and Rw, shunt the base-emitter junctions of parasitic transistors; this introduces some differences between the parasitic pnpn structure in CMOS (Fig. 1b) and the classical SCR (Fig. 1c) : (a) the holding current increases, because as Rs and Rw decrease, part of the current is diverted from the base of the parasitic transistors; (b) lateral currents which can flow in n-substrate and p-well due to overvoltages, electrical noise or ionizing radiations, develop a voltage drop across Rs and Rw. As a result, B-E junctions of parasitic transistors can become forward biased, thus firing latch-up.

Technological solutions (4-6) and layout rules (2) have been proposed to increase latch-up hardness by reducing both the current gain of parasitic transistors and the value of shunting resistances Rs and Rw.

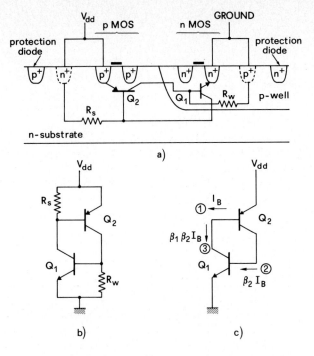

Fig. 1 a) cross section of a p-well, junction insulated CMOS inverter structure showing parasitic elements; b) first order equivalent circuit of parasitic pnpn structure; c) regeneration in the parasitic structure without shunting resistance.

3. SEM VOLTAGE CONTRAST TECHNIQUES FOR LATCH-UP CHARACTERIZATION

3.1 The Digital Differential Voltage Contrast technique

When a DC biased device is observed in the SEM, voltage changes on the IC surface cause local electric fields which modulate the energy and the trajectories of the secondary electrons. Consequently, the SEM image will show a voltage dependent contrast; higher positive potentials are usually coded as black, lower (and ground) as white. In the following we will refer to this static technique as "Conventional Voltage Contrast" (CVC).

The latch-up condition induces marked voltage drops in those semiconductor areas which are crossed by the high current flow. Latch-up sites can therefore be identified if a technique is available for observing the potential distribution on the semiconductor surface, which is covered by thermal oxide and passivation (7). This cannot be achieved by conventional voltage contrast. In fact, MOS devices must be observed at low primary beam energy ($\leqslant$ 2 keV) in order to avoid charge injection and trapping in the gate oxide, with consequent threshold shift. At such low energies primary electrons are stopped within the outrmost layers of the insulator, and no information is available about the potential status on substrate and diffused regions.

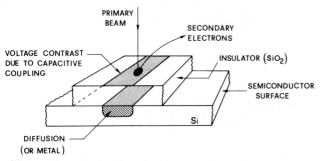

Fig. 2 Mechanism of capacitively-coupled voltage contrast.

Shortly after the device bias is switched on, however, a potential distribution, reproducing the potential pattern in the underlying semiconductor, is present on the oxide surface due to capacitive coupling (Fig. 2). The potential distribution on the oxide surface rapidly decays; therefore, to obtain a meaningful voltage contrast image at low beam energies, it is necessary to switch the device bias on and off in synchronism with the beam scan. In this way it is possible to observe voltage contrast on metal lines and on semiconductor areas covered by the various insulating layers (thermal oxide, passivation, ...). Owing to the superimposition of voltage contrast with topography and chemical contrast contribution, however, interpretation of the micrographs may be difficult, and observation of small potential differences may be impaired.

A digital SEM technique, hereafter called Digital Differential Voltage Contrast (DDVC), was therefore developed, combining the principles of capacitively-coupled voltage contrast with digital image acquisition and topography subtraction.

The experimental apparatus for DDVC is shown in Fig. 3. Digital control of X-Y deflection and A/D conversion of detector signal are achieved by means of a TRACOR NORTHERN TN 1310 column automation system. To obtain topography and chemical contrast subtraction a "differential modulus" was built (8).

System operation can be summarized as follows: field image is digitized in N x M points, where $0 \leqslant N, M \leqslant 4096$. For each point the device bias is switched on and the detector signal is repeatedly sampled and A/D converted, storing the result in R1 register. Then the bias is switched off, and the operation is repeated, storing the result in R2 register. R1 will therefore contain both voltage and chemical-topographical contrast,

while R2 contains only topographical and chemical information. By subtracting the content of R2 from R1, a digital voltage contrast image free from topographic and chemical contribution is achieved. The R1-R2 datum is stored in the R3 regiser, D/A converted and used as brightness signal for the TV display.

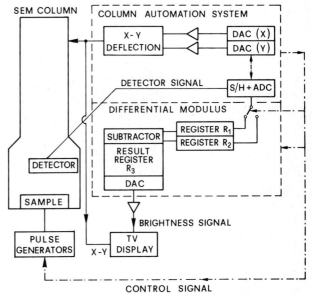

Fig. 3 Block diagram of the DDVC system consisting of the SEM, the column automation system and the differential modulus.

Fig. 4 compares the various types of voltage contrast described above.Fig. 4a shows the SEM micrograph of an unpassivated, unbiased CMOS device, while Fig. 4b shows the conventional voltage contrast image of the same device biased at Vdd = 5 V. As can be seen only the voltage contrast on the uncovered metal lines is visible, i.e. grounded lines appear white in the micrograph, while metallizations at higher potentials appear black. However, no information is available about the potential of semiconductor diffused areas, covered by thermal oxide.

Fig. 4c shows the capacitively-coupled voltage contrast image of the device, obtained by switching the supply voltage Vdd on and off, and blanking the SEM electron beam synchronously during the off period. A pseudostatic voltage contrast on p-well and substrate is obtained because the voltages on the silicon substrate are capacitively coupled through the thermal oxide to the surface. However, owing to the superimposition of voltage contrast with topographic and compositional information on the micrograph, a great difference in contrast between metal lines and diffused areas is visible even when the metallization and the semiconductor are at the same

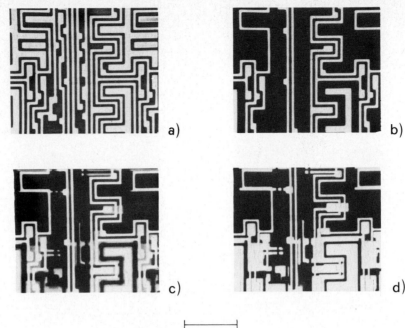

├──────────┤
100 μm

Fig. 4 a) SEM micrograph of an unpassivated CMOS IC with no
bias ; b) conventional voltage contrast image of the same
device, biased; c) capacitively coupled voltage contrast
image; d) digital differential voltage contrast image (with
topography subtraction). Higher potentials are coded as dark.

potential; voltage information from the diffused areas may thus be compromised.

Fig. 4d shows the DDVC image of the device. Thanks to topographical and chemical contrast subtraction, potentials on both p-well and n substrate can now be easily seen and the apparent potential difference between grounded metal lines and p-well is effectively cancelled.

3.2 Stroboscopic voltage contrast techniques

To follow time evolution of periodic potentials on the surface of integrated circuits stroboscopic sampling techniques must be used (9). They consist in illuminating the specimen with a pulsed electron beam in synchronism with the periodic event appearing on the specimen surface.

The e-beam strikes the IC surface only during a time interval τ much shorter than the period of the voltage waveforms on the IC surface. The voltage contrast signal detected therefore refers only to the time interval τ when the beam is on. The duration of the e-beam pulse therefore represents the temporal resolution of the technique.

Two measurement set-ups were employed: (a) voltage contrast stroboscopic imaging, which enables both temporal and spatial evolution of latch-up to be followed and critical layout areas to be identified; (b)

voltage contrast waveform observation, which enables evolution of potentials on selected points of the semiconductor surface in the latched areas to be correlated.

A Cambridge Stereoscan 250 Scanning Microscope equipped with a Lintech SSEM 2 sampling system was used (Fig. 5).

In the voltage contrast stroboscopic imaging mode the primary electron beam strikes the IC always at the same phase of the cycle while scanning the IC in a raster fashion. The image obtained shows the voltage distribution on the device surface at that particular phase of the test waveform only. Thus a sequence of voltage contrast micrographs at various phases represents the dynamic behaviour of the circuit under test.

For voltage contrast waveform observations the stationary beam is positioned at a fixed point of the IC under test. The phase of the e-beam pulse (i.e. the sampling phase) is swept by the delay unit through one or more complete periods of the Vdd signal, so that the whole cycle of the waveform to be measured is repeatedly sampled.

A voltage change at the measurement point causes a shift in secondary electron energy distribution. This energy shift is measured by a spectrometer which allows only electrons with sufficient energy to reach the detector, thus giving a secondary electron current Ib. This current is compared with a reference voltage in a comparator and is kept constant in a feedback loop by varying the voltage barrier of the electron spectrometer. In this way a linear correspondence between local voltage at the measurement point and voltage contrast signal is obtained.

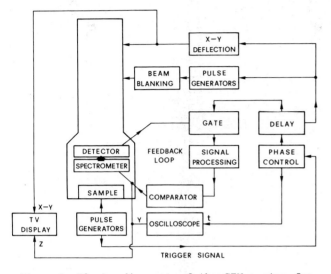

Fig. 5 Block diagram of the SEM system for stroboscopic imaging and waveform recording, consisting of the e-beam blanking unit, the voltage measuring collector, the delay unit and phase shifter.

4. EXPERIMENTAL RESULTS

Latch-up failure analysis mainly consists of layout characterization and electrical measurements, identification of latch-up current paths in steady-state by DDVC, observation of latch-up time evolution by stroboscopic voltage contrast techniques. We report here two examples of analysis, referring to a 6 µm gate array (4.1) and a 7 µm full custom device (4.2).

4.1 6 µm gate array

The first example refers to a semicustom CMOS IC, 6 µm, polysilicon gate, p-well, oxide insulated technology. Latch-up is forced by applying a negative current pulse to an output of the device with $I_{out} \geq 26$ mA for $t > 1$ µs and with a measured $V_{out} \leq -1$ V at Vdd = 5 V.
The layout of the zone involved in the phenomenon is described in Fig. 6a; as shown in Fig. 6b, the output circuit is made up of two inverters. Fig. 6c shows the equivalent circuit of parasitic transistors identified in Table I.
Fig. 7 shows supply voltage Vdd, supply current Idd and output voltage Vout waveforms as measured at IC terminals by an oscilloscope. Vout is the voltage applied the output of the device.

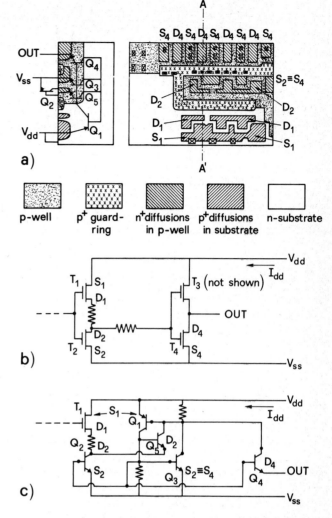

Fig. 6 a) Layout of output circuit of the 6 µm gate array (metal lines are omitted) and cross section along the aa' axis showing parasitic elements; b) electrical circuit of double output inverter; c) parasitic elements circuit.

TABLE I

PARASITIC TRANSISTOR	BASE	EMITTER	COLLECTOR
Q_1 (PNP)	S_1	n-sub	p-well
Q_2 (NPN)	S_2	p-well	D_2
Q_3 (NPN)	$S_2=S_4$	p-well	n-sub
Q_4 (NPN)	D_4	p-well	n-sub
Q_5 (NPN)	D_2	p-well	n-sub

S and D refer respectively to source and drain of MOS transistors.

TABLE I : Parasitic elements in the 6 µm gate array

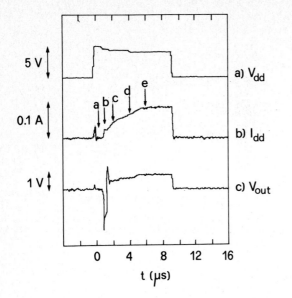

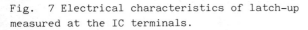

Fig. 7 Electrical characteristics of latch-up measured at the IC terminals.

A large increase in the supply current Idd can be observed as a consequence of latch-up firing. The steady-state is reached about 5 μs after the appliance of the pulse, with a current absorption of about 80 mA.

Fig. 8a and 8b show the DDVC images of the device in the non latched and latched state respectively. As a consequence of latch-up the potential of the p-well region immediately below D2 is increased, and therefore appears dark, and a voltage drop can be observed between this area and the p-well regions near Vss contact, which appear whiter and whiter as the Vss contact is approached. The p$^+$ guard-ring fails to hold the p-well zone at ground potential because of the large currents flowing. Symmetrically, the adjacent substrate region appears whiter.

For a better understanding of the firing mechanism the circuit has been analyzed by means of stroboscopic voltage contrast; let us refer to Fig. 7 for electrical conditions and timing.

Fig. 9 shows the sequence of stroboscopic voltage contrast images taken at times indicated by a, b, c, d, e in Fig. 7b. Fig. 9a refers to the normal operating condition of the device. When the firing pulse is applied, a voltage drop in the substrate takes place, as can be observed in Fig. 9b. In the following, Fig. 9c, the p-well potential rises and the drain D2 goes to a low potential status, while D1, separated from D2 by a polysilicon resistance (see Fig. 6b), remains in the high state. A progressive darkening of the p-well is observed on increasing the time (Figs. 9d and 9e),

beginning in the area where the p-well is not protected by the p$^+$ guard-ring.

The observed behaviour can be explained as follows, with the help of Figs. 6, 7 and 9 : immediately after the pulse has been applied, parasitic transistor Q4 turns on and causes a voltage drop in substrate (Fig. 9b). This voltage drop in substrate turns on Q1, causing an increase in p-well potential and the consequent switching on of Q2. Therefore D2 becomes grounded (Fig. 9c) and a low impedance path between Vdd and Vss is formed by the pMOS transistor T1 and the bipolar transistor Q2.

The latch-up condition in steady-state is maintained by two pnpn paths, including Q1-Q5-Q2 transistors, and, possibly, Q1-Q3.

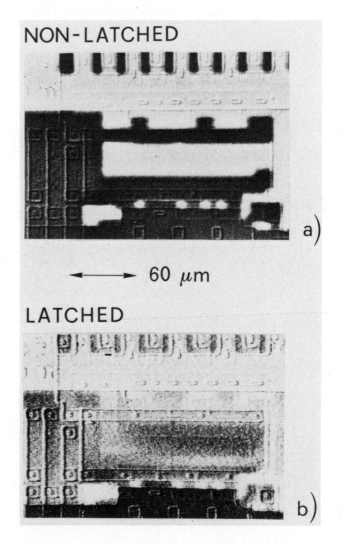

Fig.8 a) DDVC image of output circuit in the non-latched state; b) in the latched state.

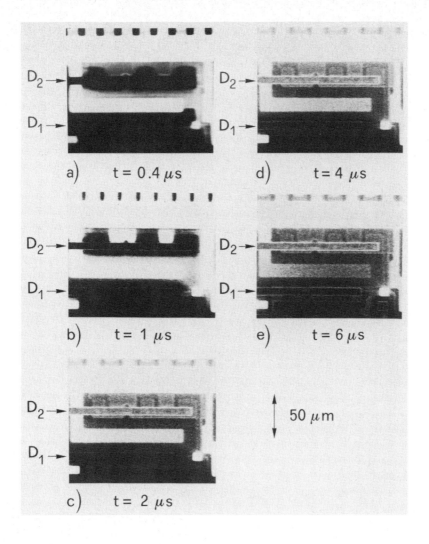

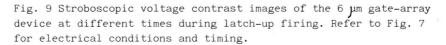

Fig. 9 Stroboscopic voltage contrast images of the 6 μm gate-array device at different times during latch-up firing. Refer to Fig. 7 for electrical conditions and timing.

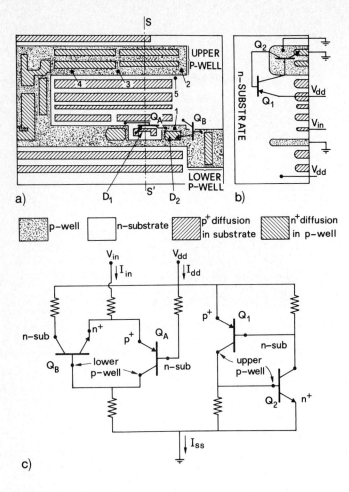

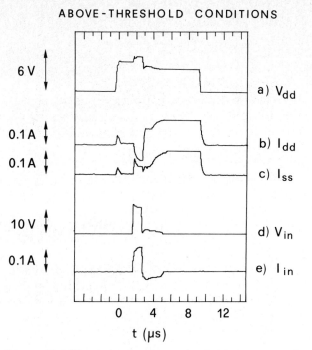

Fig. 10 a) Layout of the analog block of the 7 μm full-custom device; metal lines are omitted; b) cross section along the SS' axis and equivalent circuit of the pnpn path in steady-state; c) equivalent circuit of parasitic elements. "Upper" and "lower" p-wells refer only to the relative position in the drawing.

4.2 7 μm full-custom device

The second example refers to an analog block in a full-custom CMOS IC, manufactured with a standard 7 μm, metal-gate, p-well, junction insulated technology.
In this case latch-up is forced by applying a positive voltage pulse to a specific input. Threshold current for a pulse duration of 0.5 μs was 120 mA, with V_{in} = 10 V and V_{dd} = 5 V. The layout of the block is shown in Fig. 10a. The positive pulse forward biases the p+n protection diode D_1. When D_1 is forward biased, but the injected current is not enough to fire latch-up, the parasitic current decays and the device recovers. On the contrary, if the input pulse current is raised above the threshold value, latch-up occurs and the I_{dd} current still continue to increase to a latched steady-state afer the pulse is removed.
Fig. 11 depicts the behaviour of

Fig. 11 Electrical characteristics measured at the IC terminals in above-threshold conditions.

supply voltage V_{dd}, supply current (I_{dd} and I_{ss}), input voltage V_{in} and input current I_{in} as measured at the IC terminals. Thrre phases can be observed: (a) for t ~ 2.8 μs the input pulse is on and I_{dd} is reversed; (b) in the transient condition, 2.8 μs ≤ t ≤ 6.5 μs a positive increase in I_{dd} is measured, while I_{ss} remains lower; I_{in} current sinking from th device input into the 50 ohm termination of pulse generator is observed; (c) for t ≥ 6 μs a steady-state condition is reached; a decrease in V_{dd} is observed and both I_{dd} and I_{ss} markedly increase and become equal, whereas I_{in} has returned to zero. This suggests that different current paths could be active in transient and steady-state conditions.
Figures 12a and 12b report DDVC images of the device in the non-latched and latched state respectively. For a better understanding, in Fig. 13a and 13b magnified images of the latched area are reported. In the latched state a marked voltage drop is observed in the substrate region indicated by arrows in Figs. 12 and 13. As a consequence of latch-up this substrate zone becomes whiter, while the p-well immediately beside it becomes darker. The central region, where the highest voltage drop is observed, appears crossed by the largest current flow. In this steady-state condition the latch-up appears, therefore, to be sustained by transistors Q1 and Q2 shown in Fig. 10.
To study temporal and spatial

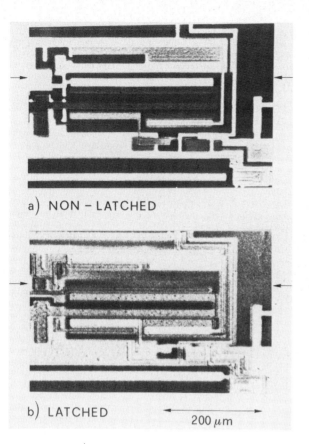

Fig. 12 a) DDVC image of analog block in the non latched state; b) in the latched state.

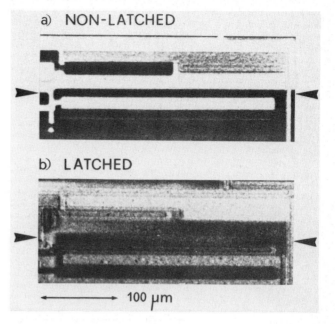

Fig. 13 Enhanced DDVC image of the latched area in the non-latched (a) and latched (b) state.

evolution of latch-up stroboscopic techniques were applied. We had recourse to stroboscopic voltage contrast imaging first.

In the following we will refer to Fig. 10 for layout description and parasitic element identification, and to Fig. 11 for timing and electrical conditons.

Fig. 14 shows stroboscopic voltage contrast micrographs of the analog block taken during the firing of the phenomenon. When the input pulse is applied, Fig. 14b, the p-well adjacent to the diode D1, hereafter called "lower p-well", reaches a high potential. Shortly after the input pulse is turned off, Fig. 14c, the potential in the "lower p-well" decays, while the upper part of the p-well ("upper p-well") has not yet reached the latched condition and still holds a low potential. A voltage increase at the extreme right of the "upper p-well" takes place at t = 4.2 μs, as shown in Fig. 14d; on increasing the time the area involved in latch-up widens. A marked increase in potential of the whole p-well area is observed when steady-state is reached, Fig. 14e, along with maximum current absorption.

Stroboscopic voltage contrast observations were repeated in "under threshold" conditions. The main feature observed during the transient supply current

absorption was a marked increase in the "lower p-well" voltage; no effect was visible in the "upper p-well".

To correlate voltage contrast analysis with electrical measurements at the IC terminals, critical points were identified within the IC layout, and stroboscopic waveform recording was performed.

Figs. 15 and 16 show Vdd, Idd and Iin waveforms measured at the IC terminals, together with stroboscopic voltage waveforms taken in the "lower p-well", point 1, in the "upper p-well", points 2-4, and in the n-substrate point 5. Figure 15 refers to "under-threshold" conditions, while Fig. 16 to "above-threshold". Point positions are identified in the layout of Fig. 10a.

As can be seen in Fig. 15, the application of the input pulse causes a large potential increase in the lower p-well. After the input pulse is turned off, the lower p-well potential does not return to zero immediately, but slowly decays in about 2 μs. The voltage waveform in the lower p-well point 1 has the same behaviour of the Idd transient current absorption and of Iin current sinking from the device into the 50 ohm termination of the voltage generator, Figs. 15c, d, e.

Furthermore, during the transient current absorption, the upper p-well remains practically at ground potential, Fig. 15e.

In the "above threshold" conditions, i.e. when the input pulse is so high that latch-up is initiated, Fig. 16, a marked potential increase in the upper p-well is

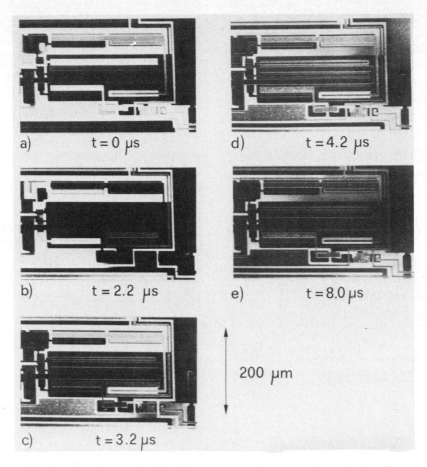

Fig. 14 Stroboscopic voltage contrast micrographs of the 7 μm full-custom device at different times during latch-up firing. Refer to Fig. 11 for electrical conditions and time indications.

observed and point 2 (at the extreme right of the upper p-well in Fig. 3 a) reaches a high potential value faster than points 3 and 4.

Furthermore, supply current absorption in steady-state appears to be correlated with the potential increase in the upper p-well.

The potential of the lower p-well in the latched steady-state is low, Fig. 16d , thus confirming that the lower p-well is not involved in the steady-state latch-up path.

    In conclusions, the latch-up triggering mechanism could be explained as follows:

(a) during the input pulse (Vin > Vdd) QA is ON. The voltage of the "lower p-well", which is QA collector, is forced to a high value due to the large current flowing;

(b) after the input pulse, QA turns off. Owing to charge storage, however, the potential of the "lower p-well" decays slowly. In this situation QB is in the active region, having the base ("lower p-well") at a higher potential than the emitter (n^+ diffusion of diode D2). Switching on QB causes the transient increase in supply

current Idd and the presence of the negative input current Iin;

(c) if the voltage drop caused in the n-substrate by the transient switching on of QB is sufficiently high, B-E junction of transistor Q1 is forward biased, thus firing latch-up (Fig. 16). In latch-up steady-state conditions QA and QB are off; the latch-up current flows through the Q1 – Q2 path which involves only the "upper p-well" (Figs. 14 and 16).

    In conclusion, input parasitic transistors QA and QB are responsible for the latch-up firing, even though they are not directly involved in the latch-up path in steady-state, which includes only Q1 and Q2.

5. DISCUSSION AND CONCLUSIONS

    Two SEM voltage contrast techniques were set up for analysis of latch-up in CMOS integrated circuits. Problems studied include: (a) identification of latch-up current paths in steady-state condition by means of the DDVC technique; (b) observation of the time evolution of latch-up from the

UNDER-THRESHOLD CONDITIONS

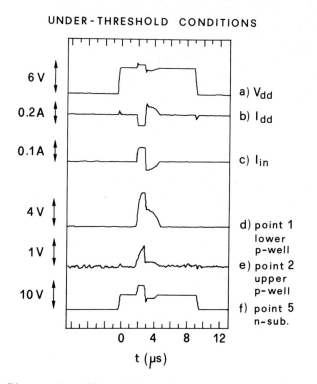

ABOVE-THRESHOLD CONDITIONS

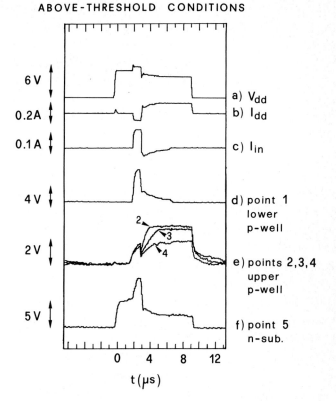

Fig. 15 Vdd, Idd and Iin waveforms as measured at the IC terminals in the under-threshold conditions, compared with stroboscopic voltage waveforms in points 1, 2, 5. See Fig. 3a for localization of the various points.

Fig. 16 Vdd, Idd and Iin waveforms as measured at the IC terminals in the above-threshold conditions, compared with stroboscopic voltage waveforms in points 1, 2, 3, 4, 5. See Fig. 10a for localization of the various points.

firing event to the final condition by means of stroboscopic voltage contrast technique.

The DDVC technique yields high quality voltage contrast images of the IC surfaces, free of topographic and chemical contrast, with good signal to noise ratio; furthermore, thanks to low beam energy and capacitive coupling, passivated devices can be observed without either causing permanent damage or altering the electrical operation of the IC, and in particular its latch-up behaviour. Stroboscopic voltage contrast technique enables detailed analysis of latch-up triggering mechanisms and of its dynamic evolution from the firing event to the steady-state condition. The high temporal ($\sim$ 2 ns) and voltage ($\sim$ 20 mV) resolution achievable in the waveform recording mode can provide a valuable tool for the experimental confirmation of simulation models of latch-up phenomena. Only dynamic observation techniques allow a complete understanding of latch-up triggering and an effective correction of IC layout; in fact, as we show by means of the second example presented, the latch-up firing point may be substantially different from the site where latch-up is sustained in steady-state

conditions.

In conclusion, a SEM system equipped for DDVC and stroboscopic voltage contrast has proved to be most useful for the diagnosis of latch-up problems in CMOS ICs, and represents a valid aid for IC layout correction.

Acknowledgement

We would like to thank Massimo Vanzi (Telettra S.p.A.) for DDVC observations and for comments and suggestions.

REFERENCES

(1) J. Fiebiger, CMOS, a designer's dream with the best yet to come, Electronics, April 5, 1984, 113-115.

(2) D.B. Eistreich, The physics and modeling of latch-up in CMOS integrated circuits, Stanford Electronics Laboratories Technical Report No. G-201-9, 1980.

(3) P. Brambilla, F. Fantini, G. Mattana, Updating of CMOS reliability, Microelectronics and Reliability, vol. 23, no. 4, 1983, 761-769.

(4) D.B. Eistreich, A. Ochoa and R.W. Dutton,
An analysis of latch-up prevention in CMOS
ICs using an epitaxial-buried layer process,
IEDM Tech. Dig., 1978, 230.

(5) M. Sugino, L.A. Akers, M.E. Rebeschini,
Latch-up free Schottky barrier CMOS, IEEE
Trans. Electron Devices, vol. ED-30, 1983,
110-118.

(6) R.D. Rung, C.J. Dell'Oca and L.G. Walker,
A retrograde reference p-well for higher
density CMOS, IEEE Trans. Electron Devices,
vol. ED-28, 1981, 1115-1119.

(7) S.M. Davidson, Latch-up and timing
failure analysis of CMOS VLSI using electron
beam techniques, IEEE 21st Ann. Proc. Int.
Reliability Physics Symp. 1983, 130-137.

(8) C. Canali, F. Fantini, M. Giannini, A.
Senin, M. Vanzi, E. Zanoni, A SEM based
system for a complete characterization of
latch-up in CMOS ICs, to be published in
Scanning, 1986.

(9) E. Menzel, E. Kubalek, Fundamentals of
electron beam testing of integrated circuits,
Scanning, vol. 5, 1983, 103-122.

BIOGRAPHIES OF AUTHORS

Fausto Fantini received the degree in
electronic engineering from the University of
Bologna, Italy, in 1971. After serving the
Army, in 1973 he joined the Quality and
Reliability Department of Telettra S.p.A. .
He is author of a number of papers about
reliability physics of integrated circuits.

Manuela Giannini received the degree in
Physics from the University of Bologna,
Italy, in 1983 and joined Olivetti S.p.A. the
same year. She works on Scanning Electron
Microscopy and Electron Beam Testing
techniques applied to VLSI analysis.

Giovanni Simeone received the degree in
Physics from the University of Bari, Italy,
in 1984; the same year he joined Centro Studi
e Applicazioni Tecnologie Avanzate (CSATA),
Bari, Italy. He is currently working on
electron microscopy techniques for failure
analysis of integrated circuits.

Enrico Zanoni received the degree in Physics
from the University of Modena, Italy, in
1982. He obtained a two-year research grant
from Marelli Autronica S.p.A., to study
reliability problems of automotive
electronics. In 1985 he joined the Electronic
Engineering Department of the University of
Bari, where he is currently a full-time
researcher. He is active in the field of
realiability physics of electronic devices.

Giuseppe Enrico was born in Scarmagno
(Torino), Italy. In 1974 he joined Olivetti
S.p.A., where he works on integrated circuits
characterization and failure analysis.

RELIABILITY MODELLING AND PREDICTION I

Reliability Technology — Theory & Applications
J. Møltoft and F. Jensen (Editors)
© Elsevier Science Publishers B.V. (North-Holland), 1986

NEW METHODS IN RELIABILITY ANALYSIS

A. Bendell
Head of Department of Mathematics, Statistics and Operational Research
Trent Polytechnic,and Services Ltd., Nottingham, U.K.

Whilst reliability engineering has made major progress and achieved important results over the last two decades, as in all disciplines there is some tendency towards conservatism. This is partially justifiable since, unfortunately, the reliability literature still contains many new methodologies which are, in fact, badly-founded. However, one bad aspect of this conservatism has been the tendency for reliability engineers to employ established methodologies on a "black-box" basis, neither verifying the validity of intrinsic assumptions, employing sensitivity analysis, nor considering alternatives. In this paper the author introduces some new methods in reliability analysis which are, however, well established in other areas of statistical application. In particular, emphasis is given to an Exploratory Data Analysis approach, in which reliability data is explored for appropriate structure to employ in the analysis. Methods such as Proportional Hazards Modelling, Multivariate Techniques and Time Series Analysis are employed. The author's recent experience of applying these various techniques in reliability work is described.

1. INTRODUCTION

Reliability analysis has a short history and an expanding present; a lot has been, and is, going on. In such circumstances, it is no surprise that two particular patterns have emerged. On the one hand, we are battered by allegedly important developments in reliability modelling, and on the other hand (and partly in consequence) professional practice tends to be conservative.

Many of the proposed and published developments in models, particularly the more "Heathcote Robinson"-ish, will not in the long run survive nor prove important; they will be written-off as "easy publications". In contrast, however, certain broad developments in modelling approach will be important conceptually. Even more importantly, certain emerging statistical data analysis techniques must have a major part to play in the future of reliability analysis. It is the purpose of this paper to identify and describe some of the techniques which fall into this later category. Some comments will also be made about relevant developments in models at the more theoretical level.

The current state of affairs in statistical reliability is very much a dichotomy between the theory and practice of the subject. Whilst the theoretical literature has tended to concentrate on aspects and issues of little real interest to the practitioners and which indeed are not practical for practical use, the practitioners have tended to imbed long established but crude methodologies on a "black-box" basis, paying little attention to the validity of the assumptions on which they are based nor the robustness of the methods and results obtained. Against this backdrop, one needs to isolate both those broad developments in modelling approach that will be important conceptually to the theoretical side, and the useful data analysis techniques that are long overdue for standard practice.

The former identification is very much in the nature of "separating the wheat from the chaff", whilst the latter is becoming increasingly urgent in order to both squeeze as much information out of the data as possible, and to avoid the currently commonplace mistakes resulting from the blind application of standard practice.

In subsequent sections we discuss the statistical data analysis techniques emerging in the reliability field, whilst in this last paragraph of this section we briefly comment on some conceptually important developments in theoretical reliability modelling. One overview of this modelling area was provided by Bendell [1] who considered in particular the trend towards a multilevel or continuous concept of reliability. More detail of this is provided in Bendell and Humble [2]. This multilevel work is in contrast to most of the theoretical multilevel literature which has instead been concerned with proving formal results for marginally varying definitions of multilevel coherent systems, rather than establishing a physically plausible multilevel component degradation model, e.g. Barlow and Wu [3], El-Neweihi, Proschan and Sethuraman [4], Griffith [5], Butler [6], Natvig [7]. See also Bendell and Ansell [8], and Bendell [9]. Other important conceptual developments in modelling identified by Bendell [1, 9, 10] are concerned with other directions to systematically liberalise the restrictive assumptions of the classical dichotomic reliability model. These extensions include allowance for (stochastic) inertias in component operation (Bendell, [11]), multiple failure modes and time scales (e.g. Bendell and Humble, [12]), multiple functions, time varying and stochastic systems structure and the like. There is a clear need for such extensions to be treated in a systematic manner, rather than haphazardly at a low level of generality and in the context of specific system configuration. Unfortunately, many such "extensions" in the IEEE Transactions on Reliability and Microelectronics and Reliability suffer from this drawback. They have often added to the confusion, rather than helped, the development of general models.

2. EXPLORATORY DATA ANALYSIS

As stated at the beginning, however, in the author's opinion it is the developments in statistical data analsyis techniques that are currently more important to the development of reliability analysis, than developments in reliability theory. One major aspect of this is the movement towards the use of the Exploratory Data Analysis (E.D.A.) approach in reliability.

Exploratory data analysis techniques have now become standard in statistical methodology. Rather than relying upon formal test procedures the approach is to apply differing and often conflicting analyses to the data, preferably making few assumptions. The idea is to search the data for unsuspected pattern and abnormalities rather than just attempt to confirm a-priori assumptions. The seminal book is Tukey [13].

In the reliability field (whether hardware or software), because of the particular characteristics of the data structure it is the approach rather than the usual techniques which are most appropriate. An introduction to the application of e.d.a. in the reliability area is provided by Bendell and Walls [14]. See also Tomikawa et al [15]. In a sense, elements of this approach may also be found in the work of Ascher and Feingold [16]. Walls and Bendell [17] provide extensive examples of applying such e.d.a. methods to diverse reliability data. Bendell [18] and Walls and Bendell [19] discuss its application to the analysis of software reliability.

The simplest reliability applications of e.d.a. consist of simple graphs and simple models and statistical tests applied to interfailure (or other) data at various levels of disaggregation by mode of failure, cause of failure, equipment or component classification, environmental conditions, etc. Examples are cumulative failure against time plots, fitted Non Homogeneous Poisson Process models with accompanying simulation bands, observed against expected plots, Laplace tests for trend, correlograms, etc.

The application of these methodologies quite extensively to date has revealed some surprising results. Typically, reliability field data behaves much less like standard theory than one would be led to believe, with times between failures not only having a non-constant hazard rate (i.e. being non-exponential) but also being non-stationary, i.e. exhibiting an increasing or decreasing trend; reliability growth or decline. Further, serial correlation and cyclical behaviour are frequently in evidence between interfailure times, and other dependencies are also common between components, and between interfailure and repair times. Problems of outliers and non-homogeneity of data sets are also commonplace. See Walls and Bendell [17].

3. EXPLANATORY FACTORS

Another major step being made in reliability analysis is the increasing awareness of the importance of explanatory factors in ascertaining the reliability of equipment or components and the developments in the use of specialised statistical methods to incorporate these. Such explanatory factors may correspond in hardware applications to variations in specifications, temperature, pressure, other environmental and operational conditions, material or design changes, etc.

In one sense the effects of certain explanatory factors have been incorporated into reliability analysis for some time by their inclusion as π-factors etc. in MIL-STD-217, the SRS databank, and the like. However, whilst certain explanatory factors are allowed for in this way, other potentially important ones are not, and the existence or not of such effects, the way they are to be incorporated, and their magnitude have all been determined a-priori from the point of view of the user's reliability problem. In contrast, the new awareness of the importance of explanatory factors upon reliability has focussed upon methodologies which enable any significant explanatory factors to be identified from field data, and the nature and magnitude of their effects estimated. The most important advent in this area has been the development and application of Proportional Hazards Modelling (PHM), although this in itself has led to an increasing interest in conceptually related, but simpler, methodologies, such as Logistic modelling.

Proportional Hazards Modelling identifies the effects of various explanatory variables or factors which may be associated with variations in the life length of equipment. Factors such as temperature, pressure, material, use, etc. may be included. Repairable as well as non-repairable systems may be studied. Data may be censored or uncensored. The technique is a powerful method for decomposing the variation in life lengths into orthogonal factors, identifying the significant ones, and reconstituting the model for prediction purposes.

The technique largely owes its origins to the seminal paper which Professor D.R. Cox read to the Royal Statistical Society in March 1972. At that time, he described the likely applications of Proportional Hazards Modelling as being "in industrial reliability studies and in medical statistics". From its introduction the technique has generated a great amount of interest, although until very recently mainly in medical fields. A vast array of research papers have dealt with theoretical considerations and developments of the basic model. Why is it, however, that recorded applications of the model to data have until recently been almost entirely associated with medical data? The answer must be very largely because of the familiar lack of qualified statisticians working in practical reliability.

The recent attempts at Proportional Hazards Modelling in the reliability literature very much reinforce the view. The papers have appeared to date almost entirely in Conference Proceedings or internal reports, with the implication that they were not initially refereed; e.g. Davis et al [20], Nagel and Skrivan [21], Booker et al [22] , Ascher [23] , Jardine [24], Dale [25] , Jardine and Anderson [26] . The data contained in these papers is diverse, including motorettes (Dale), marine gas turbines and ships sonar (Ascher), valves in Light Water Reactor nuclear generating plants (Davis et al), and aircraft engines (Jardine and Anderson). However, the proportional hazards models employed are, in general, basic and are not developed to take specific account of the complexities of structure arising in reliability data. Further, the data sets chosen to illustrate the methodology in these papers have often been inappropriate for the analysis carried out. Indeed, sometimes the data has so little structure or the models are so inappropriate that proportional hazards modelling yields almost no useful information; see e.g. Ascher [23] . It is also essential that assessments of the appropriateness and the fit of the proportional hazards model and the explanatory variables are made, but these have not always been reported in the recent reliability literature, possibly giving a misleading impression; see Jardine and Anderson [26] .

Apart from these conference papers and internal reports, there is very little in the current open literature dealing specifically with proportional hazards modelling for reliability data. The major books such as Kalbfleisch and Prentice [27] and Lawless [28] are more general and deal largely in terms of medical data. However, an important discussion of some aspects of Proportional Hazards Modelling in Reliability is contained in the recent review paper by Lawless [29] and in the contributions by its various discussants.

A systematic study of the application of proportional hazards modelling to reliability data is currently under way at Trent Polytechnic. To date the methodology has been successfully applied to diverse data in railway applications, gas and electricity supply, consumer durables, vehicles, valves, software, computer hardware, etc. Other studies are under way. See Bendell [18, 30] , Wightman and Bendell [31], Bendell et al [32] , Argent et al [33] .

The fundamental equation on which the proportional hazards model is based is an assumed decomposition of the hazard function for an item of equipment into the product of a base-line hazard function and an exponential term incorporating the effects of a number of explanatory variables or covariates, $z_1, \ldots, z_k$. That is

$$h(t; z_1, \ldots, z_k) = h_0(t) \exp(\beta_1 z_1 + \cdots + \beta_k z_k). \quad (1)$$

The β_i's are unknown parameters of the model defining the effects of each of the explanatory variables. These need to be estimated from the data and tested to see whether each explanatory variable really has an effect in explaining the variation in observed failure times. These effects are assumed to act multiplicatively on the hazard rate, so that for different values

of the explanatory variables, that is say different temperature and pressure conditions, etc., the hazard functions are proportional to each other over all time t. It is this property that gives the technique its name.

The base-line hazard function $h_0(t)$ represents the hazard function that the equipment would experience if the covariates all take the base-line value zero, which may correspond, depending on what the covariate is, either to a natural zero such as a zero temperature, or an arbitrary zero corresponding, say, to an arbitrary time point from which to measure installation date. There are essentially two ways of modelling this base line hazard. Either a parametric regression model may be assumed so that say a Weibull hazard may be assumed for $h_0(t)$, or a distribution-free approach may be taken under which no particular form is assumed for $h_0(t)$, but its form is estimated from that data. The later approach was first suggested by Cox [34] , is more common, and is sometimes taken as synonymous with the term proportional hazards modelling. Indeed, it is a major advantage of proportional hazards modelling that one does not have to assume a specified form for the base-line hazard when this may be difficult due to the confusing effects of the covariates. Instead, the model provides a distribution-free estimate of the base-line hazard function which can be used to check-out the appropriateness of standard distributional forms. The fact that in the author's experience standard forms such as the Weibull or constant hazard rate of the exponential often do then turn out to be appropriate is highly reassuring. But it is not necessary to assume a given form a-priori. However, in a number of the recent reliability papers this has been done, although sometimes with a little justification: Nagel and Skrivan [21] , Jardine and Anderson [26] .

The detail of the methodology is not developed here, but the interested reader is referred to Kalbfleisch and Prentice [27] . The usual method first iteratively estimates the effects of the covariates $\beta_1, \beta_2, \ldots, \beta_k$ using the so-called method of partial likelihood. Essentially, this technique employs the usual method of scoring baaed upon a Taylor expansion for each step in the iteration, starting with initial values of zero. Once the estimates converge, tests of whether each explanatory variable has any significant effect are based upon the asymptotic Normality of the estimates. Thus either Normal tests or the related chi-squared tests may be used. Chi-squared tests can also be conducted for the explanatory ability of the whole set of covariates included, based upon the likelihood ratio statistic, which compares the likelihood under the fitted parameter values and under the assumption that they are all zero, or upon the sums of squares of the standardised estimates.

A stepwise procedure may be incorporated into the proportional hazards approach, whereby in its backwards version non-significant explanatory factors are excluded one at a time and the model re-run until all facts are significant. Alternatively, a forward approach may be taken whereby new variables are sequentially forced into the model.

Having estimated $\beta_1, \beta_2, \ldots, \beta_k$ the method obtains a distribution-free estimate of the base line hazard function based upon discrete hazard contributions at each of the times at which failures were actually observed to occur. This can then be compared to standard distributional forms such as the Weibull.

The implementation of Proportional Hazards Modelling necessitates purpose-built computer routines, and we have developed what is now an extensive suite. More limited routines are available within commercial packages such as BMDP or SAS, and via GLIM.

One final point is particularly worth making here on explanatory factors. I am now frequently approached by engineers who believe that they have an appropriate scenario for Proportional Hazards Analysis. Usually this is so. Occasionally, however, what they are interested in is in fact a simpler situation for which the analysis methods, such as Logistic modelling, have been around for a long time, although not used in reliability.

Logistic analysis is appropriate for a situation in which the response variable is binary in nature, such as the failure or non-failure during a test. It is a well established methodology in numerous areas of statistical application, and is well-documented in the literature, e.g. Cox [35], Nelder and McCulloch [36]. My experience of applying it in a reliability setting is good. Corresponding to equation (1) for the proportional hazards model the logistic model postulates

$$\rho = \frac{exp(\beta_0 + \beta_1 Z_1 + \cdots + \beta_k Z_k)}{1 + exp(\beta_0 + \beta_1 Z_1 + \cdots \beta_k Z_k)}, \quad (2)$$

where ρ is the proportion (or probability of) failing. The reason for this formulation is complex and not discussed here, but alternatives do exist. See Cox [35]. In contrast to Proportional Hazards Modelling, the ordinary statistical method of maximum likelihood can be used to estimate the parameters in (2), and this may be applied e.g. by the use of GLIM.

4. MULTIVARIATE METHODS

Related to the possible importance of explanatory factors in describing the failure patterns of equipment by techniques such as PHM, is the realisation that, without the conventional over-simplification, the pattern of equipment failures is a complex phenomena best conceptualised in a multi-dimensional space.

Such a conception is not new; Buckland [37] proposed it in the context of the reliability of nuclear power plants. It has, however, been neglected in terms of practical implementation. Our work in this area at Trent is at an early stage of development, but is promising; Walls and Bendell [38] and Libberton et al [39],

report a study applying this approach to automatic fire detection systems on a large industrial site. Some attention is also being paid to its application to software reliability problems under the REQUEST project of the ESPRIT programme, and by my own team as part of the Software Reliability Modelling project under the Alvey Programme.

Multivariate methods are aimed at attempting to reduce the dimensionality of complex multi-dimensional data in order to ease its exploration and modelling, and to carry out such exploration and modelling to ease prediction. Useful multivariate techniques include principal component analysis, cluster analysis and principal co-ordinate analysis. See e.g. Everitt [40], Mardia et al [41].

5. TIME SERIES

The final main group of methods at which we look is the group of Time Series Methods, and in particular Box Jenkins Forecasting Techniques (Box and Jenkins [42]). Whilst widely applied in varied statistical application areas for some considerable time, these methods have again been neglected in the reliability field. The main exception is the work of Nozer Singpurwalla and his co-workers at the George Washington University, e.g. Singpurwalla [43-45], Crow and Singpurwalla [46], Horigome et al [47], Singpurwalla and Soyer [48]. A good justification for the use of Time Series techniques in the reliability area is provided by Dale and Harris [49]. A time series corresponds to an indexed or time-spaced set of measurements, so that consecutive times between failures, for example, form such a series. Time series methods are concerned with identifying and modelling such sequential behaviour for prediction purposes. As well as allowing for features in time-between-failure data such as increasing or decreasing trend (reliability growth or decline), such methods enable the convenient inclusion of cyclical fluctuations which are often observed in reliability data and which may correspond to weather effects, imperfect repair, stages of software testing, etc.; Walls and Bendell [17].

In the literature Singpurwalla and his associates have considered the approaches of several time series methodologies. These include an empirical Fourier analysis approach to search for cycles in software data in Crow and Singpurwalla [46]; transfer function models to describe the relationship between the running times and down times of a nuclear power plant in Singpurwalla [45]; a random co-efficient 1st order autoregressive model in Horigome et al [47] and its ramifications in Singpurwalla and Soyer [48].

At Trent, research is continuing into the use of time series methods in reliability and, in particular, the potential of the Box-Jenkins time series approach is being investigated. The Box-Jenkins approach is an essentially data-based, flexible tool with an in-built theory of forecasting. This approach, discussed in Box and Jenkins [42], aims to fit an appropriate model to the data, chosen and adapted from the structure exhibited by the data itself. The two basic tools for identifying structure are the correlogram, and the partial correlogram, and a suitable model is then constructed on the basis of the type of structure displayed in each of these. For example, the typical structures displayed include the tendency for the (partial) serial correlations to "cut-off" to zero at a particular lag or for the (partial) serial correlations to decay towards zero as the lag increases. Further discussions of these criteria are given in Box and Jenkins [42] .

While exploratory in nature, the Box-Jenkins time series approach requires skill and experience to identify an appropriate model. Reliability data, in particular, presents additional hazards to the unwary as it is typically noisy, e.g. Walls and Bendell [19] . It may be that in order to provide improved models for the statistical dependencies observed in reliability data that some developments of the Box-Jenkins concept are required. In the statistical literature, for example, Lawrance, Lewis and their associates are developing further time series models which are based on exponential, gamma and other assumptions, rather than normality. These models follow the Box-Jenkins concepts, but may potentially provide more suitable models for the noisy reliability data; see e.g. Lawrance and Lewis [50] and Lewis [51] . Of course, it may be that the noisy behaviour of the data is simply reflecting changes in factors that are external to the equipment, such as environmental conditions. If this is the case then it is possible to model the effects of such explanatory factors on the behaviour of the equipment as discussed in Section 3.

More detailed discussion of the application of time series methods to reliability data is given in Walls and Bendell [52] .

6. FINAL COMMENTS

This paper has highlighted some relatively new methods in reliability assessment that the author believes will be important in the (near) future. His overview of the development of reliability assessment and the barriers to progress are given in Bendell [53, 54] . He has discussed other developments in methodology, and progress with the remaining problems, at length elsewhere; e.g. Bendell and Ansell [55] consider fault-tree analysis and Markov modelling. Perhaps most important in this respect, however, are developments in data collection activities and data banks (Bendell [56]) as well as the need for more work on industrial test design (Baines [57]).

In both the long and short run, however, the methodologies to be applied by reliability analysts must depend upon the computer package and educational facilities available to them. Currently specifically in the reliability field both of these are extremely poor; Bendell [53, 54] . It is here progress must be made.

REFERENCES

[1] Bendell, A. A Generalised Semi-Markov Reliability Model, Ph.D. Thesis, C.N.A.A. (Sheffield City Polytechnic)(Aug. 1982).

[2] Bendell, A. and Humble, S. A reliability model with states of partial operation, Naval Research Logistic Quarterly, 32, 1985, 509-535.

[3] Barlow, R.E. and Wu, A.S. Coherent systems with multi-state components. Math.Operat. Res. 3, 1978, 275-281.

[4] El-Neweihi, E., Proschan, F. and Sethuraman, J. Multistate coherent systems. J. Appl. Prob. 15, 1978, 675-688.

[5] Griffith, W.S. Multistate reliability models. J. Appl. Prob. 17, 1980, 735-744.

[6] Butler, D.A. A complete importance ranking for components of binary coherent systems, with extensions to multi-state systems. Naval Res. Logist. Quart. 26, 1979, 565-578.

[7] Natvig, B. Two suggestions on how to define a multistate coherent system. Adv. Appl. Prob. 14, 1982, 434-455.

[8] Bendell, A. and Ansell, J. The Incoherency of Multistate Coherent Systems, Proceedings of the Fourth National Reliability Conference, Vol. 2. 1983, pp 5B/5/1-12. Invited for republication in Reliability Engineering, 8, 165-178 (1984).

[9] Bendell, A. The Classification of Reliability Models, in E. Lauger and J. Moltoft (eds.). Reliability in Electrical Components and Systems, North Holland 1982, 89-95.

[10] Bendell, A. A Classification System for Reliability Models, IEEE Transactions on Reliability, R-33, 1984, 160-164.

[11] Bendell, A. The Effect of Stochastic Time Delays on the Reliability of Isolated Impulse Systems, Naval Research Logistics Quarterly, 30, 1983, 537-551.

[12] Bendell, A. and Humble, S. Optimal Replacement for Systems that Operate and Idle. Journal of the Operational Research Society 32, 1981, 875-884.

[13] Tukey, J.W. Exploratory Data Analsyis, Addison-Wesley (1977).

[14] Bendell, A. and Walls, L.A. Exploring reliablity data, Qual. and Rel. Eng. Int., 1, 1985, 35-51.

[15] Tomikaiwa, T., Ueyama, T., Hatagi, K. and Masuda, A. An example of exploratory failure analysis. Proc. 14th Symposium on Reliability and Maintainability, Union of Japanese Scientists and Engineers, 29-31 May 1984, 66-67.

[16] Ascher, H.E. and Feingold, H. Repairable Systems Reliability: Modelling, Inference, Misconceptions and Their Causes, Marcel Dekker, New York.

[17] Walls, L.A. and Bendell, A. The structure and exploration of reliability field data; what to look for and how to analyse it, Proc. 5th National Reliability Conference, Birmingham, 1985, 5NRC/5B/1-18. Invited for republication in Reliability Engineering (1986).

[18] Bendell, A. The use of E.D.A. techniques for software reliability assessment and prediction, Proc. NATO A.S.I. on the Challenge of Advanced Computing Technology to Systems Design Methods, July 1985. To be published by Springer Verlag, 1986.

[19] Walls, L.A. and Bendell, A. Exploring software reliability data, in Pergamon-Infotech State-of-the-Art Report on Software Reliability. (1986).

[20] Davis, H.T., Campbell, K. and Schrader, R.M. Improving the Analysis of LWR Component Failure Data, Los Alamos Scientific Laboratory Report, LA-UR 80-92 (1980).

[21] Nagel, P.M. and Skrivan, J.A. Software Reliability: Repetitive Run Experimentation and Modelling, Boeng Computer Services Co. Report BCS-40366, NASA Report No. CR-165836 (1981).

[22] Booker, J., Campbell, K., Golldman, A.G., Johnson, M.E. and Bryson, M.C. Application of Cox's Proportional Hazards Model to Light Water Reactor, Los Alamos Scientific Laboratory, Report LA-8834-SR (1981).

[23] Ascher, H. Regression analysis of repairable systems reliability. In Electronic Systems Effectiveness and Life Cycle Costing, J.K. Skwirzynski ed., Springer-Verlag, 119-133 (1983).

[24] Jardine, A.K.S. Component and system replacement decisions. In Electronic Systems Effectiveness and Life Cycle Costing, J.K. Skwirzynski ed., Springer-Verlag, 647-654 (1983).

[25] Dale, C. Application of the proportional hazards model in the reliability field, Proceedings of the Fourth National Reliability Conference, 5B/1/1-9 (1983).

[26] Jardine, A.K.S. and Anderson, M. Use of concomitant variables for reliability estimations and setting component replacement policies, Proceedings of the 8th Advances in Reliability Technology Symposium, B3/2/1-6 (1984).

[27] Kalbfleisch, J.D. and Prentice, R.L. The Statistical Analysis of Failure Time Data, Wiley (1980).

[28] Lawless, J.F. Statistical Models and Methods for Lifetime Data, Wiley (1982).

[29] Lawless, J.F. Statistical methods in reliability (with discussion), Technometrics 25, 1983, 305-335.

[30] Bendell, A. Proportional hazards modelling in reliability assessment. Reliability Engineering, 11, 1985, 175-183.

[31] Wightman, D.W. and Bendell, A. The practical application of proportional hazards modelling in reliability, Proc. 5th National Reliability Conference, Birmingham 1985, 5NRC/2B/1-16. Invited for republication in Reliability Engineering (1986).

[32] Bendell, A., Walley, M., Wightman, D.W. and Wood, L.M. Proportional hazards modelling in reliability analysis - an application to brake discs on High Speed Trains, Quality and Reliability Engineering International, 2, 1986, No. 1.

[33] Argent, S.J., Manning, P.T., Ryan, S.G., Bendell, A., Marshall, J. and Wightman, D.W., Proportional hazards modelling in the analysis of transmission failure statistics, Proc. 5th Euredata Conference, Heidelberg, 1986.

[34] Cox, D.R. Regression models and life tables (with discussion). J.R. Stat. Soc. B., 34, 1972, 187-202.

[35] Cox, D.R., The Analysis of Binary Data, Methuen (1970).

[36] Nelder, J.A. and McCulloch, R.J., Generalised Linear Models, Chapman-Hall, (1983).

[37] Buckland, W.R. Reliability of nuclear power plants; statistical techniques for analysis, Task Force on Problems of Rare Events in the Reliability of Nuclear Power Plants, JRC ISPRA 8-10 June 1976; CSNI Report No. 10 (1976).

[38] Walls, L.A. and Bendell, A. The use of multivariate methods in reliability, Royal Statistical Society Multivariate Study Group Annual Workshop, 1985.

[39] Libberton, G.P., Bendell, A., Walls, L.A., and Cannon, A.G. Reliability data collection and analysis for automatic fire detection systems on a large industrial site, Proc. Seminar on Data Collection and Analysis for Reliability Assessment, I. Mech. E., 1986.

[40] Everitt, B., Cluster Analysis, Halsted Press, (1980).

[41] Mardia, K.W., Kent, J.T. and Bibby, J.M. Multivariate Analysis, Academic Press, (1979).

[42] Box, G.E.P. and Jenkins, G.M., Time Series Analysis: Forecasting and Control, Holden-Day, London (1976).

[43] Singpurwalla, N.D., Time Series Analysis of Failure Data, Proceedings of the Annual Reliability and Maintainability Symposium, 107-112 (1978).

[44] Singpurwalla, N.D., Estimating Reliability Growth (or Deterioration) Using Time Series Analysis, Naval Research Logistics Quarterly, 25, 1978, 1-14.

[45] Singpurwalla, N.D., Analysing Availability Using Transfer Function Models and Cross Spectral Analysis, Naval Research Logistics Quarterly, 27, 1980, 1-16.

[46] Crow, L.H. and Singpurwalla, N.D., An Empirically Developed Fourier Series Model for Describing Software Failures, IEEE Transactions on Reliability, R-33, No. 2, 1984, 176-183.

[47] Horigome, M., Singpurwalla, N.D. and Soyer, R., A Bayes Empirical Bayes Approach for (Software) Reliability Growth, Computer Science and Statistics: Proceedings of the 16th Symposium on the Interface, editor Lynne Billard, North Holland, Amsterdam, 57-56 (1984).

[48] Singpurwalla, N.D. and Soyer, R., Assessing (Software) Reliability Growth Using a Random Coefficient Autoregressive Process and its Ramifications, IEEE Transactions on Software Engineering (1985).

[49] Dale, C.J. and Harris, L.N., Approaches to software reliability prediction, Proc. Ann. Rel. Maint. Symp. (1982).

[50] Lawrance, A.J. and Lewis, P.A.W., The Exponential Autoregressive-Moving Average EARMA (p,q) Process, Journal of the Royal Statistical Society, Series B, 42, No. 2, 1980, 150-161,

[51] Lewis, P.A.W. (to appear), Some Simple Models for Continuous Variate Time Series Analysis, Water Resources Journal.

[52] Walls, L.A. and Bendell, A., Time series methods in reliability, Proc. 9th Advances in Reliability Technology Symposium, (1986).

[53] Bendell, A., Assessing the Future of Reliability: I the Basis for Growth, Safety and Reliability, (1985).

[54] Bendell, A., Assessing the Future of Reliability: II Towards the Year 2000, Safety and Reliability (1986).

[55] Bendell, A. and Ansell, J.I., Practical aspects of fault-tree analysis and the use of Markov reliability models, Proc. 5th National Reliability Conference, Birmingham, 1985, 5NRC/4B/1-8.

[56] Bendell, A., Collecting, Analysing and Applying Reliability Data in the Process Industries; an Overview, Proc. Seminar on Data Collection and Analysis for Reliability Assessment, I. Mech. E., 1986.

[57] Baines, A., Present Position and Potential Developments: Some Personal Views on Industrial Statistics and Operational Research (with discussion), J.R. Statist. Soc. A., 147, 1984, 316-326.

Dr. Tony Bendell is Head of Department of Mathematics, Statistics and Operational Research at Trent Polytechnic, Nottingham, where he leads a team of some 40 academic staff, including a developing reliability group. He has been active in reliability for 15 years, and also runs his own consultancy company, Services Ltd.

Reliability Technology — Theory & Applications
J. Møltoft and F. Jensen (Editors)
© Elsevier Science Publishers B.V. (North-Holland), 1986

BEWARE OF THE HAZARD RATE MODELS

Antoni DRAPELLA

The Technical University of Gdańsk

30-952 Gdańsk, POLAND

Application miscarriages ot he hazard rate models are presented. To explain this fact, necessary and sufficient conditions /nsc/ for the hazard rate to be of a bathtub shape are formulated in terms of the failure density function. Next, nsc for the failure density function to be of a bimodal shape are formulated in terms of the hazard rate function. It was revealed that hazard rate models generate mostly unimodal distributions being in contrary to empirical data. In view of this fact strong criticism is leveled at the hazard rate models. A simple method to convert unimodal distribution into a bimodal one is proposed.

0. INTRODUCTION

After the reliability theory had been originated, the flat-bottomed hazard rate function /commonly called bathtub curve/ has been estabilished as a failure pattern. Since over thirty years a number of hazard rate models intended for describing bathtub failure pattern was proposed.

Recently a bimodal failure density function was put forward as a new failure pattern [8], [9]. The aim of this paper is to argue reliability engineers out of the old failure pattern and to support theoretically a new one. The following question is sure to arise at this moment: did the technological progress result in change of the failure pattern or, may be our knowledge got improved and developed.

1. NOTATIONS AND BASIC FORMULAE

t   time-to-failure
$h(t)$ hazard rate function
$R(t)$ survival function
$F(t)$ cumulative distribution function
$f(t)$ failure density function
a   scale parameter of a time-to-failure distribution
b, c   shape parameters
k   hazard rate level parameter

$f(t) = -R(t)$
$h(t) = f(t)/R(t)$ /1.1/

$R(t) = \exp\left(-\int_{0}^{} h(u)\,du\right)$ /1.2/

$R(t) = 1 - F(t)$ /1.3/
$y = \ln\ln(1 - \ln R)$ axis of the M1 /Makeham/ probability paper.

2. A REVIEW AND APPLICATIONS OF THE HAZARD RATE MODELS

Several hazard rate models are gathered in Table 1. These models can be divided into two cathegories: 1/ "monolitic" models having a compact form,

2/ "hybrid" models having a compound form. To accept these models we ought to test them against applicability. It is a very meaningful fact that reliability literature offers no wider evidence to support the old failure pattern with real life-time data. Let us consider typical life-time data presented in [2]. The models M1, M2, M4, H1 and H2 were employed to describe data mentioned above. Fig. 1 shows empirical cumulative distribution function plotted on the Makeham probability paper. A considerable discrepancy observed in Fig. 1 is a typical one for all hazard rate models employed.

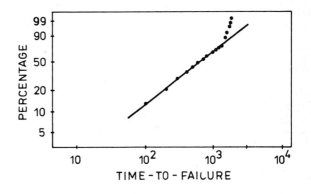

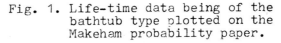

Fig. 1. Life-time data being of the bathtub type plotted on the Makeham probability paper.

3. ON MUTUAL RELATIONS BETWEEN $h(t)$ AND $f(t)$

In view of fact mentioned in the preceding section it is reasonable to investigate relations between hazard rate and failure density function. In this way we shall try to explain application miscarriages as well as to answer question asked in section 0.

Consider conditions under which $h(t)$ has a minimum. Differentiating both sides of /1.1/ with respect to t we get

$$h'(t) = f'(t)/R(t) + h^2(t) \qquad /3.1/$$

$$h''(t) = f''(t)/R(t) + (f'(t)/R(t))^2 +$$
$$\qquad\qquad + 2h(t)h'(t) \qquad /3.2/$$

A point tb is a bottom of the bathtub curve if:

$$f'(tb) \cdot R(tb) + f^2(tb) = 0 \qquad /3.3/$$

$$f''(tb) \cdot R(tb) + f'(tb) \cdot f(tb) > 0 \qquad /3.4/$$

These conditions can be verbally expressed by saying that f(t) has to be a decreasing and convex downward function in a certain neighbourhood of tb. In addition probability to survive the time longer than tb must be sufficiently high. Fig. 2 shows two time-to-failure distributions having considerably different failure density function. Each of this distribution fulfils the conditions formulated above.

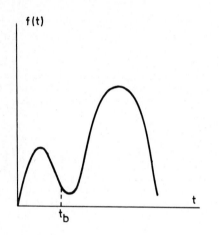

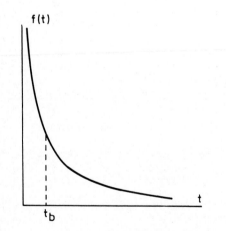

Fig. 2. Two considerably different distributions having bathtub hazard rate functions.

Now consider conditions under which modes of f(t) exist expressed in terms of h(t). Rewriting /1.1/ to the form

$$f(t) = h(t) \cdot R(t) \qquad /3.5/$$

and substituting /1.2/ into /3.5/ we get

$$f(t) = h(t) \cdot \exp\left(-\int_0^t h(u)du\right) \qquad /3.6/$$

Differentiating both sides of /3.6/ with respect to t we get

$$f'(t) = (h'(t) - h^2(t)) \cdot R(t) \qquad /3.7/$$

Let tm be a mode. Necessary and sufficient conditions are as follows

$$h'(tm) - h^2(tm) = 0 \qquad /3.8/$$

$$h''(tm) - 2h^3(tm) < 0 \qquad /3.9/$$

These conditions can be verbally expressed by saying that hazard rate must be an increasing function in a certain neighbourhood of tm. In addition it is preferable for h(t) to be a convex upward function. If h(t) is a convex downward function its convexity can not be too strong.

For illustrative purpose we shall consider the piecewise linear hazard rate function

$$f(t) = \begin{cases} ho - 2h1^2 \cdot (t - T1) & 0 \leqslant t \leqslant T1 \\ ho & T1 < t < T2 \\ ho + 2h2^2 \cdot (t - T2) & T2 \leqslant t \end{cases}$$
$$/3.10/$$

Starting from a general relation /3.5/ substituting /1.2/ into /3.5/, next substituting /3.10/ into /3.5/ we obtain the following expression for the failure density function being of interest

$$f(t) = \begin{cases} (ho - 2h1^2(t - T1)) \cdot \\ \quad \cdot \exp(-(ho - 2h2^2 T2)t - h2^2 t^2) \\ ho\,\exp(-h1^2 T1^2 - hot) \\ (ho + 2h2^2 t) \cdot \exp(-h1^2 T1^2 + \\ \quad + h2^2 T2(2t - T2) - hot - h2^2 t^2) \end{cases}$$

Fig. 3 shows two hazard rate functions and corresponding density of failure function. It should be notted that two bathtub curves generate two considerably different time-to-failure distribution. Only one of them, namely the first /bimodal/ is observed in practice.

Now we make use of the expression /3.8/ to M1 /Makeham/ distribution. In this case

$$h'(t) - h^2(t) = (1 - (1 - b)/(bz)) \cdot$$
$$\exp(-z) - 1 < 0 \quad \text{for all } t \qquad /3.12/$$
where $z = (t/a)^b$

In order to create a bimodal distribution let us introduce the "hazard rate level" parameter k

$$h(t) = (k \cdot b/t) \cdot (t/a)^b \cdot \exp(t/a)^b \qquad /3.13/$$

The condition under which extremes has

Table 1
The hazard rate models

Label	Formula	Reference
M1	$(b/t) \cdot (t/a)^b \cdot \exp((t/a)^b)$	[1]
M2	$(b/t) \cdot (t/a)^b / (1 - (t/a)^b)$	[7]
M3	$(b/t)(1/\Gamma(c))(t/a)^{bc} \cdot \exp((t/a)^b)$	[7]
M4	$k \cdot ((d - sqr(1 + t/a)) \, sqr(1 + t/a))^{-1}$	[10]
H1	$k \cdot (1 + t/a)^{-1} + (2/a) \cdot (t/a)$	[11]
H2	$(k/(2t)) \cdot (sqr(t/a1))^{-1} + ((1 - k)/a2) \cdot \exp(t/a2)$	[3]
H3	$(k/a1) \cdot \tanh(t/a1) + (1 - k) \cdot (b/t) \cdot (t/a2)^b \cdot \exp(-(t/a2)^b)$	[4]

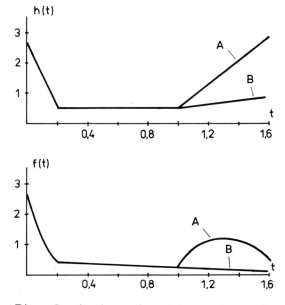

Fig. 3. A piecewise linear hazard
rate model and correspon-
ding failure density
function.

the following form

$$(1 - ((1 - b)/(b \cdot zm))) \cdot \exp(-zm) = k \qquad /3.14/$$

where $zm = (tm/a)^b$

It can be proved that equation /3.14/
may have two roots if $k < 1$. These
roots tm1, tm2 /where tm1 < tm2/ determine positions of a minimum and the second mode, respectively.

4. CONCLUSIONS

It is relatively easy to find a simple
formula which gives us the so called
bathtub curve. However it does not
mean that a useful time-to-failure
distribution has been found in this
way. Existing hazard rate models possess mostly unimodal failure density

function. In contrast empirical distributions /old ones as well as new ones/
show bimodal functions. A misconception
of the hazard rate models lies in this
that a flat-bottomed hazard rate function is a secondary phenomenon which
takes its rise from the bimodal nature
of the time-to-failure distribution.
Attention must be drawn to the fact
that the relation between the hazard
rate level and how this function rises
within the wear-out period is critical
to create bimodal distribution. Conditions $h'(te) = 0$ and $h''(te) > 0$ are not
sufficient to achive it. The bathtub
hazard rate being only a segment extracted from cannot be a starting point to
create useful life-time models.

5. SUPPLEMENT

Two weakness of the modified Makeham
distribution can be pointed out
1/ the first mode is fixed at $t = 0$
2/ "weak" and "strong" subpopulations
   are not reflected by appropriate
   sets of parameters.
The survival function of the modified
Makeham distribution is

$$R(t) = \exp(-k(\exp(t/a)^b - 1)) \qquad /5.1/$$

The HFR distribution has been derived
from /5.1/. The survival function has
the following form

$$R(t) = \exp(-k \exp((t/a)^b - (t/a1)^{-b1})) \qquad /5.2/$$

where
a, b – main scale and shape parameters,
a1, b1 – auxiliary scale and shape
         parameters, respectively.

The first set of parameters is ascribed
to "strong" subpopulation, the second
set is ascribed to "weak" subpopulation.

The HFR distribution has been success-

fully applied to describe results of
accelerated tests of hybrid IC [5], [6]

REFERENCES

[1] Barlow, R.E. and Proshan, F., Stati-
    stical theory of reliability and
    life testing /Holt, Rinehart and Wi-
    nston, New York, 1975/
[2] Calabro, S., Reliability princi-
    ples and practice /Mc-Graw-Hill,
    New York, 1962/
[3] Dhillon, B.S., A hazard rate model,
    Microel. Reliab. R-29 /1979/ 150
[4] Dhillon, B.S., A new hazard rate
    function, Microel. Reliab. /1978/
[5] Drapella, A., Reliability predic-
    tion of the hybrid I.C. with the
    H.E.R. probability distribution,
    in Proc. of the 3rd International
    Microel. Conf., Tokyo, Japan. May
    1984
[6] Drapella, A., Prediction of the re-
    liability of the hybrid IC on the
    ground of accelerated test results,
    5-th European Hybrid Microelectro-
    nics Conf., Stresa, Italy, May 1985
[7] Firkowicz, Sz., Statystyczne bada-
    nie wyrobów /PWN, Warszawa, 1970/.
[8] Jensen, F. and Petersen, N.F., Burn-
    -in. An engineering approach to the
    design and analysis of burn-in pro-
    cedures /Wiley, Chichester, 1982/.
[9] Møltoft, J., Behind the "bathtub"
    curve. A new model and its conse-
    quences, Microel. Reliab. 3 /1983/.
    489-500.
[10] Muth, E.J., Moment expressed in
     terms of the hazard function and
     applications, Microel. Reliab.
     /1974/ 469-471.
[11] Urban, H., A reliability distribu-
     tion with increasing, decreasing,
     constand bathtub failure rates,
     Technometrics 1 /1980/.

RELIABILITY MODELLING
AND PREDICTION II

Reliability Technology — Theory & Applications
J. Møltoft and F. Jensen (Editors)
© Elsevier Science Publishers B.V. (North-Holland), 1986

HOW TO IDENTIFY A "BATHTUB" DISTRIBUTION

Magne AARSET
Storebrand-Norden Ins. Co., Oslo, Norway

The total time on test plot (TTT plot) was introduced by Barlow and Campo {6} as a tool for analysing failure data. In this paper we shall study two test statistics based on this plot for testing if a random sample is generated from a life distribution with constant versus "bathtub"-shaped failure rate.

1. "BATHTUB" DISTRIBUTIONS

The failure rate constitutes a basic concept in reliability theory. If the life distribution is continuous, which very often can be assumed, the failure rate uniquely determines the life distribution.

A class of life distributions arising naturally in reliability situations may be constructed by assuming the failure rate initially decreasing during the infant mortality phase, next constant during the so-called "useful life" phase, and, finally increasing during the so-called "wear-out" phase. In reliability literature such failure rates are said to have a "bathtub" shape {8}.

2. THE TOTAL TIME ON TEST CONCEPT

Let $T_1, T_2, \ldots, T_n$ be a random sample from a life distribution F (i.e., a distribution function with $F(0^-)=0$) with finite mean, and denote the ordered sample $T_{n:1}, T_{n:2}, \ldots, T_{n:n}$. The observations may correspond to times to failure of n devices put on life test at time 0. Now

$$\sum_{j=1}^{r} T_{n:j} + (n-r) T_{n:r}$$

is defined to be the total time on test until the r-th failure. Define

$$U_r = \left(\sum_{j=1}^{r} T_{n:j} + (n-r) T_{n:r} \right) / \sum_{j=1}^{n} T_j$$

$(r=1,2,\ldots,n)$ and $U_0=0$. The plot of $(r/n, U_r)$ $(r=0,1,\ldots,n)$, where consecutive points are connected by straight lines, is known as the TTT plot {6}.

The TTT transform H_F^{-1} of a life distribution F is defined by {5}

$$H_F^{-1}(t) = \int_0^{F^{-1}(t)} (1-F(u)) \, du \quad ; \; 0 \leq t \leq 1,$$

where $F^{-1}(t) = \inf\{u:F(u) \geq t\}$. Under quite general conditions U_r will approach $\phi_F(t) = H_F^{-1}(t)/H_F^{-1}(1)$ when $n \to \infty$ and $r/n \to t$ {5}. Barlow and Campo {6} therefore suggested a comparison of the TTT plot to graphs of (scaled) TTT transforms for model identification.

Probability plotting methods are widely used in applied statistics. This plot has the benefit of being both easy to make and, because it is scale invariant, it requires only one plot. This plot may be compared to graphs on transparencies of scaled TTT transforms corresponding to different life distributions. In this way it is possible on one figure to compare, for instance, how a Weibull(2,2) and a Gamma(3,1) distribution fits the data.

An important feature of these data plots is that incomplete data can be analysed and that there is a theoretical basis for such an analysis ; see {1},{6} and {10} .

Some useful properties of the TTT transform are given below {6}.

(i)   There is a one-to-one correspondence between life distributions and their TTT transforms.

(ii) $H_F^{-1}(1) = \mu_F$, the mean of F.

(iii)   For the exponential distributions $\phi_F(t)=t$.

(iv)   If F is absolute continuous and strictly increasing

$$\frac{d}{dt} H_F^{-1}(t) = 1/r(F^{-1}(t))$$

for almost all t∈(0,1), where the failure rate is defined by $r(t) = f(t)/(1-F(t))$ $(t≥0)$.

(v) A life distribution F is IFR,"increasing failure rate" (DFR, "decreasing failure rate") if and only if the scaled TTT transform is concave (convex) for $0≤t≤1$.

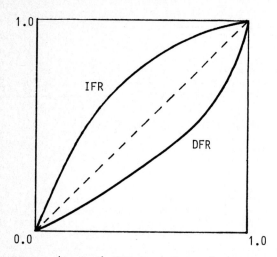

Fig. 1 (Scaled) TTT transform of an exponential distribution (dotted diagonal), an IFR distribution and a DFR distribution

(vi) A life distribution can naturally be defined as heavy-tailed if $\phi_F'(1)=∞$ and light-tailed if $\phi_F'(1)=0$ {13}.

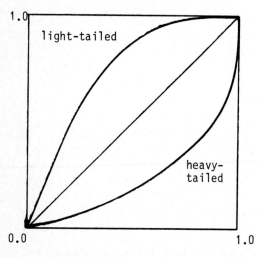

Fig. 2 (Scaled) TTT transform of a heavy-tailed distribution and a light-tailed distribution.

From (v) we see that the TTT transform of a life distribution with "bathtub"-shaped failure rate will be as illustrated in fig. 3.

When the TTT plot is compared with TTT transforms for model identification it is important to be aware of the following fact; under exponentiality, the TTT plot is likely to initially lie below the diagonal {6}. The analyst must therefore be careful not to confuse "infant mortality" with possible initial expo-

nentiality.

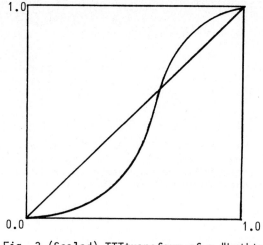

Fig. 3 (Scaled) TTTtransform of a "bathtub" distribution

3. A CROSSINGS TEST

Bergman {9} suggested the following procedure for testing exponentiality against the class of distributions with "bathtub"-shaped failure rate. Introduce

$$V_n = \min \{r≥1 \ :U_r≥r/n\},$$

$$M_n = \max \{r≤n-1:U_r≤r/n\},$$

$$G_n = V_n + n - M_n \ ,$$

and reject the hypothesis of exponentiality if G_n is large. The motivation for this test is that when the distribution has a "bathtub"-shaped failure rate, then we may expect both V_n as well as $(n-M_n)$ to be large.

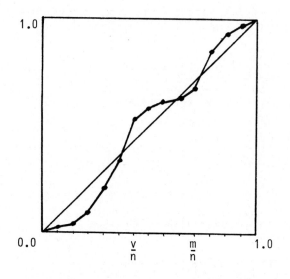

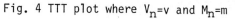

Fig. 4 TTT plot where $V_n=v$ and $M_n=m$

Under the hypothesis of exponentiality, the distribution of G_n is derived by Aarset {2} to be

$$P(G_n=i) = \sum_{v=1}^{i-1} \frac{(n-1)!}{(v-1)!\,(n-i+1)!\,(i-v-1)!} \, (v/n)^{v-1}$$

$$((n-i)/n)^{n-i+1}((i-v)/n)^{i-v-1}(1/v(i-v))$$

$(i=2,3,\ldots,n-1)$ and

$$P(G_n=n+1) = 2\,(n+1)^{n-2}/\,n^{n-1} \ .$$

In table 1 numerical values of $P(G_n \geq n-k)$ are given for selected values of k and n, and in appendix I a FORTRAN-program which calculates this distribution is given.

Aarset {2} also derived the asymptotic distribution (see table 2)

$$\lim_{n\to\infty} nP(G_n=n-k) = 2\,\frac{k^{k+1}}{(k+1)!}\,\exp(-k)$$

$(k=-1,1,2,\ldots,n-2)$. In practical use accuracy can be gained by use of the exact value for $P(G_n=n+1)$, instead of the complete asymptotic approximation (see table 2).

To illustrate the use of this test statistic we shall now consider an example. The failure data concerns an A/C Generator, and may be found in table 1 in {11}. These data are heavily censored, but will here be analysed as if they were a complete set of observed failure times.

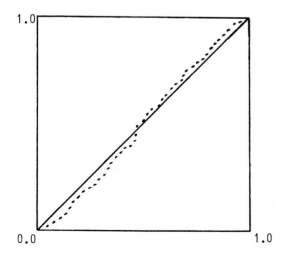

Fig. 5 TTT plot based on failure data for an A/C Generator

We observe that $G_{229}=230$, an event which occur with probability 0.024. This is to say that the hypothesis of exponentiality is to be rejected, and the life distribution expected to have a "bathtub"-shaped failure rate, whenever the level of significance is chosen greater than or equal to 0.024.

4. AN AREA-BASED TEST

Barlow and Doksum {7} proved that a test which rejects exponentiality in favor of an IFR distribution when essentially the signed area between the TTT plot and the diagonal was large, asymptotically is minimax. A test based on the signed area is of course disadvantageous to discover a "bathtub" distribution, but a test based on the (strictly positive) area might be better.

Such a test statistic can be constructed from the TTT process introduced by Barlow and Campo {6}. For absolute continuous F the TTT process is defined by

$$T_n(t) = \sqrt{n}\,\{\,\phi_n(\tfrac{r}{n}) - \phi_F(t)\}$$

where $(r-1)/n<t\leq r/n$, $1\leq r\leq n$ and $T_n(0)=0$.

Wishing also to benefit from some of the enormous literature on goodness of fit tests, Aarset {3} proposed the test statistic

$$R_n = \int_0^1 T_n^2(t)\,dt.$$

Under exponentiality

$$R_n = \sum_{r=1}^{n} \phi_n(\tfrac{r}{n})\,(\phi_n(\tfrac{r}{n})-(2r-1)/n\,) + n/3$$

and will furthermore asymptotically be equally distributed as the Cramer - von Mises statistic {3}. From this relation Aarset {3} concludes that

$$\lim_{n\to\infty} P(R_n\leq x) = 1 - 1/\P \sum_{j=1}^{\infty} \int_{(2j-1)^2\P^2}^{(2j)^2\P^2} y^{-1}\,e^{-xy/2}$$

$$(-1)^{j-1}((-\sqrt{y})/\sin\sqrt{y})^{1/2}\,dy$$

$(x\geq 0)$. The Cramer - von Mises statistic has been extensively studied (see e.g. {12}) and this asymptotic distribution is tabulated e.g. in {4}.

If we calculate R_n from the A/C Generator data we get $R_n=0.318$. Furthermore, from table 1 in {4}, we get $P(R_n\geq 0.318) \cong 0.12$.

5. COMMENT ON THE PROPERTIES OF R_n AND G_n

Many TTT-based procedures for testing exponentiality against interesting classes of life distributions have been suggested; see e.g. {6} {9} and {14}. Most of these tests have little power against "bathtub" distributions, though. Under simulations it seems like tests based on G_n or R_n more often reveals "bathtub" distributions.

Observe also that R_n will be large if the data seems to support a distribution with first IFR failure rate, then constant and finally DFR failure rate (the "opposite" of a "bathtub" distribution). If we let T_i (i=1,2,...,n) denote e.g. claims an insurance company gets after fires, it often seems realistic to assume that T_i will follow a distribution with such a failure rate.

It is not known to the author if these tests have any optimal properties, but that will (hopefully) be explored in papers to come.

6. APPENDIX I

Using G_n as a test statistic the most interesting part of its distribution is $P(G_n=n-k)$ (k=-1,1,2,...). We easily get

$$P(G_n=n-k) = \frac{n!}{n^n} \frac{k^{k+1}}{(k+1)!} \sum_{v=1}^{n-k-1} \frac{v^{v-1}}{v!} \frac{(n-k-v)^{n-k-v-1}}{(n-k-v)!}$$

(k=1,2,...,n-2). This distribution will be calculated by the following FORTRAN-program, where n=10 and k=3.

```
      INTEGER V
      N=10
      K=3
      RN=N
      RK=K
      FACT1=1.0
      FACT2=1.0
      FACT3=1.0
      FACT4=1.0
      SUM  =0.0
C
      DO 10 I1=1,N
      R1=I1
   10 FACT1=FACT1*RK/RN
      DO 11 I2=1,K+1
      R2=I2
   11 FACT2=FACT2*RK/R2
         DO 30 V=1,N-K-1
         RV=V
            DO 21 I3=1,V-1
            R3=I3
   21       FACT3=FACT3*RV/(R3+1.0)
            DO 22 I4=1,N-K-V-1
            R4=I4
   22       FACT4=FACT4*(RN-RK-RV)/(R4+1.0)
         SUM=SUM+FACT3*FACT4
         FACT3=1.0
   30    FACT4=1.0
      PROB=FACT1*FACT2*SUM
      WRITE(*,99) N,K,PROB
   99 FORMAT(1X,' P(GN=',I2,'-',I2,')= ',F8.5)
      END
```

Table 1 $P(G_n \geq n-k)$

	n			
k	10	50	100	200
-1	.42872	.10348	.05303	.02685
1	.46698	.11091	.05673	.02869
2	.51102	.11843	.06041	.03051
3	.56003	.12564	.06389	.03222
4	.61577	.13257	.06718	.03383
5	.68139	.13927	.07030	.03534
6	.76202	.14579	.07329	.03677
7	.86578	.15217	.07617	.03814
8	1.00000	.15846	.07895	.03945

Table 2. $P(G_n=n-k)$, exact and approximated

	n			
	100		200	
k	Exact	Approx.	Exact	Approx.
-1	.05303	.05437	.02685	.02718
1	.00370	.00368	.00184	.00184
2	.00368	.00361	.00182	.00180
3	.00348	.00336	.00171	.00168
4	.00329	.00313	.00160	.00156

REFERENCES
{1} Magne Aarset, Grafisk analyse av levetids-data basert på "Total Time on Test" med hovedvekt på problemstillinger fra pålitelighetsteorien,Cand. real. thesis, University of Oslo,1982 (in Norwegian).
{2} Magne Aarset, The null distribution for a test of constant versus "bathtub" failure rate,Scand. J. Statist.,vol 12,no. 1,1985, 55-61.
{3} Magne Aarset, How to identify a "bathtub" hazard rate,submitted for publication,1986.
{4} T.W. Anderson and D.A. Darling,Asymptotic theory of certain "goodness of fit" criteria based on stochastic processes, Ann. Math. Statist.,vol 23,1952,193-212.
{5} R.E. Barlow, D.J. Bartholomew,J. Bremner and H.D. Brunk,Statistical inference under order restrictions,Wiley,New York,1972.
{6} R.E. Barlow and R. Campo, Total time on test processes and applications to failure data analysis,In Reliability and fault tree analysis (ed. Barlow,Fussel and Singpurwalla),SIAM,Philadelphia,1975,451-481.
{7} R.E. Barlow and K. Doksum,Isotonic tests for convex orderings,Proc. 6th Berkeley Symp. Math. Statist. and Prob.,1972,293-323.
{8} R.E. Barlow and F. Proschan, Statistical theory of reliability and life testing,

probability models,Holt,Rinehart and Winston,New York,1981.

{9} Bo Bergman,On age replacement and the
      total time on test concept,Scand. J. Sta-
      tist.,vol. 6,1979,161-168.

{10} B. Bergman and B.Klefsjø, The total time
      on test concept and its use in reliability
      theory,Operations Research,vol. 32,no. 3,
      May-June 1984.

{11} V.E. Castellino and N.D. Singpurwalla,On
      the forecasting of failure rates-A detail-
      ed analysis,Technical Memorandum,The
      George Washington University,1973.

{12} James Durbin,Distribution theory of tests
      based on the sample distribution function,
      Regional Conference Series in Applied
      Mathematics,vol. 9,1973.

{13} Bengt Klefsjø,On aging properties and
      total time on test transforms,Scand. J.
      Statist.,vol. 9,1982,37-41.

{14} Bengt Klefsjø, Some tests against aging
      based on the total time on test transform,
      Comm. Statist. A-Theory-Methods,vol. 12,
      1983,907-927.

BIOGRAPHY

Magne Aarset was born in 1955. In 1982 he received a Cand. Real. in Mathematical Statistics from the University of Oslo, Norway. He
was employed at the institute of Mathematical
Statistics at the University of Trondheim-NTH,
1980-1983. Since 1983 he has been employed by
Storebrand-Norden Ins. Co. and primarily been
working with reliability analysis of industrial
risks. He also give lectures in Mathematical
Statistics at Oslo Ingeniør Høyskole.

Reliability Technology — Theory & Applications
J. Møltoft and F. Jensen (Editors)
© Elsevier Science Publishers B.V. (North-Holland), 1986

HUMAN ERROR ANALYSIS OF STANDBY SYSTEMS

Balbir S. Dhillon and Subramanyam N. Rayapati
Department of Mechanical Engineering
University of Ottawa
Ottawa, Ontario K1N 6N5
CANADA

This paper presents two newly developed mathematical models to perform reliability and availability analysis of two and three unit standby systems with human errors. Standby systems' reliability, availability, mean time to failure (MTTF) and variance of time to failure formulas are developed. System steady state availability, reliability and MTTF plots are shown for both the models. These plots clearly demonstrate the impact of human errors on standby systems' availability, reliability and MTTF.

1. INTRODUCTION

Numerous engineering systems are interconnected by human links. In the earlier reliability analysis, attention was directed only to equipment and reliability of the human element was neglected. Williams {1} recognized this shortcoming in the late 1950's and pointed out that realistic system reliability analysis must include the human aspect as well. Ever since then there has been an increasing interest in human-initiated equipment failures and their effect on system reliability. According to Meister {2} about 20-30 percent of failures, directly or indirectly are due to human errors. These are mainly due to wrong actions, maintenance errors, misinterpretation of instruments, and so on.

Human error is defined as a failure to perform a prescribed task (or the performance of a prohibited action), which could result in damage to equipment and property or disruption of scheduled operations. In real life most systems require some human participation irrespective of the degree of automation. Therefore predicting equipment reliability without considering human reliability will not present a true picture of that reliability. Recent studies pertaining to human reliability can be found in references {3-7}.

This paper presents the reliability and availability analysis of two and three unit standby systems with human errors. Models I and II deal with two and three non-identical unit standby systems, respectively. In both the models, a unit can fail either due to a hardware failure or due to a human error. As soon as the operating unit fails, the standby unit is switched into operation. For this purpose, it is assumed that the switchover mechanism is perfect and the switchover is instantaneous. The system failure can occur only when all the system units are non-operative (failed). The failed system is repaired back to its normal operational state (i.e., state 0). The failed system repair times are assumed to be arbitrarily distributed. Furthermore, in both the models the system is repaired back to its normal operational mode from its partially failed states (i.e., states 1 and 2). This paper makes use of the supplementary variable

method {8}, the Markov method and the Laplace transform technique.

2. ASSUMPTIONS

The following assumptions are associated with both the models:

(i) All system units are non-identical.

(ii) Failures are statistically independent.

(iii) Unit hardware failure and human error rates are constant.

(iv) Unit repair rate is constant (from states 1 and 2 to state 0).

(v) A repaired unit is as good as new.

(vi) Failed system repair times are distributed arbitrarily.

(vii) Switchover mechanism is perfect and switchover is instantaneous.

(viii) The failure rate of a unit in standby mode is zero.

3. MODEL I: TWO NON-IDENTICAL UNIT STANDBY SYSTEM

This model represents a two non-identical unit cold standby system with human errors. At time $t = 0$, unit 1 is switched into operation and unit 2 is kept as a standby. The operational unit can fail either due to a hardware failure or due to a human error. The failed unit is repaired from system states 1 and 2 to state 0. System fails only when both units are non-operative. The failed system is repaired back to its normal operational state . The state space diagram of Model I is shown in Figure 1. The numerals in circles denote the state numbers. The following symbols are associated with Model I: j is the jth state of the system; $j = 0$ (unit 1 operating, unit 2 on standby), $j = 1$ (unit 1 failed due to a hardware failure, unit 2 operating), $j = 2$ (unit 1 failed due to a human error, unit 2 operating), $j = 3$ (both units failed due to hardware failures), $j = 4$ (both units failed: unit 1 due to a hardware failure and unit 2 due to a human error),

$j = 5$ (both units failed: unit 1 due to a human error and unit 2 due to a hardware failure), $j = 6$ (both units failed due to human errors).

$P_j(x,t)$ is the probability density (with respect to repair time) that the failed system is in state j and has an elapsed repair time of x, for $j = 3,4,5,6$. $P_j(t)$ is the probability that the system is in state j at time t, for $j = 0,1,2,3,4,5,6$. $\mu_j(x)$ and $g_j(x)$ denote the repair rate and probability density function of repair times, respectively, when the system is in state j and has an elapsed repair time of x, for $j = 3,4,5,6$. λ_i is the ith unit constant hardware failure rate, for $i = 1,2$. λ_{h_i} is the ith unit constant human error rate, for $i = 1,2$. μ_j is the constant repair rate from state j to state 0, for $j = 1,2$. S is the Laplace transform variable.

3.1 Availability analysis

The system of differential equations and boundary conditions associated with Model I is:

$$\frac{dP_0(t)}{dt} + (\lambda_1 + \lambda_{h_1})P_0(t) = \sum_{i=1}^{2} \mu_i P_i(t)$$

$$+ \sum_{i=3}^{6} \int_0^\alpha P_i(x,t)\mu_i(x)\, dx \qquad (1)$$

$$\frac{dP_1(t)}{dt} + (\lambda_2 + \lambda_{h_2} + \mu_1)P_1(t) = P_0(t)\cdot\lambda_1 \qquad (2)$$

$$\frac{dP_2(t)}{dt} + (\lambda_2 + \lambda_{h_2} + \mu_2)P_2(t) = P_0(t)\cdot\lambda_{h_1} \qquad (3)$$

$$\{\frac{\partial}{\partial x} + \frac{\partial}{\partial t} + \mu_j(x)\}P_j(x,t) = 0, \text{ for } j=3,4,5,6 \qquad (4)$$

$$P_3(0,t) = P_1(t)\cdot\lambda_2 \qquad (5)$$

$$P_4(0,t) = P_1(t)\cdot\lambda_{h_2} \qquad (6)$$

$$P_5(0,t) = P_2(t)\cdot\lambda_2 \qquad (7)$$

$$P_6(0,t) = P_2(t)\cdot\lambda_{h_2} \qquad (8)$$

At time $t = 0$, $P_0(0)=1$ and $P_1(0)=P_2(0)= P_j(x,0)=0$, for $j=3,4,5,6$. By solving the above system of differential equations with the aid of Laplace transforms yield the following Laplace transforms of the state probabilities:

$$P_0(S) = \frac{1}{D(S)}, \qquad (9)$$

where $D(S) = (S + a_1) - \frac{\lambda_1}{S + a_2}\{\mu_1 + \lambda_2\, g_3(S)$

$+ \lambda_{h_2}g_4(S)\} - \frac{\lambda_{h_1}}{S+a_3}\{\mu_2 + \lambda_2 g_5(S) + \lambda_{h_2}g_6(S)\}$,

$a_1 = \lambda_1 + \lambda_{h_1}$, $a_2 = \lambda_2 + \lambda_{h_2} + \mu_1$, $a_3 = \lambda_2 + \lambda_{h_2} + \mu_2$,

$$g_j(x) = \mu_j(x)\, e^{-\int_0^x \mu_j(x)\, dx} \quad \text{and}$$

$$g_j(S) = \int_0^\infty e^{-Sx}\, g_j(x)\, dx, \text{ for } j=3,4,5,6.$$

$$P_1(S) = \frac{\lambda_1}{S + a_2}\cdot P_0(S) \qquad (10)$$

$$P_2(S) = \frac{\lambda_{h_1}}{S + a_3}\cdot P_0(S) \qquad (11)$$

$$P_3(S) = \frac{\lambda_1\lambda_2}{S(S + a_2)}\{1 - g_3(S)\}P_0(S) \qquad (12)$$

$$P_4(S) = \frac{\lambda_1\,\lambda_{h_2}}{S(S + a_2)}\{1 - g_4(S)\}P_0(S) \qquad (13)$$

$$P_5(S) = \frac{\lambda_{h_1}\lambda_2}{S(S + a_3)}\{1 - g_5(S)\}P_0(S) \qquad (14)$$

$$P_6(S) = \frac{\lambda_{h_1}\,\lambda_{h_2}}{S(S + a_3)}\{1 - g_6(S)\}P_0(S) \qquad (15)$$

The Laplace transform of the point-wise availability of the two-unit standby system is

$$AV(S) = \sum_{i=0}^{2} P_i(S) = \frac{S^2 + a_4 S + a_5}{(S + a_2)(S + a_3)D(S)}, \qquad (16)$$

where $a_4 = a_1 + a_2 + a_3$, $a_5 = a_2(a_3 + \lambda_{h_1}) + \lambda_1 a_3$.
The steady state availability of the two-unit standby system is

$$AV_{SS} = \lim_{S\to 0}\{S\, AV(S)\} \qquad (17)$$

3.2 Special case Model

This section presents the special case analysis of Model I when the system repair time distributions are described by

$$g_j(x) = \mu_j\, e^{-\mu_j x}, \text{ for } j = 3,4,5,6.$$

With the aid of equations (9) through (15), the Laplace transforms of the state probabilities of the special case model are:

$$P_0(S) = \frac{\sum_{i=1}^{7} B_i\, S^{7-i}}{S \sum_{j=1}^{7} c_j\, S^{7-j}}, \qquad (18)$$

where $B_1 = 1$, $B_2 = a_6 + a_{12}$, $B_3 = a_7 + a_6 a_{12} + a_{13}$,
$B_4 = a_7 a_{12} + a_6 a_{13} + a_{14}$, $B_5 = a_7 a_{13} + a_6 a_{14} + a_{15}$,
$B_6 = a_7 a_{14} + a_6 a_{15}$, $B_7 = a_7 a_{15}$, $c_1 = B_1$, $c_2 = a_{16}$,
$c_3 = a_{17} - (\lambda_1\mu_1 + \lambda_{h_1}\mu_2)$, $c_4 = a_{18} - (a_{33} + a_{38})$,
$c_5 = a_{19} - (a_{34} + a_{39})$, $c_6 = a_{20} - (a_{35} + a_{40})$,
$c_7 = a_{21} - (a_{36} + a_{41})$, $a_6 = a_2 + a_3$, $a_7 = a_2 a_3$,
$a_8 = \mu_3 + \mu_4$, $a_9 = \mu_3\mu_4$, $a_{10} = \mu_5 + \mu_6$, $a_{11} = \mu_5\mu_6$,
$a_{12} = a_8 + a_{10}$, $a_{13} = a_9 + a_8 a_{10} + a_{11}$,
$a_{14} = a_9 a_{10} + a_8 a_{11}$,
$a_{15} = a_9 a_{11}$, $a_{16} = B_1 a_1 + B_2$, $a_{17} = B_2 a_1 + B_3$,
$a_{18} = B_3 a_1 + B_4$, $a_{19} = B_4 a_1 + B_5$, $a_{20} = B_5 a_1 + B_6$,
$a_{21} = B_6 a_1 + B_7$, $a_{22} = B_7 a_1$, $a_{23} = a_3 + a_{10}$,
$a_{24} = a_3 a_{10} + a_{11}$, $a_{25} = a_3 a_{11}$, $a_{26} = a_2 + a_8$,

$a_{27} = a_2 a_8 + a_9$, $a_{28} = a_2 a_9$, $a_{29} = \mu_1 a_8 + \lambda_2 \mu_3 + \lambda_{h_2} \mu_4$,

$a_{30} = a_2 a_9$, $a_{31} = \mu_2 a_{10} + \lambda_2 \mu_5 + \lambda_{h_2} \mu_6$, $a_{32} = a_3 a_{11}$,

$a_{33} = \lambda_1 (\mu_1 a_{23} + a_{29})$, $a_{34} = \lambda_1 (\mu_1 a_{24} + a_{23} a_{29} + a_{30})$,

$a_{35} = \lambda_1 (\mu_1 a_{25} + a_{24} a_{29} + a_{23} a_{30})$,

$a_{36} = \lambda_1 (a_{25} a_{29} + a_{24} a_{30})$, $a_{37} = \lambda_1 a_{25} a_{30}$,

$a_{38} = \lambda_{h_1} (\mu_2 a_{26} + a_{31})$, $a_{39} = \lambda_{h_1} (\mu_2 a_{27} + a_{26} a_{31} + a_{32})$,

$a_{40} = \lambda_{h_1} (\mu_2 a_{28} + a_{27} a_{31} + a_{26} a_{32})$,

$a_{41} = \lambda_{h_1} (a_{28} a_{31} + a_{27} a_{32})$, $a_{42} = \lambda_{h_1} a_{28} a_{32}$.

$$P_1(S) = \frac{\sum_{i=1}^{6} E_i \, S^{6-i}}{S \sum_{j=1}^{7} c_j \, S^{7-j}} \; , \tag{19}$$

where $E_1 = \lambda_1$, $E_2 = \lambda_1 (a_3 + a_{12})$,

$E_3 = \lambda_1 (a_3 a_{12} + a_{13})$, $E_4 = \lambda_1 (a_3 a_{13} + a_{14})$,

$E_5 = \lambda_1 (a_3 a_{14} + a_{15})$, $E_6 = \lambda_1 a_3 a_{15}$.

$$P_2(S) = \frac{\sum_{i=1}^{6} F_i \, S^{6-i}}{S \sum_{j=1}^{7} c_j \, S^{7-j}} \; , \tag{20}$$

where $F_1 = \lambda_{h_1}$, $F_2 = \lambda_{h_1} (a_2 + a_{12})$,

$F_3 = \lambda_{h_1} (a_2 a_{12} + a_{13})$, $F_4 = \lambda_{h_1} (a_2 a_{13} + a_{14})$,

$F_5 = \lambda_{h_1} (a_2 a_{14} + a_{15})$, $F_6 = \lambda_{h_1} a_2 a_{15}$.

$$P_3(S) = \frac{\sum_{i=1}^{5} G_i \, S^{5-i}}{S \sum_{j=1}^{7} c_j \, S^{7-j}} \; , \tag{21}$$

where $G_1 = \lambda_1 \lambda_2$, $G_2 = \lambda_1 \lambda_2 (\mu_4 + a_{23})$,

$G_3 = \lambda_1 \lambda_2 (\mu_4 a_{23} + a_{24})$, $G_4 = \lambda_1 \lambda_2 (\mu_4 a_{24} + a_{25})$,

$G_5 = \lambda_1 \lambda_2 \mu_4 a_{25}$.

$$P_4(S) = \frac{\sum_{i=1}^{5} H_i \, S^{5-i}}{S \sum_{j=1}^{7} c_j \, S^{7-j}} \; , \tag{22}$$

where $H_1 = \lambda_1 \lambda_{h_2}$, $H_2 = H_1 (\mu_3 + a_{23})$,

$H_3 = H_1 (\mu_3 a_{23} + a_{24})$, $H_4 = H_1 (\mu_3 a_{24} + a_{25})$,

$H_5 = H_1 \mu_3 a_{25}$.

$$P_5(S) = \frac{\sum_{i=1}^{5} I_i \, S^{5-i}}{S \sum_{j=1}^{7} c_j \, S^{7-j}} \; , \tag{23}$$

where $I_1 = \lambda_{h_1} \lambda_2$, $I_2 = I_1 (\mu_6 + a_{26})$,

$I_3 = I_1 (\mu_6 a_{26} + a_{27})$, $I_4 = I_1 (\mu_6 a_{27} + a_{28})$,

$I_5 = I_1 \mu_6 a_{28}$;

and

$$P_6(S) = \frac{\sum_{i=1}^{5} J_i \, S^{5-i}}{S \sum_{j=1}^{7} c_j \, S^{7-j}} \; , \tag{24}$$

where $J_1 = \lambda_{h_1} \lambda_{h_2}$, $J_2 = J_1 (\mu_5 + a_{26})$,

$J_3 = J_1 (\mu_5 a_{26} + a_{27})$, $J_4 = J_1 (\mu_5 a_{27} + a_{28})$,

$J_5 = J_1 \mu_5 a_{28}$.

The Laplace transform of the point-wise availability of the two-unit standby system is given by

$$AV(S) = \sum_{i=0}^{2} P_i(S) = \frac{\sum_{i=1}^{7} D_i \, S^{7-i}}{S \sum_{j=1}^{7} c_j S^{7-j}} \; , \tag{25}$$

where $D_1 = 1$, $D_2 = a_4 + a_{12}$, $D_3 = a_5 + a_4 a_{12} + a_{13}$,

$D_4 = a_5 a_{12} + a_4 a_{13} + a_{14}$, $D_5 = a_5 a_{13} + a_4 a_{14} + a_{15}$,

$D_6 = a_5 a_{14} + a_4 a_{15}$, $D_7 = a_5 a_{15}$.

With the aid of equation (25), the steady state availability of the two-unit standby system is

$$AV_{SS} = \lim_{S \to 0} \{SAV(S)\} = \frac{D_7}{c_7} \tag{26}$$

The plots of equation (26) are shown in Figure 2. These plots clearly demonstrate the impact of human error rates λ_{h_1} and λ_{h_2} on the standby system steady state availability for the specified values of the parameters. From these plots it is evident that the system steady state availability decreases with the corresponding increasing values of human error rates λ_{h_1} and λ_{h_2}, respectively.

By utilizing the equations (18) through (24) and with the aid of the relationship $P_i = \lim_{S \to 0} \{S P_i(S)\}$, for $i = 0,1,2,\ldots,6$; the steady state probabilities of the special case model are:

$$P_0 = B_7/c_7 \tag{27}$$
$$P_1 = E_6/c_7 \tag{28}$$
$$P_2 = F_6/c_7 \tag{29}$$
$$P_3 = G_5/c_7 \tag{30}$$
$$P_4 = H_5/c_7 \tag{31}$$
$$P_5 = I_5/c_7 \tag{32}$$
$$P_6 = J_5/c_7 \tag{33}$$

3.3 Reliability Analysis with Repair

By setting $\mu_j(x) = 0$, for $j = 3,4,5,6$ in Figure 1 the following differential equations were obtained with the aid of Markov method:

$$\frac{dP_0(t)}{dt} + (\lambda_1 + \lambda_{h_1})P_0(t) = P_1(t) \cdot \mu_1 + P_2(t) \cdot \mu_2 \tag{34}$$

$$\frac{dP_1(t)}{dt} + (\lambda_2 + \lambda_{h_2} + \mu_1)P_1(t) = P_0(t) \cdot \lambda_1 \tag{35}$$

$$\frac{dP_2(t)}{dt} + (\lambda_2 + \lambda_{h_2} + \mu_2)P_2(t) = P_0(t) \cdot \lambda_{h_1} \tag{36}$$

$$\frac{dP_3(t)}{dt} = P_1(t) \cdot \lambda_2 \tag{37}$$

$$\frac{dP_4(t)}{dt} = P_1(t) \cdot \lambda_{h_2} \tag{38}$$

$$\frac{dP_5(t)}{dt} = P_2(t) \cdot \lambda_2 \tag{39}$$

$$\frac{dP_6(t)}{dt} = P_2(t) \cdot \lambda_{h_2} \tag{40}$$

At time $t=0$, $P_0(0)=1$ and $P_i(0)=0$ for $i=1,2,\ldots,6$. From equations (34) through (36) with the aid of Laplace transforms we get

$$R_R(S) = \sum_{i=0}^{2} P_i(S) = \frac{S^2 + K_4 S + K_5}{S^3 + K_1 S^2 + K_2 S + K_3}, \tag{41}$$

where $R_R(S)$ is the Laplace transform of the reliability of the two-unit standby system with repair,

$K_1 = a_1 + a_6, K_2 = a_1 a_6 + a_7 - (\lambda_1 \mu_1 + \lambda_{h_1}\mu_2)$,
$K_3 = a_1 a_7 - (\lambda_1 \mu_1 a_3 + \lambda_{h_1}\mu_2 a_2)$, $K_4 = K_1$,
$K_5 = a_7 + \lambda_1 a_3 + \lambda_{h_1}a_2$.

The mean time to failure of the two-unit standby system with repair is given by

$$MTTF_R = \lim_{S \to 0} R_R(S) = \frac{K_5}{K_3} \tag{42}$$

The variance of time to failure of the two-unit standby system with repair is

$$\sigma_R^2 = -2 \lim_{S \to 0} R_R'(S) - (MTTF_R)^2$$
$$= \{2(K_2 K_5 - K_3 K_4) - K_5^2\}/K_3^2, \tag{43}$$

where $R_R'(S)$ denotes the derivative of $R_R(S)$ with respect to S.

3.4 Reliability Analysis Without Repair

By setting $\mu_1 = \mu_2 = 0$ in equations (34) through (36) and solving for $P_0(S)$, $P_1(S)$ and $P_2(S)$ results in

$$R(S) = \sum_{i=0}^{2} P_i(S) \tag{44}$$
$$= \frac{S + \lambda_1 + \lambda_2 + \lambda_{h_1} + \lambda_{h_2}}{S^2 + (\lambda_1 + \lambda_2 + \lambda_{h_1} + \lambda_{h_2})S + (\lambda_1 + \lambda_{h_1})(\lambda_2 + \lambda_{h_2})},$$

where R(S) is the Laplace transform of the reliability of the two-unit standby system without repair.

With the aid of equation (44) the reliability of the two-unit standby system without repair is given by

$$R(t) = (1 - K_6) e^{-(\lambda_1 + \lambda_{h_1})t} + K_6 e^{-(\lambda_2 + \lambda_{h_2})t}, \tag{45}$$

where $K_6 = \frac{\lambda_1 + \lambda_{h_1}}{\lambda_1 + \lambda_{h_1} - (\lambda_2 + \lambda_{h_2})}$.

The plots of equation (45) are shown in Figure 3. These plots indicate that the system reliability decreases with the corresponding increasing values of the human error rates λ_{h_1} and λ_{h_2}.

The mean time to failure of the two-unit standby system without repair is given by

$$MTTF = \lim_{S \to 0} R(S) = \frac{\lambda_1 + \lambda_2 + \lambda_{h_1} + \lambda_{h_2}}{(\lambda_1 + \lambda_{h_1})(\lambda_2 + \lambda_{h_2})}. \tag{46}$$

The plots of equations (42) and (46) are shown in Figure 4. These plots clearly show the effect of human error rate λ_{h_2} on the system mean time to failure with and without repair for the specified values of the model parameters.

The variance of time to failure of the two-unit standby system without repair is

$$\sigma^2 = -2 \lim_{S \to 0} R'(S) - (MTTF)^2$$
$$= \frac{(\lambda_1 + \lambda_{h_1})^2 + (\lambda_2 + \lambda_{h_2})^2}{(\lambda_1 + \lambda_{h_1})^2(\lambda_2 + \lambda_{h_2})^2}, \tag{47}$$

where $R'(S)$ denotes the derivative of R(S) with respect to S.

4. MODEL II: THREE NON-IDENTICAL UNIT STANDBY SYSTEM

This model is concerned with a three non-identical unit cold standby system with human errors. At time $t=0$, unit 1 is switched into operation whereas units 2 and 3 are kept as cold standbys. The operational unit may fail either due to a hardware failure or due to a human error. The system can fail only when all the units are in failed condition. The failed system is repaired back to its normal operational state (i.e., state 0). The partially failed system is repaired back to its normal operating state 0 from states 1 and 2 only. The state space diagram for Model II is shown in Figure 5. The numerals in circles denote the state numbers.

The following are the notations associated with Model II: i is the ith state of the system; $i=0$(unit 1 operating, units 2 and 3 are on standbys), $i=1$(unit 1 failed due to a hardware failure, unit 2 operating, unit 3 on standby), $i=2$(unit 1 failed due to a human error, unit 2 operating, unit 3 on standby), $i=3$(units 1 and 2 failed due to hardware failures, unit 3 operating), $i=4$(unit 1 failed due to a hardware failure, unit 2 failed due to a human error, unit 3 operating), $i=5$(unit 1 failed due to a human error, unit 2 failed due to a hardware failure, unit 3 operating), $i=6$(units 1 and 2 failed due to human errors, unit 3 operating), $i=7$(all three units failed due to hardware failures), $i=8$(three units failed: units 1 and 2 due to hardware failures, unit 3 due to a human error), $i=9$(three units failed: units 1 and 3 due to hardware failures, unit 2 due to a human error), $i=10$(three units failed: unit 1 due to a hardware failure, units 2 and 3 due to human errors), $i=11$(three units failed: unit 1 due to a human error, units 2 and 3 due to hardware failures), $i=12$(three units failed: units 1 and 3 due to human errors, unit 2 due to a hardware failure), $i=13$(three units failed: units 1 and 2 due to human errors, unit 3 due to a hardware failure), $i=14$(all three units failed due to human errors)

$P_i(x,t)$ is the probability density (with respect to repair time) that the failed system is in state i and has an elapsed repair time of x, for $i=7,8,9,\ldots,14$. $P_i(t)$ is the probability that the system is in state i at time t, for $i=0,1,2,\ldots,14$. pdf denotes the probability density function. $\alpha_3(x)$ and $h_3(x)$ denote the repair rate and pdf of repair times, respectively, when the failed system is in state 7 and has an elapsed repair time of x. $\alpha_4(x)$ and $h_4(x)$ denote the repair rate and pdf of repair times, respectively, when the failed system is in state 8 and has an elapsed repair time of x. $\alpha_5(x)$ and $h_5(x)$ denote the repair rate and pdf of repair times, respectively, when the failed system is in state 9 and has an elapsed repair time of x. $\alpha_6(x)$ and $h_6(x)$ denote the repair rate and pdf of repair times, respectively, when the failed system is in state 10 and has an elapsed repair time of x. $\alpha_7(x)$ and $h_7(x)$ denote the repair rate and pdf of repair times, respectively, when the failed system is in state 11 and has an elapsed repair time of x. $\alpha_8(x)$ and $h_8(x)$ denote the repair rate and pdf of repair times, respectively, when the failed system is in state 12 and has an elapsed repair time of x. $\alpha_9(x)$ and $h_9(x)$ denote the repair rate and pdf of repair times, respectively, when the failed system is in state 13 and has an elapsed repair time of x. $\alpha_{10}(x)$ and $h_{10}(x)$ denote the repair rate and pdf of repair times, respectively, when the failed system is in state 14 and has an elapsed repair time of x. S is the Laplace transform variable. λ_j is the constant hardware failure rate of jth unit, for $j=1,2,3$. λ_{h_j} is the constant human error rate of jth unit, for $j=1,2,3$. α_1 is the constant repair rate from state 1 to state 0. α_2 is the constant repair rate from state 2 to state 0.

4.1 Availability analysis

The system of differential equations (and associated boundary conditions) associated with Model II is:

$$\frac{dP_0(t)}{dt} + (\lambda_1 + \lambda_{h_1})P_0(t)$$

$$= P_1(t) \cdot \alpha_1 + P_2(t) \cdot \alpha_2 + \int_0^\infty P_7(x,t)\alpha_3(x)dx$$

$$+ \int_0^\infty P_8(x,t)\alpha_4(x)dx + \int_0^\infty P_9(x,t)\alpha_5(x)dx$$

$$+ \int_0^\infty P_{10}(x,t)\alpha_6(x)dx + \int_0^\infty P_{11}(x,t)\alpha_7(x)dx$$

$$+ \int_0^\infty P_{12}(x,t)\alpha_8(x)dx + \int_0^\infty P_{13}(x,t)\alpha_9(x)dx$$

$$+ \int_0^\infty P_{14}(x,t)\alpha_{10}(x)dx \tag{48}$$

$$\frac{dP_1(t)}{dt} + (\lambda_2 + \lambda_{h_2} + \alpha_1)P_1(t) = P_0(t) \cdot \lambda_1 \tag{49}$$

$$\frac{dP_2(t)}{dt} + (\lambda_2 + \lambda_{h_2} + \alpha_2)P_2(t) = P_0(t) \cdot \lambda_{h_1} \tag{50}$$

$$\frac{dP_3(t)}{dt} + (\lambda_3 + \lambda_{h_3})P_3(t) = P_1(t) \cdot \lambda_2 \tag{51}$$

$$\frac{dP_4(t)}{dt} + (\lambda_3 + \lambda_{h_3})P_4(t) = P_1(t) \cdot \lambda_{h_2} \tag{52}$$

$$\frac{dP_5(t)}{dt} + (\lambda_3 + \lambda_{h_3})P_5(t) = P_2(t) \cdot \lambda_2 \tag{53}$$

$$\frac{dP_6(t)}{dt} + (\lambda_3 + \lambda_{h_3})P_6(t) = P_2(t) \cdot \lambda_{h_2} \tag{54}$$

$$\left\{ \frac{\partial}{\partial x} + \frac{\partial}{\partial t} + \alpha_3(x) \right\}P_7(x,t) = 0 \tag{55}$$

$$\left\{ \frac{\partial}{\partial x} + \frac{\partial}{\partial t} + \alpha_4(x) \right\}P_8(x,t) = 0 \tag{56}$$

$$\left\{ \frac{\partial}{\partial x} + \frac{\partial}{\partial t} + \alpha_5(x) \right\}P_9(x,t) = 0 \tag{57}$$

$$\left\{ \frac{\partial}{\partial x} + \frac{\partial}{\partial t} + \alpha_6(x) \right\}P_{10}(x,t) = 0 \tag{58}$$

$$\left\{ \frac{\partial}{\partial x} + \frac{\partial}{\partial t} + \alpha_7(x) \right\}P_{11}(x,t) = 0 \tag{59}$$

$$\left\{ \frac{\partial}{\partial x} + \frac{\partial}{\partial t} + \alpha_8(x) \right\}P_{12}(x,t) = 0 \tag{60}$$

$$\left\{ \frac{\partial}{\partial x} + \frac{\partial}{\partial t} + \alpha_9(x) \right\}P_{13}(x,t) = 0 \tag{61}$$

$$\left\{ \frac{\partial}{\partial x} + \frac{\partial}{\partial t} + \alpha_{10}(x) \right\}P_{14}(x,t) = 0 \tag{62}$$

$$P_7(0,t) = P_3(t) \cdot \lambda_3 \tag{63}$$

$$P_8(0,t) = P_3(t) \cdot \lambda_{h_3} \tag{64}$$

$$P_9(0,t) = P_4(t) \cdot \lambda_3 \tag{65}$$

$$P_{10}(0,t) = P_4(t) \cdot \lambda_{h_3} \tag{66}$$

$$P_{11}(0,t) = P_5(t) \cdot \lambda_3 \tag{67}$$

$$P_{12}(0,t) = P_5(t) \cdot \lambda_{h_3} \tag{68}$$

$$P_{13}(0,t) = P_6(t) \cdot \lambda_3 \tag{69}$$

$$P_{14}(0,t) = P_6(t) \cdot \lambda_{h_3} \tag{70}$$

At time $t=0$, $P_0(0)=1$ and all the other initial condition probabilities are equal to zero.

By solving the above system of differential equations yield the following Laplace transforms of the state probabilities:

$$P_0(S) = (S+b_2)(S+b_3)(S+b_4)/B(S), \tag{71}$$

where

$$B(S) = (S+b_1)(S+b_2)(S+b_3)(S+b_4)$$

$$- (S+b_4)(b_5S+b_6) - (s+b_3)A_1(S) - (s+b_2)A_2(S),$$

$$A_1(S) = \lambda_1\lambda_2\{\lambda_3h_3(S) + \lambda_{h_3}h_4(S)\} + \lambda_1\lambda_{h_2}\{\lambda_3h_5(S)$$

$$+ \lambda_{h_3}h_6(S)\},$$

$$A_2(S) = \lambda_{h_1}\lambda_2\{\lambda_3h_7(S) + \lambda_{h_3}h_8(S)\}$$

$$+ \lambda_{h_1}\lambda_{h_2}\{\lambda_3h_9(S) + \lambda_{h_3}h_{10}(S)\},$$

$$h_i(S) = \int_0^\infty e^{-Sx} h_i(x)\,dx,$$

$$h_i(x) = \alpha_i(x)\,e^{-\int_0^x \alpha_i(x)dx} \qquad \text{, for } i=3,4,5,\ldots,10;$$

$b_1 = \lambda_1 + \lambda_{h_1}$, $b_2 = \lambda_2 + \lambda_{h_2} + \alpha_1$, $b_3 = \lambda_2 + \lambda_{h_2} + \alpha_2$,

$b_4 = \lambda_3 + \lambda_{h_3}$, $b_5 = \lambda_1\alpha_1 + \lambda_{h_1}\alpha_2$,

$b_6 = \lambda_1\alpha_1 b_3 + \lambda_{h_1}\alpha_2 b_2$.

$$P_1(S) = \frac{\lambda_1}{S + b_2} \cdot P_0(S) \tag{72}$$

$$P_2(S) = \frac{\lambda_{h_1}}{S + b_3} \cdot P_0(S) \tag{73}$$

$$P_3(S) = \frac{\lambda_1\lambda_2}{(S + b_2)(S + b_4)} \cdot P_0(S) \tag{74}$$

$$P_4(S) = \frac{\lambda_1\lambda_{h_2}}{(S + b_2)(S + b_4)} \cdot P_0(S) \tag{75}$$

$$P_5(S) = \frac{\lambda_{h_1}\lambda_2}{(S + b_3)(S + b_4)} \cdot P_0(S) \tag{76}$$

$$P_6(S) = \frac{\lambda_{h_1}\lambda_{h_2}}{(S + b_3)(S + b_4)} \cdot P_0(S) \tag{77}$$

$$P_7(S) = \frac{\lambda_1\lambda_2\lambda_3}{S(S + b_2)(S + b_4)} \{1 - h_3(S)\} P_0(S) \tag{78}$$

$$P_8(S) = \frac{\lambda_1\lambda_2\lambda_{h_3}}{S(S + b_2)(S + b_4)} \{1 - h_4(S)\} P_0(S) \tag{79}$$

$$P_9(S) = \frac{\lambda_1\lambda_{h_2}\lambda_3}{S(S + b_2)(S + b_4)} \{1 - h_5(S)\} P_0(S) \tag{80}$$

$$P_{10}(S) = \frac{\lambda_1\lambda_{h_2}\lambda_{h_3}}{S(S + b_2)(S + b_4)} \{1 - h_6(S)\} P_0(S) \tag{81}$$

$$P_{11}(S) = \frac{\lambda_{h_1}\lambda_2\lambda_3}{S(S + b_3)(S + b_4)} \{1 - h_7(S)\} P_0(S) \tag{82}$$

$$P_{12}(S) = \frac{\lambda_{h_1}\lambda_2\lambda_{h_3}}{S(S + b_3)(S + b_4)} \{1 - h_8(S)\} P_0(S) \tag{83}$$

$$P_{13}(S) = \frac{\lambda_{h_1}\lambda_{h_2}\lambda_3}{S(S + b_3)(S + b_4)} \{1 - h_9(S)\} P_0(S) \tag{84}$$

$$P_{14}(S) = \frac{\lambda_{h_1}\lambda_{h_2}\lambda_{h_3}}{S(S + b_3)(S + b_4)} \{1 - h_{10}(S)\} P_0(S) \tag{85}$$

The Laplace transform of the point-wise availability of the three unit standby system is given by

$$AV(S) = \sum_{i=0}^{6} P_i(S) = (S^3 + b_{19}S^2 + b_{20}S + b_{21})/\Delta(S), \tag{86}$$

where $\Delta(S) = S^4 + b_{22}S^3 + b_{26}S^2 + b_{27}S + b_{28}$

$\qquad\qquad - (S + b_3)A_1(S) - (S + b_2)A_2(S)$,

$b_7 = b_1 + b_2$, $b_8 = b_1 b_2$, $b_9 = b_7 + b_3$,

$b_{10} = b_8 + b_3 b_7$, $b_{11} = b_3 b_8$, $b_{12} = b_3 + b_4$,

$b_{13} = b_3 b_4$, $b_{14} = b_2 + b_{12}$, $b_{15} = b_2 b_{12} + b_{13}$,

$b_{16} = b_2 b_{13}$, $b_{17} = b_2 + b_4$, $b_{18} = b_2 b_4$,

$b_{19} = b_1 + b_{14}$, $b_{20} = b_{15} + \lambda_1 b_{12} + \lambda_{h_1} b_{17}$

$\qquad + b_1(\lambda_2 + \lambda_{h_2})$

$b_{21} = b_{16} + \lambda_1 b_{13} + \lambda_{h_1} b_{18} + (\lambda_2 + \lambda_{h_2})(\lambda_1 b_3 + \lambda_{h_1} b_2)$,

$b_{22} = b_4 + b_9$, $b_{23} = b_4 b_9 + b_{10}$, $b_{24} = b_4 b_{10} + b_{11}$,

$b_{25} = b_4 b_{11}$, $b_{26} = b_{23} - b_5$, $b_{27} = b_{24} - (b_4 b_5 + b_6)$,

$b_{28} = b_{25} - b_4 b_6$.

The steady state availability of the three unit standby system can be obtained from the following equation:

$$AV_{SS} = \lim_{S \to 0} \{S \ AV(S)\} \tag{87}$$

4.2 Special case Model

In this section, a special case model pertaining to Model II is studied in detail when the system repair time distributions are described by $h_i(x) = \alpha_i \, e^{-\alpha_i x}$, for $i = 3, 4, 5, \ldots, 10$.

With the aid of equations (71) through (85), the Laplace transforms of the state probabilities of the special case model are:

$$P_0(S) = \frac{\sum\limits_{i=1}^{12} f_i S^{12-i}}{S \sum\limits_{j=1}^{12} e_j S^{12-j}} , \tag{88}$$

where $e_1 = d_1$, $e_2 = d_{10}$, $e_3 = d_{11}$, $e_4 = d_{12}$,

$e_5 = d_{13} - (m_{13} + m_{29})$, $e_6 = d_{14} - (b_3 m_{13} + d_{22} + b_2 m_{29} + d_{29})$,

$e_7 = d_{15} - (b_3 d_{22} + d_{23} + b_2 d_{29} + d_{30})$,

$e_8 = d_{16} - (b_3 d_{23} + d_{24} + b_2 d_{30} + d_{31})$,

$e_9 = d_{17} - (b_3 d_{24} + d_{25} + b_2 d_{31} + d_{32})$,

$e_{10} = d_{18} - (b_3 d_{25} + d_{26} + b_2 d_{32} + d_{33})$,

$e_{11} = d_{19} - (b_3 d_{26} + d_{27} + b_2 d_{33} + d_{34})$,

$e_{12} = d_{20} - (b_3 d_{27} + d_{28} + b_2 d_{34} + d_{35})$,

$f_1 = d_1$, $f_2 = d_1 b_{14} + d_2$, $f_3 = d_1 b_{15} + d_2 b_{14} + d_3$,

$f_4 = d_1 b_{16} + d_2 b_{15} + d_3 b_{14} + d_4$,

$f_5 = d_2 b_{16} + d_3 b_{15} + d_4 b_{14} + d_5$,

$f_6 = d_3 b_{16} + d_4 b_{15} + d_5 b_{14} + d_6$,

$f_7 = d_4 b_{16} + d_5 b_{15} + d_6 b_{14} + d_7$,

$f_8 = d_5 b_{16} + d_6 b_{15} + d_7 b_{14} + d_8$,

$f_9 = d_6 b_{16} + d_7 b_{15} + d_8 b_{14} + d_9$,

$f_{10} = d_7 b_{16} + d_8 b_{15} + d_9 b_{14}$,

$f_{11} = d_8 b_{16} + d_9 b_{15}$, $f_{12} = d_9 b_{16}$,

$d_1 = 1$, $d_2 = m_5 + m_{25}$, $d_3 = m_6 + m_5 m_{25} + m_{26}$,

$d_4 = m_7 + m_6 m_{25} + m_5 m_{26} + m_{27}$,

$d_5 = m_8 + m_7 m_{25} + m_6 m_{26} + m_5 m_{27} + m_{28}$,

$d_6 = m_8 m_{25} + m_7 m_{26} + m_6 m_{27} + m_5 m_{28}$,

$d_7 = m_8 m_{26} + m_7 m_{27} + m_6 m_{28}$, $d_8 = m_8 m_{27} + m_7 m_{28}$,

$d_9 = m_8 m_{28}$, $d_{10} = d_1 b_{22} + d_2$,

$d_{11} = d_1 b_{26} + d_2 b_{22} + d_3$,

$d_{12} = d_1 b_{27} + d_2 b_{26} + d_3 b_{22} + d_4$,

$d_{13} = d_1 b_{28} + d_2 b_{27} + d_3 b_{26} + d_4 b_{22} + d_5$,

$d_{14} = d_2 b_{28} + d_3 b_{27} + d_4 b_{26} + d_5 b_{22} + d_6$,

$d_{15} = d_3 b_{28} + d_4 b_{27} + d_5 b_{26} + d_6 b_{22} + d_7$,

$d_{16} = d_4 b_{28} + d_5 b_{27} + d_6 b_{26} + d_7 b_{22} + d_8$,

$d_{17} = d_5 b_{28} + d_6 b_{27} + d_7 b_{26} + d_8 b_{22} + d_9$,

$d_{18} = d_6 b_{28} + d_7 b_{27} + d_8 b_{26} + d_9 b_{22}$,

$d_{19} = d_7 b_{28} + d_8 b_{27} + d_9 b_{26}$, $d_{20} = d_8 b_{28} + d_9 b_{27}$,

$d_{21} = d_9 b_{28}$, $d_{22} = m_{14} + m_{13} m_{25}$,

$d_{23} = m_{15} + m_{14} m_{25} + m_{13} m_{26}$,

$d_{24} = m_{16} + m_{15} m_{25} + m_{14} m_{26} + m_{13} m_{27}$,

$d_{25} = m_{16}m_{25} + m_{15}m_{26} + m_{14}m_{27} + m_{13}m_{28},$

$d_{26} = m_{16}m_{26} + m_{15}m_{27} + m_{14}m_{28},$

$d_{27} = m_{16}m_{27} + m_{15}m_{28}, \quad d_{28} = m_{16}m_{28},$

$d_{29} = m_{30} + m_{5}m_{29}, \quad d_{30} = m_{31} + m_{5}m_{30} + m_{6}m_{29},$

$d_{31} = m_{32} + m_{5}m_{31} + m_{6}m_{30} + m_{7}m_{29},$

$d_{32} = m_{5}m_{32} + m_{6}m_{31} + m_{7}m_{30} + m_{8}m_{29},$

$d_{33} = m_{6}m_{32} + m_{7}m_{31} + m_{8}m_{30}, \quad d_{34} = m_{7}m_{32} + m_{8}m_{31},$

$d_{35} = m_{8}m_{32}, \quad m_{1} = \alpha_{3} + \alpha_{4}, \quad m_{2} = \alpha_{3}\alpha_{4}, \quad m_{3} = \alpha_{5} + \alpha_{6},$

$m_{4} = \alpha_{5}\alpha_{6}, \quad m_{5} = m_{1} + m_{3}, \quad m_{6} = m_{2} + m_{1}m_{3} + m_{4},$

$m_{7} = m_{2}m_{3} + m_{1}m_{4}, \quad m_{8} = m_{2}m_{4}, \quad m_{9} = \lambda_{1}\lambda_{2}(\lambda_{3}\alpha_{3} + \lambda_{h_{3}}\alpha_{4}),$

$m_{10} = \lambda_{1}\lambda_{2}\alpha_{3}\alpha_{4}b_{4}, \quad m_{11} = \lambda_{1}\lambda_{h_{2}}(\lambda_{3}\alpha_{5} + \lambda_{h_{3}}\alpha_{6}),$

$m_{12} = \lambda_{1}\lambda_{h_{2}}\alpha_{5}\alpha_{6}b_{4}, \quad m_{13} = m_{9} + m_{11},$

$m_{14} = m_{3}m_{9} + m_{10} + m_{1}m_{11} + m_{12},$

$m_{15} = m_{4}m_{9} + m_{3}m_{10} + m_{2}m_{11} + m_{1}m_{12},$

$m_{16} = m_{4}m_{10} + m_{2}m_{12}, \quad m_{17} = \lambda_{h_{1}}\lambda_{2}(\lambda_{3}\alpha_{7} + \lambda_{h_{3}}\alpha_{8}),$

$m_{18} = \lambda_{h_{1}}\lambda_{2}\alpha_{7}\alpha_{8}b_{4}, \quad m_{19} = \lambda_{h_{1}}\lambda_{h_{2}}(\lambda_{3}\alpha_{9} + \lambda_{h_{3}}\alpha_{10}),$

$m_{20} = \lambda_{h_{1}}\lambda_{h_{2}}\alpha_{9}\alpha_{10}b_{4}, \quad m_{21} = \alpha_{7} + \alpha_{8}, \quad m_{22} = \alpha_{7}\alpha_{8},$

$m_{23} = \alpha_{9} + \alpha_{10}, \quad m_{24} = \alpha_{9}\alpha_{10}, \quad m_{25} = m_{21} + m_{23},$

$m_{26} = m_{22} + m_{21}m_{23} + m_{24}, \quad m_{27} = m_{22}m_{23} + m_{21}m_{24},$

$m_{28} = m_{22}m_{24}, \quad m_{29} = m_{17} + m_{19},$

$m_{30} = m_{17}m_{23} + m_{18} + m_{19}m_{21} + m_{20},$

$m_{31} = m_{17}m_{24} + m_{18}m_{23} + m_{19}m_{22} + m_{20}m_{21},$

$m_{32} = m_{18}m_{24} + m_{20}m_{22} \ .$

$$P_{1}(S) = \frac{\lambda_{1}}{S + b_{2}} \cdot P_{0}(S) \tag{89}$$

$$P_{2}(S) = \frac{\lambda_{h_{1}}}{S + b_{3}} \cdot P_{0}(S) \tag{90}$$

$$P_{3}(S) = \frac{\lambda_{1}\lambda_{2}}{(S + b_{2})(S + b_{4})} \cdot P_{0}(S) \tag{91}$$

$$P_{4}(S) = \frac{\lambda_{1}\lambda_{h_{2}}}{(S + b_{2})(S + b_{4})} \cdot P_{0}(S) \tag{92}$$

$$P_{5}(S) = \frac{\lambda_{h_{1}}\lambda_{2}}{(S + b_{3})(S + b_{4})} \cdot P_{0}(S) \tag{93}$$

$$P_{6}(S) = \frac{\lambda_{h_{1}}\lambda_{h_{2}}}{(S + b_{3})(S + b_{4})} \cdot P_{0}(S) \tag{94}$$

$$P_{7}(S) = \frac{\lambda_{1}\lambda_{2}\lambda_{3}}{(S + b_{2})(S + b_{4})(S + \alpha_{3})} \cdot P_{0}(S) \tag{95}$$

$$P_{8}(S) = \frac{\lambda_{1}\lambda_{2}\lambda_{h_{3}}}{(S + b_{2})(S + b_{4})(S + \alpha_{4})} \cdot P_{0}(S) \tag{96}$$

$$P_{9}(S) = \frac{\lambda_{1}\lambda_{h_{2}}\lambda_{3}}{(S + b_{2})(S + b_{4})(S + \alpha_{5})} \cdot P_{0}(S) \tag{97}$$

$$P_{10}(S) = \frac{\lambda_{1}\lambda_{h_{2}}\lambda_{h_{3}}}{(S + b_{2})(S + b_{4})(S + \alpha_{6})} \cdot P_{0}(S) \tag{98}$$

$$P_{11}(S) = \frac{\lambda_{h_{1}}\lambda_{2}\lambda_{3}}{(S + b_{3})(S + b_{4})(S + \alpha_{7})} \cdot P_{0}(S) \tag{99}$$

$$P_{12}(S) = \frac{\lambda_{h_{1}}\lambda_{2}\lambda_{h_{3}}}{(S + b_{3})(S + b_{4})(S + \alpha_{8})} \cdot P_{0}(S) \tag{100}$$

$$P_{13}(S) = \frac{\lambda_{h_{1}}\lambda_{h_{2}}\lambda_{3}}{(S + b_{3})(S + b_{4})(S + \alpha_{9})} \cdot P_{0}(S) \tag{101}$$

$$P_{14}(S) = \frac{\lambda_{h_{1}}\lambda_{h_{2}}\lambda_{h_{3}}}{(S + b_{3})(S + b_{4})(S + \alpha_{10})} \cdot P_{0}(S) \tag{102}$$

The Laplace transform of the point-wise availability of the three unit standby system is

$$AV(S) = \sum_{i=0}^{6} P_{i}(S) = \frac{\sum_{i=1}^{12} g_{i} S^{12-i}}{S \sum_{j=1}^{12} e_{j} S^{12-j}}, \tag{103}$$

where $g_{1} = d_{1}$, $g_{2} = d_{1}b_{19} + d_{2}$,

$g_{3} = d_{1}b_{20} + d_{2}b_{19} + d_{3},$

$g_{4} = d_{1}b_{21} + d_{2}b_{20} + d_{3}b_{19} + d_{4},$

$g_{5} = d_{2}b_{21} + d_{3}b_{20} + d_{4}b_{19} + d_{5},$

$g_{6} = d_{3}b_{21} + d_{4}b_{20} + d_{5}b_{19} + d_{6},$

$g_{7} = d_{4}b_{21} + d_{5}b_{20} + d_{6}b_{19} + d_{7},$

$g_{8} = d_{5}b_{21} + d_{6}b_{20} + d_{7}b_{19} + d_{8},$

$g_{9} = d_{6}b_{21} + d_{7}b_{20} + d_{8}b_{19} + d_{9},$

$g_{10} = d_{7}b_{21} + d_{8}b_{20} + d_{9}b_{19}, \quad g_{11} = d_{8}b_{21} + d_{9}b_{20},$

$g_{12} = d_{9}b_{21}.$

The steady state availability of the three-unit standby system is given by

$$AV_{SS} = \lim_{S \to 0} \{SAV(S)\} = \frac{g_{12}}{e_{12}} \tag{104}$$

The plots of equation (104) are shown in Figure 6. The Figure 6 plots clearly demonstrate the impact of human error rates $\lambda_{h_{1}}$ and $\lambda_{h_{2}}$ on the standby system steady state availability for the specified values of the model parameters.

With the aid of equations (88) through (102) and the relationship $P_{i} = \lim_{S \to 0} \{S\,P_{i}(S)\}$, for $i = 0,1,2,\ldots,14$; the steady state probabilities of the three unit standby system are:

$$P_{0} = f_{12}/e_{12} \tag{105}$$

$$P_{1} = \lambda_{1}f_{12}/b_{2}e_{12} \tag{106}$$

$$P_{2} = \lambda_{h_{1}}f_{12}/b_{3}e_{12} \tag{107}$$

$$P_{3} = \frac{\lambda_{1}\lambda_{2}f_{12}}{b_{2}b_{4}e_{12}} \tag{108}$$

$$P_{4} = \frac{\lambda_{1}\lambda_{h_{2}}f_{12}}{b_{2}b_{4}e_{12}} \tag{109}$$

$$P_{5} = \frac{\lambda_{h_{1}}\lambda_{2}f_{12}}{b_{3}b_{4}e_{12}} \tag{110}$$

$$P_{6} = \frac{\lambda_{h_{1}}\lambda_{h_{2}}f_{12}}{b_{3}b_{4}e_{12}} \tag{111}$$

$$P_{7} = \frac{\lambda_{1}\lambda_{2}\lambda_{3}f_{12}}{b_{2}b_{4}\alpha_{3}e_{12}} \tag{112}$$

$$P_{8} = \frac{\lambda_{1}\lambda_{2}\lambda_{h_{3}}f_{12}}{b_{2}b_{4}\alpha_{4}e_{12}} \tag{113}$$

$$P_{9} = \frac{\lambda_{1}\lambda_{h_{2}}\lambda_{3}f_{12}}{b_{2}b_{4}\alpha_{5}e_{12}} \tag{114}$$

$$P_{10} = \frac{\lambda_{1}\lambda_{h_{2}}\lambda_{h_{3}}f_{12}}{b_{2}b_{4}\alpha_{6}e_{12}} \tag{115}$$

$$P_{11} = \frac{\lambda_{h_1} \lambda_2 \lambda_3 f_{12}}{b_3 b_4 \alpha_7 e_{12}} \qquad (116)$$

$$P_{12} = \frac{\lambda_{h_1} \lambda_2 \lambda_{h_3} f_{12}}{b_3 b_4 \alpha_8 e_{12}} \qquad (117)$$

$$P_{13} = \frac{\lambda_{h_1} \lambda_{h_2} \lambda_3 f_{12}}{b_3 b_4 \alpha_9 e_{12}} \qquad (118)$$

$$P_{14} = \frac{\lambda_{h_1} \lambda_{h_2} \lambda_{h_3} f_{12}}{b_3 b_4 \alpha_{10} e_{12}} \qquad (119)$$

4.2   Reliability Analysis with Repair

By setting $\alpha_i(x) = 0$ for $i = 3,4,5,\ldots,10$ in Model II and solving the resulting system of differential equations obtained with the Markov method yields

$$R_R(S) = \sum_{i=0}^{6} P_i(S) = \frac{\sum_{i=1}^{4} u_i \, S^{4-i}}{\sum_{j=1}^{5} v_j \, S^{5-j}} , \qquad (120)$$

where $R_R(S)$ is the Laplace transform of the reliability of the three-unit standby system with repair,

$u_1 = 1$, $u_2 = b_1 + b_{14}$, $u_3 = b_{15} + x_5 + b_1 x_6$,

$u_4 = b_{16} + x_5 x_6$, $v_1 = x_1$, $v_2 = x_1 b_4 + x_2$,

$v_3 = x_2 b_4 + x_3$, $v_4 = x_3 b_4 + x_4$, $v_5 = x_4 b_4$, $x_1 = 1$,

$x_2 = b_9$, $x_3 = b_{10} - b_5$, $x_4 = b_{11} - b_6$,

$x_5 = \lambda_1 b_3 + \lambda_{h_1} b_2$, $x_6 = b_4 + \lambda_2 + \lambda_{h_2}$.

The mean time to failure of the three-unit standby system with repair is given by

$$\text{MTTF}_R = \lim_{S \to 0} R_R(S) = \frac{u_4}{v_5} \qquad (121)$$

The variance of time to failure of the three-unit standby system with repair is

$$\sigma_R^2 = -2 \lim_{S \to 0} R'_R(S) - (\text{MTTF}_R)^2$$

$$= \frac{2(u_4 v_4 - u_3 v_5) - u_4^2}{v_5^2} , \qquad (122)$$

where $R'_R(S) = \dfrac{d \, R_R(S)}{ds}$.

4.3   Reliability Analysis Without Repair

By setting $\alpha_1 = \alpha_2 = \alpha_i(x) = 0$,

for $i = 3,4,5,\ldots,10$ in Model II and solving the resulting system of differential equations obtained with the Markov method leads to

$$R(S) = \sum_{i=0}^{6} P_i(S) = \frac{S^2 + w_1 S + w_2}{S^3 + w_1 S^2 + w_2 S + w_3} , \qquad (123)$$

where $R(S)$ is the Laplace transform of the reliability of the three-unit standby system without repair,

$w_1 = b_1 + b_4 + y$, $w_2 = y(b_1 + b_4) + b_1 b_4$,

$w_3 = b_1 b_4 y$, $y = \lambda_2 + \lambda_{h_2}$.

With the aid of equation (123), the reliability of the three-unit standby system without repair is given by

$$R(t) = z_1 e^{-b_1 t} + z_2 e^{-yt} + z_3 e^{-b_4 t} , \qquad (124)$$

where $z_1 = \dfrac{b_1^2 - w_1 b_1 + w_2}{(y - b_1)(b_4 - b_1)}$, $z_2 = \dfrac{y^2 - w_1 y + w_2}{(b_1 - y)(b_4 - y)}$,

$$z_3 = \frac{b_4^2 - w_1 b_4 + w_2}{(b_1 - b_4)(y - b_4)} .$$

The plots of equation (124) are shown in Figure 7. The Figure 7 plots clearly show the effect of time t and human error rate λ_{h_1} on the system reliability. It is evident from these plots that the system reliability decreases with the corresponding increasing values of t and λ_{h_1}.

The mean time to failure of the three-unit standby system without repair is

$$\text{MTTF} = \lim_{S \to 0} R(S) = \frac{w_2}{w_3} \qquad (125)$$

The plots of equations (121) and (125) are shown in Figure 8. These plots clearly demonstrate the impact of human error rate λ_{h_1} on the system mean time to failure with and without repair for the specified values of the Model II parameters.

The variance of time to failure of the three-unit standby system without repair is

$$\sigma^2 = -2 \lim_{S \to 0} R'(S) - (\text{MTTF})^2$$

$$= (w_2^2 - 2w_1 w_3)/w_3^2 , \qquad (126)$$

where $R'(S) = \dfrac{d \, R(S)}{ds}$.

5.   CONCLUSIONS

In this paper two mathematical models are developed to study the effect of human errors on two and three non-identical unit standby systems. For both the models system availability, reliability, mean time to failure and variance of time to failure formulas are developed. In addition, system steady state availability, reliability and mean time to failure plots are shown. Finally, it is contended that this study will be useful to practicing design, system, human factors and reliability engineers.

ACKNOWLEDGEMENT

The financial assistance from the Natural Sciences and Engineering Research Council of Canada is gratefully appreciated.

REFERENCES

{1} H.L. Williams, Reliability evaluation of the human component in man-machine systems, <u>Electrical Manufacturing</u>, April 1958.

{2} D. Meister, The problem of human-initiated failures, <u>Proceedings of the eighth national symposium on reliability and quality control</u>, 1962.

{3} B.S. Dhillon, On human reliability — Bibliography, <u>Microelectronics and reliability</u>, Vol. 20, 1980, 371-373.

{4} B.S. Dhillon and S.N. Rayapati, Analysis of redundant systems with human errors, <u>Proceedings of annual reliability and maintainability symposium</u>, 1985, 315-321.

{5} B.S. Dhillon and C. Singh, <u>Engineering reliability: New techniques and applications</u>, John Wiley and Sons, Inc., New York, 1981.

{6} B.S. Dhillon, <u>Reliability engineering in systems design and operation</u>, Van Nostrand Reinhold company, New York,1982.

{7} B.S. Dhillon, <u>Human reliability: with human factors</u>, Pergamon Press Inc.,New York, 1986.

{8} D.R. Cox, The analysis of non-Markovian stochastic processes by the inclusion of supplementary variables, <u>Proceedings of Cambridge philosophical society</u>, Vol. 51, 1955, 433-441.

BIOGRAPHIES

Balbir S. Dhillon, Ph.D., P.Eng.
Department of Mechanical Engineering
University of Ottawa
Ottawa, Ontario K1N 6N5
Canada

Dr. Balbir S. Dhillon is Professor of Mechanical Engineering. He has published about 150 articles on Reliability Engineering and related areas. He is on the Editorial Advisory Board (Advisory Editor) of Microelectronics and Reliability: An International Journal. Dr. Dhillon is associate Editor of International Journal of Energy Systems and Editor-at-Large for engineering books (Marcel Dekker, Inc., New York). He served as an associate editor of the 10th-13th Annual Modeling and Simulation Proceedings, published by the Instrument Society of America. He has written seven books on various aspects of System Reliability, Maintainability and Engineering Management published by Wiley (1981), Van Nostrand (1982), Butterworth (1983), Marcel Dekker (1984), etc. His first book with Dr. Singh on reliability was translated into Russian by the MIR publishers of Moscow. He is a recipient of the American Society for quality control Austin Bonis Reliability Award, Society of Reliability Engineers' Merit Award. He is a registered Professional Engineer in Ontario, and is listed in the American Men and Women of Science, Men of Achievements, International Dictionary of Biography, and Who's Who in International Intellectuals.

Balbir S. Dhillon attended the University of Wales where he received a B.Sc. in Electrical and Electronic Engineering and M.Sc. in Indus-

trial and systems engineering. He received the Ph.D. in Industrial Engineering from the University of Windsor.

S.N. Rayapati, Ph.D.
Department of Mechanical Engineering
University of Ottawa
Ottawa, Ontario K1N 6N5
Canada

Dr. S.N. Rayapati (Subramanyam Naidu Rayapati) is a Post-doctoral Fellow pursuing his research in the field of Reliability Engineering. He has published about 40 papers on Reliability Engineering and related areas. He received a M.Sc. in Statistics from Srivenkateswara University, Tirupati (India). He received the Ph.D. in Reliability Theory from Indian Institute of Technology, Bombay. His field of research interest includes cost-benefit analysis of maintainable systems, Transit System Reliability, Human Reliability, etc.

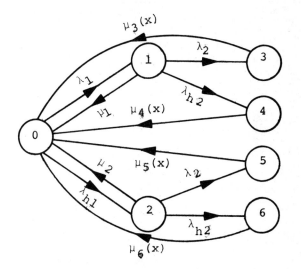

Figure 1.   State space diagram for Model I.

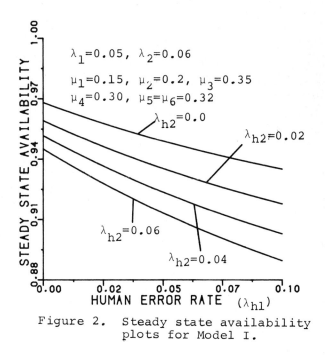

Figure 2.   Steady state availability plots for Model I.

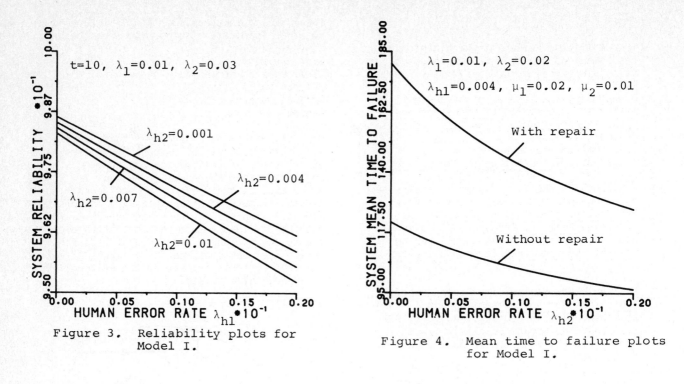

Figure 3.   Reliability plots for
             Model I.

Figure 4.   Mean time to failure plots
             for Model I.

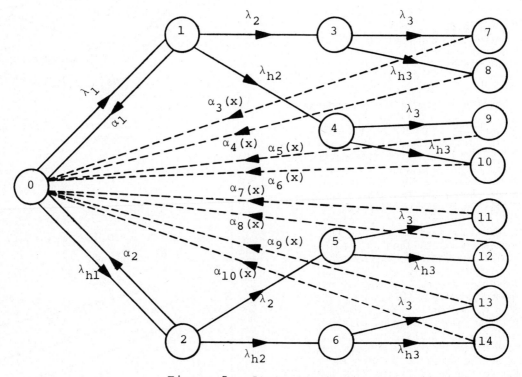

Figure 5.   State space diagram for Model II.

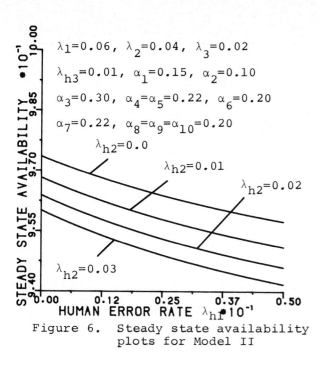

Figure 6. Steady state availability
plots for Model II

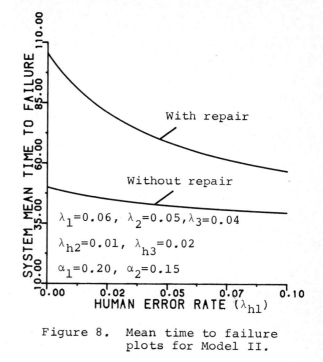

Figure 8. Mean time to failure
plots for Model II.

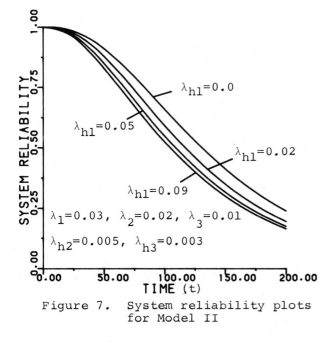

Figure 7. System reliability plots
for Model II

Reliability Technology — Theory & Applications
J. Møltoft and F. Jensen (Editors)
© Elsevier Science Publishers B.V. (North-Holland), 1986

RELIABILITY ALLOCATION FOR MECHANICAL
SYSTEMS WITH SPECIFIED RELIABILITY AND COST

Gradimir IVANOVIĆ
Faculty of Mechanical Engineering, University of Belgrade
Belgrade, Yugoslavia

Up-to-date mechanical systems are required to have high reliability characteristics. To attain them, it is necessary to take particular care (apart from laboratory and in-service reliability examination) of reliability design from the early stage of the development of mechanical system. In order to meet certain technical and economical requirements of systems, a method has been developed for the choice of "the best" reliability allocation from the aspect of achieving minimum system cost (C_{smin}) for specified system reliability (R_s). In addition, this method offers the possibility of achieving maximum system reliability (R_{smax}) for specified system cost (C_s). This method uses the Lagrange multipliers method and the multicriteria ranking method by compromise programming.

1. INTRODUCTORY CONSIDERATIONS

Today´s technical, or, to be more specific, mechanical systems are required not only to attain some technical performances but also to meet certain cost/effectiveness as well as reliability and availability requirements. To this end, special attention must be given to system reliability during the stages of design, production and exploitation.

According to some authors, until recently 90 percent of the time in design stage was devoted to system performances. However, the situation has changed and it is assessed that today performances, cost and reliability analysis, i.e. logistic support each account for one-third of the time of design stage. According to {19}, the costs are distributed as follows: about 28 percent for development and production, 12 percent for exploitation and 60 percent for logistic support. Such an approach has, in turn, changed the views of system reliability design. In other words, this means that it is not sufficient to perform in-service reliability analysis, but it is necessary to carry out certain procedures in the system design stage, i.e. system design for reliability must be performed.

Reliability design for a new system is an incomparably more complex problem than the determination of in-service reliability. Depending on the type of system being designed and on the actual conditions, the following approaches are possible in principle:

- a system is designed to comprise the elements whose realibility and economic characteristics are aither known or can be assessed accurately enough,

- a system is designed to comprise whose reliability and economic characteristics are unknown, i.e. cannot be assessed with sufficient accuracy.

Reliability design for a new system includes the allocation of system reliability to system elements which should meet specified system reliability requirements.

This problem is complex for several reasons; some of them are: different roles played by elements in performing the overall system function, different effects of failures that can occur on these elements, different operation times and different numbers of elements, different system complexity degrees, different amounts of resources required for the development of certain elements, etc. Since it is impossible to cover all these factors by allocation methods, some simplifications must be made.

Numerous models have been developed for reliability allocation purposes. Some of them rely on technical parameters, i.e. on technical and functional requirements and properties of system elements and components, whereas the other group includes the models based on cost optimization.

The allocation models based on technical parameters neglect the effects of resources required to develop a system with specified reliability but ensure the satisfaction of not only the specified reliability level but also other technical requirements (e.g. safety aspects). On the other hand, the models relying on cost optimization neglect the effects of technical requirements and properties of system elements but provide a favourable economic result. The shortcomings of both approaches may sometimes cause serious difficulties. Therefore, it is sensible in some cases to combine both approaches, {15}, {16} and {21}, i.e. to employ such allocation methods as will meet both technical and economic requirements to the highest possible extent.

This paper presents a reliability allocation method allowing the techno-economic requirements imposed on the system to be satisfied in an appropriate way. The method was developed having in mind some specific points arising in the construction of mechanical systems and the dependence between system cost and reliability. In addition, a procedure was developed which permits the choice of "the best" reliability allocation by compromise programming, where system cost and reliability represent the criterion for choosing "the best" solution.

2. DEFINITION OF CONDITIONS AND CONSTRAINTS

To permit the reliability and cost problems to be solved properly in the system design stage, the following conditions and constraints were accepted.

1. Reliability allocation should be performed for the specified values of (fig. 1):

 - system reliability level R_s
 - system operation time t_s
 - system cost C_s

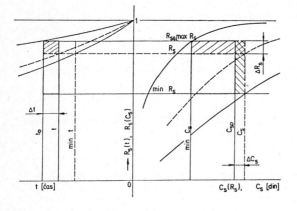

Fig. 1.

2. System reliability function $R_s(t)$ relates the reliabilities of system elements $R_i(t)$ in the following form:

$$R_s(t) \leq f(R_i(t)) \tag{1}$$

3. The total system cost C_s represents a sum of the costs of single system components C_i. This can be written as:

$$C_s \geq \sum_{i=1}^{n} C_i \tag{2}$$

4. Reliability allocation should ensure both technical and economic requirements of the system to be met. In doing this one should tend to:

 - cost minimization,
 - reliability maximization, or
 - a compromise solution between cost minimization and reliability maximization.

5. To perform reliability allocation, in order to form a reliability block diagram, one chould make a proper choice of the elements constituting the block diagram, i.e. of the reliability function.

6. Reliability allocation should be done gradually from the system to its elements, i.e. from the system to subsystems, from the subsystems to units, and so on, until components are reached, fig. 2. Proceeding in this way, one can write the reliability functions in the following forms:

 - for the systems:

$$R_s(t) = \prod_{i=1}^{n} R_i(t) \tag{3}$$

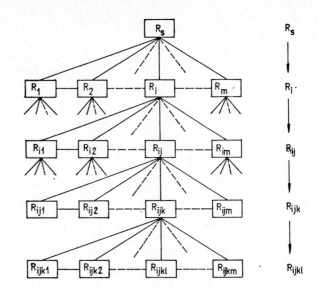

Fig. 2.

- for the subsystem:

$$R_i(t) = f_i(R_{ij}(t)) \tag{4}$$

- for the unit:

$$R_{ij}(t) = f_{ij}(R_{ijk}(t)) \tag{5}$$

- for the module:

$$R_{ijk}(t) = f_{ijk}(R_{ijk\ell}(t)) \tag{6}$$

7. The operation times of signel elements, depending on the functions of elements, may be equalt to or smaller than the system operation time, $t_i \leq t_s$.

8. The reliability of system components depends on the amount of resources required to achieve the reliability. According to {4-13}, this function may be written in the form (fig. 3):

$$C_i(R_i) = f(R_i) = K_i(\ln R_i)^{-a_i} \tag{7}$$

$$R_i(C_i) = \phi(C_i) = e^{-(K_i/C_i)^{1/a_i}} \tag{8}$$

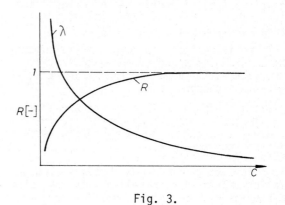

Fig. 3.

where:

$C_i(R_i)$ — cost of component i of the system required to achieve reliability R_i,

$R_i(C_i)$ — reliability of component i of the system achieved by cost C_i,

k_i, a_i — a constant for component i of the system.

9. In view of the large number of system elements, reliability allocation to the elements should be done using appropriate mathematical programming methods and optimization techniques.

10. To make decisions about "the best" reliability allocation, one should employ a compromise programming method that takes into account the satisfaction of technical and economic requirements imposed on the system.

3. MATHEMATICAL MODEL

Having in mind the results achieved in this field {10-15}, we developed a method for solving the reliability allocation problem for mechanical systems. The method relies on the following three approaches.

1. To allocate the design-specified system reliability to system elements so as to meet the technical requirements, on condition that the specified reliability is achieved at a minimum system cost.

2. To allocate the design-specified system cost to system elements so as to meet the economic requirements, on condition that maximum system reliability is obtained for the specified cost.

3. To allocate the design-specified reliability and cost of the system so as to meet particular and economic requirements.

Speaking in mathe-matical terms, these three approaches to solving the reliability allocation problem may be defined as follows (fig. 1).

The first approach is to determine the minimum system cost minC_s for the specified system reliability $R_{so}(t)$. This can be written as:

$$R_{so}(t) \leq f(R_i(t)) \tag{9}$$

subject to the constraint:

$$C_{smin} \geq \sum_{i=1}^{n} C_i \tag{10}$$

The second approach is to determine the maximum system reliability max$R_s(t)$ for the specified system cost C_{so}. This may be written in the form:

$$C_{so} \geq \sum_{i=i}^{n} C_i \tag{11}$$

subject to the constraint

$$R_{smax}(t) \leq f(R_i(t)) \tag{12}$$

The third approach represents an appropriate compromise between the previous two approaches in that equations (9) and (10) are used in reliability allocation for essential elements, whereas equations (11) and (12) are used in reliability allocation for nonessential elements (those which do not cause a system failure). This can be written as:

$$R_s \leq R_{so} (E_i=1) R_{smax}(C'_{co}(E_i<1)) \tag{13}$$

$$C_s \geq C_{smin}(R_{so}(E=1) + C'_{co}(E_i<1)) \tag{14}$$

$$C'_{so}(E<1) = C_s - C_{smin}(R_{so}(E_i<1)) \tag{15}$$

where:

R_s — minimum, compromise-defined reliability that must be achieved by the system,

C_s — maximum, compromise-defined cost of the system that must not be exceeded,

$R_{so}(E_i=1)$ — specified component reliability for elements with which $E_i=1$,

$R_{smax}(C'_{so}(E<1))$ — maximum component reliability for elements with which $E_i<1$,

$C_{smin}(R_{so}(E=1))$ — minimum component cost for elements with which $E_i=1$,

$C_{so}(E_i<1)$ — specified system cost used for allocation to components for elements with which $E_i<1$,

E_i — component essentiality factor including the effects of a component failure on system operation; if E=1, a component failure causes a system failure; if E<1, a component failure does not cause a system failure, but degrades system characteristics.

The mathematical model for determining the minimum system cost and maximum system reliability was developed on the basis of Lagrange's equation with the Lagrange multiplier of the form:

$$F(x,A) = F(x) + \sum_{i=1}^{n} \Lambda_i A_i(x) \tag{16}$$

Introducting substitutions for

$$F(x,A) = L(R,\lambda) \tag{17}$$

$$F(x) = \sum_{i=1}^{n} K_i(-\ln R_i)^{-a_i} \tag{18}$$

$$A_i(x) = \ln R_{so} \tag{19}$$

since there is only one constraint, i.e. since expression (3) can be written as

$$\sum_{i=1}^{n} \ln R_i - \ln R_{so} = 0 \tag{20}$$

the general form of Lagrange's equation for determining the minimum cost for the specified system reliability reads:

$$L(R_i, \lambda) = C_{so} + \lambda\left[\sum_{i=1}^{n} \ln R_i - \ln R_{so}\right] =$$

$$= \sum_{i=1}^{n} K_i(-\ln R_i)^{-a_i} + \lambda\left[\sum_{i=1}^{n} \ln R_i -\right.$$

$$\left. - \ln R_{so}\right] \tag{21}$$

To determine the minimum cost for the specified system reliability, this equation should be partially differentiated with respect to R_i and λ, and the results should be put to equal zero. As a result, one obtains the reliability of element i of the system in the following form:

$$R_i = \exp\left[\frac{(K_i a_i)^{1/(1+a_i)}}{\sum_{i=1}^{n}(K_i a_i)^{1/(1+a_i)}} \ln R_{so}\right] \tag{22}$$

Inserting this solution into equation (7), and then into (10), yields the minimum system cost.

To determine the maximum reliability or the minimum unreliability for the specified system cost, one should introduce the appropriate substitutions for F(X) and $A_i(X)$ into Lagrange's equation (16). As a result, the following equation is obtained:

$$L(C_i, \lambda) = \min(1-R_s) + \lambda\left(\sum_{i=1}^{n} C_i - C_{so}\right) =$$

$$= n - \max\sum_{i=1}^{n} \exp\left[-(K_i a_i)^{1/a_i}\right] +$$

$$+ \lambda\left(\sum_{i=1}^{n} C_i - C_{so}\right) \tag{23}$$

By partial differentiation of this equation with respect to C_i and λ and by putting the results to equal zero, the solution for C_i is obtained in the following form:

$$C_i = \frac{(K_i^{1/a_i} a_i^{-1})^{a_i/(1+a_i)}}{\sum_{i=1}^{n}(K_i^{1/a_i} a_i^{-1})^{a_i/(1+a_i)}} C_{so} \tag{24}$$

Inserting this solution into equation (8), and then into (12), gives the maximum reliability for the specified system cost.

To provide the best possible satisfaction of both technical and economic requirements, the allocation process may be performed in accordance with the appropriate compromise solution. The compromise solution is obtained by determining, on the basis of required system reliability R_s (fig. 1), the specified system reliability for essential system elements ($E_i=1$) in the form:

$$R_{so}' = \exp(\ln R_s\, Q(E_i=1)) \tag{25}$$

where $Q(E_i=1)$ is the allocation factor for essential elements.

Having this value of specified component reliability R_{so}', one determines the minimum system

cost using equations (22), (7) and (10). The obtained system cost is subtracted from the specified system cost C_s (fig. 1) in accordance with equation (15), and a new specified system cost $C_{so}'(E<1)$ is obtained in this way. This cost is then used to determine the maximum component reliability for E<1 by employing equations (24), (8) and (12).

4. ALGORITHM FOR CHOOSING "THE BEST" RELIABILITY ALLOCATION

The mathematical model developed for reliability allocation was used to write the computer program shown in fig. 4.

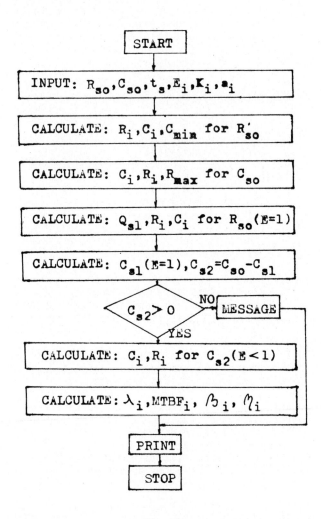

Fig. 4.

Choice of "the best" reliability allocation was performed by compromise programming, and the measure of distance from the ideal point was obtained according to the following metrics:

$$L_p(Z^*, Z) = \left\{\left(\sum_{i=1}^{n} Z_i^* - Z_i(x)\right)^p\right\}^{1/p}, \quad 1 \le p \le \infty \tag{26}$$

This metrics gave the distance between the ideal point Z^* and a point $Z(x)$ in the space of criterion functions. The parameter p plays the role of a "balancing factor" between the total utility and the maximum individual deviation. An increase in the value of p decreases the total utility, but it also decreases the individual deviation from the best value. Small values of p are used when a group utility is more than dominates the individual deviation {18}.

Choice of "the best" solution was performed according to eight criteria:

- system reliability R_s,
- component reliability of system $R_s(E_i=1)$,
- component reliability of system $R_s(E<1)$,
- system cost C_s,
- component cost of system $C_s(E_i=1)$,
- component cost of system $C_s(E_i<1)$,
- allocation factor $Q(E=1)$,
- allocation factor $Q(E<1)$.

5. RELIABILITY ALLOCATION FOR A GEAR TRAIN OF MOTOR VEHICLES

The described allocation model was used for reliability allocation of a gear train of motor vehicles. To determine the reliability block diagram, the gear train (considered as a system) was divided into five subsystems. The gear train will be in proper operation if each subsystem is in proper operation. If any subsystem fails, the gear train (as a system) will also fail. This means that these subsystems are series-connected, as shown in fig. 5. The reliability function of the gear train with five series-connected subsystems reads:

$$R_s(t) = R_1(t)\ R_2(t)\ R_3(t)\ R_4(t)\ R_5(t) \qquad (27)$$

Further development of the reliability block diagram refers to subsystems. By analyzing the operation of each of these five subsystems it was shown that units are series-connected into a subsystem, modules into a unit and components into a module. Taking this into account, the reliability functions may be written as follows:

- for the subsystem according to equation (4),
- for the unit according to equation (5),
- for the module according to equation (6).

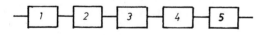

Fig. 5.

The allocation values of the reliability and cost of the gear train depending on the specified values for all variants were determined, in accordance with the algorithm, in order to choose the best solution for the components of gear train with essentiality factor E=1 and for the components with E<1.

44 alternatives were selected for the choice of "the best" solution. From the following variants:

- 16 alternatives with R_{so} ranging from 0.9 to 0.998 from variant 1,

- 12 alternatives with C_{so} ranging from 400,000 to 900,000 dinars from variant 2,

- 16 alternatives with R_{so} ranging from 0.993 to 0.9976 and C_{so} ranging from 350,000 to 700,000 dinars.

The sequence of choosing "the best" solution according to the algorithm was as follows:

1. In each variant several near - "best" solutions and "the best" solution were sought from the standpoint of eight criteria: cost minimization (3 criteria), reliability maximization (3 criteria) and allocation factor minimization (2 criteria) by:

   a) assigning the same importance to each criterion,

   b) assigning different importance to each criterion.

2. "The best" solution was sought within all alternatives from all variants, by considering a total of 44 alternatives (regardless of the variant from which a solution was taken) from the standpoint of 8 criteria, by:

   a) assigning the same importance to each criterion,

   b) assigning different importance to each criterion.

3. "The best" solution was sought within the choice of several near - "best" solutions from each variant by selecting several "best" solutions from each variant. Out of them, "the best" solution was sought (regardless of the variant from which a solution originates) from the standpoint of 8 criteria by assigning:

   a) the same importance to each criterion,

   b) different importance to each criterion.

The choice of "the best" solution in accordance with the described procedure is shown in fig. 6 (criteria with the same importance) and in fig. 7 (criteria with different importance. The curves in these figures have the following meanings:

- curve I represents the determination of minimum cost for the specified system reliability,
- curve II represents the determination of maximum reliability for the specified system cost,
- curve III represents a compromise solution from the standpoint of specified system reliability and cost.

For illustration purposes, Table 1 gives the values of cost, reliability and allocation factor of alternative 11 as well as alternatives 12 and 15, which are close to the 11th. As may be seen, alternative 11 is better than alternatives 12 and 15, because its advantage in cost is greater than the advantages of the other two alternatives in reliability.

Alternative 28 was obtained as "the best" solution from all variants by applying compromise programming. The alternative 28 represents a compromise solution (curve II in fig. 6) between

Table 1

VAR IJA NTA	BROJ ALTE RNAT IVE	R A N C	PRE DNC ST (ADVANTAGE)	CS	C E N A (COST) CS1	CS2	P O U Z D A N C S T (RELIABILITY) RS	RS1	RS2	ALOKACIONI FAKTOR QS1	QS2
I	11	1	4	415445.	404420.	11025.	.993	.9931	.9999	.9789	.0208
I	12	2	1	420817.	409719.	11097.	.994	.9941	.9999	.9789	.0208
I	15	3	10	445384.	434453.	11430.	.997	.9971	.9999	.9789	.0208

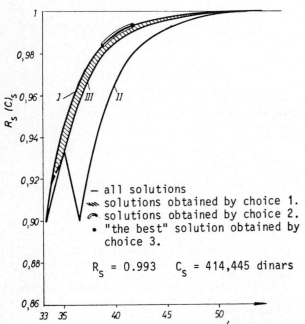

$R_S = 0.993$ $C_S = 414,445$ dinars

– all solutions
solutions obtained by choice 1.
solutions obtained by choice 2.
• "the best" solution obtained by choice 3.

$C_s (R)_s 10^4 (din)$

Fig. 6. Choice of "the best" solution with all criteria having the same importance

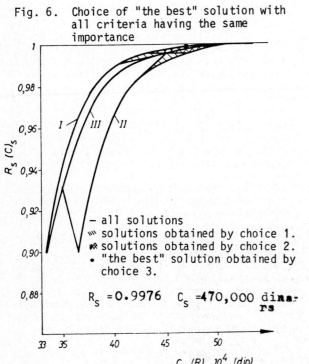

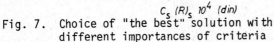

– all solutions
solutions obtained by choice 1.
solutions obtained by choice 2.
• "the best" solution obtained by choice 3.

$R_S = 0.9976$ $C_S = 470,000$ dinars

$C_s (R)_s 10^4 (din)$

Fig. 7. Choice of "the best" solution with different importances of criteria

the technical and economic requirements imposed on the system, since minimum cost for specified reliability is determined for system components with $E_i=1$, whereas maximum reliability for specified cost is determined for system components with $E_i<1$.

Alternative 28 is the result of the choice of "the best" solution from the standpoint of assigning the same and different importance to the criteria of cost (CS, CS1, CS2), reliability (RS, RS1, RS2) and allocation factor (QS1, QS2) for system reliability allocation (the gear train in this case). The alternative 28 incorporates in a particular was:

- cost minimization for the specified system reliability for elements that are essential to system operation $E_i=1$,

- reliability maximization for the specified system cost for elements that are not essential to system operation $E_i<1$,

- technical requirements imposed on the system, since a system component for essential elements is determined on the basis of technical requirements imposed on the system through factor Q E=1

- economic requirements imposed on the system, since the reliability of nonessential elements is determined through the specified system cost with reliability maximization.

The solution obtained represents system reliability allocation for the specified reliability and cost of the system. The allocated values of cost, reliability, failure rate - (for the case of exponential distribution), parameters β, η and MTBF (for the case of Weibull distribution) for 169 components of the gear train are given in Table 2.

CONCLUSIONS

Studies of reliability design for technical systems represents a very important area of engineering activities in the present stage of industrial development. These activities should be directed towards the largest possible reliability improvement of technical and, thus, mechanical systems. This would, in turn, result in higher human safety. Investments in this field would be highly rational, since economic issues would be solved in parallel wit solving reliability problems.

The results obtained indicate that the method developed in this paper can be applied to reli-

Table 2

CS= 470000 DIN			RS=0.9976		TS= 100000.		
CS(E.EQ.1)= 452771 DIN		RS(E.EQ.1)=0.9982		CS(E.EQ.1)=0.7482			
CS(E.LT.1)= 17228 DIN		RS(E.LT.1)=0.9994		CS(E.LT.1)=0.2515			

BROJ DELA NAZIV (No.) (NAME)	CENA /DIN/ (COST)	POUZDA- NOST (RELIABILITY)	FAKTOR ALOKACIJE	LAMBDA *1.E-6 /1/KM/	BETA	ETA *1.E3 /KM/	T-SRED. *1.E3 /KM/
010100001 KUCISTE (ROKS)	2986	0.999997	0.00106	0.256E-04	1.0	************	
					2.0	62463	55342
					3.0	7307	6525
					4.0	2499	2264
					5.0	1312	1205
					6.0	854	793
020200005 VRATILO GL (SHAFT)	56825	0.999764	0.09781	0.236E-02	1.0	423403	423403
					2.0	6506	5765
					3.0	1617	1444
					4.0	806	730
					5.0	531	487
					6.0	402	373
020302024 ZUPCANIK 3 (GEAR)	27860	0.999890	0.04564	0.110E-02	1.0	907316	907316
					2.0	9525	8439
					3.0	2085	1862
					4.0	975	884
					5.0	618	568
					6.0	456	423
020303011 CAURA ODST (CAPISLE)	626	0.999999	0.00035	0.834E-05	1.0	************	
					2.0	109470	96990
					3.0	10621	9485
					4.0	3308	2997
					5.0	1643	1508
					6.0	1030	956

ability allocation in the stage of designing mechanical systems. All factors affecting reliability allocation should be studies in detail and due attention should be given to each of them. Satisfaction of all techno-economic criteria imposed on the system should be attained, with particular attention devoted to the elements whose failures cause a system failure.

The developed mathematical model for reliability allocation by determining minimum cost for the specified reliability has some advantages over the allocation model for determining maximum reliability for the specified cost. Since variant 3 employs both mathematical models and since it satisfies in an appropriate way the techno-economic system requirements, it is obvious that both models should be used for allocation purposes.

Compromise programming method should be used to make a decision in choosing "the best" reliability allocation solution. "The best" solution is obtained very rapidly by this method. In addition, it should be emphasized that the ranking procedure yields a percent advantage of each solution over the next from the list, which also facilitates making the decision about "the best" solution.

REFERENCES

{1} Todorović J., Zelenović D., System Effectiveness in Mechanical Engineering (in Serbo-Croat), Naučna knjiga, Beograd, 1981.

{2} Kapur K.C., Lamberson L.R., Reliability in Engineering Design, John Wiley and Sons, New York, 1977.

{3} Belman E.R., Dreyfus E.S., Applied Dynamic Programming, Princeton, New Jersey, 1976.

{4} Kulakov N.N., Zagorinko A.S., Vibor modeli staimosti, primenjaenih v ekonomiko matematičeskih isledovanijah povišega nadežnosti, Izd. "Nauka", Novosibirsk, 1968.

{5} Tillman A., Frank, Hwang Ching-lai, Fan Liang-Tseng and Lai C. Keeting, Optimal Reliability of a Complex System, IEEE Transactions on Reliability, Vol. R-19, No. 3, 1970.

{6} Kulshrestha K.D., Gupta C.M., Use of Dynamic Programming for Reliability Engineers, IEEE Transactions on Reliability, Vol. R-22, No. 5, 1973.

{7} Misra Behari Krishna, Ljubojević D. Milan, Optimal Reliability Design of a System: A New Look, IEEE Transactions on Reliability, Vol. R-22, No. 5, 1973.

{8} Aggarwal K.K., Gupta S.J., On minimizing the cost of reliable systems, IEEE Transactions on Reliability, Vol. R-24, 1975.

{9} Tillmann A. Frank, Hwang Shing-Lai, Kuo Way, Optimization Techniques for System Reliability with Redundancy - A Review, IEEE Transactions on Reliability, Vol R-26, No. 3, 1977.

{10} Fan Tseng Liang, Wang Sen Chiu, Tillman A. Frank, Hwang Lai Ching, Optimization of Systems Reliability, IEEE Transactions on Reliability, Vol. R-16, No. 81, 1967.

{11} Tillman A. Frank, Hwang Ching-Lai, Kuo Way, Determining Componented Redundancy for Optimum System Reliability, IEEE Transactions on Reliability, Vol. R-26, No.3, 1977.

{12} Hwang L.C., Tillmann A.F., Kuo W., Reliability Optimization by Generalized Lagrangian-Function and Reduced-Gradient Methods, IEEE Transaction on Reliability, Vol. R-28, No. 4, 1979.

{13} Aggarwal K.K., Chopra C.Y., Bajwa S.J., Reliability Evaluation by Network Decomposition, IEEE Transaction on Reliabaility, Vol. R-31, No. 4, 1982.

{14} Todorović J., Ivanović F., Zuordnung der Zuverläsigkeit für Kraftfahrzeugenbaugruppen, 3. Fachtagung Materialäkonomie in Automobilbau, Karl-Marx-Stadt, 1982.

{15} Todorović J., Ivanović G., Reliability Allocation Problem in Technical Systems (in Serbo-Croat), SYM-OP-IS, H. Novi, 1981.

{16} Ivanović G., Reliability Allocation Problem for Motor Vehicles, Conference on Science and Motor Vehicles, Opatija, 1983.

{17} Zeleny M., Compromise Programming, Multiple, Criteria Decision Making, University of South Carolina Press, Columbia, 1973.

{18} Opricović S., Multicriteria Optimization: Comrpomise Programming (in Serbo-Croat), XXII-th Yugoslavien Conference ETAN, Zadar, 1978.

{19} Jovičić S., An Approach to Determining System Effectiveness, Technical Paper, Vol. XXX, No. 6, Beograd, 1980. (in Serbo-Croat).

{20} Todorović J., Ivanović G., The Optimization of Reliabilities Allocation by EFTES-method, VI-th Int. Conf. on Prod. Research, Novi Sad, 1981

BIOGRAPHY

Mr. Gradimir Ivanović was born in 1947. He graduated from the Faculty of Mechanical Engineering, Belgrade University, Yugoslavia, where he took a M.A. degree of technical sciences in 1978. Mr. Ivanović is working at the Motor Vehicles Department on the problems of Effectiveness Maintainability, Availability, and Reliability of technical systems. In the field of motor vehicles and effectiveness he presented more than 30 papers. Mr. Ivanović and Mr. Stanivuković edited two books (in Serbo-Croat): Exercises in Effectiveness Technical Systems with Background Theoretical (Faculty of Mechanical Engineering, Belgrade University, Faculty of Technical Sciences, Novi Sad University, 1978), and Exercises in Reliability Technical Systems with Background Theoretical (Faculty of Mechanical Engineering, Belgrade University, 1983).

Reliability Technology — Theory & Applications
J. Møltoft and F. Jensen (Editors)
© Elsevier Science Publishers B.V. (North-Holland), 1986

REJECTION OF UNRELIABLE ITEMS BY THEIR INFORMATIVE PARAMETERS

Anatoliy I. LUCHINO
Institute of Control Sciences
Moscow, USSR

A method is described for unreliable items rejection based on pattern recognition techniques. The underlying hypothesis is built around the fact that the information on latent defects is felt in the item parameters and characteristics. The algorithms of the developed technique are programmed in FORTRAN IV language. Results of practical implementation are presented.

1. INTRODUCTION

The present time requirements to reliability of devices, hardware, computer and control systems cause the necessity of developing the methods ensuring the advancement of reliability of their elements and components (elements base). Reliability of the elements base (of semiconductor devices, integrated circuits and other items) has essentially improved, but its level does not yet satisfy the system development engineers. Further improvement of the elements base reliability is possible at the cost of the construction advancement, strict control of the technological process, development of new designing principles and by means of rejecting potentially unreliable (with latent defects) items on the various stages of the technological process and from the lot. Effective rejection enables the manufacture of more homogeneous products with respect to quality and reliability and may considerably reduce the intensity of failures. Besides, rejection on the early stages of manufacture allows the avoidance of expences covering the realization of further technological operations required for potentially unreliable items.

In the course of searching the ways of increasing the reliability of items the important problem is the analysis and definition of failures depending on the causes [1].

The failures occurring in the items are to be referred to as inherent and introduced. Inherent failures (area I in Fig. 1) are caused by degradation or interaction of the materials used in a given technological process. These failures are natural and specified by the structure, production technology, operation conditions, prescribed for a given item and they are actually its properties. Inherent failures are defined by the

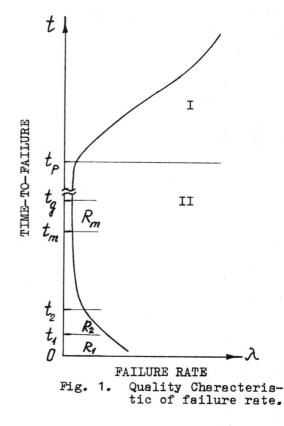

Fig. 1. Quality Characteristic of failure rate.

physical and chemical processes, which can be reproduced, studied and described. Item reliability, determined on the basis of this type failure mechanisms is an ideal or theoretical reliability. Evidently in the case of an ideal technological process the item failures are conditioned by their own limitations and interactions of the materials used in items production. These failures can be simulated in the corresponding test structures. The kinetics of the degradation evolution may be studied and the failure mechanisms parameters can be defined together with the forecasting of the resource t_p (time of failure occurence, caused by aging) of items.

Introduced failures (area II in Fig.1) are conditioned by the disturbances in the technological process, nonobservance of maintenance norms, insufficient control of quality in all manufacturing stages and such failures are not referred to as item properties. Introduced failures cannot be practically reproduced and they are characterized by a random distribution of faults. Therefore the description of physical models simulating such failures is factually impossible due to the complexity of a mechanisms description, random nature and rare occurrence of failures. Because of the above reasons, the reliability forecasting for the items containing introduced failures is actually impossible with the help of the described approach.

2. THE STATEMENT OF THE PROBLEM

It becomes obvious from the characteristic of the failure rate $\lambda = f(t)$

(Fig. 1) that failure-free performance of items within a time interval less than t_g is determined by introduced failures. Thus the basic problem of increasing the reliability of an items lot consists in detection and rejection of potentially unreliable items from the lot.

There exist two principally different ways for detecting potentially unreliable items. The first one involves the idea that faulty items tend to failures and they are very sensitive to overloading. Having this in view, the items are subjected to rejection testing under load (not less than maximum permissible load according to the technical specifications) the duration of which is shorter compared with the term of items application in real objects. The results of rejection tests will show that all or at least the greater part of faulty items will fail and it will increase reliability of the total lot of items, subjected to such tests. Successful realization of rejection tests requires the correct definition of a kind, magnitudes and duration of load action having in mind that reliable items should not be damaged. It is a complicated problem which can be solved if one has the data on the inherent failure mechanisms of items. The knowledge of the inherent failure mechanisms makes it possible to choose the method, regimes and duration of rejection tests with no significant reduction of a reliable items resource. The main disadvantage of rejection tests is a considerable increase of cost (in a number of cases – in tens of times) of items.

The other way of detecting potentially unreliable items consists in the analysis of failure physics. In this case, if physical causes, failure mechanisms

and faults are known, one should determine the parameters and characteristics feeling the faults introduced into the item. If the values of informative parameters set, which were measured in definite conditions, exceed the established limit, such an item is to be excluded from the lot. Optimal boundary values are to be determined in this case only in combination with the realization of reliability tests within a given time interval (in a particular case, a test for a guaranteed longevity t_g). Contrary to the method of rejection tests this method of item rejection provides for valid data with lesser material expences within shorter terms and with no reduction of a reliable items resource.

3. THE SELECTION OF A SOLUTION METHOD

In this paper the problem of detection and rejection of potentially unreliable items is solved by means of the second approach. The author proposes the developed and investigated effective method of items rejection, based on the hypothesis that the faults introduced in an item should be felt in its parameters and characteristics, measured in the initial moment or time interval. With the help of this method one can solve the problem of individual reliability prediction for items of any kind, for which stochastical relation between reliability and informative parameters set values is revealed.

Variety of failure types and dispersal in various parameters and characteristics of the data on the item behaviour creates the necessity of an information processing method which would make it possible to use all available data conserning the item and construct a forecasting (prediction) model. If item reliability is viewed as a function of the set of initial values of its informative parameters and characteristics (henceforth – "informative parameters") then this problem can be solved with the help of the pattern recognition method the essence of which is as follows. Let the items of some general totality, described by the informative parameters set initial values be subdivided into the Classes A and B. Using training (statistical) lots q and h of the items belonging to the classes A and B one should define to which of these classes some new item, not included into the training lots, belongs. In the k -dimensional space of informative parameters χ , each item of this general totality is characterized by the vector $\chi_j = (x_{1j}, x_{2j}, \ldots, x_{kj})$ and represented as a certain point (a vector end) $\chi_j (j = 1, 2, \ldots, \infty)$ of this space, classes A and B as certain sets of the points Q and H, and the training lots as certain subsets q and h of points of these sets. The problem of defining the pattern (class), to which a new item represented as the point X_i (i-number of a new item), is to establish to which of the sets Q of H this point belongs if it is known that $q \in Q$ and $h \in H$.

Thus, the geometrical problem of the pattern recognition training consists in constructing (in the space of informative parameters X) a hypersurface separating the sets Q and H specified only by the subsets q and h.

If the unknown function $f(X)$ is introduced which with $X_j \in Q$ takes on the value +1 and with $X_j \in H$, -1 (i.e. the sign of $f(X)$ indicates whether X_j belongs to A or B), then the problem of finding the class to which the new point X_i belongs is in finding the value of the function $f(X_i)$ in that point if it is known that $f(X)$ in the points of the subsets q and h takes values equal to 1 and -1, respectively. Finding an unknown separating (forecasting) function $f(X)$ is the problem of the pattern recognition training.

It is obvious that application of the pattern recognition methods makes it possible to forecast reliability of items of the same type, manufactured under the same conditions (or sufficiently similar conditions) of manufacturing items of the training sampling which was employed for construction of the forecasting rule. Since the faults of items vary it may happen that for one and the same type of items the set of informative parameters will be not identical for items produced by different manufactures.

4. THE SELECTION OF THE METHOD FOR CONSTRUCTION OF THE FORECASTING RULE

The pattern recognition methods are widely used for individual forecasting of reliability indices in some types of electronic items. For this purpose the developed techniques utilize the methods of potential functions and statistical solutions. However these methods are difficult to be applied to forecast the reliability indices of highly reliable items due to the complexities connected with the training sampling formation since the construc-

tion of the forecasting rule according to these methods requires sufficiently great quantity of failed items.

Application variety of the pattern recognition methods resulted in the availability of a great number of pattern recognition training algorithms for particular problems solution. The abilities of these algorithms are different. They also differ in realization complexity and initial data needed for their operation. Therefore the selection of an algorithm type for pattern recognition training presents considerable difficulty each time when solving a new problem.

When solving the problem of increasing reliability of items belonging to the modern element base, in a great number of cases one deals with a limited amount of unreliable items (beginning with single units up to dozens of them) and, as a result, with different prices of two possible error types occurring in reliability forecasting. Hence it is important to obtain such a forecasting function $f(X)$ which would help to reduce an error of classifying the unreliable items (belonging to the Class B) within the training sampling to zero with a minimal error of classifying the opposite class items (belonging to the Class A) because the value of the first type error (an error of the second kind) is significantly greater. Using this criterion the problem, dealing with the synthesis of the separating function $f(X)$ between the Classes A and B, is solved by means of a specially developed piece-wise-linear approximation algorithm. In every detail this algorithm is described in paper [2]. The main advantages of this algorithm are the following: 1) it is not parametrical and hence it does not require data on the kind of probabilities density distribution of the item parameters in the classes, and it is very important when solving the problems with the limited number of items in one of the classes; 2) the absence of restrictions on the number of informative parameters and a minimal number of items in the class; 3) this algorithm is simple in operation and does not require large computer memory.

5. THE METHOD OF AN INDIVIDUAL ESTIMATION OF THE ITEM RELIABILITY

The block-diagram demonstrating the method of the item reliability individual forecasting is given in Fig. 2. For each item included in the general totality, the initial values of informative parameters are to be measured. In the K-dimentional space of these parameters, each item has a corresponding vector of their values. A part of

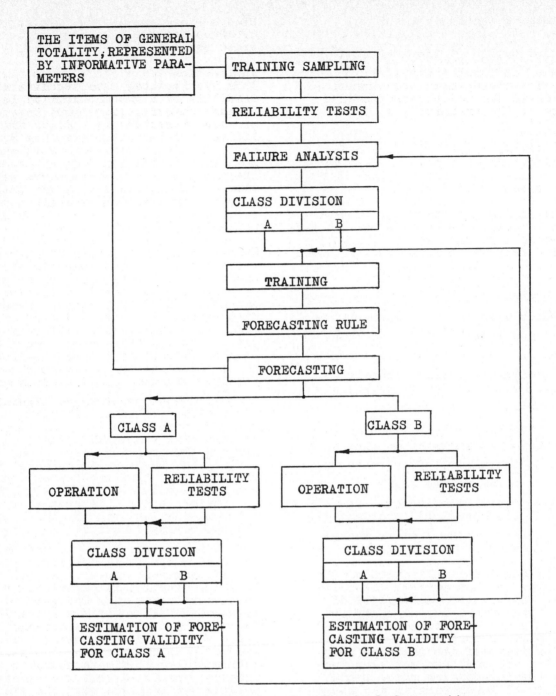

Fig. 2.   The block-diagram of individual forecasting.

these items (chosen by means of the random choice method) forms the training sampling so that it will include the representatives of the Classes A and B. The training sampling items are tested for reliability within a time-period t_g and it is the actual period for which reliability forecasting is needed. In the course of tests the longevity of each unreliable item (of the Class B) is defined. Depending on the test results the items of this sampling are subdevided into the Classes A and B. The Class A includes the items with the longevity longer than forecasting time t_g or equal to it.

The Class B contains the items whose longevity is shorter than the time t_g.

The data obtained in the tests and showing to which of the two classes each item of the training sampling belongs, depending on the vector value of informative parameters, make it possible to find the position of a separating hypersurface between the Classes A and B with the help of the piece-wise-linear approximation algorithm and thus to determine the forecasting rule.

To estimate the forecasting validity some part of the general totality items after forecasting can be tested for reliability similarly to items of the training sampling. Depending on the test results, the training sampling may be given some items which in the course of forecasting were by mistake referred to the opposite class and afterwards the forecasting rule is subjected to correction. Correction may be carried out on the basis of the data on the items which are already in operation in the modes corresponding to those of the training sampling items tests.

Correct dividing into classes requires the realization of a physical analysis of the failed items and finding the failure causes.

Having a sufficient amount of information the problem can be made more complicated by assigning the requirement of a more accurate individual forecasting of items reliability. To achieve this, one should construct forecasting rules for several time instants t_1, t_2, ..., t_m (Fig. 1). In this case in the process of forecasting it is to be defined to which of the time Classes $R_1, R_2, ..., R_m$ each item of the general totality belongs, with respect to the lower and top time limits:

$$0 - t_1, \quad t_1 - t_2, \quad \ldots, \quad t_m - t_g.$$

Depending on the assigned problem this method helps to estimate the quality or to forecast a certain property (longevity, failure-free operation, keeping quality) of an item.

6. THE MAIN RESULTS

The technique of "Forecasting of items reliability making use of informative parameters" has been developed on the basis of this method of reliability estimation. The software of this method algorithms is accomplished in Fortran-IV. The developed program means, included in the method in the form of a supplement, provides for: construction of a forecasting rule according to a given set of informative parameters and realization of forecasting; estimation of the informativeness of various sets of the initial totality parameters and selection of the optimal set of the informative parameters; realization of forecasting according to a given forecasting rule; estimation of forecasting validity.

These methods describe the order of the program means application and decoding of the results printed out in the process of operation.

7. APPLICATION

The testing of various types of electronic items and lots proved that the developed (with the corresponding data available) method makes it possible to realize a sufficiently effective rejection of potentially unreliable items. For all considered problems

the frequency of a correct forecasting for items reliability is within the range of 0.7-0.95.

In practice the sorting of more than 3000 hybrid integrated circuits of series K217 into two reliability level classes reduces the failure intensity of these items within a class with an increased reliability level by an order of magnitude. At present the rejection of potentially unreliable semiconductor laser radiators is carried out for one of the many types of radiators with the help of this method.

This method facilitates the possibility of finding the radiators of the above type whose longevity considerably exceeds the guaranteed one which is very important when solving the problem of prolonging these items resource.

REFERENCES

[1] D.N. Gretchin, An approach to estimating of the integrated circuits reliability based on the study of failure physics, <u>Technical means of control and reliability</u> problems, Nauka, Moscow, 1974, 172-178.

[2] A.I. Luchino, Construction of algorithm of the separating function synthesis and finding the approach to the choice of the initial parameters totality for an individual forecasting of items reliability, <u>STAQUAREL'84</u>, Proceedings, vol.2, Praha, 1984, 212-222.

AUTHOR

Dr. A.I. Luchino was born on 1939 November 18 in South Ural, USSR. He received the PhD degree in Elements and Arrangements of Computer and Control Systems from Institute of Control Sciences USSR Academy of Sciences in Moscow in 1977. From 1966 he is employ at the Institute of Control Sciences. His research interests include electronic means of control, statistical life testing, prediction of reliability and individual forecasting of items reliability.

Reliability Technology — Theory & Applications
J. Møltoft and F. Jensen (Editors)
© Elsevier Science Publishers B.V. (North-Holland), 1986

A SIMULATION APPROACH TO RELIABILITY PREDICTION
IN THE PRESENCE OF UNCERTAINTY

Mirko VUJOŠEVIĆ, Radivoj PETROVIĆ,
Aleksandar ŠENBORN
Mihailo Pupin Institute
Volgina 15, Belgrade, Yugoslavia

The importance and practical applicability of the analytical results of reliability theory decrease if data about system element reliabilities do not properly reflect the real situation. A high level of uncertainty is present in these data not only in the stage of designing new equipment and systems, when no experience about their behaviour in real operation is available, but also after performing a reliability test and gathering failure data for a particular period of time. This is especially true for high-reliability systems or equipment which is not in common use. A simulation model for system reliability analysis in the presence of uncertainty is described in this paper. The main parts of the simulation model are: the generators of random numbers, i.e. of the state of the components of the system under consideration, determination of system state on the basis of the states of elements, statistics keeping, statistical processing and presentation of simulation results. The simulation model is modular and allows experiments with systems of various structures and purposes to be carried out.

1. INTRODUCTION

A plenty of mathematical models and algorithms are available for system reliability evaluation. They have been developed and used by reliability engineers who are trying to obtain quantitative measures of reliability performances. In practice, however, very often, qualitative assessments (such as the existence of redundancy, the presence of components made in a particular technology or by a particular manufacturer, and the like) are preferred by a purchaser to quantitative data about reliability (e.g. the values of MTBF or inherent availability). The reason for this lies in that the former data are easy to check, whereas this is very difficult or even impossible to do for the latter. Apart from the usual engineers' distrust and fear of any probabilistic measures related to technical equipment, the reliability context contains an additional reason for their suspicion: reliability data are frequently unreliable.

The essential building blocks supporting the mathematical theory of system reliability are reliability and maintainability functions of system components, i.e. the probability distribution of time to failure and the probability distribution of time to repair. Nearly all mathematical models, whether simple or very sophisticated, have been developed on the assumption that these functions and their parameters are known. Application of the results obtained should be conditioned by a previous check as to whether the assumptions stated correspond to a real system. Although the check itself does not necessarily represnet a difficulty, since it is a standard hypothesis test, problems do arise because of a lack of failure data {1,2,3}. This is why the type of distribution and its parameters become uncertain as well. The uncertainties related to determining the type of distribution will not be treated here. Uncertainty considerations

will be limited to the simplest and the most frequent case of the application of single-parameter exponential distribution. In addition, we will consider separately the reasons for uncertainty in reliability prediction during the stage of designing new equipment and systems, in reliability evaluation after performing a reliability test and gathering failure data as well as in mission reliability prediction.

A component classification based on interval estimates is proposed in Section 3. The use of interval estimates aggravates largely the application of well-known analytical results, and one simple example in illustrated in Section 4. The simulation approach can rely on the use of either system reliability function or structure function. Both cases are illustrated by examples presented in Section 5.

2. SOURCES OF DATA UNCERTAINTY

According to the concept of integrated logistics support (ILS) {4}, it is already *in the system design stage* that care must be taken, among other issues, of reliability. A quantitative measure of reliability is required for applying the available mathematical models for life cycle cost estimation, i.e. for the evaluation, ranking and choice of different design proposals {5}. Of course, there cannot exist any failure data for a system that is being designed; what does exist is the experience about the reliabilities of components to be incorporated into the system. Using mathematical models, system reliability is predicted on the basis of component reliabilities.

Various handbooks and standards represent the main source of data for component failure rates. A selected list of these sources is provided in {6}, e.g. {7,8}, but MIL-HDBK-217 {7} is most commonly used in electronics applications. The mathematical models for predicting component failure rates given in this Handbook depend on component types (a resistor, solenoid, integrated circuit, etc.) but, in a general case, they include a variety of different factors such as environment temperature, electrical and mechanical stress, environment conditions, quality, intended application, etc. The influence of single factors on the component failure rate may amount to one order of magnitude, or more, but it is impossible to determine the exact values of these factors. Thus, the environment temperature varies significantly, with the exception of units intended to operate under laboratory conditions. For failure rate prediction, one accepts, by rule, a temperature higher than some expected mean temperature, or even a maximum expected temperature, or a temperature close to this. A certain security level is achieved in this way, because it is regarded that the predicted reliability will not be lower than the calculated. Another important piece of data is the quality factor, π_Q, {7}, whose value is difficult to determine for components that do not comply with MIL standards. The situation is similar with other factors (e.g. the number of switching operations for switches and relays), so that the error in component failure rate ranges from about 3 (one order) to 10 (two orders of magnitude).

An important source of uncertainty is the use of new components for which not enough experience is available and which are not covered by existing handbooks. All these uncertainties lead to an error of the order of 3 to 10 in determining the MTBF of the system under consideration.

There is a delusion among engineers and mathematicians that a high level of uncertainty may be eliminated by performing *a reliability test* or by gathering failure data from field operation. Enough data for performing statistical analysis will become available, they regard, if enough resources are devoted to this end. But such an amount of resources will never be available, and accurate data will, unfortunately, always be lacking. Seven different reasons for this are given in {1}. However, a single reason underlying all these problems is the so-called "tyranny of numbers". This phenomenon occurs whenever one should check the probability of very seldom events such as failures of high-reliability components. For example, to be able to check a failure rate of 10^{-6}/h with a 90% confidence interval, one must spend a thousand of hours checking 2300 components, and no failure will occur, or 9300 components, and 5 failures will occur {9}.

In reliability prediction for a system subject to a phased mission, there are two causes of uncertainty:

(a) By rule, the available algorithms yield an approximate solution only.

(b) The environmental conditions and structure for each phase may vary significantly. This is especially important for systems with redundant elements that are, rule degradable i.e. the use of standard reliability prediction methods may load to overestimates {10}.

As a result of preceding considerations, one may conclude that:

(a) In performing reliability prediction or failure rate prediction, one should use interval estimates rather than point estimates {11}.

(b) These intervals may be rather wide, one or two orders of magnitude.

The problem arising now is how to handle parameters which have no point but intervalu velues, i.e. estimates.

3. COMPONENT CLASSIFICATION

Prior to developing and using any mathematical model that would incorporate the mentioned uncertainty in system parameters, it is suitable to make a certain component classification. This can be done in two ways. The first consists of determining the confidence interval or the error factor that may occur in using the available data from handbooks or after performing statistical analyses. The second relies on pre-specified ranges to which the parameter under consideration, e.g. the failure rate, belongs. For example, in the Spares provisioning guide by the Collins Radio Comp. {12},{17} categories of failure rates are proposed. Similarly, in {13} a category table with 18 categories is presented, whereas for a higher uncertainty enviroment, according to {14}, it is sufficient to have 7 categories of failure rates, such that the range of each category is one order of magnitude.

Having made the classification, one should provide an interpretation of the membership of failure rate in a certain class. For example is the uncertainty such that the component failure rate may take any of the values from the range under consideration with equal probability, or else would some distribution other than the uniform be more suitable? Or, would the introduction of fuzzy concepts perhaps permit a more realistic interpretation of the membership? We will not discuss the fuzzy membership possibilities, but will treat the uncertainty in the failure rate by regarding the value of failure rate as a random variable with a given distribution.

4. MATHEMATICAL MODELLING

One can now proceed to the development and analysis of mathematical models. Generally, two approaches are possible: analytical and simulation.

There are very few analytical methods that require a rather complicated mathematical apparatus for the analysis of complex stochastic processes. The results are usually approximate in nature. This will be illustrated by the example of reliability prediction for a component subject to a phased mission profile.

Example 1

For each phase k, k = 1,...,K of duration τ_k,

Fig. 1, let the failure rate τ_k be a random variable. The probability that the component will surve phase k, if it functioned at the begining of the phase, is also a random quantity

$$R_k(\tau_k) = \exp(-\lambda_k \tau_k)$$
$$= \exp(-\alpha_k) \tag{1}$$

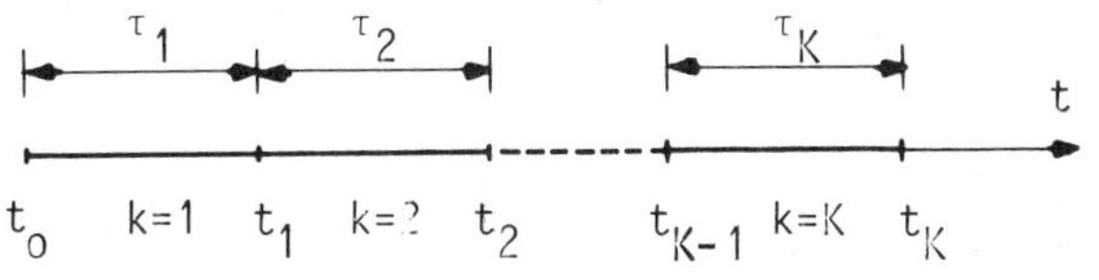

Fig. 1.

The mission reliability $R_M(t_K)$ is a product of the probabilities that the component will not fail in any of the phases when considering them separately

$$R_M(t_K) = R_1(\tau_1) \dots R_K(\tau_K)$$
$$= \exp\left(-\sum_{k=1}^{K} \alpha_k\right) \tag{2}$$

Let us denote $\alpha_M = \sum_{k=1}^{K} \alpha_k$. If $\alpha_M < 0.1$, with an accuracy higher than 10 percent one can use the approximation

$$R_M(t_K) = 1 - \alpha_M \tag{3}$$

Now, if α_k are random variables of normal distribution with the mean $\mu_{\alpha k}$ and variances $\sigma^2_{\alpha k}$, the reliability $R_M(t_K)$ also obeys the normal distribution with the mean $\mu_R = 1 - \sum_{k=1}^{K} \mu_{\alpha k}$ $\sigma^2_R = \sum_{k=1}^{K} \sigma^2_{\alpha k}$. In other cases, if the uncertainty in parameters λ_k is modelled in a different way, it is not obvious how one would analize the uncertainty in reliability of this simple system. Let us state that the same mathematical model is applicable to the analysis of series systems.

The largest number of analytical approaches start from the assumption that the time to failure has a logarithmic normal, or lognormal distribution. In other words, when the failure rate is expressed as 10^{-x}, where x is some exponent, the use of the lognormal disrribution implies that the exponent satisfies a normal distribution.

5. SIMULATION MODEL

Simulation permits more freedom in treating the uncertainty in component reliability parameters. As in other cases, the simulation model comprises the following wholes:

(a) Random number generation
(b) Establishment of system state
(c) Data accumulation
(d) Statistics
(e) Presentation of simulation results.

Only the whole listed under (b) is specific in this problem. This part of the reliability analysis model is even simpler than many other simulation models. It can be developed starting from a reliability function or from a structure function.

The *system reliability* function relates the system reliability R_S to the component reliabilities R_j, j = 1,...,J, i.e.

$$R_S = f(R_1, \dots, R_J)$$

Before defining the *structure function*, for each component C_j, j = 1,...,J, it is necessary to define a performance state indicator $x_j(t)$ as a binary random variable

$$x_j(t) = \begin{cases} 1 & \text{if } C_j \text{ functions at time t} \\ 0 & \text{otherwise} \end{cases}$$

Further, one defines a performance state indicator vector $\underline{X}(t) = \{x_1(t),\dots,x_J(t)\}$ and corresponding stochastic processes $\{x_j(t), t \geq 0\}$ and $\{X(t), t \geq 0\}$. The structure function $\overline{\Phi}$ is a binary function of binary variables $x_1,\dots,x_J$ which relates the system performance state with the performance states of its components

$$\Phi(\underline{X}) = \Phi(x_1,\dots,x_J) = \begin{cases} 1, & \text{if the system functions} \\ 0, & \text{otherwise} \end{cases}$$

Relation between the reliability function and the structure function is

$$f(R_1,\dots,R_J) = \Pr[\Phi(x_1,\dots,x_J) = 1]$$
$$= E\Phi(x_1,\dots,x_J)$$

where E denotes a mathematical expectation. In the text to follow we will present two simulation models; one of them relies on the reliability function and the other uses the structure function.

Example 2

Suppose that the reliability function $f(R_1,\dots, R_J)$ of a system intended to operate for a time t_K is known. Component failure rates λ_j, j = 1, ...,J are assumed to be random variables with known distributions according to component classification. Then the system reliability is regarded as a random variable whose distribution should be obtained by simulation.

The simulation model for this case is relatively simple and contains the following steps that should be repeated for a sufficient number of times:

1. Using random number generators, find random values for failure rates λ_j, j = 1,...,J

2. Compute $R_j = \exp(-\lambda_j t_K)$

3. Compute $R_M = f(R_1,\dots,R_J)$

Note that if the lower and upper bounds on component failure rates are determined according to the component classification, one can calculate the upper and lower bounds on system reliability by using the reliability function.

Example 3

We consider again a phased mission and use the same notation as in Example 1. But, instead of a single component, now we consider a system whose configuration varies in such a way that for each phase k there exists a different structure function. Use of the available algorithms for reliability prediction of phased missions {14,15} does not always permit the reliability function to be obtained in a suitable form. Therefore, we will first present a simulation model for determining a mission reliability assuming that failure rates are known, and then give a modification of the algorithm for the case when failure rates are uncertain.

By definition, a mission reliability is

$$R_M(t_K) = \Pr \left\{ \prod_{k=1}^{K} \phi_k \left[\underline{X}(t_k) \right] = 1 \right\}$$

For each component C_j one defines a performance state indicator in phase k, considering the phases independently

$$u_{jk} = \begin{cases} 1, & \text{if } C_j \text{ functions at the end of phase k assuming that it functioned at the begining of the phase} \\ 0, & \text{otherwise} \end{cases}$$

Then

$$x_j(t_k) = ^{st} u_{j1} \cdots u_{jk}$$

where $=^{st}$ means "stochastically equal" or "has the same distribution as".

The main steps of the simulation algorithm for reliability prediction of a phased mission are:

1. Initialization, $N_M = 0$.

2. Repeat steps 3-13 N_t times; N_t is the number of simulation runs.

3. Initialization; $j = 0$; $k = 0$.

4. Put $k \leftarrow k+1$.

5. Put $j \leftarrow j+1$.

6. For each component C_j generate the time to railure T_{jk} assuming the conditions of phase k

$$T_{jk} = - \frac{1}{\lambda_{jk}} \ln z$$

where z is a random number uniformly distributed in interval (0,1).

7. If $T_{jk} \geq \tau_k$, then $u_{jk} = 1$; otherwise $u_{jk} = 0$.

8. Compute $x_j(t_k) = u_{j1} \cdots u_{jk}$.

9. If $j < J$, return to 5.

10. Compute $\phi_k [\underline{X}(t_k)]$.

11. If $k < K$, return to 4.

12. Compute $Y = \prod_{k=1}^{K} \phi_k [\underline{X}(t_k)]$.

13. If Y=1, then $N_M \leftarrow N_M + 1$.

14. Calculate $R_M = N_M/N_T$.

If the component failure rates in different phases are treated as random variables according to the accepted classification, the described algorithm is applied several times by determining, prior to Step 1., random values λ_{jk}, $j = 1,...,J$; $k = 1,...,K$. The mission reliability is calculated for each set of values $\{\lambda_{jk}^{(n)}\}$ and, after terminating the simulation, statistical processing of results is performed.

6. CONCLUDING REMARKS

Probability theory is a tool for the treatment of the uncertainty that is immanent in all random processes. In reliability theory, which, for its largest part, may be viewed, as an application of probability theory, there exists another, higher level of uncertainty: probability laws and/or system parameters are usually not known. This paper considers the causes of such a situation and provides an attempt to obtain a quantitative assessment of reliability in uncertainty conditions by applying Monte Carlo simulation.

The simulation approach concept is very simple, as illustrated by the selected examples. Still, there remain two problems in its application. The first, which is characteristic of highly reliable and complex systems, lies in that "the tyranny of numbers" is faced in applying the simulation. This phenomenon may cause computational difficulties.

The other problem arises in the analysis and interpretation of obtained results. If upper and lower bounds of component failure rates can be determined, it is possible to calculate the lower and upper bounds of system reliability. In many applications, the MTBF of a system may be more relevant than its reliability, but in both cases difficulties may occur in the interpretation of tables of scatergrams obtained by simulation for these values.

REFERENCES

{1} A. Evans, Data, data, oh where art thou data, IEEE Transactions on Reliability, Vol. R-23, No. 5, 1974, p. 289.

{2} A. Evans, O data, data! Wherefore art thou data?, IEEE Transactions on Reliability, Vol. R-31, No. 4, oct. 1982, p. 321.

{3} J. Fragola, Comment on Editorial: O data, data! Wherefore art thou data?, IEEE Transactions on Reliability, Vol. R-32, No. 1, Apr. 1983, p. 2.

{4} B.S. Blanchard, Logistics engineering and management, New Jersey: Prentice Hall, 1981.

{5} L. Palsson, Applied life cycle cost technique, NOAK '81, The 10th nordick congress on operations research, Göteborg, Sweden, Oct. 1981.

{6} N.J. NcCormick, Reliability and risk analysis, Academic Press, New Yrok, 1981, pp. 80-90.

{7} Military standardization handbook, Reliability prediction of electronic equipment MIL-HDBK-217C, (DoD USA, Available from Naval Publications and Forms Center, Philadelphia, PA 19120), 1979 and updates.

{8} IEEE Guide to the collection and presentation of electrical, electronic, and sensing component reliability data for nuclear power generating stations, IEEE Std. 500-1977, Institute of electrical and electronic engineers, 1977.

{9} Military standardization hadbook, Life test sampling procedures for established levels of reliability and confidence in electronic parts, MIL-STD-690A, Department of Defence, USA.

{10} J.D. Esary, The effect of modelling depth on reliability prediction for systems subject to a phased mission profile, in J.B. Fussell and G.R. Burdick (eds.), Nuclear systems reliability and risk assessment SIAM, pp. 562-583, 1977.

{11} R.A. Evans, Point estimates-no, interval estimates-yes, IEEE Transactions on Reliability, Vol. R-23, No. 5, 1974, p. 289.

{12} Collins Radio Company, Spares provisioning guide, Cedar, Rapids, IA, 1974.

{13} H. Malec and D. Steihorn, A new technique for depot and sub-depot spares, IEEE Transactions on Reliability, Vol. R-29, No. 4, 1980, pp. 381-386.

{14} R. Petrović, A. Šenborn, M. Vujošević, Spares allocation in the presence of uncertainty, European Journal of Operational Research, Vol. 11, pp. 77-81, 1982.

{15} J.D. Esary, H.Ziehms, Reliability analysis of phased missions, in R.E. Barlow, J.B. Fussell and N.D. Singpurwalla (eds), Reliability and fault tree analysis, SIAM, pp. 213-236, 1975.

{16} M. Vujošević, D. Meade, Reliability evaluation and optimization of redundant dynamic systems", IEEE Transactions on Reliability, Vol. R-34, No. 2, pp. 171-174, 1985.

QUALITY AND
FIELD RELIABILITY

Reliability Technology — Theory & Applications
J. Møltoft and F. Jensen (Editors)
© Elsevier Science Publishers B.V. (North-Holland), 1986

INFLUENCE OF QUALITY OF MANUFACTURE ON ELECTRONIC EQUIPMENT AND SYSTEM RELIABILITY

Urs Ender, Siemens-Albis AG, Zurich
Wolfgang Gerling, Siemens AG, Munich

Summary

For electronic equipment and systems a coincidence of major defects is observed after electrical tests and during operational use. From this fact a relationship is derived between quality of manufacture and reliability in operation. Quality of manufacture is measured by the defect level of PBA's resp. by their frequency of repair.

Practical experience with an electronic communication system demonstrates that an improvement of manufacturing quality by a factor of 5 has increased reliability by about the same degree.

1. PROBLEM

The established methods of predicting electronic equipment and system reliability are based on the failure rate figures of the applied components and connections. They take into account as parameters the operational and environmental stresses acting on the components, the complexity of components and their quality related to the measures taken for quality assurance and screening. However, a corresponding influence of the manufacturing quality of the equipment or system supplier is not taken into consideration, probably on the assumption that all defects will be detected by electrical tests performed at the board and system levels and will be eliminated without subsequent detrimental effects. But the observation of

- escapes (undetected defects)
- secondary defects (caused by a primary defect or failure)
- coincidence of focal points of defects in electrical tests an system operation

tell a different story.

2. ESCAPE RATE

Printed board assemblies (or PBA's) - typical components of modern electronic systems - contain microelectronic devices and interconnections that, due to their complexity, are a potential source of defects. Electrical testing therefore requires thousands of individual steps and mearurements to be made. The performance of a complete functional check is, however, not technically and economically feasible. A certain number of defects will thus go unnoticed, even when highly sophisticated test programs are used (fig. 1).

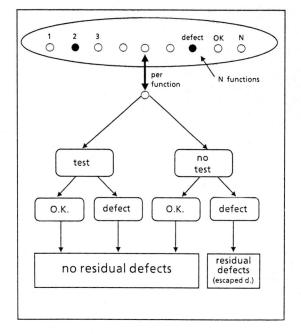

Model for Probability of Undetected Defects

Fig. 1

The escape rate is determined by the test program's fault coverage and by latent defects, which do not impair the proper functioning of the unit that is being tested.

- Test programs can detect hard faults. Test programs for complex components or boards do not provide 100 % coverage of this type of fault.

- Intermittent and transient defects are extremely difficult to identify.

- Soft parameter faults, which cause improper functioning only when specific boundary conditions are present, are also difficult to detect without variation of test conditions (voltage level, temperature).

- Latent defects, which are inactive at the time of test but will eventually result in failure are another kind of escapes.

The defects of interest are stochastic defects, since most systematic defects were already eliminated in the development phase. Statistical analyses show that in practice the escape rate is approximately proportional to the number of defects found. This means that, as the quality of manufacture improves, i.e. the number of defects found during electrical testing decreases, the defects existing in the delivered products will decrease as well.

3. SECONDARY FAILURES DUE TO INDUCED DEFECTS

Electrical faults such as defective or incorrectly inserted components, shorts, opens, etc. often change the operating conditions of other components. The consequences for components indirectly affected by this type of fault may be as follows:

a) effects are not evident or expected

b) one or more components are destroyed

c) one or more components are still functional, but are pre-damaged by an induced defect and thus have a higher probability as secondary failures

Case c) is of special importance to electronic equipment and system reliability. The more faults have to be located and removed during the test phases, the more probable it is that pre-damaged components are contained in the delivered products (see fig. 2).

Model for the failure probability of originally latent defects and additionally induced (latent) defects

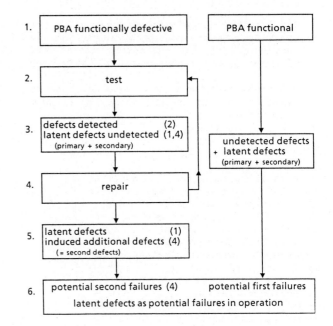

Fig. 2

The question is whether and to what extent pre-damage is to be expected in practice.

Let us consider the following model:

If induced defects as per items c) did not exist, PBA's which needed repairing at some previous stage would have the same reliability as those which did not. If induced defects do exist, the reliability of repaired PBA's would be reduced. It is, of course, essential that repairs are carried out competently.

The procedure to be taken is as follows:

- recording of the failures in large batches of PBA's operation over a prolonged period of time

- statistical prediction of the second failures that can be expected on the assumption that repair of first failures will not affect reliability

- comparison of the number of second failures observed with that of the number predicted.

- determination of the operating time that elapsed between the first failure and the second failure.

The results of the investigation are shown in fig. 3 and 4.

Result:

- the average failure rate of repaired PBA's is increased about fivefold

- the second failures are typical early failures as characterized by their physical defects which caused the failure, i.e. the unexpected failures are secondary failures caused by induced defects.

Influence of Defects induced by Rework on Operational Reliability

PBA - type	number PBA's	failures	2. failures expected	2. failures observed	σ
analog converter	9500	323	5	35	7
exchange system	50 000	600	8	40	5
μP - control unit	5000	215	5	53	10
subscriber line circuit	13 000	242	5	11	2

$$\sigma = \frac{\text{second failures observed}}{\text{second failures expected}}$$

Fig. 3

Influence of Defects Induced by Rework on Operational Reliability

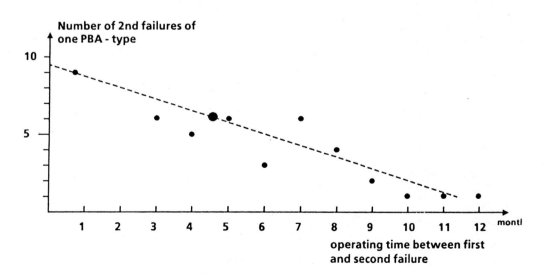

Fig. 4

4. COINCIDENCE OF FOCAL POINTS OF FAILURES IN ELECTRICAL TESTS AND SYSTEM OPERATION

The statistical analysis of failures detected by electrical testing and of components that failed in system operation clearly shows some focal points of faults. The quality of manufacture established by electrical testing and the observed operational reliability of equipment and systems are determined to a great extent by relatively few components with excessively high failure rates. The causes of these dominant defects, or 'weak points', can be traced to the component manufacturer or the system manufacturer, i.e. they are the result of circuit design, PBA production process, or overstressing.

It is decisive for the relationship between quality of manufacture and operational reliability that the majority of weak points detected at the board and system levels are of the same kind, i.e. they govern to a great extent both quality of manufacture and reliability. Fig. 5 gives an overview of the causes determined for weak points in the last few years. The majority of causes influences quality of manufacture as well as reliability.

As far as orignal component defects are concerned, the weak points do not relate to specific component functions or technologies. They are (temporary) weaknesses in some specific types from known sources. Examples of such defects are chip damage, improper bonding, contamination, and leaks in hermetic seals.

Defects resulting from board manufacture include ESD-damage, soldering heat damage, mecha-

Fig. 5

Effect of failure causing faults found as major problems at PBA - testing and system operation on quality of manufacture and reliability in operation

Failure causing defect	Q	R
component (IC) :		
- chip damage	X	X
- defective contact	X	X
- chip surface contamination	X	X
- defects on metallization	X	X
- current density too high		X
- Au / Al - intermetallic formation		X
- parameter drift	X	X
- package leaks		X
- insufficient qualification	X	X
printed circuit board assembly :		
- circuit design fault	X	X
- parts storage conditions	X	X
- ESD - damage	X	X
- soldering heat damage	X	X
- mechanical damage, faulty part insertion	X	X
- faulty testing	X	X
- climatic stress, electrical overload		X
- secondary failures caused by primarily failed parts	X	X
- service and maintenance faults		X

nical damage, incorrectly inserted components, and overloading. As latent defects they may give rise to failure at a later stage.

The question as to the extent to which manufacturing quality and system reliability are governed by these weaknesses remains open for the time being. It can be answered only once the results of experience gained in the field have been analysed. The following case study is presented as an example.

5. CASE STUDY

Since 1977 extensive measures have been introduced by Siemens-Albis AG in Zurich to improve the manufacturing quality of its electronic systems. The results are given in fig. 6.

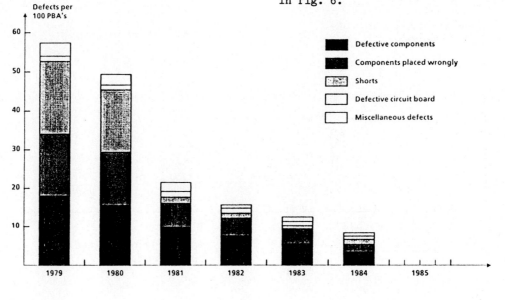

Fig. 6

The figures include all defects detected in the various test steps, ranging from in-circuit testing to system operation. The quality of manufacture achieved was satisfactory. More than 90 % of the PBA's passed all tests without a single defect being found. The figures for 1984 indicate the production volume:

Number of PBA's tested approx. 135 000

Number of types (PBA's) approx. 1 500

containing the following components:

Integrated circuits	3.4×10^6
Small signal semiconductores	8.5×10^6
Capacitors	4.8×10^6
Resistors	10.1×10^6
Total	26.8×10^6

The test results for components taken from the total production volume are given below, using the fully electronic ECS 400 communication system as an example. This system has been manufactured since about 1978, and more than 1000 systems are currently in service.

Quality of manufacture 1984, ECS 400:

Number of PBA's tested 3.9×10^4

Defects found:

Defective components	2'869
Incorrectly inserted components	1'164
Shorts	259
Printed board defects	123
Total	4'415

Fig. 7

ECS 400 - Quality of Manufacture 1984

Tested printed circuit board assemblies	39'000
Defects detected by electrical testing	4'415

Defect breakdown :

component defective	2'869
components placed wrongly	1'164
shorts	259
defective circuit board	123

Component defects :

components	internal code	number of defects	percentage of defective components
Integrated Transistor Crosspoint	V3724-Z7-X1	1415	
*capacitor	B32540-C3682-K	220	
*stab. diode	V3706-Z1-X1	122	
*diode	U169-H2-X2	105	⌐ 80 %
Z-diode	U169-E505-X210	103	
hybride	-	93	
IC	V3721-Z273-X3	88	
capacitor	V2600-Z7-X350	87	
transistor	V3324-Z159-X3	50	
131 different types of components about 800 different types		586	⌐ 20 % \ 0

*in 1984 detected new dominant defects

Defective components are clearly the dominant type of fault. Fig1. 7 lists the defective components grouped by effected type.

The table shows that few component types with disproportionately low volumes cause by far the greatest number of defects. Comparing the figures for 1984 with those for 1983 indicates that some very prominent focal points of fault have made their first appearance, while others have vanished as a result of the measures that were introduced, or are no longer of any importance.

The reliability profile of the electronic ECS 400 communications systems is given in fig. 8.

Fig. 8

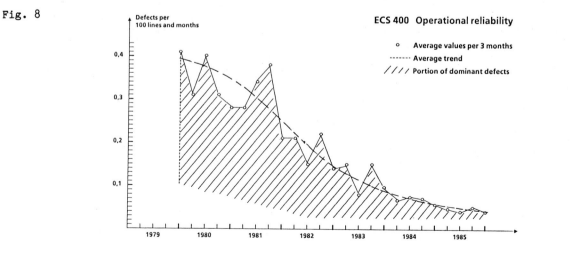

ECS 400 Operational reliability

○ Average values per 3 months
----- Average trend
//// Portion of dominant defects

It shows the average values measured at intervals of 3 months, the approximate mean values and the approximate portion of focal points of faults.

The present operational reliability of the system is very good, but there is room for improvement. Fig. 9 shows the share of focal points of fault in 1984. Eight component types with very low numbers account for 70 % of the system's operational reliability. At least six were identical with those previously or currently encountered in PBA manufacturing.

It is to be noted that the system (ECS 400 in 1984) is not dealt with as an example but is representative of all the equipment and systems that have been manufactured and put into service by Siemens-Albis in recent years. On the basis of this experience it can be safely maintained that effective measures aimed at achieving and maintaining a high quality of manufacture are absolutely necessary to ensure the operational reliability of electronic systems.

6. COMPARISON TO RELIABILITY PREDICTION MODELS

The reliability results obtained hardly can compare with those obtained by established forecast methods (e.g. MIL HDBK217D)

- The observed failure focal points are other than would be expected from forecasts based on component failure rates.

- The failure focal points are generally characterized by excessive component failure rates that are not related to technology or complexity.

- Besides secondary failures ocure due to repair-induced defects to a significant extent.

- After elimination or minimization of the focal points the component failure rates observed in application are considerably lower than those used as basic figures in prediction models.

It is obvious that 'quality of manufacture' should be incorporated in the reliability prediction models. From this concept of weak-point analysis a close relationship can be derived between manufacturing quality and reliability. Despite the high level of reliability achieved, this procedure still points out further possible improvements.

7. CONCLUSION

Practical experience shows that a close connection exists between the quality of manufacture and the operational reliability of electronic systems. The most important reasons for this are:

- Coincidence of focal points of failures in electrical tests and operation.

- Secondary failures due to defective parts, manufacturing faults, and repair actions

- Escapes during electrical testing (at a level still lower than the failure causes mentioned before)

The measures taken to improve the quality of manufacture will also increase the operational reliability. Conversely, high reliability can only be achieved through controlled quality of manufacture.

ECS 400 - Operational Reliability 1984

PBA's in service	147'000
PBA's failed in 1984	1'326
Defective components	1'150

Breakdown of failed components:

type	internal code	number of defects	portion of defect components
* Z-diode	U169-E505-X210	174	
EPROM 16 K	V3724-Z4-X1	169	
* Integrated Transistor Crosspoint	V3724-Z7-X1	115	
* CMOS-NAND	V3724-Z40-X111	86	⌐ 68 %
CPU	V3724-Z8080-X1	62	
* CMOS-trigger	V3724-Z40-X931	59	
* hybride c.		58	
* stab. diode	V3706-Z1-X1	55	
105 different types of components about 950 different types		372	⌐ 32 %

* Dominant defect identical to dominant defect in manufacture in 1984 or before

Fig. 9

Urs Ender, graduate engineer, was born 1929 in Niederwil, Switzerland. After studying electrical engineering, he joined the R&D section of Siemens-Albis AG in Zurich in 1953. Since 1977 he has been the chief of the quality assurance organization of that Company. He was also the chairman of the Electronics Group of the SAQ (Schweizerische Arbeitsgemeinschaft für Qualitätsförderung) from 1980 to 1984.

Wolfgang H. Gerling was born in Solingen, Germany, in 1936. He received Dipl. Ing. and Dr. Ing. degrees in 1962 and 1968 from the Technical University at Aachen, where he was assistant and technical head of the Institute for Electrical Engineering Materials.

In 1970 he joined SIEMENS Components Division, where he is manager of the IC-Reliability Engineering Department.

COMPUTER AIDED
RELIABILITY ENGINEERING

Reliability Technology — Theory & Applications
J. Møltoft and F. Jensen (Editors)
© Elsevier Science Publishers B.V. (North-Holland), 1986

Advances in Computer-Aided Prediction of Reliability.

R. Nitsch

Siemens AG. Zentrallaboratorium für Nachrichtentechnik
München, Germany

A computer-aided prediction of hardware failure rates for communication equipment includes the evaluation of the temperature acceleration factor as well as a method to deal with the failure rate during the non-operating period of the device. A model for computing these effects is presented; the results are compared with field data. Preliminary failure rates of non-stressed components are discussed.

1. Introduction.

In the past reliability engineering frequently consisted in evaluating the field experience and subsequently eliminating the detected flaws. This method, often known as "trial by customer", fails at the latest when the production process is modified before the field experience can be evaluated, as it happens nowadays more often due to accelerated innovation.

Future-proof reliability engineering must, therefore, start at the very beginning of the development of a new product. Since at this stage no samples yet exist for an experimental investigation of reliability, the operating behavior expected in later operation must be estimated by means of failure rate prediction. The results of this prediction are required for:

- Checking at an early stage of development whether or not the customer specification with respect to reliability has been met.

- Obtaining a basis for calculating the availability and necessary redundancies in the system.

- Detecting, as early as possible after initial use, any flaws in the components used or the design or the production process by comparing the prediction with field data.

Most published prediction methods are implemented in three steps:

- Retrieving the base failure rate for the device from an appropriate database. These base failure rates are valid for standard temperature, stress and environment.

- Modifying the base failure rate according to the actual values of electric load, temperature, operating time etc.

- Summing up the device failure rates for a module or total system.

In complex electronic systems, economic failure rate prediction is only possible with the aid of a computer [1]. It is consequently important to insure that the algorithms needed for modifying the base failure rates are suitable for computer processing. It is advisable to use the same computing model for all component families.

The programs used at Siemens AG for computer-aided prediction obtain the failure rates and the standard temperature as well as other necessary data from SN 29 500, a company standard on electric component reliability. The results of many thousands of predictions made previously have shown that:

- Correct modification of the failure rate to meet actual stress parameters is at least as important as the use of correct base failure rates.

- Of the many factors affecting the base failure rates, temperature and intermittent operation are of particular significance.

This paper deals with the progress achieved in evaluating the two parameters mentioned last.

2. Temperature dependance of failure rates.

It has already been known for many years that the failing of components depends on temperature. Especially the failure rate of semiconductors depends heavily on the junction temperature of the device. It is generally assumed that this dependance follows Arrhenius law. Unfortunately, however, the most important constant in this law, the activation energy, is quite different for various failure mechanisms. Computing models may range between the following extreme cases: [2,3]:

- Only one mean activation energy, relatively low in most cases, is used to deal with various failure mechanisms.

  Disadvantage: the rise of the failure rate at high temperatures is underestimated.

- The temperature dependance is determined separately for every failure mechanism.

  Disadvantage: it is not known precisely what share each of the many possible mechanisms has in the total failure rate.

At Siemens, a compromise procedure is applied for the prediction programs. It is the same for all components and uses two different activation energies to represent two groups of failure mechanisms. The following formulae and designations are used:

$$\lambda = \lambda_{BASE} \cdot \Pi_T$$

λ :   Failure rate

λ_{BASE} : Base failure rate

as listed in the failure rate handbook

Π_T :   Temperature acceleration factor

$$\Pi_T = A \cdot e^{Ea_1 \cdot X} + (1-A) \cdot e^{Ea_2 \cdot X}$$

$$X = 11605 \cdot (1/(T_{ST}+273) - 1/(T_J+273)) \; 1/V$$

A :   Fraction of failures due to Ea_1

Ea_1 : Activation energy eV, lower value

Ea_2 : Activation energy eV, higher value

T_{ST} : Standard junction temperature °C

as listed in the failure rate handbook

T_J :   Junction temperature °C

$A = 0,9$ $Ea_1 = 0,3$ eV $Ea_2 = 0,8$ eV $T_{JST} = 45$ °C

In Fig. 1, the temperature acceleration factor for LSTTL-IC is plotted vs. the junction temperature. The dotted curve represents an earlier algorithm in which a uniform activation energy Ea=0,4 eV was assumed.

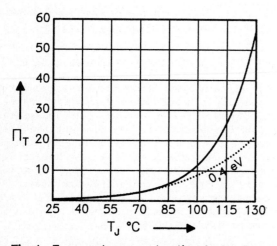

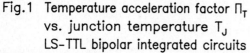

Fig.1   Temperature acceleration factor Π_T vs. junction temperature T_J LS-TTL bipolar integrated circuits

By using a second, higher activation energy, higher failure rates are obtained at elevated temperatures. This corresponds well with the empirical evidence that hot integrated circuits highly contribute to electronic device failures.

The lower activation energy Ea=0,3 eV is characteristic of oxide and silicone defects. The higher value of Ea=0,8 eV reflects surface charge, electromigration, contamination of the chip surface and similar effects. It is assumed that the failure mechanisms with the highest activation energies have already been eliminated before use by a suitable burn-in. The fraction A, like the activation energy, has been derived from component manufacturers reports. Some examples are given in [4,5,6].

The junction temperature required in the formulae shown above can be either calculated or measured. Calculation is possible if the power consumption and thermal resistance junction-ambient are known. The following data for the thermal resistance junction-ambient of the 40-pin plastic package frequently used for integrated circuits were obtained from various manufacturers:

Table 1.

Manufacturer	S	R_{thja} °C/W
A	2	62-73
B	4	50
C	3	65
D	2	92-100
E	1	75
F	1	38-44
G	3	100
H	3	110-116

Source of information S:

1:   Written information by manufacturer
2:   Data book, type or family
3:   Data book, package
4:   Non written information by vendor

IC – Package dual in line, Plastic, 40 pins
Thermal resistance junction-ambient R_{thja}

Even considering that a part of these differences is due to technological causes, such as different material for the lead frame and the package, the range of values revealed was rather unexpected. Let us hope that future manufacturers data books will describe the thermal behaviour of integrated circuits with sufficient accuracy for every type or at least for every package technology.

The thermal resistance junction-ambient also depends on the velocity of the ambient air. The corresponding relationship is shown in Fig. 2 for dual-in-line IC packages with 4-40 pins as published in [7]. The values refer to plastic materials with a thermal conductivity of 0,65 W/(mK). After the velocity of the air has been entered, the thermal resistance is corrected by the computer program.

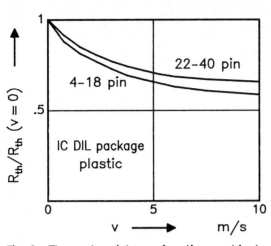

Fig. 2 Thermal resistance junction-ambient
vs. velocity of surrounding air

The junction temperature must be measured in case the heat sink or package used have unknown thermal properties or if the power consumption or its fluctuations vs. time are unknown. The junction temperature must usually be determined indirectly by measuring the surface temperature of the package, for which two procedures are available:

- Measurement with a thermocouple. The thermal conditions may be considerably affected by the probes, especially where the devices under investigation are small.

- Measurement with infrared thermography. This obviates any falsification due to measuring equipment. It is the method of choice for accesible objects.

Conversion of the surface temperature to the junction temperature is based on experience and still represents a problem today.

3. Failure rate for intermittent operation or storage.

The manuals giving failure rates for various components usually contain data for continuous operation. Many widely used electronic devices, such as telephone equipment and television sets are, however, operated only intermittently, for instance only a few hours per day. Spare parts or spare modules are stored out of action until required.

Experience has shown that components also fail when not in operation, however less frequently than under load. This is evident since some failure mechanisms are active without electrical load, for example corrosion. There is a deplorable lack of exact data on this complex of problems. Earlier data on this subject give failure rates during storage which are higher than those currently determined under operating conditions and, therefore, are no longer applicable [8]. One of the following preliminary procedures has hitherto been adopted:

- The failure rate is set equal to zero when the device is out of action.

  Disadvantage: failures of spare parts, spare modules and infrequently operated devices cannot be predicted.

- The failure rate during the out-of-action phase is obtained by multiplying the failure rate at continuous operation under service conditions with a constant factor which is the same for all components. A typical value for this factor is 0,1.

  Disadvantage: an increased junction temperature under load leads not only to an higher failure rate during service but, paradoxically enough, also to an increased failure rate during storage.

- Intermittent operation is neglected and continuous operation assumed.

  Disadvantage: the predicted failure rate is too high.

In order to obtain consistent results, it is suggested to estimate the out-of-action failure rate, i.e. the failure rate during storage, using the following steps:

- Applying the formulae shown in section 2, the base failure rate of the component is converted to the failure rate of the stressed component at storage temperature:

$$\lambda_{TSTORE,S} = \lambda_{BASE} \cdot \Pi_{TSTORE}$$

$\lambda_{TSTORE,S}$: Failure rate at storage temperature stress applied

λ_{BASE} : Base failure rate as listed in the failure rate handbook

Π_{TSTORE} : Temperature acceleration factor converting the base failure rate to failure rate at storage temperature

- The failure rate during storage is computed using the residual factor:

$$\lambda_{STORE} = \lambda_{TSTORE,S} \cdot Rf$$

λ_{STORE} : Failure rate at storage temperature no stress applied

Rf : Residual factor as listed in the failure rate handbook

- To get results consistent with experience different residual factors are chosen for various component technologies. The following preliminary values are suggested:

Table 2

■ Electrolytic capacitors	Rf = 2
■ Dry piles, batteries electrochemical devices	Rf = 1
■ Semiconductors	Rf = 0,2
■ Capacitors, resistors	Rf = 0,1
■ Disc drives, printers Electromechanical devices	Rf = 0,05
■ Inductance, electric motors Cathode ray tubes, bulbs	Rf = 0,01

Residual factor Rf, preliminary values

– The failure rate for intermittent operation is computed using the intermittent duty factor:

$$\lambda_{IO} = \lambda_{CO} \cdot If + \lambda_{STORE} \cdot (1 - If)$$

λ_{IO} :   Failure rate, intermittent operation

λ_{CO} :   Failure rate, continuous operation

If :   Intermittent duty factor
           fraction of time during which
           the device is stressed

Numerical example:

A module operated 24 hours per day contains a power transistor to which load is applied for one hour per day. During this one hour of operation a junction temperature of 100 degrees C is measured. During the remaining 23 hours the temperature decreases to 25 degrees C, since the transistor is disconnected from the power supply:

λ_{BASE} = 120 fit   Power transistor, metal can

T_{ST} = 100 °C   Standard junction temperature

T_{JCO} = 100 °C   Junction Temperature
                                 continuous operation

Failure rate, continuous operation:

λ_{CO} = 120 fit

T_{STORE} = 40 °C   Storage temperature

Π_{TSTORE} = 0,151   Temperature acceleration factor
                                      converting the base failure rate
                                      to failure rate at 40 °C

A = 0,1   constant used

Ea_1 = 0,3 eV   constant used

Ea_2 = 0,8 eV   constant used

Rf = 0,2   Residual factor semiconductor

Failure rate, storage:

$$\lambda_{STORE} = 120 \cdot 0,151 \cdot 0,2 = 3,6 \text{ fit}$$

If = 1/24 = 0,042 Intermittent duty factor

Failure rate, intermittent operation:

$$\lambda_{IO} = 0,042 \cdot 120 + 0,958 \cdot 3,6 = 8,5 \text{ fit}$$

This example shows that the predicted failure rate can vary by 1 to 2 orders of magnitude if temperature and intermittent operation are correctly considered.

The computing procedure described for intermittent operation should be used only when both the turn-on and the turn-off time are long in comparison with the thermal time constant of the component.

4. Comparison of prediction and field data.

Subscriber line modules are used in telephone switching systems. Some of the components mounted on these modules are stressed only during calls. The failure rate was initially predicted without and subsequently with consideration of temperature and intermittent operation and then compared with field data obtained in two sucessive half-year periods:

Table 3.

Failure rate prediction Standard junction temperature and continuous operation assumed	7 506 fit
Field data 6.84 – 12.84	5 100 fit
Field data 12.84 – 5.85	4 304 fit
Failure rate prediction Real junction temperature and intermittent operation considered	4 747 fit

Predicted failure rate compared with field data
Fit: One failure per 10^9 operating hours

5. Quality list.

As shown in sections 2 and 3 the knowledge of junction and ambient temperature as well as the intermittent duty factor are absolutely essential for a correct prediction of failure rates. The computer must, therefore, be able to read the necessary values for each component during prediction. This is done by storing the data in a data set termed the "quality list". An extract from the computer printout of such a "quality list" is shown in Fig. 3. The abbreviations used have the following meanings:

Adr   Addres of device in database

Cf   Correction factor, considers effects
             not discussed in this paper

D	Code of database	
Dta	Difference between ambient temperature of device and mean ambient temperature of the module. In this column the results of temperature measurements are entered	
Fn	Foot note, code for remarks	
If	Intermittent operation factor as discussed in this paper	
PmW	Power dissipation, mW	
Sf	Stress factor, considers the supply voltage, not discussed in this paper	
Te	Increase of junction temperature due to power dissipated in the semiconductor as measured or calculated	

It is not necessary to enter individual data for components whose failure rate does not depend on the mode of operation. Fortunately, those are the majority of all components used in a system.

REFERENCES

[1] Nitsch R., Computer-Aided Prediction of Reliability in E.Lauger and J.Möltoft (eds.), Reliability in electrical and electronic components and systems, North Holland, Amsterdam, 1982, p.399-403.

[2] MIL-HDBK 217D, January 1982, US DOD

[3] F.M.Wurnik, Quality assurance system and reliability testing of LSI circuits Microelectron. Reliab. Vol 23, No.4, p. 709-715, 1983

[4] Reliability Report, NEC Electronics (Europe)

[5] Reliability Monitor Program Intel Corporation, 1982

[6] EPROM Reliability Data Summary, Reliability report, Intel Corporation, 1982

[7] Gerling/Wörther, Wärmewiderstand Integrierter Schaltungen, Siemens AG, 1984 B WIS QA 3/31, Company report

[8] Bajenescu T.I., Zuverlässigkeit elektronischer Komponenten, (VDE-Verlag Berlin-Offenbach, 1985)

Rudolf Nitsch received the Dipl.Ing. degree in physics from the Technical University in Wien, Austria, and joined the Siemens AG in 1953. Currently he is Deputy Director in the "Zentral-Laboratorium für Nachrichtentechnik" in München, Germany. For about twenty years he was engaged in the development of telephon switching networks and relays with sealed contacts. Now he is working on methods of quality and reliability engineering, especially in the field of communication systems.

Pcs	Family circuit code	Item/Type	D	Adr	Dta	Te	PmW	Sf	If	Cf	Fn
5	IC	Q67000-J973-S20	1	288	0	0	0	1,00	1,00	1,00	0
1	IC	Q67000-A2001-S20	1	908	0	0	13	1,00	1,00	1,00	12
3	IC	Q67000-H962-S20	1	81	0	0	0	1,00	1,00	1,00	0
4	IC	Q67100-Y234-S31	1	210	0	0	0	2,40	1,00	1,00	8
6	IC	Q67000-J713-S20	1	106	0	0	0	1,00	1,00	1,00	0
1	IC	Q67000-A1121-S20	1	213	0	0	12	1,00	1,00	1,00	12
4	IC	Q67100-Z140-S20	1	640	0	0	0	1,00	1,00	1,00	0
3	IC	Q67000-A1537-S20	1	215	0	0	0	1,00	,10	1,00	2
1	IC	Q67000-H1287-S30	1	223	0	0	5	1,00	1,00	1,00	12
15	DIO	Q62702-A496-S50	1	233	0	0	0	1,00	,10	1,00	2
8	ELKO	ELKO ALU	2	899	0	0	0	,70	1,00	1,00	4
36	KOND	KOND KER-HDK-VS	2	888	0	0	0	1,00	1,00	1,00	0
8	WIDNETZ	S30814-Q156-A	1	4 899	0	0	0	1,00	1,00	1,00	0
2	INDUKTIV	B82111-E-C23	1	913	0	0	0	1,00	1,00	1,00	0
Pcs	Family circuit code	Item/Type	D	Adr	Dta	Te	PmW	Sf	If	Cf	Fn

Fig 3: Quality list, extract, computer printout

Reliability Technology — Theory & Applications
J. Møltoft and F. Jensen (Editors)
© Elsevier Science Publishers B.V. (North-Holland), 1986

RELIABILITY AND SAFETY ENGINEERING ON THE PC - WHAT ARE THE LIMITS -

Hiromitsu Kumamoto
Department of Precision Mechanics
Faculty of Engineering
Kyoto University
Kyoto 606, JAPAN

Ernest J. Henley
Department of Chemical Engineering
University of Houston
Houston, TX 77004

ABSTRACT

This paper first overviews reliability and safety engineering, and then demonstrates existing programs with worked examples to show how they cover a subset of the field. Relative merits, demerits, and prospects of the PC-based approach are discussed.

INTRODUCTION

Personal computers are now widely used for clerical tasks, word processing, games, hobbies, etc. However, engineering designs and analyses are being done on mainframes or mini-computers, and PC applications to these professional fields are still embryonic. It appears, however, that in the near future the PC will compete with mainframes in solving engineering and scientific problems, including chemical plant simulations and syntheses, structural designs, medical diagnoses, control systems design, symbolic manipulation of mathematical equations, computer-aided instructions, etc. This trend is now established, even though it represents a challenge for managers of large computing centers.

This paper discusses the state of the art of PC applications to reliability and safety engineering. The field is first outlined and basic tasks are identified in Section 2. Typical PC programs are demonstrated in Section 3, with worked examples to illustrate how they currently cover a subset of the field. Section 4 discusses relative merits and demerits of PC applications. References (1) and (2) provide more detailed information.

RELIABILITY AND SAFETY ENGINEERING

Typical steps in the analysis are:

Step 1.1. Hazard identification. This identifies types and locations of hazards such as explosions, poison releases, fires, falls, collision, production/service interrupts, etc.

Step 1.2. Preliminary hazard analysis. This scripts brief scenarios consisting of triggering events, hazardous conditions, and potential accidents and their effects. Also considered are corrective measures, and prototype system configurations.

Step 2.1. Detailed causal modeling. This utilizes the system schematic developed thus far, and applies a logic, tabular, or graphic modeling technique such as FTA, FMEA, FMEA-CA, RBD (Reliability Block Diagram), CCD (Cause-Consequence Diagram), ET (Event Tree), Markov diagram, operability study MORT (Management Oversight and Risk Tree), etc.

Step 2.2. Qualitative/Quantitative analysis. Models like FT, ET, RBD, FMEA-CA, and Markov transition diagram are analyzed qualitatively and/or quantitatively. For instance, minimal cut sets and path sets for FT or RBD diagrams are generated and analyzed. System reliability parameters such as interval reliability, availability, mean time to failure and repair, expected number of failures are quantified from basic reliability data for hardware components and humans.

Step 2.3. System improvement. Redundancies, inspection/maintenance policies, protective systems, and fault diagnoses, etc. are introduced to improve system reliability and safety performance. The resulting system schematic is again analyzed (Step 2.1) and the iterations proceed.

Step 3.1. Consequence analysis and detailed simulation. Potential accidents and production/service interrupts are analyzed using as boundary conditions the environment. Meteorological conditions, population densities, shock waves, flame fronts, thermodynamic processes, control system dynamics and time lags, etc. are all considered in this step.

Step 3.2. Design evaluation. Several alternative system designs are evaluated and ranked according to subject expert criteria, including comparisons with other societal risks. Utility theories and multi-objective optimizations provide formal approaches.

Current PC applications are most advanced in activities relating to Step 2.2, "Qualitative/ Quantitative Analysis" because of the established algorithmic procedures. The same applies to Step 3.1, "Consequence Analyses and Detailed Simulation". Other steps, however, include intelligent capabilities specific to humans. Automation of fault tree construction is still an open question, since this is considered an art; hazard identification, scenario writing, and system improvement, etc. are also largely based on human expertise; system evaluation requires expert judgement except for formal algorithmic methods. One promising approach to these non-classical problems is to utilize artificial intelligence and database technologies. This innovation is illustrated in Section 3 by 1) expert system fault diagnosis and 2) human reliability quantification. Section 3 also discusses reliability/safety programs, now available in PC version (MOCUS, PATHCUT, CONVER, SPOCUS, SAMPLE, BACFIRE, RELICS, SCEH, MARKOV, HEUR 1, HEUR 2, NLB, NONLIN, M-EXPERT, M-THERP).

PC-BASED RELIABILITY AND SAFETY ANALYSIS

Fault-Tree Anaylsis

Figure 1 shows a reaction system in which the temperature increases with the feed rate of flow-controlled stream A. Heat is removed by water circulating through a water-cooled exchanger. Normal reactor temperature is 200°F, but a catastrophic runaway will start if this temperature reaches 300°F. In view of this situation:

1) The reactor temperature is monitored (by TE/TT-714).
2) Rising temperature is alarmed at 225°F (see Horn).
3) An interlock shuts off stream A at 250°F, stopping the reaction (SV-1).
4) The operator can initiate the interlock by punching the panic switch (NC).

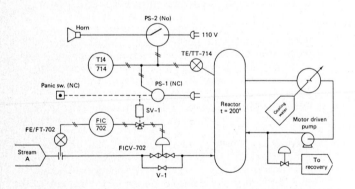

Figure 1. Chemical Reactor Safety System.

The fault tree for the top event, "runaway reaction", includes 2 AND gates, 13 OR gates, and 13 basic events, and is shown in (1).

Qualitative Reliability Analysis

A path set (i.e. system success mode) of a fault tree is a collection of basic events, and if none of the events in the set occur, the top event is guaranteed not to occur. A minimal path set is a path set, such that, if any basic event is removed from the set, the remaining events collectively are no longer a path set.

A cut set (i.e. system failure mode) is a collection of basic events; if all of these basic events occur, the top event is guaranteed to occur. The minimality is defined similarly to the min path criteria.

Example 1: MOCUS, PATHCUT, and CONVER.
It is sometimes difficult to generate all the min cuts for a large fault tree. A more efficient way is to first generate path sets, and then convert them into minimal cut sets. Using MOCUS, we find three path sets:

C3, C4, C5, C6, E1, E2, E3, E5
C3, C8, C10, E4, E5
C3, C8, C12, C11, C13, C14, E3, E4, E5.   (1)

These three are minimal since there are no supersets.

Given the three (minimal) path sets of (1), PATHCUT or CONVER yield 41 minimal cut sets. It turns out that the min cuts include the two single-event cut sets E5 and C3; a serious potential problem which should be corrected. Figure 2 shows a reasonable modification which eliminates the one-event cut sets. Here valve XV-714 is added in front of FICV-704, and SV-1 is relocated.

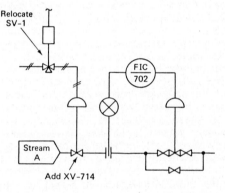

Figure 2. Flow sheet modification.

Example 2: MOCUS, PATHCUT, and CONVER.
With the fault tree modification reflected in Fig. 2, the minimal path sets according to MOCUS are:

C3, C4, C5, C6, C7, E1, E2, E3, E5
C2, C8, C10, E4
C2, C8, C11, C12, C13, C14, E3, E4 .   (2)

Given the three minimal cut sets of (2), PATHCUT or CONVER yields 60 minimal cut sets. The revised version of the chemical plant has no single-event cut sets.

Quantitative Analysis

Assume basic event data such as failure rates, mean times to repair is available, and constant failure probabilities exist. Consider the original version of the chemical reactor safety system which has 41 minimal cut sets. SPOCUS yields: 1) basic event information such as unavailability, availability, unreliability, reliability, and expected number of failures; 2) cut set information identical to the basic event information; 3) top event information as in 1) and 2); 4) basic event importance; and 5) cut set importance. Since the top event is a runaway reaction, neither unavailability nor availability is a good measure for system quantification. The expected number of failures and the reliability are measures of merit. The program execution shows that the runaway reaction occurs 0.63 times a year.

Example 3: SPOCUS. Consider next the revised version of the chemical reactor safety system. Assume the same basic event data as in Example 2. Given the 60 minimal cut sets, SPOCUS quantifies the system. The program execution shows that the runaway reaction occurs 0.02 times a year, which is much smaller than the 0.63 times a year for the original design.

Sensitivity Analysis

Assume that the runaway reaction causes a small temperature increase which only effects the product quality. Then, unavailability (or availability) provides information on how frequently production interrupts occur during a plant operation time interval. The SPOCUS output indicates that system unavailability is 0.000032, i.e., 876 x 0.000032 = 2.7 hours of average interrupts per year.

The basic event reliability data in Example 3, however, are nominal values, and hence are subject to uncertainty. The resulting uncertainty at the system level can be evaluated using the SAMPLE code.

Example 4: SAMPLE. Assume that the failure rates of basic events are random variables with log-normal distributions with appropriate medians and 90% error factors. Consider, for instance, basic event C5. The median of the component unavailability is 0.0021, which is the same as that derived from the nominal failure rate and mean time to failure. The 90% error factor of 10 indicates that the 90% lower and upper bounds of the component unavailability are 0.00021 and 0.021, respectively.

A function subroutine must be provided to return the value of system unavailability, given a sample vector of component unavailabilities; there are 17 two-or less-event cut sets in the 41, and the system unavailability is approximated by a sum of products over these cuts. We see, for instance, from the SAMPLE output that the 90% confidence upper bound of system unavailability is 4.29 x 10^{-4}, more than 10 times larger than the point estimate of 3.2 x 10^{-5} calculated by SPOCUS. If the upper bound is too large, then another iteration on the system design improvements is desirable.

Common Cause Analysis

Consider the hardware minimal cut sets for a fault tree which contains no secondary failures. Such trees can be more easily constructed, and are smaller in size than those with secondary failures. For each hardware component, secondary events such as impact, temperature, vibration, errors of operation, test, installation, manufacture, etc., i.e., common cause candidates, are assigned. Also assigned are the physical locations of components. For instance, some

components are susceptible to impact at location A whereas others are similarly affected at location B. BACFIRE can identify minimal cut sets subject to common causes.

BLOCK DIAGRAMS ANALYSIS

Consider a tail gas quench and clean up system. The system is designed to: 1) decrease the temperature of a hot gas by water quench; 2) saturate the gas with water vapor; and 3) remove solid particles entrained in the gas. The reliability block diagram is given as Fig. 3, where the booster fan (A), either quench pump (B or C), the feedwater pump (D), either circulation pump (E or F) and the filter system (G) must be operating successfully for the system to work.

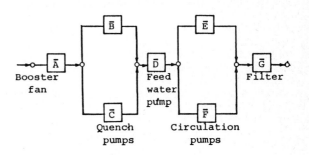

Figure 3.   Reliability block diagram of tail gas quench and clean up system.

Example 5: RELICS. Given the reliability block diagram with mean times to failure and repair for the blocks, RELICS will yield: 1) minimal path sets of the diagram; 2) minimal cut sets of the diagram; 3) steady state system availability (or reliability); 4) system mean time to failure, 5) system mean time to repair; and 6) importance ranking of blocks.

We see from the program output that: 1) system availability is 0.938; 2) system MTTF is 190 (hours), i.e., 8 days; and 3) system MTTR is 12 (hours). In other words, every 8 days, on the average, the system fails. When system failure occurs, it takes 12 hours (average) to repair it. We observe a low reliability for the original system design. The "importance factor" in the output data suggests which blocks are more critical to system failure. Block G, i.e., "filter system" is the system component which must be modified for better system design.

Example 6: RELICS. Let us introduce a redundant configuration of two filters. The filter system now fails only if both filters fail simultaneously. RELICS is used to evaluate the improved design. We note that the new filter is added to the original design to form a parallel redundancy. The performance of the tail-gas-quench-cleanup system is now improved: 1) system availability is 0.983; 2) system MTTF is 1422 (hours), i.e., 59 days; and 3) system MTTR is 24 (hours) i.e., 1 day. System availability and MTTF are improved considerably, while MTTR is doubled. The increased MTTR results from the redundant configuration, i.e., we have more filters to repair.

Example 7: SCHE. SCHE converts the reliability block diagram into a fault tree, i.e., a logic tree for system failure. The tree can be used as a check list to search for the cause of system failure. The first candidate of the cause is the filter G, then the circulation pumps, and so on. Programs for fault trees can be applied to reliability block diagrams via SCHE.

MARKOV ANALYSIS

Example 8: MARKOV. Figure 4 is a MARKOV transition diagram for a subsystem of a nuclear reactor (3). Node 0 represents successful operation of the subsystem, and node F a critical failure. Other nodes correspond to partial failures or repair and maintenance activities. There are a number of routes from node 0 to F. MARKOV models the transition diagram by 53 mother-child relationships together with the transition rates. Two output functions are defined, and the probabilities of nodes 0 to F are calculated as "Normal" and "Failure", respectively. The analysis shows that the probability of critical failure continuously increases as time evolves. The subsystem must be refreshed before the critical failure probability reaches a specified level.

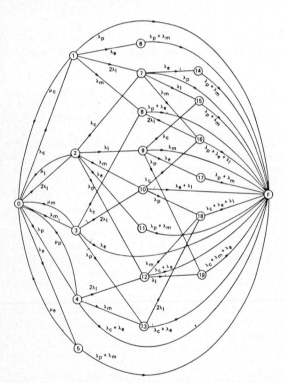

Figure 4.   Transition diagram for nuclear reactor sub-system.

OPTIMIZATION

Discrete Optimization

Example 9: HEUR 1 and HEUR 2. In Example 6, we changed the single-filter system into a two-filter parallel redundancy. This is an allocation problem which is usually formulated as an integer-programming problem. We now optimize the number of parallel components for each subsystem (booster fan, quench pump, feedwater pump, circulation pump, or filter). There are five decision variables, IX(1) to IX(5). The objective function to be maximized is the system

availability. The constraints are the maximum total cost of system components and maximum numbers of components for subsystems. The total money available is assumed to be $500.00. The number of components for each subsystem is specified as less than or equal to 4.

Starting with the non-redundant system configuration, heuristic integer programming software (HEUR 1 and HEUR 2) reaches the optimal solution: IX(1) = 4, IX(2) = IX(3) = IX(4) = 2, IX(5) = 4. The $500.00 invested. Booster fans and filters are the components protected most by the 4-component parallel configuration.

Example 10: NLB. NLB is a rigorous optimization tool which finds the optimal solution by an implicit enumeration procedure. The optimal solution is (4, 2, 2, 2, 4), the same as in Example 9. NLB, however, generates successively inferior solutions. We see that the third best solution is (2, 2, 2, 2, 4) with an availability 0.99939, which is close to the optimal availability 0.99940. The third optimum may be more practical because it only requires two booster fans.

Continuous Optimization

Example 11: NONLIN. Let us further increase the system availability by improving the component availabilities in the block diagram of Example 9. A component's availability increases when it has a longer MTTF. However, more reliable components cost more. We optimize the MTTF's within a cost constraint while minimizing system unavailability. Starting at an initial guess, NONLIN reaches the optimal solution. The reliability of the circulation pump and quench pump are improved, while booster fan, feedwater pump, and filter are compromised. The increased MTTF's are attained, for instance, by better operating environment, maintenance, etc. NONLIN is constrained, nonlinear optimization software based on a multiplier method and a flexible polyhedron search which does not require derivatives.

Example 12: NONLIN. Of the N components, N–F survived beyond final observation time T, and of the F failures, k_i failed in interval $[t_{i-1}, t_i]$, for i = 1 to NF, the number of observation intervals. Here $t_0 = 0$, $t_{NF} = T$ $k_1 + k_2 + ... + k_{NF} = F$. For data of this type, the negative log-likelihood function $J(x_1, x_2)$ can be obtained analytically, where x_1 is Weibull scale parameter > 0, and x_2 is Weibull shape parameter > 0. The maximum likelihood estimates x_1^* x_2^* are the values of x_1, x_2 which minimize $J(x_1, x_2)$. Assume grouped frequency data for the failure times of the cloth, which is a component of the centrifuge. Assume also the failure data for bearings in (4).

The inequality constraints $x_1 > 0$ and $x_2 > 0$ are satisfied by suitable variable transforms, and we now have unconstrained optimization problem. The solutions are: 1) cloth, $x_1^* = 13.7$, $x_2^* = 1.04$; and 2) bearing, $x_1^* = 18.1$, and $x_2^* = 3.05$. The importance parameter is the Weibull shape parameter x_2 since this determines whether or not the component wears out. As seen from the optimal solution, the estimate of x_2 was 1.04 for the cloth, thus indicating that wearout was not the cause of failure. The optimal replacement policy for the cloth is therefore one of replacing the component only upon failure. The other shape parameter indicates that wearout of the bearing occurs and thus preventive replacement is the correct maintenance strategy. Prior to the analysis, the bearing was replaced about every 4 months. Failure replacement is costly due to unexpected production losses. Thus a preventive replace-ment policy reduces the loss.

FAULT DIAGNOSIS BY EXPERT SYSTEM TECHNIQUE

An expert system has an inference machine and conversational capability which work on a knowledge base created by subject experts. The machine and the conversational capability are usually stand-alone systems which can operate on a variety of knowledge bases. In a fault diagnosis application, if-then rules can be used to organize the expert knowledge regarding fault isolation. The process of successive classification (or refinement) of system states can be formulated by the following form of the rule.

if [system state i] **and** (observable facts)
then [system state j] .

For example

if [enough coolant flow in loop A] **and** (there ia a high outlet water temperature in loop A)
then [heat exchanger is ineffective]
if [heat exchanger is ineffective] **and** (there is a low outlet temperature in loop B)
then [there is heat exchanger fouling and hence poor heat transfer] .

The rules show how states can be decomposed into more definitive states as new facts are brought to light.

M-EXPERT (Micro Expert) is programmed in the relational database manipulation language of K-Man (Knowledge Manager) or dBASE. This language was chosen partly because of the availability of crucial information (meta-knowledge and various databases) other than the rules themselves.

We illustrate the expert system approach by an application to a cooling system of the ship engine shown in Fig. 5. At engine startup, the warm-up steam valve opens and the fresh water is heated by the stream. At steady-state operation of the engine, the valve which controls the steam flow is closed and the engine is cooled by the fresh water loop which, in turn, is cooled by the sea water in the heat exchanger loop. The expansion tank is used to replenish the fresh water coolant. The sea water is pumped from the sea into the loop through the strainer which removes foreign materials such as seaweed and fish. The control valve is provided to control the flow of the sea water in accordance with the temperature of the fresh water loop. The sea water is discharged to the sea through the discharge valve.

The abnormal event which triggers the investigation of cause is an engine overheat at steady-state operation. For simplicity we assume that the engine itself does not become a cause.

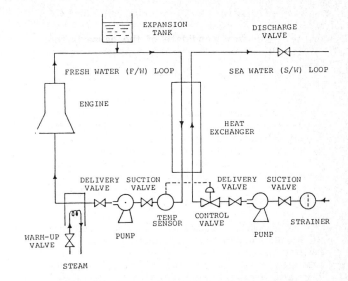

Figure 5. A cooling system of ship engine.

There are 16 possible causes including blockage of the strainer. The problem is to extract the one cause which fits the plant data. A total of 22 rules are used (5). The fault isolation expert system is similar in structure to the MYCIN medical diagnosis tool. Figure 6 shows a very brief summary of a fault-diagnosis dialogue between the M-EXPERT and an operator.

```
Q1   ENGINE ITSELF IS NOT THE CAUSE ?   TRUE
Rule 1 deduces that:
COOLING SYSTEM IS RESPONSIBLE FOR OVERHEAT

Q2   PRESSURE GAUGE READING OF F/W IS LOW ?   FALSE

Q3   PRESSURE GAUGE READING OF F/W IS NORMAL ?   TRUE
Rule 10 deduces that:
ENOUGH COOLANT CIRCULATES IN F/W LOOP

Q4   WARM-UP STEAM VALVE IS OPEN ?   FALSE

Q5   TEMPERATURE READING OF F/W AT EXIT OF HEAT
     EXCHANGER IS HIGH ?   TRUE
Rule 12 deduces that:
HEAT EXCHANGER WORKS POORLY

        .....        .....        .....        .....
        .....        .....        .....        .....

Q12  PRESSURE GAUGE READING OF SUCTION VALVE OF S/W
     COOLANT PUMP IS LOW ?   TRUE
Rule 19 reduces that:
BLOCKAGE OF STRAINER OR CLOSED SUCTION VALVE OF S/W
COOLANT IS RESPONSIBLE FOR OVERHEAT
```

Figure 6. Computer-operator diaglog of M-EXPERT.

HUMAN RELIABILITY QUANTIFICATION

The THERP (Technique for Human-Error Rate Prediction) which was first developed and publicized by Swain, Rook et al., at the Sandia Laboratory in 1962, was later used in the WASH-1400 study, and since then it has been improved to the point where it is regarded as the most powerful and systematic methodology for quantification of human reliability (6).

Human errors, which may be defined as deviations from assigned tasks, often appear as basic events in fault trees. A THERP analysis begins by decomposing a human task into a sequence of unit activities. Possible deviations are postulated for each unit activity. A so-called HRA (Human Reliability Analysis) event tree construct is then used to visualize the normal sequence of unit activities together with the deviations. The event tree thus becomes a collection of chronological scenarios associated with the human task. Each limb of the event tree represents either the normal execution of a unit activity or an omission or commission error related to the activity. The human error for the task can be defined by a subset of the terminal nodes of the event tree. The human reliability is calculated after probabilities are assigned to the limbs of the event tree. The events described by the limbs can be statistically dependent.

The M-THERP (Micro THERP) program automates human reliability quantification. It is written in the K-Man database manipulation language with spread sheet options. Major functions of the program include: 1) construction of the human-error probability database; 2) event-tree definition; 3) human-error probability assignment to the limb by retrieval from the database; 4) probability modifications by stress levels, human dependencies, and recovery factors such as alarms; 5) human reliability quantification by the event tree; and 6) sensitivity analysis of alternatives such as digital vs. analog readings. Figure 7 shows a dialogue for specifying the unit activity "initiate cooldown", an element of the task, "responding to loss of coolant accident". Other unit activities are specified in a similar way, and then the human error rate for the task is calculated.

```
***** 5     Initiate Cooldown        *****
*******************************************

***** SEARCH FOR BASIC PROBABILITY ******
Select ERROR TYPE
C -- Commission;  O --- Omission: ANS ? O

Omission in valve operation (Y or N) ?  N

43   Procedure with checkoff provisions.
        short list =< 10 items
44   Procedure with checkoff provisions.
        long list > 10 items
45   Procudure with no checkoff provisions
        short list =< 10 items
46   Procudure with no checkoff provisions
        long list > 10 items
47   Performance of simple arithmetic
        calculations

Select a number ?  46

Is the stress level of this task the same
   as the last one (Y or N) ?  N

***** STRESS LEVEL *****
A --- very low (no decision involved)
B --- optimum
C --- moderately high (possible damage)
D --- extremely high (death or injury)

Select a level ?  C

1 --- Step-by-step task
2 --- Dynamic task

Select a task ?  1

Input recovery factor ?  ANUNCIATOR
```

Figure 7. Dialogue to specify a unit activity.

RELATIVE MERITS AND DEMERITS

Processing Speed

The PC's are slower than up-to-date mainframes that are shared by a moderate number of log-on users. The slow speed poses problems in the following cases.

1) A large number of computations such as in the numerical integrations in MARKOV and the Monte Carlo simulations in SAMPLE are required. Typical computer times would range from 20 minutes to several hours an IBM CP with an 8087 chip.
2) A problem-specification subroutine must be written, compiled, and linked with the main program every time a new problem is encountered. This situation is typical for nonlinear optimization software such as NONLIN, and HEUR 1. Compiler and linker speeds becomes a factor.
3) Database application programs such as M-EXPERT require frequent access to floppy disks, and processing speed degrades. The so-called memory disks significantly improve disk access.

PC response times in these three cases remain tolerable for problems of moderate size. It should be noted that the PC's are rapidly catching up in speed with older mainframes.

Memory size: This does not seem to pose any serious problems, provided that compilers or interpreters accept "large" multi-dimension arrays in a flexible way.

Cost performance: A general-purpose PC compares favorably in cost with special purpose TSS terminal for mainframes. The PC world is extremely competitive, and a variety of programs are becoming available at relatively low prices. The CPU time for the PC is free. The PC's have better cost performance than the mainframes although this depends on the problems to be solved.

Scientific subroutine packages: Complete sets of subroutine packages such as Gaussian sweepout, Fourier transforms and numerical integration are available for the mainframes. These packages must be transferred to the PC's if software reliability and development efficiency are to be attained.

Hardware and software reliability: PC programs directly downloaded from the mainframes usually have high reliabilities. It take time and competition for software developed exclusively for the Pc's to enjoy comparable reliabilities. It is still risky to use PC hardware as a critical element of an on-line plant control system.

Portability: The programs can be demonstrated anywhere, including underdeveloped countries, where maintainability is a problem.

Man-Machine interface: Colors, graphs, spread sheets, function keys, screen editors, multi-windows, and the mouse contribute to the PC's better man-machine interface.

<u>Mental aspects</u>: Human desires for territorial imperatives and freedom play as important a role for scientists and engineers as for other people. The PC's are always available, while mainframes are sometimes unavailable or under maintenance. A telephone call, registration number and passwords are required to log-on the mainframe, but the PC is activated by simply turning on the power with an autoexec file. The mainframe account output, registration form, and other administrative reports sometimes stress the user. Users affect each other in the mainframe environment, and various security mechanisms exist, causing additional stress to the users, especially when the response is slow, log-on is frequently rejected, or an automated time-out features is implemented.

PC-based reliability and safety engineering is a challenging field, and much remains to be done in addition to downloading conventional programs. Packages must be developed to replace the intelligence capability of humans. A return to qualitative treatments occurs when symbols, concepts, heuristics, satisfactions, expertises, explanations, relations, symbol and picture databases, etc. play crucial roles.

REFERENCES

1. Henley, E. J. and Kumamoto, H., Reliability Engineering and Risk Assessment, (Prentice-Hall Inc., Englewood Cliffs, N. J., 1981).
2. Henley, E. J. and Kumamoto, H., Designing for Reliability and Safety Control, (Prentice-Hall Inc., Englewood Cliffs, N. J., 1985).
3. Gokcek, O., Bazovsky, I., and Crellin, G. L., Markov Analysis of Nuclear Plant Failure Dependencies, 1979 Proc. Annual Reliability and Maintainability Symposium, ASME, N. Y., pp. 104-109.
4. Jardine, A. K. S., Solving Industrial Replacement Problems, 1979 Proc. Annual Reliability and Maintainability Symposium, ASME, N. Y., 1979, pp. 136-141.
5. Kumamoto, H., Ikenchi, K., Inoue, K., and Henley, E. J., Application of Expert Systems to Fault Diagnosis, Chem. Eng. J., Vol. 29, 1985, pp. 1-9.
6. Swain, A. D. and Guttman, H. E., Handbook of Human Reliability Analysis with Emphasis on Nuclear Power Plant Applications, NUREG/CR-1278, Washington, D. C., 1980.

ACKNOWLEDGEMENT

The authors wish to acknowledge the support of the National Science Foundation under grant CBT-8501024.

Reliability Technology — Theory & Applications
J. Møltoft and F. Jensen (Editors)
© Elsevier Science Publishers B.V. (North-Holland), 1986

IMPLEMENTING RELIABILITY CALCULATIONS IN A PORTABLE COMPUTER

Claudio BENSKI
Merlin Gerin
Dept. Recherches Générales
38050 Grenoble Cedex, FRANCE

The operating system of a computer which allows a non hierarchical structure of multiple programs presents new possibilities in the realm of reliability calculations. This would not be sufficient without good accuracy and multiple interfaces for its potential to be fully and easily realized. This article describes the applications that have been implemented in a HP 71 scientific computer to solve, in a portable environment, some of the typical reliability problems that have challenged reliability engineers using desktops and mainframes for statistical calculations.

INTRODUCTION

Statistical computer packages have now existed for many years, {1} and {2}, and some of them have been used with success for calculations related to reliability problems. In general they are either subroutine type, like IMSL {3} or programs that execute a specific type of statistical treatment upon the user's data {4}. In the former case a mainframe computer or at least a minicomputer must be available. In the latter case desktop computers with a disk operating system are used. A few remarks are in order here concerning these procedures. First, these routines and programs have a general statistical approach in that they are not specifically addressed to reliability calculations. For example one finds Least Square fitting routines, more rarely sampling plans for attributes or variables, but no MTBF sampling plans or Weibull Maximum Likelihood Estimates (or other estimates for that matter) and of course no confidence intervals for reliability predictions. It is very likely that an advanced user could use however these programs and adapt them to the particular problem at hand but the point is that it would require a considerable programming effort and in many instances the result will only be satisfying to the specific user. Second, all the commercially available packages require a disk operating system and this implies that the user is restricted to a fixed computer site. We would like to suggest that the system described below adds significant computing power to the reliability engineer as it solves the problems mentioned above with a unit the size of a paperback book.

SYSTEM DESCRIPTION

The HP 71 computer from Hewlett-Packard is a handheld computer whose microprocessor can address half a megabyte of memory. The 64 Kbyte operating system is ROM resident, can be user expanded in RAM, and Math and Forth/Assembly language ROM's are available allowing for very fast matrix inversions.

In addition, these ROM's implement complex data types, fast computation of gamma functions, numerical integration, function roots, Fast Fourier Transforms of complex data points and language extensions to the 240 word resident Basic language. The operating system allows the CMOS-RAM memory to be partitioned in independent segments and thus create a multicatalog structure, each capable of generating its own particular computing environment. In addition, subroutine calling, to disk or memory resident programs, in the same or in a different partition, is possible in a Fortran like fashion. That is, parameter passing by reference or value is allowed with no nesting limit. Recursive subroutine calling is also possible. Practically this means that a single file can contain all the necessary subroutines for statistical hypothesis testing as will be seen below. This file can in turn be resident in a memory partition dedicated to a specific application along with programs and data or text files. These partitions can be considered as independent application modules, can be downloaded to ROM or EPROM memories and be protected from computer crashes or careless editing. Any program, implanted in the same or in a different partition, can gain access to these subroutines. Keyboard calling is also an easy alternative, useful for one shot applications or for debugging purposes. Furthermore, variables can be global or local to an environment meaning that a sophisticated memory management system is at work.

External mass memory support is possible but, as has been shown, it is not essential and is mostly used for archival purposes in this case. The whole system is self-contained and totally autonomous. The user can transport it to different factory points, perform some local data acquisitions and do the corresponding reliability calculations, then go somewhere else for further work. Although downloading data to another computer is quite easy, this was seldom found to be needed due to the powerful mathematical functions already built-in and the fairly large amount of resident memory available for programs as well as data.

Accuracy and dynamic range are also exceptional: all internal calculations are carried out internally to 15 digit accuracy for the mantissa. The basic exponent range is from -499 to +499 but the Math ROM exponent range uses a -50000 to +50000 dynamic to prevent overflows, for instance in root finding routines. Notice that this is a significant improvement over the standard computing accuracy of most desktop computers. It is particularly convenient to compute, for instance, the integrals involved in the conditional estimation of the Weibull distribution parameters and their confidence intervals {5} or to carry out Markov chains calculations without loss of precission.

Interfacing the computer to external peripherals is basically done through the HPIL interface {6} and, by means of available converters, to any peripheral using RS232-C, IEEE 488 or 16 bit parallel interfaces. The HPIL interface has several features which make it particularly well suited for a portable computing device: it has low power requirements, it has a closed loop structure allowing a single interface to link up to 30 main peripherals (over 900 with extended addressing) and, presently, very adequate 5 Kbytes/second transmission rates are possible.

The compact nature of this computer also meant some sacrifices: it only has a one line scrollable display and the keyboard is too small for extensive work. These handicaps can however be easily overcome by connecting an external keyboard or a terminal, through the HPIL-RS232-C converter, thus transforming the computer in a full blown desktop system. This is actually necessary for important program development jobs.

SOFTWARE IMPLEMENTATION

Since the aim of this article is to expose the benefits of the non hierarchical structure of this RAM based operating system we will show how a user would use a procedure to estimate the two parameter Weibull distribution using Maximum Likelihood Estimates (MLE) and the corresponding confidence intervals {7} as well as the probability to refuse the hypothesis that the data come from such a distribution {8}. No tables have to be consulted since all the standard statistical distribution and their inverses are resident in a single file as has been explained above. The raw data is the only requirement. This is in contrast with some statistical packages which require some form of table reading, an error prone and time consuming activity.

Many devices are known to have a failure distribution which follows Weibull's law:

$$f(t) = \frac{\beta}{\alpha} \left(\frac{t}{\beta}\right)^{\beta-1} \exp\left[-\left(\frac{t}{\alpha}\right)^{\beta}\right] \qquad (1)$$

where $\beta > 0$ and $\alpha > 0$ are the parameters that must be estimated from the data. One popular technique for this is the MLE. This procedure requires solving the following equation to determine $\tilde{\beta}$, the parameter estimate of β:

$$\frac{\sum_{i=1}^{r}{}^{*} t_i^{\tilde{\beta}} \log t_i}{\sum_{i=1}^{r}{}^{*} t_i^{\tilde{\beta}}} - \frac{1}{\tilde{\beta}} - \frac{1}{r} \sum_{i=1}^{r} \log t_i = 0 \qquad (2)$$

where the Σ^{*} indicates the sum over all recorded failure times t_i, there are r failures, plus (n-r) times the last recorded failure time t_r. The sample is of size n.

Once eq. (2) is solved the value of $\tilde{\alpha}$, the MLE of α, can be easily obtained by:

$$\tilde{\alpha} = \left[\frac{1}{r} \sum_{i=1}^{r}{}^{*} t_i^{\tilde{\beta}}\right]^{1/\tilde{\beta}} \qquad (3)$$

Eq. 2 must be solved by iterative techniques. In our case the Math ROM contains a fast function root finder which can easily be incorporated in an application program. Thus finding α and β is very straightforward: two BASIC program lines are enough once eq. 2 has been programmed.

However, since the user usually ignores wether the data can be considered as coming from a Weibull distribution, it is essential that the program give an indication of the probability to refuse this hypothesis. Mann, Scheuer and Fertig {8} have proposed a method to test this and the discrimination statistic is closely approximated by Fisher's distribution function. The main program calls this function in the statistical distribution file (DISTAT) by its name with parameter passing as follows:

CALL FISHER(M,N1,N2,P) IN DISTAT

M is the statistic value, N1 and N2 are the degrees of freedom and, on return from the subroutine call, the program finds the probability in the variable P. The FISHER routine uses a well known series expansion to compute P {9} in a few seconds. All calls to any other subroutine follow the same structure. Thus, when one of the two degrees of freedom is bigger than 200, a Khi squared approximation is used and when both are bigger than 200, the normal approximation is used {9}. Both of these functions are also contained in the DISTAT file along with their inverses.

The system allows language extensions to the normal BASIC language, somehow like the root finding routine mentioned above. It is possible to write these functions in assembly langauge.

The statistical distribution functions would then become part of the BASIC language with a significant gain in speed over the subroutine calling method. The new functions would be part of a language extension file (called LEX file) and the invoking syntax would be, for example:

P=FISHER(M,N1,N2)

The parsing and decompiling portion of the operating system would recognize this syntax and jump immediately to the execution code {10}. This would be specially desirable for very long routines like the ones necessary to implement the calculation of conditional confidence intervals of the Weibull parameters as explained in {5}. For our purposes this approach was not deemed necessary since the whole procedure takes less than a minute. In fact the confidence intervals for the parameters can be obtained by using Bain and Engelhardt's approximate method {7}. This is done in our program by invoking the inverse Khi squared, inverse normal and inverse Student distributions in response to the user's desired confidence level. The program computes also the MLE of the mean, standard deviation, median and mode of Weibull's distribution as well as the reliability function along with its lower confidence limit for a time span covering between 95% and 5% reliabilty. The following is a typical computer output for this program:

MLE for Weibull data
DATA: 23 failure times on 23 devices

t_1	=	17.88
t_2	=	28.92
t_3	=	33.00
t_4	=	41.52
t_5	=	42.12
t_6	=	45.60
t_7	=	48.48
t_8	=	51.84
t_9	=	51.96
t_{10}	=	54.12
t_{11}	=	55.56
t_{12}	=	67.80
t_{13}	=	68.64
t_{14}	=	68.64
t_{15}	=	68.88
t_{16}	=	84.12
t_{17}	=	93.12
t_{18}	=	98.64
t_{19}	=	105.12
t_{20}	=	105.84
t_{21}	=	127.92
t_{22}	=	128.04
t_{23}	=	173.40

Probability to refuse the hypothesis of the Weibull nature of the data: 75.09%.

ALPHA = 81.88
Inf and Sup of ALPHA at 90%:
[68.15 , 98.37]

BETA = 2.102
Inf and Sup of BETA at 90%:
[1.49 , 2.60]

Mean = 72.52
Std. Dev. = 36.25

Median = 68.78
Mode = 60.22

10% fail before t = 28.07
Inf of this at 90%: 20.32

10% survive after t = 121.75
Inf of this at 90%: 106.82

t	R(t)	Inf at 90%
19.65	0.95	0.91
33.56	0.86	0.78
47.47	0.73	0.63
61.38	0.58	0.48
75.29	0.43	0.33
89.20	0.30	0.21
103.11	0.20	0.12
117.02	0.12	0.06
130.93	0.07	0.03
144.84	0.04	0.01

The DISTAT file takes 3 Kbytes and the Weibull program occupies slightly less than 4 Kbytes. This compactness is possible thanks to the multiple program nature of the operating system since, during the computation of some of the statistical distribution functions, calls can be made to the others, typically Normal and Khi squared functions, to shorten the program code.

In addition to the Weibull program, it is apparent that other programs will normally coexist such as an equivalent MTBF confidence interval program for Type I or II censored samples, sampling plan routines and regression fitting programs, producing very complete Analysis of the Variance tables with significance calculations, tests of normality {11}, and a test for checking the constant failure rate hypothesis {12}.

Our programs and routines have been checked against some of the published literature and accuracy has always been comparable to the best known mainframe results. For example, the well known Longley data {13} is known to be a particularly troublesome set due to its correlated nature. It has recently been used to check the accuracy of statistical packages {14}. We have least square fitted it using our own regression program (MLR) and the Math ROM matrix functions. In Table 1 we compare the results of two mainframe programs, Sas and Glim, to MLR:

TABLE 1
Accuracy of SAS, Glim and MLR

Param.	Sas	Glim	MLR
1	.0150602	.015060	.0150617
2	-.035819	-.0358108	-.358189
3	-.020202	-.0202023	-.0202022
4	-.010332	-.0103323	-.0103323
5	-.051104	-.0511071	-.0511049
6	1.829151	1.829150	1.829147
7	3482.259	3482.256	3482.250

Since these data have been hand calculated {14} to exact values it is possible to compare the relative accuracy of the parameters as calculated by the three different programs. The results in ppm (parts per million) are presented in table 2.

Table 2
Relative accuracy in ppm.

Param.	Sas	Glim	MLR
1	113	126	13
2	6	11	3
3	15	0	5
4	29	0	0
5	2	61	18
6	0	1	2
7	0	1	3

Our MLR program gives, in addition to the parameters and basic statistics, the significance of the t value for each parameter, the intercorrelation matrix, the Analysis of the Variance table with its F value and its significance.

This again is easily implemented by using the DISTAT file and taking advantage of the utility routines it contains any time a statistical distribution is required. In fact, the basic computer has a built-in series of operations to perform basic statistics on up to 15 variables which greatly accelerates and simplifies program execution.

CONCLUSIONS

We have shown that a non hierarchical multicatalog operating system provides many advantages for statistical life data analysis. Several programs can coexist, performing different tasks and calling all the needed routines, resident in a separate file, and eventually calling each other. This structure reslults in a very compact software system which is particularly well suited for a portable computer: routines are shorter, faster and do not need to rely on an external mass memory support. Data files resulting from physical measurements can reside in the computer's memory and be accessed by different programs for analysis, e.g. one program can determine that the data can be considered to come from a constant failure rate population and give the probability to refuse this hypothesis and another program can access the data to esimate confidence intervals for the failure rate. Both programs will have accessed the same statistical distribution file to compute probabilities. The external disk will be used only for backup and archival purposes.

Although many portable computers exist at this moment, we have found that accuracy, RAM based multicatalog operating system and ease of interfaceing of the HP 71 computer put it in a class on its own for reliability calculations.

REFERENCES

{1} R. W. BERGER, What do you look for when you are looking for software, Quality Progress, March 1984, 28-53.

{2} A. W. DICKINSON, The support of a statistics program library, J. of Quality Technology, vol. 6, no. 2, April 1974, 84-94.

{3} International Mathematical and Statistical Library, IMSL Inc., 7500 Bellaire Blvd. Houston, Texas 77036-5085.

{4} J. CARPENTER, D. DELORIA and D. MORGANSTEIN, Statistical software for microcomputers, Byte, April 1984, 234-264.

{5} J. F. LAWLESS, Confidence interval estimation for the Weibull and extreme value distributions, Technometrics, vol. 20, no. 4, Nov. 1978, 355-368.

{6} Electronic Design, Dec. 24, 1981.

{7} L. BAIN, M. ENGELHARDT, Simple approximate distributional results for confidence and tolerance limits for the Weibull distribution based on Maximum Likelihood Estimators, Technometrics, vol. 23, no. 1, Feb. 1981, 15-20.

{8} N. MANN, E. SCHEUER and K. W. FERTIG, A new goodness of fit test for the two parameter Weibull or extreme value distributions, Commun. Stat., vol. 2, 1973, 383-400.

{9} M. ABRAMOWITZ and I. STEGUN, Handbook of Mathematical Functions, National Bureau of Standards, 1964.

{10} HP 71 Internal Design Specification, Hewlett-Packard, vol. 1, Dec. 1983.

{11} S. S. SHAPIRO and R. S. FRANCIA, An approximate analysis of variance test for normality, Journal of the American Statistical Association, vol. 67, no. 337, 1972, 215-216.

{12} C. W. BRAIN and S. S. SHAPIRO, A regression test for exponentiality: censored and complete samples, Technometrics, vol. 25, Feb. 1983, 69-76.

{13} J. CHAMBERS, Computational Methods for Data Analysis, John Wiley & Sons, 1977.

{14} P. LACHENBRUCH, Statistical programs for microcomputers, Byte, Nov. 1983, 560-570.

BIOGRAPPHY

Claudio Benski received his Licenciado in Physics degree from the University of Buenos Aires in 1967 and his Ph. D. in Physics from Brandeis University in 1972. He worked as an invited scientist at the Institut de Physique Nucléaire at Orsay and the Centre d'Etudes Nucléaires de Grenoble in France before joining Merlin Gerin in 1978 where he is in charge of reliability activities at the corporate research department. He has taught reliability courses for Merlin Gerin engineers and at the Institut de Sciences Nucléaires in Grenoble.

Reliability Technology — Theory & Applications
J. Møltoft and F. Jensen (Editors)
© Elsevier Science Publishers B.V. (North-Holland), 1986

RELTREE - A FAULT TREE CODE FOR PERSONAL COMPUTERS

Ulf Berg
RELCON AB
Box 137
S-182 12 Danderyd
Sweden

A computer code for fault tree documentation and analysis has been developed for use on IBM or compatible personal computers. The code includes an interactive graphic fault tree construction module, an editor for input of reliability data bases, and an analysis module. The analysis module can determine minimal cut sets, calculate top event unavailability, basic event importance and time-dependent reliability characteristics for the top event. The program also includes printer and plotter routines for documentation of fault trees, data bases and analysis results. The program is completely menu-driven and is easy to use.

1. INTRODUCTION

RELTREE is a commercially available software for fault tree construction and analysis, which has been developed for IBM (or compatible) personal computers (PCs).The program has been developed by RELCON AB in cooperation with Control Data AB. In Sweden and Norway the program is marketed under the name CYBERNET*EXPRESS Reliability.

The RELTREE software includes several program modules, and is completely menu-driven. Thus, you will only have to start up RELTREE from the operating system, and then all program modules are accessed through the menus.

There is one program module for preparation of fault tree input (fault tree construction). The fault trees are built interactively using a graphic representation of the tree on the screen. This means that not only the fault tree logic, but also the graphical layout of the tree is defined while the tree is being built up. This is a considerable advantage, because all the information needed for both analysis and documentation is defined at the same time. The fault tree may be documented in several ways: 1. A traditional fault tree listing, 2. a page-by-page semi-graphical printer-plot, and 3. a report quality fault tree plot using a pen-plotter.

Another program module is used for preparation of data bases. These data bases are defined by the user and may contain reliability data for different component types, failure modes etc. The data bases may also include descriptive texts (such as "motor-operated valve fails to open") for each entry. A data base may then be connected to a fault tree in such a way that when the fault tree is constructed, reliability data and descriptive texts are automatically assigned to basic events if a matching entry is found in the data base. A systematic coding scheme for basic events will, of course, have to be used if this feature is to be utilized.

An analysis module is provided which performs the following types of analysis:

(a) Determination of minimal cut sets (MCS) and calculation of top event unavailability.

(b) Calculation of basic event importance according to the Fussell-Vesely importance measure.

(c) Calculation of time-dependent reliability characteristics for a set of time points.

There is also a directory module which allows you to list the various types of files created by RELTREE, to help you keep track of your fault trees, data bases and various result output files.

The following sections will describe the different program modules in more detail, starting with the fault tree input module.

2. FAULT TREE INPUT MODULE

2.1 The Standardized Fault Tree Format

In RELTREE the fault trees are input page by page, where each page has a fixed, standardized format. Each page is organized as a 10 rows by 6 columns "grid", where you may put an event in anyone of the 60 fixed positions. A fault tree page (produced by the plotter routine in RELTREE) is shown in fig. 1.

The 10 vertical rows are numbered 0 - 9 and the 6 columns are numbered 0 - 5. You may therefore refer to an event by specifying the page and the vertical and horizontal position coordinates. This property is used when defining gates, as will be described in the next section.

When a fault tree does not fit into one single page, a transfer is made to another page where the fault tree continues. The idea of splitting fault trees into pages is widely used by fault tree analysts, since it is the best way to make large fault trees suitable for presentation in reports. The standardized format described here has been used very extensively in the probabilistic risk assessments which have been carried out for Swedish nuclear power plants. The format has proven to be very

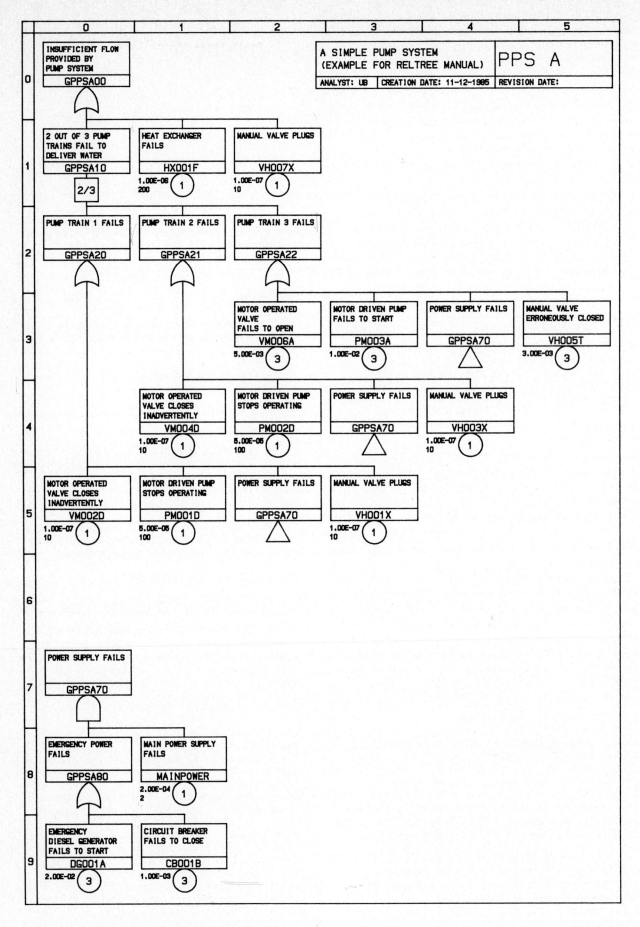

Fig. 1. A sample fault tree plot.

useful and easy to work with, and is much liked by both analysts and by those who review the fault trees.

2.2 Gates

Three different types of logical gates may be defined in RELTREE: AND-gate, OR-gate and K-out-of-N-gate (K/N-gate). The K/N-gate is failed when at least K of the N inputs to the gate are failed. The maximum values of K and N are limited by the fixed fault tree format, which allows a maximum of 6 inputs to a gate.

Each event in the fault tree must have a unique identifier, a label. Since a gate can be defined only once, each gate is uniquely defined by its location in the tree. This property is used by RELTREE to generate gate labels automatically. The labels are defined as "GxxxPPPPvh", where

```
G    = first character in gate labels
xxx  = fault tree id (3 characters)
PPPP = page identification (1 - 4 characters)
v    = vertical coordinate on page (0 - 9)
h    = horizontal coordinate on page (0 - 5)
```

When constructing a fault tree, you start by defining a 3 character fault tree identification (preferably something that indicates what the fault tree is for, e.g. a system name abbreviation). The fault tree in fig. 1 has the id "PPS". Then, as each page is defined, you will give it a page identification, 1 - 4 characters long. The page id of the example fault tree page is "A". Now, for each gate that you define on page "A" in the fault tree "PPS", the complete label is given by the position where you put the gate.

When working with the program, you will have the graphic lay-out of a fault tree page on the screen. When you want to define a gate, you just have to move a cursor to the position where you want the gate, press a key (which defines gate type), and the the gate is there, complete with a unique label.

Obviously, the standardized format and the automatic gate labeling has several advantages: Gate definition is very simple and is made quickly; the risk of input errors in gate definition is minimized; a particular gate can easily be found in a fault tree since the gate label defines the page and position. The last point is very useful when defining transfers as we shall see in the next section.

2.3 Transfers

When there is not enough space available at a particular location in the tree to continue development of the fault tree logic, you can make a transfer to a gate located somewhere else where there is free space. This may be another location on the same page or on another page. That is, a transfer is a reference to gate - located elsewhere - where the tree continues. Transfers are also used when you want a gate to be input to more than one gate. There is no limitation in the number of repetitions of a particular transfer, that is you may use the same transfer as many times as you wish in the tree.

When fault trees become large and span over many pages, it may sometimes be rather awkward to search through several fault tree pages to find the page where the logic continues. When using the gate labeling scheme described above, however, this problem is solved. To make a transfer you just enter the label of the gate you want to make a transfer to. Since this label defines page and position, you can easily find your way through the fault tree, even if it contains many pages and transfers.

2.4 External Transfers

The transfers described above are transfers within one particular fault tree. RELTREE also allows you to transfer to another fault tree. These transfers are called external transfers. Use of external transfers makes it possible to construct several fault trees, which can be documented and analyzed separately, and then linked together and evaluated as one tree. The linking, or merging, of trees is performed as a pre-processing step before analysis.

To allow you to easily evaluate a fault tree containing external transfers separately - i.e. without connecting the external fault trees - you may re-define an external transfer so that it is treated as a basic event. This is done simply by defining reliability data for the external transfer. When you decide to connect the external tree you just remove the data.

2.5 Basic Events

The basic events are the terminating events in the tree, for which no further logic development is necessary or desired. Reliability data are entered for basic events.

Basic events are defined with 1 - 10 character labels. Each unique basic event must have a unique label, but the same basic event may occur at several places in the fault tree. There is no limitation in the number of repetitions of a basic event. All character combinations are allowed in a basic event label, with one important exception: The label may not begin with "G", since this indicates a gate. If you enter a label starting with a "G", the program will assume that it is a transfer to a gate, and if the label does not follow the gate labeling scheme you will be promptly notified.

RELTREE supports three different basic event - or component - models. The models are selected by specifying the basic event TYPE. A type 1 component is repairable and has exponentially distributed lifetime and repairtime. The parameters for such a component are failure rate and MTTR (Mean Time To Repair). Type 2 is an unrepairable, periodically tested component with exponentially distributed lifetime. The parameters are failure rate and test interval. The tests are assumed to be perfect, that is the unavailability is instantaneously brought down to zero at each test, so that the unavailability curve will have a saw-tooth shape. If the test interval is selected in such a way that it is larger than the observation period of the analysis, the type 2 component will behave as a usual unrepairable component. Type 3, finally, is a component with a constant unavailability (probability per demand).

2.6 Program Operation

The fault trees are built up interactively, using a graphic representation of the tree on the screen. You start by selecting a page. This is done by entering a page identification. If the page already exists it is brought up on the screen. If it does not exist previously, the page will be created and you will start with an empty page on the screen.

You will see the full width of a page on the screen, but only half the height (for example vertical positions 0 - 4). There are special function keys that will let you scroll the tree up and down. You "move around" on a page with a cursor which is operated by the arrow keys (up, down, left, right). The cursor moves between the fixed positions in the page "grid". When you want to place an event in a particular position, you just move the cursor tho this position, and then the event is defined as follows:

(a) For gates: Press one of the pre-defined keys that specifies gate type.

(b) For basic events, transfers and external transfers: Enter the event label and press ENTER.

For all events you may enter a descriptive text (3 lines) which goes into the event "box" (see fig. 1). For basic events you may, of course, enter reliability data.

The texts and the data are not shown for each event on the page, but only for the event on which the cursor is positioned.

Simple commands (pressing one or two keys) are provided for connecting input events to a gate, removing events, changing page etc. A summary of the available commands is given on the screen, so there is no need to memorize a lot of complicated commands. Thus, the fault tree input routine is easy to learn, and the trees can be built up very quickly and efficiently. Error checking is performed on all input to minimize the risk of input errors.

3. DATA BASE PREPARATION MODULE

RELTREE includes a data base editor, which allows you to enter generic reliability data, along with descriptions, for different component types, failure modes etc. Such a data base may be connected to a fault tree, so that the data base is searched for matching data when basic events are input to the fault tree. In this way, basic event data can be assigned automatically, thus making fault tree input quicker and at the same time reducing the risk of input errors. The descriptions (e.g. "PUMP FAILS TO START") are also fetched from the data base and are entered as event descriptions in the fault tree.

Specific, user-defined, positions of the basic event labels are searched for in the data base to find out if there is a matching data item. These positions are specified when a data base is created. Then, for each item in the data base, a label is entered which follows some systematic coding scheme (defined by you). Usually, the coding scheme will cover component type and failure mode.

4.   ANALYSIS MODULE

4.1 Minimal Cut Set Analysis

In the minimal cut set (MCS) analysis the MCS of the fault tree are determined, and based on these MCS the top event unavailability is calculated.

The MCS are generated according to a top-down algorithm which is essentially similar to the well-known MOCUS algorithm. The performance in terms of capacity and execution speed has been improved by the use of modularization and an effecent cut set minimization technique.

Before starting to generate the MCS the fault tree is restructured to a more optimal form for analysis, and all modules of the tree are found. A module is a combination of events that are independent from the rest of the fault tree, i.e. none of the events exists somewhere else in the tree. A module can consist of basic events and/or other modules. Thus, a large fault tree structure may be a module. If all events in the tree are non-replicated, i.e. occur only once in the tree, then the whole fault tree is a module.

The advantage of finding the modules is that when the MCS are to be determined, one may first treat all modules as if they were basic events and generate the MCS in terms of modules. Since the modules replace a more complex logic, the amount of computation needed to generate the MCS is usually drastically reduced. The MCS which have been generated for the modularized tree are called module-MCS from here on.

When the module-MCS have been determined, it is a simple matter of substituting modules with their inputs to arrive at the final list of MCS in terms of the original basic events of the tree.

When generating the module cut sets, a cutoff value may be used. This cutoff value is supplied by the analyst for each analysis. All cut sets found to have an unavailability lower than this cutoff are neglected. The use of a cutoff value may reduce the amount of computation significantly. For large and complex trees it may be necessary. An upper bound for the error introduced by neglecting cut sets is given in the output from the MCS-analysis.

In addition to the determination of MCS, the top event unavailability is calculated in the MCS-analysis. The calculation will be performed for a specific time point, which is given as input to the analysis. If there are no time dependent basic events in the fault tree (i.e. all basic events are type 3, probability per demand), this time point will, of course, be meaningless.

Any gate in the tree may be specified as the top event for a particular analysis. The default top gate is, however, the first gate in the tree, i.e. the first gate that was defined when the fault tree was constructed.

The result from the analysis is written to a file and contains a summary of the numerical results, and a list of the minimal cut sets, ordered according to unavailability. A sample printout is shown in fig. 2.

4.2 Importance Analysis

The importance analysis module in RELTREE calculates the Fussell-Vesely measure of importance for basic events. This importance measure is defined as: The probability that event i contributes to the occurrence of the top event, given that the top event has occurred.

When the program performs this calculation, it reads the module-MCS from a file produced by the MCS-analysis program. Next, all unavailabilities are calculated for the evaluation time point given as input. Finally, the importance values are calculated based on these unavailabilities. The importance values therefore corresponds to a given time point.

The result from the analysis is written to a file and contains a list of basic events and importance values, sorted in importance order. A sample print-out is shown in fig. 3.

4.3 Time Dependent Calculations

In the time dependent analysis several parameters are calculated for the top event at a series of time points. In this way the variation of the parameters over a given time interval may be studied. The calculated parameters are:

(a) Unavailability. The probability that the system is in its failed state at time T.

(b) Unconditional failure intensity. The expected number of failures per unit time at time T.

(c) Conditional failure intensity. The probability of a failure per unit time at time T, given no failures at time T.

(d) The expected number of failures during the time interval from 0 to T.

(e) The probability of at least one failure in the time interval from 0 to T.

A sample print-out from a time-dependent analysis is shown in fig. 4.

4.4 Merging Fault Trees

By using external transfers (see section 2.4), you may connect several fault trees, so that these separate trees are converted to one larger, inter-connected tree. This operation is called fault tree merging.

If you wish to analyze a tree which consists of several separate - but connected - trees, proceed in the following way: Use external transfers in the individual fault trees to define the desired connection between the trees. When all the individual fault trees are completed, select the "MERGE" option from the RELTREE analysis menu, and specify the name of the top fault tree (of those trees that are to be merged). The program will now generate a fault tree input file which contains the merged tree, but the original individual trees will be retained. The merged tree may then be analyzed just like any other tree.

4.5 Main-Frame Analysis

If desired, the fault tree input files used by RELTREE may rather easily be converted to input files for other fault tree analysis codes. The RELTREE input files are written in ASCII-format, and it is fairly simple to write a routine that reads a RELTREE file and performs the necessary conversions. Using some communication software, one may then transmit such a file to a main-frame computer. In this way, the fault trees may be constructed and stored on the PC and, if desired, analyzed on a powerful main-frame. Note, however, that the capacity and performance of the RELTREE analysis module should be sufficient unless the fault trees are extremely large and complex.

5. CAPACITY AND PERFORMANCE

Each individual fault tree constructed in the graphic input module must contain no more than 200 gates, 200 basic events, 75 transfers, 20 external transfers and 50 pages.

Each data base may contain up to 99 data entries (with both reliability data and descriptive texts).

If several fault trees are merged into one, the merged tree may contain up to 1000 gates and 1000 basic events. This is also the fault tree size limit of the analysis module. The maximum number of cut sets that can be handled in the MCS-analysis is 5000.

The performance of a fault tree analysis program is hard to measure, since the run times depend more on the complexity of the tree structure than on measurable quantities such as the number of events in the tree. In spite of this, table 1 should give some indication of the type of performance that may be expected from the MCS-analysis module in RELTREE. The table shows the run times in seconds for five fault trees of different size and complexity. The run times given are total times from initiation of analysis until the result file has been written to disk, and correspond to the use of an IBM PC-XT or compatible. A PC-AT would be at least a factor of two faster.

The "no. of MCS evaluated" in the table is the number of MCS used by the program when calculating the top event probability.

```
┌─────────────────────────────────────────────────────────────────────┐
│ ─────────────────────────────────────────────────────────────────── │
│                                                                       │
│ Fault tree: PUMPSYS                            Serial no.=    1       │
│                                                                       │
│ A SIMPLE PUMP SYSTEM (EXAMPLE FOR RELTREE MANUAL)                     │
│ ─────────────────────────────────────────────────────────────────── │
│                                                                       │
│ Top event: GPPSA00                                                    │
│                                                                       │
│ Top event unavailability            =     4.124E-04                   │
│ Time point for calculation (hrs)    =        10000.                   │
│                                                                       │
│ Cutoff value used                   =     0.000E+00                   │
│ Max. cutoff error                   =     0.000E+00                   │
│                                                                       │
│ Number of Boolean Indicated Cut Sets =           77                   │
│                                                                       │
│ Number of MCS evaluated             =           31                    │
│ Number of MCS listed                =           31                    │
│                                                                       │
│ Listed MCS represents   100.0000 % of top event unavailability        │
│                                                                       │
│                                                                       │
│ MINIMAL CUT SETS ORDERED BY UNAVAILABILITY                            │
│ -------------------------------------------                           │
│                                                                       │
│   1. 2.000E-04   HX001F                                               │
│   2. 4.975E-05   PM003A      PM002D                                   │
│   3. 4.975E-05   PM003A      PM001D                                   │
│   4. 2.488E-05   PM001D      VM006A                                   │
│   5. 2.488E-05   PM002D      VM006A                                   │
│   6. 2.475E-05   PM002D      PM001D                                   │
│   7. 1.493E-05   VH005T      PM001D                                   │
│   8. 1.493E-05   VH005T      PM002D                                   │
│   9. 7.997E-06   DG001A      MAINPOWER                                │
│  10. 1.000E-06   VH007X                                               │
│  11. 3.998E-07   MAINPOWER   CB001B                                   │
│  12. 1.000E-08   PM003A      VH003X                                   │
│  13. 1.000E-08   VM004D      PM003A                                   │
│  14. 1.000E-08   PM003A      VH001X                                   │
│  15. 1.000E-08   VM002D      PM003A                                   │
│  16. 5.000E-09   VH003X      VM006A                                   │
│  17. 5.000E-09   VH001X      VM006A                                   │
│  18. 5.000E-09   VM006A      VM002D                                   │
│  19. 5.000E-09   VM006A      VM004D                                   │
│  20. 4.975E-09   PM001D      VM004D                                   │
│  21. 4.975E-09   VH003X      PM001D                                   │
│  22. 4.975E-09   VM002D      PM002D                                   │
│  23. 4.975E-09   PM002D      VH001X                                   │
│  24. 3.000E-09   VM002D      VH005T                                   │
│  25. 3.000E-09   VM004D      VH005T                                   │
│  26. 3.000E-09   VH005T      VH001X                                   │
│  27. 3.000E-09   VH005T      VH003X                                   │
│  28. 1.000E-12   VM002D      VH003X                                   │
│  29. 1.000E-12   VH003X      VH001X                                   │
│  30. 1.000E-12   VM004D      VM002D                                   │
│  31. 1.000E-12   VH001X      VM004D                                   │
│                                                                       │
└─────────────────────────────────────────────────────────────────────┘
```

Fig. 2. An MCS-analysis print-out.

```
┌─────────────────────────────────────────────────────────────────────┐
│                                                                       │
│   ─────────────────────────────────────────────────────────────      │
│   Fault tree: PUMPSYS                        Serial no.=    1         │
│                                                                       │
│   A SIMPLE PUMP SYSTEM (EXAMPLE FOR RELTREE MANUAL)                    │
│   ─────────────────────────────────────────────────────────────      │
│                                                                       │
│   Top event: GPPSA00                                                  │
│                                                                       │
│   Top event unavailability       = 4.124E-04                          │
│   Time point for calculation (hrs) =   10000.                         │
│                                                                       │
│                                                                       │
│   VESELY-FUSSELL BASIC EVENT IMPORTANCE                               │
│   -------------------------------------                               │
│                                                                       │
│   BASIC EVENT  IMPORTANCE  FAIL.RATE   MTTR      TEST INT.   T.T.F.T. │
│   ----------   ----------  ---------   ----      ---------   -------- │
│                                                                       │
│   HX001F        .48        1.00E-06   2.00E+02   0.00E+00   0.00E+00  │
│   PM002D        .28        5.00E-05   1.00E+02   0.00E+00   0.00E+00  │
│   PM001D        .28        5.00E-05   1.00E+02   0.00E+00   0.00E+00  │
│   PM003A        .24        1.00E-02/d 0.00E+00   0.00E+00   0.00E+00  │
│   VM006A        .12        5.00E-03/d 0.00E+00   0.00E+00   0.00E+00  │
│   VH005T       7.23E-02    3.00E-03/d 0.00E+00   0.00E+00   0.00E+00  │
│   MAINPOWER    2.03E-02    2.00E-04   2.00E+00   0.00E+00   0.00E+00  │
│   DG001A       1.93E-02    2.00E-02/d 0.00E+00   0.00E+00   0.00E+00  │
│   VH007X       2.42E-03    1.00E-07   1.00E+01   0.00E+00   0.00E+00  │
│   CB001B       9.67E-04    1.00E-03/d 0.00E+00   0.00E+00   0.00E+00  │
│   VH003X       5.56E-05    1.00E-07   1.00E+01   0.00E+00   0.00E+00  │
│   VM004D       5.56E-05    1.00E-07   1.00E+01   0.00E+00   0.00E+00  │
│   VH001X       5.56E-05    1.00E-07   1.00E+01   0.00E+00   0.00E+00  │
│   VM002D       5.56E-05    1.00E-07   1.00E+01   0.00E+00   0.00E+00  │
│                                                                       │
└─────────────────────────────────────────────────────────────────────┘
```

Fig. 3. An importance analysis print-out.

```
┌─────────────────────────────────────────────────────────────────────┐
│                                                                       │
│   Top event: GPPSA00                                                  │
│                                                                       │
│   TIME-DEPENDENT RELIABILITY CHARACTERISTICS                          │
│   ------------------------------------------                          │
│                                                                       │
│   T       = Time (hrs)                                                │
│   Q(T)    = Unavailability at time T                                  │
│   W(T)    = Unconditional failure intensity (expected number of       │
│             failures per hour at time T)                              │
│   L(T)    = Conditional failure intensity (probability of failure     │
│             per hour at time T, given no failure at time T)           │
│   E(O,T)  = Expected number of failures in the interval (O,T)         │
│   F(T)    = Probability of at least one failure in the interval (O,T) │
│                                                                       │
│                                                                       │
│   T       Q(T)       W(T)       L(T)       E(O,T)     F(T)            │
│   -       ----       ----       ----       ------     ----            │
│                                                                       │
│      0.   0.000E+00  2.907E-06  2.907E-06  0.000E+00  0.000E+00       │
│    100.   2.117E-04  3.217E-06  3.218E-06  3.062E-04  3.062E-04       │
│    200.   3.096E-04  3.330E-06  3.331E-06  6.336E-04  6.335E-04       │
│    300.   3.575E-04  3.371E-06  3.373E-06  9.687E-04  9.685E-04       │
│    400.   3.822E-04  3.387E-06  3.388E-06  1.307E-03  1.306E-03       │
│    500.   3.954E-04  3.392E-06  3.393E-06  1.646E-03  1.645E-03       │
│    600.   4.028E-04  3.394E-06  3.395E-06  1.985E-03  1.983E-03       │
│    700.   4.071E-04  3.395E-06  3.396E-06  2.324E-03  2.322E-03       │
│    800.   4.096E-04  3.395E-06  3.397E-06  2.664E-03  2.661E-03       │
│    900.   4.111E-04  3.395E-06  3.397E-06  3.003E-03  3.000E-03       │
│   1000.   4.119E-04  3.395E-06  3.397E-06  3.343E-03  3.338E-03       │
│                                                                       │
└─────────────────────────────────────────────────────────────────────┘
```

Fig. 4. A time-dependent analysis print-out.

All of the fault trees are taken from nuclear power plant PRAs are system reliability studies. They may be considered as rather typical nuclear system fault trees. The fault trees 4 and 5 in the table are examples of trees that models several inter-connected systems. In fact, tree number 5 describes a complete core-melt sequence.

Ulf Berg is a consultant with RELCON AB where his activities include development and application of methods in the field of risk and reliability analysis. He recieved the M.S. degree in mechanical engineering from Chalmers University of Technology, Sweden in 1980. For a few years he worked as a research engineer at the nuclear engineering departments at Chalmers and at the Royal Institute of Technology, where he participated in several re-eliability analysis projects for the Swedish nuclear industry. These projects included application of fault tree and event tree analyses, common cause failure analysis and data analysis. He has also developed computer codes for fault tree analysis, uncertainty evaluation and Bayesian data analysis.

Table 1
Run times for some nuclear power plant PRA fault trees

Fault tree number	Number of gates	Number of basic events	Top event unavailab.	Cutoff unavailab.	Max cutoff error	No. of MCS evaluated	Run time (sec.)
1	65	107	9.3E-3	0	0	197,489	175
2	120	82	5.9E-3	1.0E-7	1.7E-5	2,084	125
3	85	73	3.3E-4	1.0E-8	5.0E-6	91	155
4	507	705	6.8E-2	1.0E-7	3.7E-5	1,500	245
5	201	205	3.0E-10	1.0E-14	1.0E-11	1.193E6	250

LATE PAPERS

B.H. Clemmensen
Methods for Spare Parts Calculations

A. Salomon
Mechanical Reliability - A Practical Approach

A. Schornagel
Optimum Preventive Maintenance for Complex Systems
with Interdependent Equipment Units

Reliability Technology — Theory & Applications
J. Møltoft and F. Jensen (Editors)
© Elsevier Science Publishers B.V. (North-Holland), 1986

METHODS FOR SPARE PARTS CALCULATIONS

Bent Haunstrup CLEMMENSEN
KTAS, Copenhagen Telephone Company
Denmark

This paper presents and compares basic principles and models for spare part stores. Simple calculation methods are demonstrated which enable estimation of the consequences of different structures, turnaround times etc.

1. INTRODUCTION

Availability is one of the most basic and important reliability measures. In order, however, to achieve a certain objective concerning availability several other measures must be taken into account. Some of these measure are shown in fig.1.

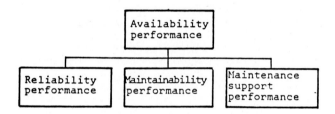

Fig.1. Survey of availability measures.

While the reliability and maintainability measures primarily refer to the design of the equipment, the maintenance support measures depend on the service organization. An important duty is to procure spare parts in a fast and economical way. This problem has increased due to increasing complexity and specialization of components and PCBs. At the same time the ac- tive repair time has decreased considerably due to failure location (fault isolation) sy- stems etc. Thus the logistic delay including the time to procure spare parts plays an important role concerning the availability performance.

The spare parts costs, including organization of stocks, handling and transporting, need consideration of the maintenance philosophy in order to achieve an optimum between costs and availability. This is hardly ever to obtain due to the numerous unknown parameters involved. It is possible, however, to make simple and easy to understand models, which can clarify some of the most important mechanisms of a spare part system. As such the consequences of different structures, turnaround times etc. can be estimated. We can "play" with the system.

2. STRUCTURE MODELS

A structure model describes the normal flow of spare parts to and from systems in operation, various stores, repair workshop etc. and as such the model reflects the maintenance organization. Most structures are in principle based on one of the following three models:

- Local stores with a repair workshop, one level system (Fig.2)

- Local stores with one (or more) central store(s) and a repair workshop, hierarchical system (Fig.3)

- Local store, self-contained. Mostly for equipment furnished with spare component parts sufficient for a certain time (Fig.4)

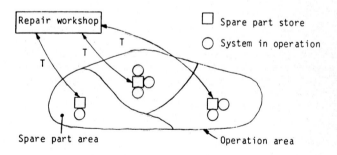

Fig.2. One level storage systems

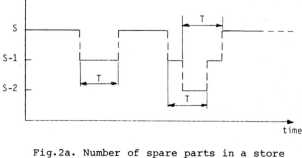

Fig.2a. Number of spare parts in a store versus time (example)

The schematic procedure for <u>local stores with a repair workshop</u> is shown in fig.2. The neccesary spare parts are (immediately) taken from the local store and the faulty parts are sent to the repair workshop for repair/replacement. After a while the local store receives repaired/new parts. The total time from shipment to receipt is called the <u>turnaround time</u> (T). An example of the behaviour of a store is shown in fig.2a.

In case of lack of critical spare parts in a local store emergency procedures may be established, such as loan from a neigbourhood local store or from the repair workshop. These procedures are relatively simple to establish in practice, but difficult to introduce in mathematical models.

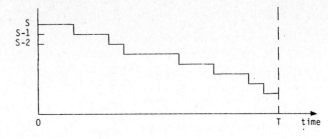

Number of spare parts in store

Fig.4a. Number of spare parts in the store versus time (example)

The last situation, a self-contained store, is typical for an equipment left "alone" for a relative long operating time. In this case shown in fig. 4, sufficient spare parts shall be available during the operating time (T). Thus T is no longer a turnaround time. An example of the behaviour of this kind of store is shown in fig.4a.

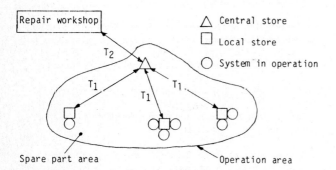

Fig.3. Two level storage system

The schematic procedures for local stores with one central store and a repair workshop are shown in fig.3. This is a typical hierarchical system because the spare parts have to pass the central store. In this case two turnaround times are relevant, one for the time between the local and the central store (T_1), generally short, and another for the time between the central store and the repair workshop (T_2), generally long. Also in this case emergency procedures may be established.

The two level system can with approximation be treated as two independent one level systems with different turnaround times.

3. MATHEMATICAL MODELS

3.1 General

Many models have been suggested during time, some of them very complicated, few of them, however, are in practical use. All relevant models are based on random variables and require statistical methods, assumptions, simplification and approximations. The assumptions are discussed below. The general model requires as a minimum the following input parameters:

- The failure rate pr. item (λ)

- The number of items in operation (N)

- The acceptable shortage risk (α)

- The turnaround time (T)

The <u>failure rate</u> is usually assumed constant and <u>equal</u> to the predicted value. Neither of these assumptions, however, is usually true. In fact <u>replacement rate</u> should be used instead, considering the often large amount of "Non Found Failures" (typical 20-40%).

The acceptable <u>shortage risk</u> is often traditionally 1% but <u>should be</u> chosen in the light of the importance of the item, the turnaround time, the maintenance organization, the possibility of loan and emergency procedures and redundancy, if any. Eventhough it might be simple to calculate the number of spare parts corresponding to a stated risk, it may be difficult to choose the relevant risk level.

Any model implies many not fulfilled assumptions and many doubtful input values. It is therefore recommended - especially in the beginning of a planning procedure for spare parts stores - to use simple models in which different consequences can be easily recognized, such as economy.

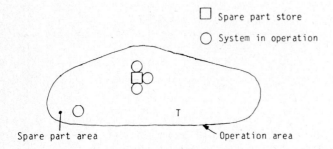

Fig.4. Local self-contained store.

3.2 The Poisson method

The Poisson method is a simple and very gene-rally used method (ref. {1}, {5}, {7}) for calculating the neccesary number of spare parts for a one level storage system.

The method may also be used for a two or more level storage system by calculating each level indepent of the others. This is an approxima-tion and care should be taken when evaluating the total shortage risk, (see ref. {7}).

The assumptions for the Poisson method are the following:

λ = failure/replacement rate is constant (or the time between failures is exponentially distributed).

T = turnaround time is constant

The probability of finding r or less failures P(r) is:

$$P(r) = e^{-\rho} \sum_{i=0}^{r} \rho^{i}/i! = 1 - \alpha \qquad (1)$$

where

ρ = $N\lambda T$ = the expected total number of failu-res of N items during the time period T at failure rate λ

α = the probability of finding more than r failures = accepted shortage probability

When r is calculated by iteration using (1) as the lowest value satisfying

$$P(r) \geqslant 1 - \alpha$$

then the initial amount of spare parts S is

$$S = r + B$$

Where B = 1 when T is the turnaround time according to fig. 2 or 3 and B = 0 when T is the operating time according to fig. 4.

Table 1 and graph 1 show r versus ρ at dif-ferent α levels. By means of the table or the graph it is very simple to get a quick estima-te of different maintenance policies.

Table 1: Values $\rho = \rho(r, \alpha)$ for cumulative Poisson distribution

$$Pr(X \leq r) = \sum_{i=0}^{r} \frac{\rho^{i}}{i!} e^{-\rho} = 1 - \alpha$$

r	100·α (%)					
	.1	.5	1.0	5.0	10.0	20.0
0	.0010	.0050	.0101	.0513	.105	.223
1	.0454	.103	.149	.355	.532	.824
2	.191	.338	.436	.818	1.10	1.54
3	.429	.672	.823	1.37	1.74	2.30
4	.739	1.08	1.28	1.97	2.43	3.09
5	1.11	1.54	1.79	2.61	3.15	3.90
6	1.52	2.04	2.33	3.29	3.89	4.73
7	1.97	2.57	2.91	3.98	4.66	5.38
8	2.45	3.13	3.51	4.70	5.43	6.43
9	2.96	3.72	4.13	5.43	6.22	7.29
10	3.49	4.32	4.77	6.17	7.02	8.16
11	4.04	4.94	5.43	6.92	7.83	9.03
12	4.61	5.58	6.10	7.69	8.65	9.91
13	5.20	6.23	6.78	8.46	9.47	10.79
14	5.79	6.89	7.48	9.25	10.30	11.68
15	6.41	7.57	8.18	10.04	11.14	12.57
16	7.03	8.25	8.89	10.83	11.98	13.47
17	7.66	8.94	9.62	11.63	12.82	14.37
18	8.31	9.64	10.35	12.44	13.67	15.27
19	8.96	10.35	11.08	13.25	14.53	16.17
20	9.62	11.07	11.83	14.07	15.38	17.08
21	10.29	11.79	12.57	14.89	16.24	17.99
22	10.96	12.52	13.33	15.72	17.11	18.90
23	11.65	13.26	14.09	16.55	17.97	19.81
24	12.34	14.00	14.85	17.38	18.84	20.72
25	13.03	14.74	15.62	18.22	19.72	21.64
26	13.73	15.49	16.40	19.06	20.59	22.56
27	14.44	16.25	17.17	19.90	21.47	23.48
28	15.15	17.00	17.96	20.75	22.35	24.40
29	15.87	17.77	18.74	21.59	23.23	25.32
30	16.59	18.53	19.53	22.44	24.11	26.24
31	17.32	19.30	20.32	23.30	25.00	27.17
32	18.05	20.08	21.12	24.15	25.89	28.09
33	18.78	20.86	21.92	25.01	26.77	29.02
34	19.52	21.64	22.72	25.87	27.66	29.95
35	20.26	22.42	23.53	26.73	28.56	30.88
36	21.00	23.21	24.33	27.59	29.45	31.81
37	21.75	24.00	25.14	28.46	30.34	32.74
38	22.51	24.79	25.96	29.33	31.24	33.67
39	23.26	25.59	26.77	30.20	32.14	34.60
40	24.02	26.38	27.59	31.07	33.04	35.54
41	24.78	27.18	28.41	31.94	33.94	36.47
42	25.54	27.99	29.23	32.81	34.84	37.41
43	26.31	28.79	30.05	33.69	35.74	38.34
44	27.08	29.60	30.88	34.56	36.65	39.28
45	27.85	30.41	31.70	35.44	37.55	40.22
46	28.62	31.22	32.53	36.32	38.46	41.15
47	29.40	32.03	33.36	37.20	39.36	42.09
48	30.18	32.85	34.20	38.08	40.27	43.03
49	30.96	33.66	35.03	38.96	41.18	43.97
50	31.74	34.48	35.87	39.85	42.09	44.91
$u_{1-\alpha}$	3.090	2.576	2.326	1.645	1.282	.8416

For higher values of ρ, use formula: $r = \rho + u_{1-\alpha}\sqrt{\rho} - 0.5 + \frac{1}{8} u_{1-\alpha}^{2}$,

where r is rounded up to the nearest integer and $u_{1-\alpha}$ is the (1-α)-fractile of the cumulative Normal distribution

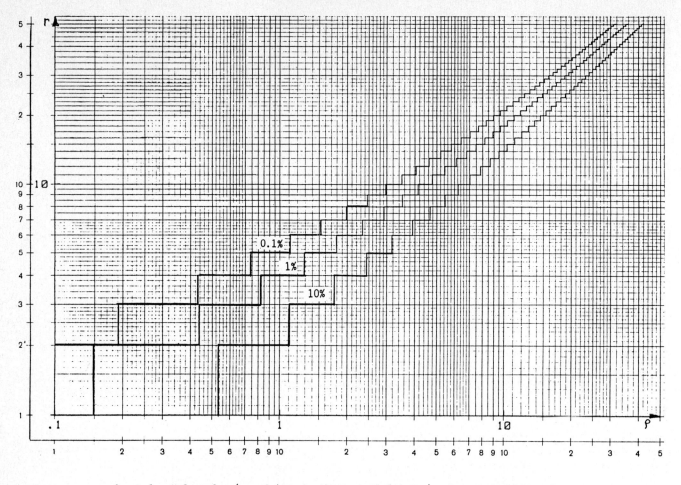

Graph 1 : Value of r (rounded up to the nearest integer) versus parameter ρ for
cumulative Poisson distribution (see Table 1), α = 0.1%, 1%, 10%

3.2 The Markov method

The Markov method is in family with the Poisson method and is believed to be more accurate. It is usually applicable for one level storage system, (se ref. {3}, {6}).

The assumptions for the Markov method is the following:

λ = failure/replacement rate is constant

T = turnaround time is exponentially distributed with mean T ($1/T = \mu$ = constant repair rate). This means in fact that the number of repairmen is always large enough to avoid significant delay due to lack of repairmen.

The behaviour of the storage system may be described by the state diagram with transitions due to the utilization and repair of the spare parts. From the state transition diagram, fig. 5 the steady state probabilities for each state are obtained. Using these probabilities the expected long-run fraction of demands for spare parts filled with delay, due to shortage of spare parts in the store can be calculated.

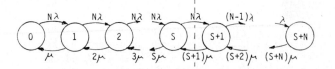

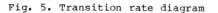

Fig. 5. Transition rate diagram

A more detailed description may be found in ref. {1}, {2} and {3}. The derivation and calculation are quite complicated.

Table 2 shows for comparison some α risks calculated according to Poisson and Markov for small and large N and for small and large values of ρ = NλT. It is seen that except for very small N values there is no significant difference between Poisson and Markov, and Poisson still being on the safe side. It may be mentioned that for N = 1 the α risk can be calculated according to the Erlang loss formula (wellknown within telecom). These examples may show that for most practical situations the Poisson method is fully adequate.

ρ = N λ T	N	SHORTAGE PROBABILITY α % Spare parts S = r + 1						
		1	2	3	4	5	6	7
.05	1	4.78	.12					
	10	4.87	.12					
	500	4.88	.12					
	Poiss.	4.88	.12					
1.00	1			6.52	1.54	.31		
	10			7.78	1.85	.36		
	100			8.00	1.89	.37		
	500			8.03	1.90	.37		
	Poiss.			8.03	1.90	.37		
2.0	1					3.67	1.21	.34
	10					5.00	1.59	.44
	100					5.24	1.65	.45
	500					5.26	1.66	.45
	Poiss.					5.27	1.66	.45

Table 2. Comparison of shortage probabilities α % calculated by Erlang
loss formula (N=1), Markov method
(N=1-500) and Poisson distribution (Poiss.).

3.4 The simulation method (Monto-Carlo)

Computer simulation may be applied for complicated storage structures and other statistical models that cannot be calculated analytically.

The statistical assumptions may be whatever,
but normally the model, where subsequent times
between failures are exponentially distributed
(failure rate is constant) and turnaround time
is constant, should be used.

All generated events are put into a que, where
they can be selected in time order. After several runs store counters are used to calculate the mean shortage risk.

The simulation method requires many assumptions and initial values and it is time consuming.

Simulation is mostly recommended in cases
where approximation methods have been used in
the first planning phase and then a calculation of the true (often combined) risks is
wanted.

For further details see ref, {1} and {8}.

3.5 Other methods

Many other methods have been suggested, for
example based on logistics or cost optimizing,
que theory or allocation of different risks
according to failure consequenses (e.g. con-
gestion). The last method is described in ref.
{4}.

Generally most of the methods cannot be recommended, because they require many assumptions,
are very complex and often only result in a
"local optimum" not taking care of an overall
point of view.

4. CONCLUSIONS

The provisioning of spare parts requires careful consideration during the planning procedu-

re. It is important to keep an overview of all
the aspects and to use a relative transperent
calculation method just as an approach. The
final general judgement must not be left to a
calculation method.

The simple Poisson method demonstrated here
and in more details in ref. {1} has proved its
practical applicability and is generally used
by telecom manufactures and telecom administrations.

In fig.6 is as an example shown the relative
cost of spares in at store for a typical PCB
versus the number in operation. It is seen
that the relative cost decreases rather slowly
with the number in operation. Thus the important conclusion can be drawn that a local
store is costly to establish for covering a
few number of units in operation. Unfortunately many computer systems do in fact have unique PCBs, which may spoil optimizing.

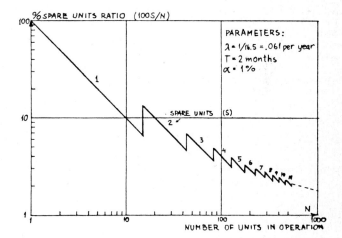

Fig.6. Relative cost (spare units ratio)
and number of spares versus number
of units in operation. Example.

Acknowledgement

This paper is based on ref. {1}, which is
prepared by the Working Group, NT Reliability
Methodology (NT/R) under the Nordic Telecommunications Administrations. The tables and
graphs etc. have been elaborated by Mr. E.Nou-
sianen, Finnish Telecom Research Institute.

REFERENCES:

{1} Nordic Telecommunications Administrations:
Spare parts calculation, WG NT/R-SPEC-
E 4.4, November 1985.

{2} Barlow, R.E. and Proschan, F.: Statistical
theory of reliability and life testing,
probability models. (Holt, Rinehart and
Winston Inc., 1975).

{3} Ericsson Public Telecommunication: Dimen-
sioning of spare parts (Booklet 1984).

{4} Tigerman, B: Methods for calculation of adequate number of spare parts for transmission systems, Ericsson Review 48 (1971) : 4, 145-156.

{5} O'Neil, F.J.: Spare equipment: How much is enough?, Bell Lab. Rec. 54 (May 1976) 5, 135-138.

{6} Bell, E.M., Kwiatkowski, C., Ross, C.E.J.: Computer aids for Reliability prediction and spares provisioning, Electrical Comm. 54 (1979) 2, 136-142.

{7} Malec, H.A., Steinhorn, D: A new technique for depot and sub-depot spares, IEEE Trans. on Reliability, R29 (Dec.1980) 5, 381-386.

{8} Murray, L.R., Morris, R.S.: Spare/repair parts provisioning recommendations, Proc. Ann. Reliability and Maintainability Symposium, 1979, 224-230.

Author

Bent Haunstrup Clemmensen is M.Sc. in Electronic Engineering. Formerly he worked with military projects within the industry and is now a project manager in the Research and Development department, Reliability and Environmental Engineering of the Danish Telecommunication Administration, KTAS (Copenhagen Telephone Company). He is a member of IEC TC 56, WG3 and member/chairman of the WGs, NT/Environment and NT/Reliability Methodology (Nordic Telecommunications Administrations).

Reliability Technology — Theory & Applications
J. Møltoft and F. Jensen (Editors)
© Elsevier Science Publishers B.V. (North-Holland), 1986

MECHANICAL RELIABILITY,
A PRACTICAL APPROACH

Arne Salomon
Radiometer A/S
Copenhagen NV, Denmark

In electro-mechanical products the prevailing failures reported from the field originate mostly from mechanical parts. The paper describes basic principles in the design and testing of mechanical systems in order to achieve high reliability. It is described how failure data from the field are used to improve the product.

1. INTRODUCTION

Modern instruments are often highly complex, using a mixture of both mechanical and electronic components. Analyzing failures reported from the field show that more than 80% of all failures originate from mechanical parts.

There is no simple way to ensure a high reliability of mechanical parts. On the contrary: a careful planning for high reliability is essential through-out the total manufacturing process.

This paper describes, in a practical way, some basic principle on how to ensure higher reliability. It is, of course, not a handbook in good engineering. Rather does the paper look upon the problem from a quality assurance point of view.

2. RANDOM FAILURES

In equipment with only electronic components, it is generally accepted that failures occur randomly. The mean time between the failures (MTBF) is considered to be constant.

When planning a life test this means that instead of testing one instrument over say 24 months, we might as well test 24 instruments for one month. If our statistical assumption is correct, we will experience the same number of failures. This is because the accummulated test time is the same.

With electronic components the test time could even be shortened by increasing the temperature. The approximate increase in failure rate with respect to temperature is stated in various handbooks. If the temperature is increased to correspond to an acceleration factor of 3, it means that we - in the above mentioned example - will only need 8 instruments in one month for the testing.

3. WEAR-OUT FAILURES

Also mechanical components may have a low random failure rate up to the point where the wear-out starts. In contrast to most electronic components, however, the wear-out of mechanical components usually starts much earlier.

The problem with the testing of mechanical parts is, that the wear-out can only be found through a test lasting the actual time to wear-out.

As it is the case with electronic components, an acceleration of the time to wear-out on mechanical parts is not easy to perform. The only certain way to shorten the test time is to run the test for 24 hours a day - thus leaving out the resting time which most mechanical parts have in practical use.

The time required for a wear-out test is illustrated in Fig. 1. It shows the accumulated percentage of failing parts with respect to the calendar time.

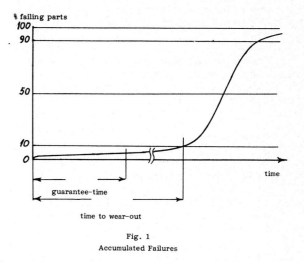

Fig. 1
Accumulated Failures

The curve shows that the wear-out time starts much later than the time, when the product is under guarantee. This is because normally one would not like the parts fo fail just after the guarantee period has run out. Rather would one expect the mechanical components to last 3 to 10 times longer than the guarantee period.

If this requirement cannot be met, preventive service should be prescribed - either in the form of lubrication, adjustment and the like or by replacement of critical parts.

From the result of the wear-out test we are interested in finding out when the parts start to fail - not when all have failed.

Sumarizing, we therefore define the wear-out time as the <u>average time when 10% of a certain part or subassembly has failed.</u>

4. DESIGN-PRINCIPLES

Good engineering practice is still the fundamental basis for a reliable mechanical construction. Below are given some guidelines, mainly related to reliability.

1) <u>Low Complexity</u>. Mechanical constructions, which are optimized with respect to low cost, are generally more likely to become unreliable if just one part is out of tolerance. The interaction between the parts is complex. One should try to keep this complex interaction as low as possible.

2) <u>Parts with Proven Reliability</u>. Only parts with a proven reliability should be used. The wear-out time under the expected load must be specified. If this is not possible, the parts should be exposed to a life-test.

3) <u>Written Specifications</u>. It is essential that the manufacturer - internal or external - refer to written specifications for materials, parts and processes. This is often overlooked when buying plastic or rubber parts. Any changes should be accepted only on the basis of a revised specification.

4) <u>Well-known Manufacturing Processes</u>. A step forward in process technology usually leads to more advanced constructions and to lower production cost. At the same time, however, this also leads into fields with little proven experience with respect to reliability. Rather costly experiences, where the influence of various process parameters are studied, seem to be the only way to overcome this problem.

5) <u>Few Corrections</u>. The purpose of introducing changes in drawings, specifications etc. is, of course, to improve the product or make it cheaper to produce. Unfortunately, it is not always so easy to see if the correction will influence the reliability in a negative way. If there is any doubt, the life test should be repeated.

5. SETTING UP A TEST PLAN

Before starting af life test it is important to set up a test plan. Assuming it to be a normal life test, the following should be considered:

1) <u>Time Duration</u>. It must be estimated how many hours a day the parts are running in average, when used by the customer. The specifications for the instrument must show what the expected wear-out time is. From these data the duration of the test time can be planned.

2) <u>Definition of a Failure</u>. It must be defined what a failure is: whether it is a complete break down, whether tolerances are outside a certain specification etc. It must also be considered how the failures are going to be detected - manual inspection or failure detectors. A recorder may sometimes be adviseable to show slow variations.

3) <u>Registration Intervals</u>. Failures should be registered at gradually increasing time intervals, so that these intervals form an approximate logarithmic scale. In this way most information can be gained when plotting the failures on Weibull-paper.

4) <u>Load and Environment</u>. In order to obtain realistic test data the acutal load and environment should be simulated as far as possible. If the environment is normal indoor use it is adviseable to vary the temperature in a climatic chamber. Maximum load normally simulates the worst condition for mechanical parts, whereas for electro-mechanical parts containing contacts, these should preferably be tested under the same conditions as when they are used in the instrument.

5) <u>Bump and Vibration</u>. Apart from the above mentioned items, which mainly refer to the life-test, other tests should also be carried out. Especially the bump-test in all planes is important. This test could be supplied with a vibration test. Only on the basis of such tests the final packing can be designed.

6. EVALUATING THE TEST DATA

It is important to gain as much information from a life test as possible. The following steps are suggested:

1) <u>Plot failures on Weibull-paper</u>. This graphic depiction is a very strong tool in estimating the failure development. For further details about the Weibull-distribution, please refer to the quite extensive litterature.

2) <u>Analyze Failures</u>. The failing parts should be taken apart to find the cause of the failure. This is often a quite cumbersome job - but necessary in order to improve the reliability.

3) <u>Revise Specifications</u>. Suggested improvements often result in tighter specifications or in a revision of the production method. Be sure that the written specifications are revised, and consider whether a renewed test is necessary.

4) <u>Define Critical Parameters</u>. Certain tolerances are critical with respect to the reliability of the product. If possible, a test-method should be defined to test these tolerances - directly or indirectly - in order to check new parts. An example: measurement of noise-level to check the play in moving parts.

7. FEED-BACK FROM THE FIELD

It is important to receive data on failing parts from the field. The data can be used to verify if the reliability goal has been met, and to identify the dominating failing parts.

In order to improve a product on the basis of information from the field, three conditions must be fulfilled:

1) <u>Information on Failures</u>. Reports on failures together with the defective parts must be returned to the company. Copy of the servicereports from a

selected market-segment is often used to solve this problem.

2) <u>Computing the Failure Distribution</u>. For each failure-report the time of failure-free operation must be computed. The time is calculated as the difference between the service date and the sales date. For a specific product the accumulated number of failures should be drawn as shown in Fig. 1. This curve will show whether the reliability goal is met or not.

Furthermore, the different failing parts should be registered in the order of sequence, showing the most frequently failing parts first. It is quite often so, that few parts contribute to 80 - 90% of all failures.

Form the feed-back data a wear-out curve should be drawn for each one of these few prevailing failures. Concentrating on improving these few parts will often greatly improve the reliability of the total instrument.

3) <u>Corrective Action</u>. The failure-reports should be discussed in a group with representatives from different departments, and the task of correcting the failures should be delegated to responsible persons.

Such a group must meet regularly and also follow up on the pending matters.

In theory the analyzing of failures from the field can seem quite simple. In practice, however, it often takes some years to establish an effective feed-back system.

Failures are inevitable, especially in new products. The success of a company is therefore also dependent on how quickly it reacts to correct the failures.

8. CONCLUSION

This paper has been dealing with ways to ensure high reliability on mechanical products. The actual design of reliable products has not been discussed - the intention has been to analyze the problems from a quality assurance point of view.

How to organize the various quality activitees has not been dealt with. There is, howeer, no doubt that an independent quality function is necessary in order to coordinate the activities. The quality function also plays an important role in demonstrating to the management that the considerable effort is well worth the money.

Reliability Technology — Theory & Applications
J. Møltoft and F. Jensen (Editors)
© Elsevier Science Publishers B.V. (North-Holland), 1986

OPTIMUM PREVENTIVE MAINTENANCE FOR COMPLEX SYSTEMS
WITH INTERDEPENDENT EQUIPMENT UNITS

Aat SCHORNAGEL*
Mathematical Centre, Amsterdam, The Netherlands

We outline how the theory of generalised Markov decision processes can be used to determine
optimum preventive maintenance strategies for complex systems if the units are interdependent
(correlated) (statistically, economically and/or in the evaluation of their state). By way
of illustration we give an example of a system with two units but only one repair facility.
The units are correlated (interdependent) solely in terms of the repair facility: if both
units require repair or maintenance at the same time, one has to wait.

1. INTRODUCTION

In industry as a whole a growing concern for
the cost effectiveness of plant maintenance is
discernible. Although low as a percentage of
capital replacement value, maintenance costs
may form an important part of the total operating budget.

Most maintenance strategies encountered in
practice are of the "control-limit" or "age
replacement" type: do preventive maintenance
as soon as an indicator of the equipment condition (e.g. number of starts or run hours)
reaches a certain threshold value t* and carry
out breakdown maintenance (repair) whenever a
failure occurs before t*.

A large proportion of the maintenance literature employs mathematical models and techniques
and deals with finding an optimum maintenance
interval t* for a single unit of equipment
[1-3]. With the growing complexity of industrial plants these methods are no longer appropriate due to the correlation between equipment
units. Setting an optimum maintenance interval
t* for each individual unit does not necessarily yield the most cost-effective solution for
the whole plant.

Three different kinds of equipment correlation
may be distinguished:

- Performance correlation. The evaluation of
  the state (part of the performance process)
  of a unit depends on that of other units. For
  example, in a redundant parallel system a
  stand-by unit starts operation on failure of
  another unit. Another example is a series system, where non-failed units are put on suspended animation (or stand-by) on failure of
  one unit in the series. A third example is a
  system with a shortage of repair facilities,
  so that a queueing situation may occur before
  repair.

- Economic correlation. Failure, repair and
  maintenance costs and rates of return on
  operating equipment may depend on the actual

state of other units. For example, with load-
sharing parallel equipment the rate of return
on each unit depends on the number of operating units.

- Statistical correlation. The reliability
  characteristics (e.g. lifetime and repair
  time distributions) depend on the condition
  of other units. Also units may be statistically correlated due to common environmental
  factors.

A well-known maintenance strategy for complex
systems is block replacement, applied to a
group of equipment units. However, such a
strategy does not answer the question as to
what condition the plant as a whole must be in
for preventive maintenance on a certain unit to
be worth while. This paper describes a unified
method for finding an optimum preventive maintenance strategy for a complex system with interdependent (correlated) units. ("Optimum"
here means the most cost-effective in the long
run.) The equipment condition indicator is run
hours and the maintenance strategy is of the
control-limit type in the sense that for each
unit an optimum t* is calculated which is now a
function of the condition of (all) other units
in the system.

Compared to earlier papers on preventive maintenance strategies for systems with correlated
units (cf [3-6]) our model assumptions are
less restrictive.

We begin with a more formal statement of the
problem, followed by an example. This example
is then used as an aid in the explanation of
the mathematical methods. In fact we give a
general outline of how the theory and computational methods of generalised Markov decision
processes can be used. Finally, some numerical
results for the example are presented, from
which the implications of practical implementation of the method can be quantified.

* Present adress: Koninklijke/Shell-Laboratorium, Amsterdam,
  P.O. Box 3003, 1003 AA Amsterdam, The Netherlands.

2. STATEMENT OF THE PROBLEM

We consider a system with n units which, once
operating, are subject to random failures. After a failure and (possibly) some waiting time
the unit is repaired and restored to the "as
good as new" state. Upon completion of the repair process and (possibly) a stand-by period
the unit enters the next cycle of operating
(phase 1), waiting (phase 2), repair (phase 3)
and stand-by (phase 4) periods (see Fig. 1).

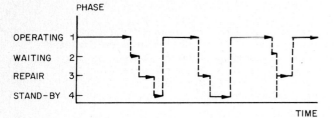

Fig. 1   Example of performance of a unit.

The state of each unit is specified by the pair
(k,s) where k is the phase number and where s
is either the time already spent in that phase
(first state model) or the time still to be
spent (residual occupation time) in that phase
(second state model). If the first state model
is used then, for each unit i, the time (to be)
spent in a certain phase is a random variable,
the distribution of which is specified by its
conditional transition rate $q_k^i(s; x_i, i \neq j)$,
i.e. the probability that unit i will have a
transition from phase k to phase k+1 (mod 5)
between time s and time s+Δs, given that it
entered phase k at time 0 and given that it
has spent s time units in phase k, equals
$q_k^i(s; x_i, i \neq j) \cdot \Delta s + o(\Delta s)$. In other words, phase
transitions may depend on the states of all
units in the system. With k=1 (the operational
phase) q_k^i represents the well-known failure
rate of unit i, which is now a function of the
state of all units in the system (this means
statistical correlation).

In model 1 the occupation times in each phase
may not depend on the actual occupation times
in the previous phase. If, for example, repair
times do depend on the failure time, then model 2 can be used. The time to be spent in a
phase k by unit i has a probability distribution which depends on the state of the whole
system when unit i enters phase k, and, hence,
on the time spent in phase k-1 (mod 5) by
unit i.

Note 1
If unit i has a maximum repair time r_i then the
use of the second state model enables the
state of a unit to be represented by a scalar
s.s>0 means that the unit has been operating
for s time units (phase 1). s=$-r_i$ means that
the unit is waiting for repair (phase 2).
$r_i < s < 0$ means that the unit is being repaired
with residual repair time $-s$ (phase 3) and s=0
represents the stand-by phase (phase 4).

Selection of the first or the second state
model for each individual unit and phase guarantees that the statistical laws governing the
evaluation of the future state of the system as
a whole depend on the present state of the system only and not on how this state was arrived
at. This lack of memory (Markov property) is
essential in order for the theory of generalised Markov decision processes to be applicable.

For the stochastic model described above the
following cost structure is defined. For each
state $(x_1, .., x_n)$ of the system there is a
cost rate $c(x_1, .., x_n)$ for being in that
state; a negative cost rate implies a positive
return. Further, there are immediate costs incurred at a phase transition of a unit:
$b_k^i(s; x_j, j \neq i)$ is the cost of a transition of
phase k to k+1 (mod 5) at unit i. Observe that
cost functions may depend on the states of all
units in the system (economic correlation).

The stochastic (Markov) process describing system state evaluation can be controlled by a decision-maker. The decision-maker may interrupt
the process and thereby enforce a phase transition for one or more units. Without loss of
generality we shall only consider enforced
transitions of phase 1 (operating) to phase 2
(waiting) or to phase 3 (repair) if there is no
waiting time before repair. These transitions
represent preventive maintenance: an operating
unit is directed to a waiting place or even to
the repair facility without having suffered
from a failure. The cost of a phase transition
caused by a preventive decision is denoted by
$p_k^i(s; x_j, j \neq i)$ similar to $b_k^i(s; x_j, j \neq i)$.

As we let the units be interdependent a decision to interrupt (stop) a unit's operational
period may depend on the state of all units in
the system. Because a unit i can never reach
the state (1,s), s>t*, if at t* the unit is
taken out of operation preventively, a preventive maintenance strategy for the whole system
is described by the n functionals $t_i^*(x_j, j \neq i)$,
representing the age of unit i (depending on
the state of all other units) at which preventive maintenance is carried out, i.e. the phase
transition 1$\longrightarrow$2 is enforced. $t_i^*(x_j, j \neq i)$
(i=1,...,n) represents an n-dimensional age replacement or control-limit preventive maintenance strategy. It indicates when preventive
maintenance should be carried out and on what
unit the work has to be done.

The objective is to find that n-dimensional age
replacement strategy which minimises the expected long-run average cost per unit time for
the system.

Note 2
In many practical applications there will be
a return (reward) rate connected to units being
in operation, i.e. the cost rate associated
with those unit states will be negative. It is
then also possible to speak of long-term optimisation of the average return on the system.

We claim that for the class of maintenance
problems described above the theory of generalised Markov decision processes can be applied
as a unified approach. We now give an example,
which we shall then use to illustrate the theory.

3. EXAMPLE

In this section we give a fairly simple example of two units which are correlated in terms of performance as they share one repair facility. This means that a queueing situation for the repair facility may arise. The units are neither statistically nor economically correlated. For both units we have the following parameters

- Failure rate $\lambda_i(s)$. This corresponds to the conditional transition rate $q_i^1(s; x_j, j \neq i)$ mentioned previously. Because the units are not statistically correlated we have immediately an explicit expression for the unit lifetime distribution:

$$F_i(s) = 1 - \exp\left(-\int_0^s \lambda_i(u)du\right)$$

- Each unit has a fixed repair time r_i
- When operating the return rate is $C_i(s)$, where s is the age of unit i (in terms of the time since it last entered the operational phase)
- Upon unit failure there are immediate costs $B_i(s)$, where s is the time to failure. We assume that repair costs, $M_i(s)$, are also incurred upon unit failure. In terms of the previous section we have
  $b_i^1(s; x_j, j \neq i) = B_i(s) + M_i(s)$
- If a unit is taken out of operation preventively, only repair costs $M_i(s)$ have to be paid. The maintenance time is equal to the repair time r_i

A discrete time version of this model is analysed in [7].

For this example the state of a unit can be represented by a scalar; see note 1 above. The system state can be represented by the pair (s_1, s_2) $(s_i > -r_i)$ where s_i is the state of unit i. In fig. 2 we show a maintenance strategy for this example, together with a possible realisation, which is explained in table 1. The hatched area can never be reached because only one repair facility is available.

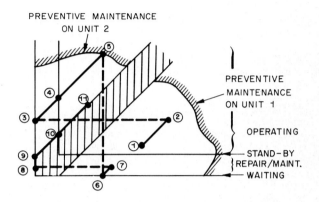

Fig. 2 Example of preventive maintenance strategy and possible realisation of the process.

Table 1

Example of realisation of the decision process with associated costs and return rates

Fig. ref.	State	Unit 1	Unit 2	Costs	Return rate
(1)	(x,y)	in operation	in operation	–	$C_1(x)+C_2(y)$
(2)	$(x+s,y+s)$	fails	in operation	$B_1(x+s)$	$C_2(y+s)$
(3)	$(-R_1,y+s)$	repair starts	in operation	$M_1(x+s)$	$C_2(y+s)$
(4)	$(0,y+s+R_1)$	starts operating	in operation	–	$C_1(0)+C_2(y+s+R_1)$
(5)	$(u,y+s+R_1+u)$	in operation	preventive maintenance decision	–	$C_1(u)$
(6)	$(u,-R_2)$	in operation	starts preventive maintenance	$M_2(y+s+R_1+u)$	$C_1(u)$
(7)	$(u+t_1-R_2+t)$	fails	being maintained	$B_1(u+t)$	–
(8)	$(-R_1,-R_2+t)$	waiting	being maintained	–	–
(9)	$(-R_1+R_2-t,0)$	repair starts	starts operating	$M_1(u+t)$	$C_2(0)$
(10)	$(0,R_1)$	starts operating	in operation	–	$C_1(0)+C_2(R_1)$
(11)	(w,R_1+w)	in operation	in operation	–	$C_1(w)+C_2(R_1+w)$

4. THE GMDP FORMULATION

The self-repeating process of operating, waiting, repair, stand-by can be seen as a controlled Markov drift process. This implies that the well-known theory of Markov decision processes (MDP, cf [8]) cannot be used. In this section we show how the model can be formulated as a "generalised Markov decision process" [9]. The GMDP theory is less accessible than that of MDPs and stochastic dynamic programming but it can be applied perfectly well to our maintenance problem. As we wish to avoid too general terms we shall confine ourselves to the example described in Section 3 above.

In ref. [9] a GMDP is specified by a list of seven elements. The first four give the essence of a GMDP.

A GMDP may be seen as a natural process with superposed decisions. The natural process should be a strong Markov process with stationary transition probabilities. Decisions are interruptions of the natural process. Superposition of the decisions onto the natural process should result in the original decision model. Or, conversely, the original decision model should be decomposed into a natural process which is not disturbed by external interruptions (decisions) and decisions at given states of the natural process. In our example we take as the natural process the original process without any decision to start repair on a unit, either after a failure or preventively. This implies that the natural process will always end in the absorbing state $(-r_1,-r_2)$ where both units have failed and are not under repair (i.e. both units are waiting), provided that the lifetime distributions are not defective.

Elements 4 and 5 in ref. [9] deal with the cost/reward structure of the GMDP. For each accessible state of the natural process the following functions must be calculated:

$k_0(s_1,s_2)$ = the total expected costs (and/or profits) incurred by the natural process up to and including the time at which state $(-r_1,-r_2)$ is reached, if the natural process starts in state (s_1,s_2)

$t_0(s_1,s_2)$ = the total expected time to state $(-r_1,-r_2)$ being reached, if the natural process starts in (s_1,s_2)

$k_1(s_1,s_2;i)$ = the cost of deciding to repair unit i (after a failure or preventively) plus $k_0(-r_1^+,s_2)$ (if $i=1$) or $k_0(s_1,-r_2^+)$ (if $i=2$), i.e. plus the k_0 costs associated with the state entered after the decision to repair a unit. $-r_i^+$ represents the state of unit i just after the start of a repair period

$t_1(s_1,s_2;i)$ = $t_0(-r_1^+,s_2)$ (if $i=1$) or $t_0(s_1,-r_2^+)$ (if $i=2$), i.e. the t_0 time associated with the state entered after the decision to repair a unit.

The above four functions can easily be derived by summing (integrating), over all possible realisations of the natural process ending in state $(-r_1,-r_2)$, the total costs incurred

(or time length) associated with that realisation multiplied by the corresponding probability (density).

Because the cost/reward structure is additive the k functions can be obtained by looking at each unit individually and adding the results. For example, if $s_1,s_2 > 0$, then:

$$k_1(s_1,s_2;1) = R_1(s_1) + k_0(-r_1^+,s_2) =$$
$$= R_1(s_1) + \int_0^\infty [B_1(x) + R_1(x) + \int_x^x C_1(u)du]\, dF_1(x)$$
$$+ \int_0^\infty [B_2(s_2+x) + R_2(s_2+x) + \int_0^x C_2(s_2+u)du]$$
$$dF_2(s_2+x\,|\,s_2)$$

Here $F_i(s_i+x\,|\,s_i)$ is the conditional probability that unit i fails before the age $s+x$, given that it has survived s time units.

Obviously, the time functions are not additive. For example, for $s_1>0$ and $-r_2<s_2<0$

$$t_0(s_1,s_2) = -s_2 + F_1(s_1-s_2\,|\,s_1) \int_0^\infty xdF_2(x) +$$
$$+ [1-F_1(s_1-s_2\,|\,s_1)].t_0(s_1-s_2,0)$$

and for $s_1,s_2>0$

$$t_0(s_1,s_2) = \int_0^\infty [1-F_1(s_1+x\,|\,s_1).F_2(s_2+x\,|\,s_2)]dx$$

Let us now denote by D_z the set of all system states (s_1,s_2) for which a given maintenance/repair strategy z prescribes a repair. By $z(s_1,s_2)=i$ we denote that strategy z prescribes a repair decision on unit i in the system state (s_1,s_2). Of course, $s_i>0$ or $s_i=-r_1$ $(i=1,2)$. With $z(s_1,s_2)=0$ no repair will take place. In any reasonable strategy all states $(s_1,-r_1)$ with $s_1 \geqslant 0$ and $(-r_1,s_2)$ with $s_2 \geqslant 0$ will be repair decision states because in these states one unit is waiting for an available repair facility. All states (s_1,s_2) with $s_i>-r_i$ and at least one $s_i \geqslant 0$ where $z(s_1,s_2)=i$ $(i=1,2)$ are states where z prescribes preventive repair (maintenance) on a unit.

It now follows from theorem 2 in ref. [9] that the long-term average return per unit time associated with strategy z, g_z, can be derived from a set of functional equations embedded on the set of states where a repair is prescribed. For our example we have:

$$v(s_1,s_2) = \begin{cases} k(s_1,s_2;1) - g_z t(s_1,s_2;1) + v(-r_1^+,s_2) & \text{if } z(s_1,s_2)=1 \\ k(s_1,s_2;2) - g_z t(s_1,s_2;2) + v(s_1,-r_2^+) & \text{if } z(s_1,s_2)=2 \\ E[v(T_z(s_1,s_2))] & \text{if } z(s_1,s_2)=0 \end{cases}$$

Here $k(s_1,s_2;i)$ equals $k_1(s_1,s_2;i) - k_0(s_1,s_2)$, $t(s_1,s_2;i)$ equals $t_1(s_1,s_2;i) - t_0(s_1,s_2)$, $T_z(s_1,s_2)$ is the first state where z prescribes a repair starting from state (s_1,s_2) where $z(s_1,s_2)=0$ and $Ev(T_z(s_1,s_2))$ is the expected value of v at $T_z(s_1,s_2)$, $T_z(s_1,s_2)$ itself being governed by the statistical failure laws of the units. The values $v(s_1,s_2)$ play the same role as the relative values in MDPs (cf [8]).

The set of functional equations (1) can be decreased if it is noted that:

$$v(s_1,s_2) = k(s_1,s_2;1)-g_z t(s_1,s_2;1)+v(-r_1^+,s_2)$$
$$\text{if } z(s_1,s_2)=1 \text{ and } s_1>0$$

and

$$v(-r_1,s_2)= k(-r_1,s_2;1)-g_z t(-r_1,s_2;1)+v(-r_1^+,s_2)$$

Hence there is a simple relation between $v(s_1,s_2)$ and $v(-r_1,s_2)$ if $z(s_1,s_2)=1$ and $s_1>0$. This implies that the set of functional equations (1) can be embedded on the states $(-r_1,s_2)$, $s_2 \geqslant -r_2$ and $(s_1,-r_2)$, $s_1 \geqslant -r_1$ only. According to ref. [9] these functional equations can be solved after adding a normalising equation, e.g. $v(0,0)=0$.

Now that we are able to calculate the long-term average return of a strategy z the question arises as to how to find the optimum strategy, i.e. the one which has the highest long-term average return associated with it. In ref. [9] an iterative method is described to find the optimum strategy z*. This method provides two ways of improving a strategy z. The first is called the cutting operation, which involves considering whether the repair decision set Dz can be decreased. For each state (s_1,s_2) for which $z(s_1,s_2)=i$ $(s_i>0)$ it is considered whether it is profitable to delay the preventive maintenance decision Δs time units, of course with the risk of having a failure of unit i in that small period. It is profitable if the total expected return in that small period Δs is greater under the delayed preventive maintenance case than under the original strategy z.

The other way in which a strategy can be improved is called the policy improvement operation. Here it is analysed whether bringing forward a preventive maintenance decision over a small period Δs would improve the strategy z. In this way either the cutting operation or the policy improvement operation may improve the strategy at a certain point. Of course, with each policy improvement or cutting operation preventive maintenance on either unit is allowed. The results of a cutting operation and a policy improvement operation are depicted schematically in fig. 3, parts a and b respectively. Fig. 3c gives an example of what an improved strategy may look like if at all states of D_z the policy improvement and cutting operation have been applied.

The optimum strategy can be derived by alternately solving the set of functional equations (1) for the present strategy and improving that strategy by policy improvement and/or cutting operations until no further improvement is possible. When converged, the algorithm has produced both the optimum maintenance strategy z* and the associated long-term average return per unit time g*.

Note
Due to the continuity of the state space a discretisation should be used to solve the set of functional equations (1) and to apply the cutting and policy improvement operations.

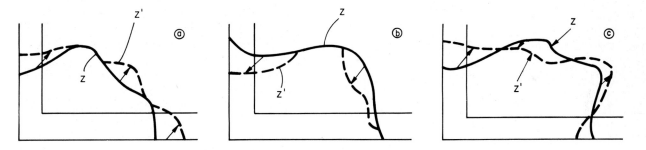

Fig. 3 Result of (a) the cutting operation, (b) the policy improvement operation and (c) both cutting and policy improvement. z is the present strategy, z' the improved strategy.

5. NUMERICAL RESULTS FOR THE EXAMPLE

The GMDP method has been implemented in a computer program (in PASCAL) for the example given in section 3. It is beyond the scope of this paper to give an extensive overview of the sensitivity of the optimum preventive maintenance strategy and the maximum average return to all system parameters. A detailed description of how the example can be modelled as a GMDP, together with extensive sensitivity results, will be reported in [10]. Our general impression is that both the optimum strategy and the associated average return are more sensitive to the mean lifetime, repair time and failure costs of a unit than to the return rates and repair costs.

To give an idea of the effect of correlating performance in the manner shown in our example,

we have depicted in fig. 4 the optimum strategy, together with the initial strategy and that after four iterations, for a system with parameters as given in table 2. In this example we calculated an optimum average return on the system of 9 units of revenue per unit time. If the equipment units were taken as being independent of each other, the optimum long-term average return would be less than 7.5, a difference of almost 20 %. However, the long-term average return would have increased by almost 20 % if a second repair facility had been installed, thus decoupling the units so that no time was lost by waiting. This kind of argument gives an impression of how much one should be prepared to pay for one more repair facility.

Table 2

Parameters of the model leading to the optimum strategy z* in fig. 4

Parameter		Unit 1	Unit 2
Maintenance/repair time	R_i	2	1
Time to failure distr.	$F_i(t)$	$1 - \exp(-0.05\ t)$	$1 - \exp(-0.1\ t^{1.3})$
Failure cost	$B_i(t)$	$9 + 0.55\ t$	2.20
Maintenance/repair costs	$M_i(t)$	$2 + 0.01\ t^2$	$0.5 + 0.35\ t$
Rate of return	$C_i(t)$	$8.5 - 0.55\ t$	$8.1 - 0.58\ t$

In fig. 4 the preventive maintenance strategy for unit 1 does not depend on the residual repair time of unit 2 if the latter is in the process of being repaired, and vice versa. Indeed, because for the optimum maintenance strategy neither cutting nor policy improvement operation may result in a different strategy, the optimum preventive maintenance age of unit i, given that the other unit is being repaired, follows from (cf [10])

$$\frac{d}{ds} R_i(s) + C_i(s) + q_1^i(s)\ B_i(s) = 0 \qquad (2)$$

Here $q_1^i(s)$ represents the conditional failure rate function of unit i (cf section 2). Equation (2) means that when one unit is under repair, the operation of the other unit should be stopped preventively only if the risk of continuing operation (in terms of failure costs) is higher than the expected revenues from continuing operation. This property of an optimum preventive maintenance strategy is explained by the fact that the operating unit has to wait in any case after failure or a repair decision if the other unit is under repair. Equation 2 is an example of the infinitesimal look-ahead rule (cf [8]) where an optimum control-limit strategy is such that a small change in the control limit in any direction in the state space would decrease the average return associated with that strategy.

Finally, we mention that system availability calculations can be done by taking all cost functions as being identical to 0 and by taking the return rate $C_i(t)$ of a unit as being identical to 1. In the case of Weibull-distributed unit lifetimes (shape factor 1.4) we calculated the system effectiveness for 9 combinations of unit availabilities. Here unit availability is the ratio of the unit mean time between failures (MTBF) to the sum of the MTBF and the mean time to repair (MTTR):

$A_i = MTBF_i/(MTBF_i+MTTR_i)$.

The system effectiveness is defined as:

$$E_s = \tfrac{1}{2} \cdot \sum_{i=1,2} \frac{MTBF_i}{MTBF_i+MTTR_i+MW_i}$$

where MW_i is the mean waiting time of unit i in the system due to the shortage of one repair facility. If the units were independent then

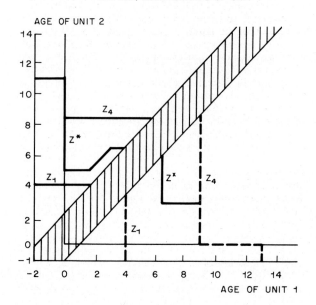

Fig. 4 Convergence to an optimum preventive maintenance strategy. Model parameters as in table 2. z_1 is the initial strategy, z_4 is the strategy after 4 iterations, $z_1=z^*$ is the optimum strategy.

there would be no waiting time in the system $(EW_i=0)$, in which case we have $E_s = (A_1+A_2)/2$. From table 3 we conclude that the effect of performance correlation on the system effectiveness is small. The system effectiveness is decreased by at most 3 % compared to the value if there were no correlation.

Table 3

Maximum system effectiveness for different combinations of unit availabilities

A_2 \ A_1	0.80	0.90	0.95
0.80	0.78	0.84	0.87
0.90	0.84	0.89	0.93
0.95	0.87	0.93	0.95

6. CONCLUSIONS

We have given a general outline of how optimisation of preventive maintenance of the age replacement type for systems with correlated equipment units can be tackled with the theory of generalised Markov decision processes (GMDP). We believe that this theory can be used for most practical cases of equipment correlation. While the GMDP method may be reported as a unifying approach, each particular application requires a check of all seven elements of the GMDP model and, in particular, the development of all k and t functions. Nevertheless, the method opens up the possibility of analysing very complicated maintenance models with correlation; hence, we expect new insights to be obtained through it.

REFERENCES

[1] K. Bosch and U. Jensen, Instandhaltungs-modelle - eine Ubersicht, OR Spektrum, vol. 5, 1983, 105-118.

[2] Y.S. Sherif and M.L. Smith, Optimal maintenance models for systems subject to failure - a review, Naval research logistics quarterly, vol. 28, 1981, 47-74.

[3] F.A. Tillman, C.L. Hwang and W. Kuo, Optimization techniques for system reliability with redundancy - a review, IEEE transactions on reliability, vol. 26, 1977, 148-155.

[4] M.J. Beckmann and R. Subramanian, Optimal replacement policy for a redundant system, OR Spektrum, vol. 6, 1984, 47-51.

[5] S. Epstein and Y. Wilamowsky, A replacement schedule for multicomponent life-limited parts, Naval research logistics quarterly, vol. 29, 1982, 685-691.

[6] K. Okumoto and E.A. Elsayed, An optimum group maintenance policy, Naval research logistics quarterly, vol. 30, 1983, 667-674.

[7] D. Stengos and L.C. Thomas, The blast furnaces problem, European journal of operational research, vol. 4, 1980, 330-336.

[8] Sheldon M. Ross, Applied probability models with optimisation applications, Holden-Day, San Francisco, 1970.

[9] G. de Leve, A. Federgruen and H.C. Tijms, A general Markov decision method I: model and techniques, Advances in applied probability, vol. 9, 1977, 296-315.

[10] A. Schornagel, Report in preparation.

Biography

The author was awarded a degree at MSc level in 1978 from Twente University of Technology in the field of Stochastic Operations Research. Until 1982 he worked at the Mathematical Centre in Amsterdam, where he studied Markov decision models, in particular controlled queueing models and preventive maintenance models. Since 1982 he has been in the employ of Shell Research B.V. at Koninklijke/Shell-Laboratorium, Amsterdam, where he is active in the field of reliability, availability and maintenance of systems.

AUTHOR INDEX